AF506294

Scientific Computation

Editorial Board

J.-J. Chattot, Davis, CA, USA
P. Colella, Berkeley, CA, USA
W. Eist, Princeton, NJ, USA
R. Glowinski, Houston, TX, USA
M. Holt, Berkeley, CA, USA
Y. Hussaini, Tallahassee, FL, USA
P. Joly, Le Chesnay, France
H. B. Keller, Pasadena, CA, USA
J. E. Marsden, Pasadena, CA, USA
D. I. Meiron, Pasadena, CA, USA
O. Pironneau, Paris, France
A. Quarteroni, Lausanne, Switzerland
   and Politecnico of Milan, Italy
J. Rappaz, Lausanne, Switzerland
R. Rosner, Chicago, IL, USA
P. Sagaut, Paris, France
J. H. Seinfeld, Pasadena, CA, USA
A. Szepessy, Stockholm, Sweden
M. F. Wheeler, Austin, TX, USA

Jean Cousteix · Jacques Mauss

Asymptotic Analysis and Boundary Layers

With 85 Figures

 Springer

Prof. Jean Cousteix
ONERA Centre de Toulouse
2, av. Edouard Belin
31055 Toulouse, France
e-mail: jean.cousteix@onecert.fr

Prof. Jacques Mauss
Institut de Mécanique des Fluides de Toulouse
(IMFT)
Université Paul Sabatier
118, Route de Narbonne
31062 Toulouse Cedex, France
e-mail: Jacques.Mauss@imft.fr

Library of Congress Control Number: 2007921677

ISSN 1434-8322
ISBN 978-3-540-46488-4 Springer Berlin Heidelberg New York

This work is subject to copyright. All rights are reserved, whether the whole or part of the material is concerned, specifically the rights of translation, reprinting, reuse of illustrations, recitation, broadcasting, reproduction on microfilm or in any other way, and storage in data banks. Duplication of this publication or parts thereof is permitted only under the provisions of the German Copyright Law of September 9, 1965, in its current version, and permission for use must always be obtained from Springer. Violations are liable to prosecution under the German Copyright Law.

Springer is a part of Springer Science+Business Media

springer.com

© Springer-Verlag Berlin Heidelberg 2007

The use of general descriptive names, registered names, trademarks, etc. in this publication does not imply, even in the absence of a specific statement, that such names are exempt from the relevant protective laws and regulations and therefore free for general use.

Typesetting: Camera-ready copy from the authors
Data conversion and production by LE-TeX Jelonek, Schmidt & Vöckler GbR, Leipzig, Germany
Cover design: eStudio Calamar, Girona/Spain

Printed on acid-free paper   SPIN: 11576716   55/3100/YL - 5 4 3 2 1 0

Preface

Two distinguished specialists, one of asymptotic methods, the other of simulation of flows with boundary layer, made an evident effort, on one side, to understand their respective disciplines and, on the other side, to progress together. The result is this very original book, which represents a significant contribution to the problem of laminar flow calculation, at high Reynolds number, with moderately separated boundary layers.

The basic tools used in this book are not new. Asymptotic methods, ordinary differential equations, fluid mechanics, Euler and Prandtl equations are the fundamental blocks on which the edifice is built. The proposed examples, examined theoretically and numerically, produce convincing results. In addition, standard applications such as the second order boundary layer theory, the triple deck model for localized separation, the effect of localized wall perturbation on plane Poiseuille flow, are revisited with this new construction.

Then, where is the novelty? It lies in the effort conducted to fully review the application of asymptotic methods for the resolution of problems involving a boundary layer. The reader is guided in this progression through twelve chapters. Chapter two to six are devoted to asymptotic methods, in general, and to their use to solve problems of ordinary differential equations containing a small parameter. Here, we have the first opportunity to find out the ingenuity of the method recommended by the authors. Attributing a loose meaning to the metaphor, let us say that, with the commonly used Method of Matched Asymptotic Expansions (MMAE), appropriate expansions play the role that a mathematician associates with intuition, while matching is a substitute to proof. Here appears a basic change in strategy: the intuition consists of guessing that one may write down a Uniformly Valid Approximation (UVA), and the substitute to the proof is that a Successive Complementary Expansion Method (SCEM) may be used to achieve the goal. Usually, we use asymptotic expansions appropriate to different regions and the matching between the expansions play a crucial role. Here, the construction of the UVA is performed abreast, i.e. without going back and forth, by substituting the assumed approximation in the equations and in the boundary conditions, and by minimizing the error in an asymptotic sense. The authors show that the same result is obtained as with the back and forth method with matching.

The conclusive argument appears after chapter seven, when the method is applied to boundary layers in fluid mechanics. Indeed, in this case, MMAE comes up against the impossibility to solve the Prandtl equations beyond the point of vanishing skin-friction. Many researchers tried to overcome the difficulty by introducing an interaction between the inviscid flow and the boundary layer, but it was necessary to abandon any asymptotic expansion while this was an underlying idea. The authors apply SCEM in order to guess a UVA proceeding according to (fractional) powers of the inverse Reynolds number (Re). This UVA is substituted in the Navier-Stokes equations and boundary conditions, and the remainders are rendered as asymptotically small as possible. The inviscid flow and the boundary layer are considered simultaneously by imposing to the two simulations to be as close as possible to each other in a region which is common to the two domains of validity. There is no need of the back and forth process and of matching conditions because these latter conditions are implicitly contained in the construction process of the UVA due to SCEM and to the asymptotic minimization of errors. The authors show that SCEM produces the same results as MMAE, if this method works, including the triple deck results with separation having a longitudinal extent of order $Re^{-3/8}$. With SCEM, the separated zone can be more extended but not up to order one, which is not surprising.

The effect of wall deformation on plane Poiseuille flow raises a difficult problem and asymptotic approximation has been thoroughly studied for large Reynolds number in laminar flows. Discarding any attempt to be exhaustive, the authors find there a splendid test shedding light on feasibility of their strategy, which leads to solving two sets of Prandtl's equations, linking each other by their pressures. The numerical results are convincing.

Any reader mastering a little bit of asymptotic machinery, or the one who tackled over some of the proposed problems, will be rewarded when reading chapter eleven on turbulent boundary layer which reveals flexibility of the strategy. He will understand how, some calculus, inspired by it, leads to an equation valid over the whole boundary layer thickness. Of course, such an equation relies on a mixing length model for the Reynolds stress. The numerics exhibits two zones with logarithmic overlapping which reduces in extent when the Reynolds number reduces towards transition. Such an overlapping, which is usually thought about as a consequence of matching, is now a consequence of the strategy, and obviously of the choice of the mixing length model for the Reynolds stress.

For the reasons given above, I find this book very attractive. I am convinced that experienced readers will share this opinion and that young researchers will find new perspectives.

Meudon, 10 march 2006

Jean-Pierre GUIRAUD
Professeur honoraire
Université Pierre et Marie CURIE
Paris

Acknowledgements

We are profoundly grateful to Tuncer Cebeci for his contribution to the preparation of this work. He reviewed the manuscript very carefully, making many constructive suggestions, and in doing so, helped us improve the book significantly. We are entirely responsible for any remaining mistakes and imperfections.

Contents

Abbreviations

AE: asymptotic expansion
EST: exponentially small term
IBL: interactive boundary layer
MMAE: method of matched asymptotic expansions
MVDP: modified Van Dyke's principle
SCEM: successive complementary expansion method
TST: transcendentally small term
UVA: uniformly valid approximation
VDP: Van Dyke's principle

1 Introduction

The history of relations between science and technique is surprising and stormy just like those of a couple combining love, hatred and necessity. Obviously, we can go into ecstasies over the marvels obtained by the thought in the study of motion from Aristotle to Einstein, passing by Galileo, Newton and Laplace. We can be also attracted by the successes of technique from the wheel to the computer passing by the astronomical telescope and the aircraft. Beyond the secular questioning about the pre-eminence of one on the other, are not science and technique the two faces of intelligence and reason?

Can the modern physics be satisfied with mathematical models which lead us to the outermost bounds of the knowledge of our macroscopic world? No, evidently, man needs to realize objects, to check theories, to experiment, to simulate, to explore. Man needs to search, to create and to understand.

Nowadays, the science of motion – the mechanics – rests on three supports which ensure its equilibrium: mathematical modelling, numerical simulation and experiment. Now, the cost of experiment, the modelling difficulty and the ever increasing power of numerical calculations disorder this beautiful structure to the detriment of reflection. The close connection between the mathematical model, constructed by the physicist, and the mathematics, sometimes very difficult, required to its resolution leads us too often to renounce the analysis of the model in favour of its numerical resolution. Obviously, mechanicists cannot wait for mathematicians to progress in the analysis of their models. However, they must prepare the path of mathematics by instilling a strict rigour in their heuristic reasoning. Many mathematical tools have been implemented from Leibniz and the advent of analysis in the too restricted world of geometry. The power of mathematics in the development of models and the search for solutions contributed to a large extent to remarkable progress in physics. Sometimes, surprising results have been obtained in what physicists call generically "the theory of approximation".

Thus, among the different tools of the theories of analysis and of approximation, the divergent series have been used for a long time. Not without reason, mathematicians took a great interest in these series. Calculated from well-defined functions, the terms of these series must contain information on the expanded functions. In general, the divergent series are nothing else than asymptotic series. The difference with a convergent series is that an asymptotic series is such that a partial sum is a better representation of the expanded function when a certain parameter is smaller. When the parameter vanishes, the function is exactly recovered with the first term of the series. When the parameter is not zero but simply small, any partial sum is an approximation of the function. The generic notation ε is often used for the small parameter. The small parameter is a determining factor in physics to reduce the considered mathematical model to a simpler model whose solution is an approximation of the solution of the initial model.

Beyond the notion of asymptotic series, this is the notion of *asymptotic expansion*, AE, and, perhaps more generally, the notion of approximation which is the core of our reflection. Like the word "theory" can have different degrees in its meaning, the word "approximation" can be interpreted in very different ways. Even if we restrict ourselves to mathematical physics, the ambiguity still exists. In contrast with recommendations required for any rigourous reasoning as formulated by Euclid, the word approximation has two different meanings. An asymptotic approximation is obtained, according to mathematicians, for values of ε as small as required by the mathematical formulation, the accuracy of the approximation being here perfectly well-defined. On the other hand, according to physicists, the approximation is sought for a given value of the parameter and its accuracy is not known in advance.

The goal of this book is to reconcile both definitions by proposing a method, the successive complementary expansion method, SCEM, which takes into account we have to solve concrete problems, while a rigourous mathematical procedure is followed. SCEM addresses the so-called *singular perturbation problems* which are the subject of the study throughout this book. In these problems, as $\varepsilon \to 0$, the solution does not tend uniformly towards the corresponding reduced problem obtained for $\varepsilon = 0$. It must be noted that the non-uniformity occurs in a domain whose dimension is smaller than the initial domain. That is why these problems are usually called *boundary layer problems*.

The non-uniformity of an approximation of the solution as a parameter is small is a mathematical problem. Now, as physicists, we are fortunate to be able to identify the known and unknown quantities to physical quantities. This fundamental piece of information on the nature of the physical problem enables us to better grasp the mathematical model. This is the case of the nondimensionalizing process with characteristic scales which enables us to determine if certain parameters are small. In fact, it is through the multiple choice offered by the physical description to nondimensionalize that singular perturbations can be suspected.

Thus, the flow around an airfoil is practically inviscid far away from the airfoil. However, for a steady incompressible flow, the governing equations are the Navier-Stokes equations in which, in dimensionless form, the only physical parameter is the Reynolds number. Now, away from the airfoil, the characteristic length scale is such that the inverse of the Reynolds number is very small compared to unity. Neglecting the terms containing the inverse of the Reynolds number, we obtain the Euler equations as if the viscosity were neglected. It is not that the fluid viscosity takes another value, it is that away from the airfoil, its influence is negligible because the velocity gradient is small enough. By contrast, near the airfoil, viscosity must be effective. This means that the characteristic length changes so that we can take into account the proximity of the wall where the viscous effects are essential. Then, the Reynolds number based on this latter length scale is no longer large. Near the airfoil, the Navier-Stokes equations reduce to the boundary layer equations. Even though this model is simpler than the Navier-Stokes model the wall conditions are satisfied.

How to construct a uniformly valid approximation, UVA, of the solution of Navier-Stokes equations by using solutions to Euler equations, valid only far away from the airfoil, and solutions to the boundary layer equations, valid only near the airfoil? That is the key question we want to answer for this particular problem. This is the main idea even if, obviously, other problems than high Reynolds number flows are considered. How to find the characteristic reduced problems and their domain of validity, how to link them and, finally, how to construct an approximation of the initial problem, are the points at issue which lead this book. Admittedly, the main domain of application is fluid mechanics but the scope of Chaps. 2–6 is very broad and can be useful to physicists and more generally to modellers faced to large or small parameters leading them to singular perturbation problems.

Chapter 2 is an introduction to these questions. Even the very simple example of the linear oscillator shows that the nondimensionalizing process of the equations is the first key which enables us to educe the nature of the mathematical models. Within this frame, the skill of the physicist to understand his topic and to model it is clearly the most powerful tool to solve it. Friedrichs' model problem, whose simplicity is such that the exact solution is immediate, is a so pedagogical model for singular perturbtion problems that the main methods of resolution are outlined with this example. In fact, the next chapters are focussed on two methods. One of them is the well-known method of matched asymptotic expansions, MMAE, the other, less known and it is seen why, is SCEM which is the heart of the rest of this book.

Chapter 3 deals with the structure of boundary layer. Generally, physical considerations give the necessary clues to find the location of the boundary layers. However, with a very simple problem, a second order linear ordinary differential equation whose exact solution is not known, we can study the location of the boundary layer as a stability problem. A few examples are given through the search for an approximation of the solution and the required corresponding boundary layer structures are studied. In all cases, we are concerned with a boundary value problem for which existence theorems are not available, contrary to initial value problems, at least locally.

In Chapter 4, mathematical definitions are stated. A deliberate choice is made to ally rigour and simplicity. After defining a total ordering on a set of order functions, the order of a function is defined. This explains why two different notations are found in the literature; Hardy's notations are devoted to order functions whereas Landau's notations are used for functions. We attach a particular importance to gauge functions which, when carefully chosen in equivalence classes for a given problem, enable us to introduce a certain "uniqueness" in AE. The notion of AE is discussed in this chapter. For many authors, an AE is regular, that is an expansion in the Poincaré acceptation. Now, an AE is more general and it is seen why this point is essential. Instead of calling it non regular, which could be misleading, it is chosen to call it *generalized* AE.

Chapter 5 is the heart of this book. Its title, *Successive complementarty expansion method* is associated with the fact that the central goal is to seek approximations to the solution of a problem and that this simple idea leads us to reconsider more standard methods from another point of view. Within the frame of MMAE, after discussing the standard notions of outer and inner expansions with a simple example, main definitions such that the notion of expansion operator and significant approximation bring the minimum required basic rules. Next, the notion of asymptotic matching is explored by comparing the respective merits of the intermediate matching set up as a rule and of Van Dyke's principle which is more systematic. From the construction of a UVA called a composite approximation, we show how a modified matching principle, MVDP, enables us to eliminate known counter-examples. This reflection on the matching, either formal or based on the notion of overlap, sometimes illusory in practice, leads us to invert the reasoning and to consider that the assumed form of the UVA must define the method of construction of the UVA. SCEM is proposed according to this idea. We show that MMAE is well-adapted when regular AEs are sufficient and that MMAE is contained in SCEM. SCEM demands a more indepth reflection than MMAE, in particular in its initialization. SCEM becomes really powerful when, for different reasons, MMAE is no longer adapted or when a UVA is necessary to analyze the problem.

Second order ordinary differential equations are discussed with SCEM in Chap. 6. Both methods, MMAE and SCEM, are systematically compared on several cases. The study of an equation whose coefficients are regular enough is performed in detail, showing the advantage of SCEM even on this example. A few singular cases are addressed specially when a logarithm appears with MMAE whereas SCEM shows that this singularity is only due to the method. In fact, the logarithmic behaviour appears only as an asymptotic behaviour of the solution as ε tends formally towards zero which is never the case in practice. All the studied examples lead us to the conclusion that the use of generalized AEs requires a more indepth reflection which is superfluous for the simplest cases but essential in cases that MMAE only cannot deal with.

Chapter 7, devoted to the study of high Reynolds number flows, is the transition from abstract theory to physical problems of great importance in fluid mechanics, the boundary layer. All the following chapters indeed concern boundary layers in fluid mechanics. It is interesting to note that the term *boundary layer* is now in use as a mathematical term. MMAE, through the MVDP, gives us a whole new insight into Prandtl's boundary layer theory and into the triple deck theory which complements it very nicely. This latter theory yields in particular a detailed analysis of certain types of separated flows. Based on an integral method, a simplified investigation of problems associated with separation is also given. This study enables us to understand why the inverse and simultaneous modes are better adapted to cope with separated flows than the direct mode. This knowledge is essential to tackle the numerical resolution.

Various degeneracies of Navier-Stokes equations are studied with SCEM in Chap. 8. The approximation starts with the Euler equations. As this model is not valid everywhere, in particular in the neighbourhood of walls where no condition is written, the approximation is complemented thanks to a generalized AE by adding a boundary layer term in order to seek a UVA. This analysis enables us to construct *interactive boundary layer* models, IBL, to first and second order. The models are analyzed according to their accuracy. In particular, for an irrotational external flow, the accuracy is sufficient to write reduced IBL models. Thanks to the generalized AEs, these IBLs provide a strong coupling between the viscous and inviscid zones. The hierarchy between these zones and also the asymptotic matching no longer exist. SCEM and the associated generalized AEs are the basis of the rational justification of IBLs.

Chapter 9 presents calculation results. First, the flow around a standard hump deforming a flat plate is calculated in the presence of a separated flow. Next, applications of an IBL method to aerodynamic flows are presented, including the flow around the trailing edge of a flat plate and the flow around airfoils with and without separation. Finally, the cases of different rotational freestreams are discussed. This is particularly important because a reduced IBL is not obtained as simply as in the case of an irrotational external flow. SCEM results are compared to Van Dyke's model and to numerical solutions of Navier-Stokes equations. It is shown that the results are closer to the Navier-Stokes model if the rotational character of the flow is weaker.

Chapter 10 is devoted to the derivation of nowadays standard theories such that Prandtl's, Van Dyke's and triple deck theories starting from the IBL models and not from the Navier-Stokes equations. It is clearly seen how the various degeneracies of Navier-Stokes equations are embedded into each other. Thanks to generalized expansions, initially we have first and second order IBLs; next, by using regular SCEM, equivalent to MMAE, we find Prandtl's, Van Dyke's second order and triple deck models.

The turbulent boundary layer is revisited with SCEM in Chap. 11. First, it is shown that, under hypotheses stemming from experimental data, the

VDP or, better, the MVDP indicates the existence of a logarithmic overlap without any apparent closure relationship. When applied to this problem, SCEM shows its ability to perform a better modelling of the physical problem. Adapted to values of ε admittedly small but far from vanishing, SCEM shows the necessity to write a closure relationship to construct a UVA. Moreover, the UVA underscores the asymptotic character of the logarithmic law formally obtained as $\varepsilon \to 0$.

In Chap. 12 a similar study as in Chap. 8 is carried out for laminar channel flows. More precisely, we consider high Reynolds number flows in a two-dimensional channel in which the flow is slightly perturbed by a small indentation of the walls for example. The wall deformation is sufficient to induce flow separation. In a channel, there is no external flow region and the regular asymptotic models for the flow perturbations are mainly based on an inviscid rotational core flow region together with boundary layers near the walls. Here, SCEM is used to produce a uniformly valid approximation of the flow. Once again, SCEM leads to a very fruitful analysis of the flow and to the formulation of an interactive boundary layer model.

The appendices bring complements while lightening the main text. At the end of each chapter, detailed problems are given to allow the reader to fully exploit the results described in the corresponding chapters.

Very detailed solutions are given at the end of the book. Certain problems are really research topics and derive from results often unpublished.

This book is the English version of the book entitled "Analyse asymptotique et couche limite" published in French. For most of the chapters, the material is the same in both versions. However, Chap. 9 has been supplemented with examples of application of an IBL method to aerodynamic flows and Chap. 12, dealing with channel flows, is entirely new. These complementary elements provide a further evidence of the efficiency of SCEM.

We do hope that this book will provide the reader with the essential elements, mathematical and practical as well, to understand and to apply the standard asymptotic methods devoted to study boundary layers. In many problems of mathematical physics, these methods form the basis of a sharp understanding of the solution structure, which is often the key for an appropriate numerical solution. In addition, we think that SCEM sheds a new light on the search for a UVA of the solution of problems encompassing a boundary layer. In its regular form, equivalent to MMAE, SCEM provides a complementary point of view of this very efficient technique. With the implementation of generalized expansions, SCEM enables us to bring a rational justification of IBL which was missing until now. Finally, we think that the goal of this work will be reached if generalized SCEM is applied to domains not addressed here. For example, in fluid mechanics, unsteady or three-dimensional boundary layers, instabilities and their control are important topics for the future.

2 Introduction to Singular Perturbation Problems

Mathematical models used in physics often lead to problems which do not have explicit solutions. Their numerical solutions become more difficult when small parameters are present or when the calculation domains are very large. In such cases, simpler models can be developed either by setting a parameter to zero or by restricting the study to a smaller domain. When a small parameter, denoted by ε, is set to zero, it is possible that the solution of the initial problem does not tend uniformly to the solution of the reduced problem as $\varepsilon \to 0$. A *singular perturbation* problem arises for which difficult mathematical questions need to be addressed.

To discuss this further, let us consider an integro-differential operator L_ε and seek a solution $\Phi_\varepsilon(\boldsymbol{x}, \varepsilon)$ of equations $\mathrm{L}_\varepsilon\left[\Phi_\varepsilon(\boldsymbol{x}, \varepsilon)\right] = 0$ where $\boldsymbol{x}$ is a variable in a domain D and where $0 < \varepsilon \leq \varepsilon_0$, ε_0 being a fixed positive number as small as desired. The parameter ε is dimensionless which implies that the whole problem is expressed in terms of dimensionless variables. Let $\mathrm{L}_0\left[\Phi_0(\boldsymbol{x})\right] = 0$ be the so-called reduced problem, supposedly a simpler problem, and let us assume that the norm $\|\Phi_\varepsilon - \Phi_0\|$ is small in the considered domain D. Using the supremum norm (see Subsect. 4.1.4), we have

$$\mathrm{Max_D} \left|\Phi_\varepsilon - \Phi_0\right| < K\delta(\varepsilon) \, ,$$

with K denoting a positive number independent of ε and $\delta(\varepsilon)$ a positive function such that

$$\lim_{\varepsilon \to 0} \delta(\varepsilon) = 0 \, .$$

If this property is satisfied, the problem is called a *regular perturbation problem* (see Problem 2-4).

In some problems, this property is not satisfied, at least in the whole domain D. A singularity can occur, generally in a domain whose dimension is smaller than D. The problem is called a *singular perturbation problem*.

The models considered in this Chapter are such that Φ_ε is known. These pedagogical problems are used to describe the main conceptual difficulties and the different classes of methods used to solve them.

2.1 Regular and Singular Problems

2.1.1 Linear Oscillator

The linear oscillator is a typical example of a regular perturbation problem.
To discuss this further, let us consider the following equation

$$
\mathrm{L}_\varepsilon\, y = \frac{\mathrm{d}^2 y}{\mathrm{d}x^2} + 2\varepsilon \frac{\mathrm{d}y}{\mathrm{d}x} + y = 0 \,,
\tag{2.1a}
$$

subject to the initial conditions

$$
y\big|_{x=0} = 0 \,, \qquad \frac{\mathrm{d}y}{\mathrm{d}x}\bigg|_{x=0} = 1 \,.
\tag{2.1b}
$$

The function $y(x, \varepsilon)$ is defined for $x > 0$ and ε is a small positive parameter
which is as small as desired. All quantities are dimensionless.

This equation models the motion of a mass in a mass-spring-damping
system when the damping is small. The meaning of "small" is important
in the subsequent analysis. Obviously, other physical problems may be of
interest, for example, the case of a small mass.

Let $y^*(t, m, \beta, k, I_0)$ be the location of the mass m as a function of time,
measured from its equilibrium location, k the spring constant and β the
damping coefficient. If the mass is set into motion from its equilibrium location with an impulse I_0, (2.1a) can be written as

$$
m\frac{\mathrm{d}^2 y^*}{\mathrm{d}t^2} + \beta\frac{\mathrm{d}y^*}{\mathrm{d}t} + ky^* = 0 \,,
\tag{2.2a}
$$

subject to the initial conditions

$$
y^*\big|_{t=0} = 0 \,, \qquad m\,\frac{\mathrm{d}y^*}{\mathrm{d}t}\bigg|_{t=0} = I_0 \,.
\tag{2.2b}
$$

Let y and x be dimensionless variables

$$
y = \frac{y^*}{L} \,, \qquad x = \frac{t}{T} \,,
$$

and L and T respectively a length and a time scale, not yet defined. As the
origin of the motion is the impulse, it is quite logical to set

$$
T = \frac{mL}{I_0} \,.
$$

With these new variables, (2.2a) in dimensionless form can be written as

$$\frac{I_0^2}{mL^2k}\frac{\mathrm{d}^2y}{\mathrm{d}x^2} + \frac{\beta I_0}{mLk}\frac{\mathrm{d}y}{\mathrm{d}x} + y = 0 \ , \tag{2.3a}$$

subject to the initial conditions

$$y\big|_{x=0} = 0 \ , \qquad \frac{\mathrm{d}y}{\mathrm{d}x}\bigg|_{x=0} = 1 \ . \tag{2.3b}$$

Two dimensionless groups appear and contain the arbitrary length L which can be defined in two ways

$$L = \frac{I_0}{\sqrt{mk}} \quad \text{or} \quad L = \frac{\beta I_0}{mk} \ .$$

If, according to the physical problem studied, the two-dimensional groups in (2.3a) are not of the same order, an *asymptotic analysis* can be used. Two cases arise:

1. If the action of the spring dominates the action of the damper, the first group $\dfrac{I_0^2}{mL^2k}$ is larger than the second group $\dfrac{\beta I_0}{mLk}$ and

$$L = \frac{I_0}{\sqrt{mk}} \quad \text{and} \quad T = \sqrt{\frac{m}{k}} \ ,$$

so that the small parameter ε is defined by

$$\varepsilon = \frac{\beta}{2\sqrt{mk}} \ .$$

It is seen below that the corresponding problem is typically a *regular perturbation problem* as far as x is bounded. This is the case of a small damping. Equation (2.3a) becomes

$$\frac{\mathrm{d}^2y}{\mathrm{d}x^2} + 2\varepsilon\frac{\mathrm{d}y}{\mathrm{d}x} + y = 0 \ . \tag{2.4}$$

According to Poincaré, the asymptotic behaviour of the solution as $\varepsilon \to 0$ can be sought as an *expansion* in powers of ε

$$y(x,\varepsilon) = y_0(x) + \varepsilon y_1(x) + \varepsilon^2 y_2(x) + \cdots \ . \tag{2.5}$$

As for a Taylor series expansion, the small dots $\cdots$ mean that the neglected terms are smaller than ε^2 and the approximation is better and better when ε is smaller and smaller.

Substituting the expansion into the initial equation and equating coefficients of like powers of ε, the following equations result for the first two powers of ε:

a) $\dfrac{d^2 y_0}{dx^2} + y_0 = 0$ with $y_0|_{x=0} = 0$, $\left.\dfrac{dy_0}{dx}\right|_{x=0} = 1$,

b) $\dfrac{d^2 y_1}{dx^2} + y_1 = -2\dfrac{dy_0}{dx}$ with $y_1|_{x=0} = 0$, $\left.\dfrac{dy_1}{dx}\right|_{x=0} = 0$.

The first problem for y_0 is the *reduced problem* which yields a solution without damping

$$y_0 = \sin x .$$

The second problem for y_1 yields a correction

$$y_1 = -x \sin x ,$$

so that an *approximation* of the solution is

$$y = (1 - \varepsilon x) \sin x + \cdots . \tag{2.6}$$

It is seen that, in any interval of finite time, $0 < x < \tau$, where τ is independent of ε, the approximation is *uniformly valid*; the correction is small. This is no longer true if the time interval becomes large; this is clearly seen by taking $\varepsilon\tau = 1$. This problem is called a *secular problem* because a singularity occurs in the expansion when the time interval is too large. The terminology comes from the study of the trajectories of planets. Solutions are obtained from perturbation methods which are valid on a small time scale, but the value of the secular terms is non realistic over time scales of the order of a century.

The comparison of the above approximation with the exact solution is enlightening. Approximation given by (2.6) is exactly the first terms of a Taylor series expansion of the exact solution

$$y(x, \varepsilon) = \dfrac{e^{-\varepsilon x}}{\sqrt{1 - \varepsilon^2}} \sin \sqrt{1 - \varepsilon^2}\, x .$$

2. In the second case, the mass is small and the length and time scales are

$$L = \dfrac{\beta I_0}{mk} \quad \text{and} \quad T = \dfrac{\beta}{k} .$$

The small parameter ε is defined by

$$\varepsilon = \dfrac{mk}{\beta^2} ,$$

and (2.3a) becomes

$$\varepsilon\dfrac{d^2 y}{dx^2} + \dfrac{dy}{dx} + y = 0 \quad \text{with} \quad y|_{x=0} = 0 , \quad \left.\dfrac{dy}{dx}\right|_{x=0} = 1 . \tag{2.7}$$

This problem is typically a *singular perturbation problem* whose study is precisely the subject of this book.

2.1.2 Secular Problem

We consider the equation

$$\mathrm{L}_\varepsilon\, y = \frac{\mathrm{d}y}{\mathrm{d}x} + \varepsilon y = 0 \,, \tag{2.8a}$$

subject to the initial condition

$$y|_{x=0} = 1 \,, \tag{2.8b}$$

and we seek its solution in the domain $x \geq 0$. Using an expansion as (2.5), we seek an approximation of y in the form

$$y(x, \varepsilon) = y_0(x) + \varepsilon y_1(x) + \varepsilon^2 y_2(x) + \cdots + \varepsilon^n y_n(x) + \cdots \,.$$

Substituting this expression into (2.8a) and equating coefficients of like powers of ε, the following successive equations result:

1. $\dfrac{\mathrm{d}y_0}{\mathrm{d}x} = 0$ with the initial condition $y_0|_{x=0} = 1$.

2. $\dfrac{\mathrm{d}y_1}{\mathrm{d}x} = -y_0$ with the initial condition $y_1|_{x=0} = 0$.

3. $\dfrac{\mathrm{d}y_n}{\mathrm{d}x} = -y_{n-1}$ with the initial condition $y_n|_{x=0} = 0$.

Collecting the solutions for $y_0, y_1, \ldots, y_n$, the result is well-known

$$y(x, \varepsilon) = 1 - \varepsilon x + \varepsilon^2 \frac{x^2}{2} + \cdots + (-1)^n \varepsilon^n \frac{x^n}{n!} + \cdots \,. \tag{2.9}$$

From the exact solution,

$$y(x, \varepsilon) = \mathrm{e}^{-\varepsilon x} \,, \tag{2.10}$$

the difficulty is clearly seen. When x becomes large, for any number of terms considered, the above expansion is no longer valid (Fig. 2.1). The salient feature is that the infinite series converges towards the exact solution for any value of ε and the partial sum is an approximation of the solution when ε is small and x is bounded. The considered expansion is a convergent series whereas the partial sum is the simplest form of an *asymptotic expansion*.

In order to transfer the singularity when x is large to the neighbourhood of the origin, the following change of variable is used

$$t = \frac{1}{x+1} \,.$$

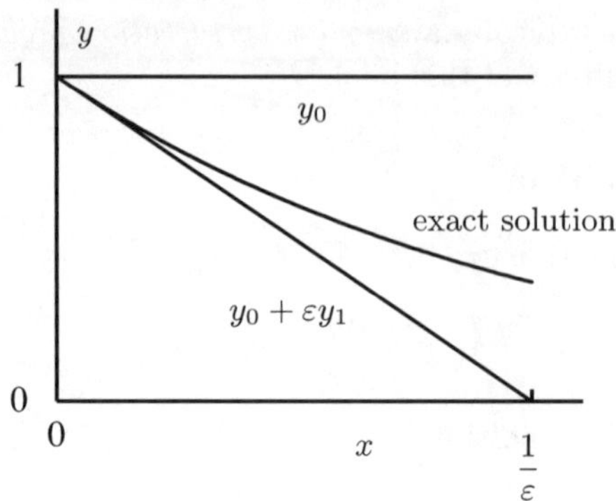

Fig. 2.1. Approximations of the solution of (2.8a) given by (2.9). The exact solution y is given by (2.10)

By setting
$$Y(t, \varepsilon) \equiv y(x, \varepsilon) \,,$$

we can write (2.8a) as

$$\mathrm{L}_\varepsilon\, Y = t^2 \frac{\mathrm{d}Y}{\mathrm{d}t} - \varepsilon Y = 0 \,, \tag{2.11a}$$

subject to the initial condition

$$Y|_{t=1} = 1 \,. \tag{2.11b}$$

A straightforward expansion,

$$Y(t, \varepsilon) = Y_0(t) + \varepsilon Y_1(t) + \varepsilon^2 Y_2(t) + \cdots \,,$$

leads to the approximation

$$Y(t, \varepsilon) = 1 + \varepsilon\left(1 - \frac{1}{t}\right) + \varepsilon^2\left(\frac{1}{2} - \frac{1}{t} + \frac{1}{2t^2}\right) + \cdots \,. \tag{2.12}$$

The successive approximations are more and more singular in the neighbourhood of the origin (Fig. 2.2). This is clear by expanding the exact solution

$$Y(t, \varepsilon) = \exp\left[-\varepsilon\left(\frac{1}{t} - 1\right)\right] \,. \tag{2.13}$$

This feature is also present in similar problems to which a special treatment can be applied. Let us consider the equation

$$\mathrm{L}_\varepsilon\, y = (x + \varepsilon y)\frac{\mathrm{d}y}{\mathrm{d}x} + y = 0 \quad \text{with} \quad y|_{x=1} = 1 \,. \tag{2.14}$$

and let us seek its solution in the domain $0 \le x \le 1$.

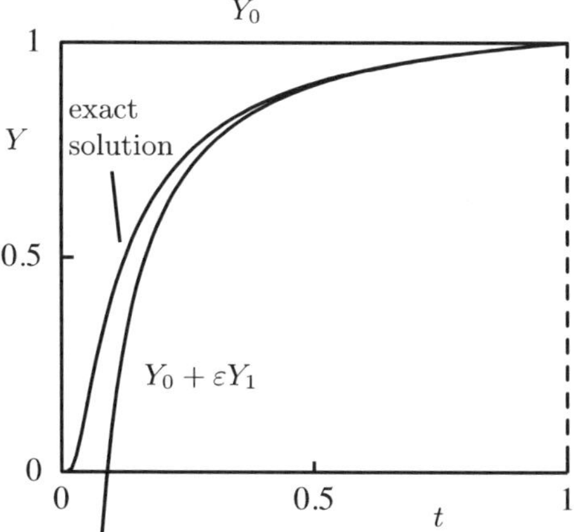

Fig. 2.2. Approximations of the solution of (2.11a) given by (2.12). The exact solution Y is given by (2.13)

The expansion

$$y(x, \varepsilon) = y_0(x) + \varepsilon y_1(x) + \cdots$$

leads to the following equations:

1. $x\dfrac{\mathrm{d}y_0}{\mathrm{d}x} + y_0 = 0$ with $y_0\big|_{x=1} = 1$.

2. $x\dfrac{\mathrm{d}y_1}{\mathrm{d}x} + y_1 = -y_0\dfrac{\mathrm{d}y_0}{\mathrm{d}x}$ with $y_1\big|_{x=1} = 0$.

The result

$$y(x, \varepsilon) = \frac{1}{x} + \varepsilon\frac{1}{2x}\left(1 - \frac{1}{x^2}\right) + \cdots \tag{2.15}$$

clearly shows that the second approximation is more singular than the first one in the neighbourhood of the origin (Fig. 2.3). The exact solution,

$$y(x, \varepsilon) = -\frac{x}{\varepsilon} + \sqrt{\frac{x^2}{\varepsilon^2} + \frac{2}{\varepsilon} + 1} \, , \tag{2.16}$$

is bounded at the origin,

$$y(0, \varepsilon) = \sqrt{\frac{2}{\varepsilon} + 1} \, ,$$

for any value of $\varepsilon > 0$. This is typical of *secular problems*.

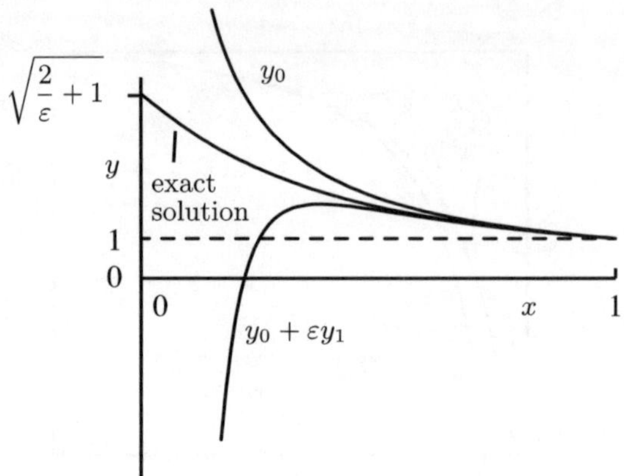

Fig. 2.3. Approximations of the solution of (2.14) given by (2.15). The exact solution y is given by (2.16)

2.1.3 Singular Problem

The archetype of a *singular problem* has been introduced by Friedrichs [36] to justify the matching between the boundary layer and the inviscid flow as proposed by Prandtl [78]. We consider the equation

$$L_\varepsilon\, y = \varepsilon \frac{d^2y}{dx^2} + \frac{dy}{dx} - a = 0 \quad \text{with} \quad 0 < a < 1 \,, \tag{2.17a}$$

subject to the boundary conditions

$$y|_{x=0} = 0 \,, \qquad y|_{x=1} = 1 \,, \tag{2.17b}$$

and we seek its solution in the domain $0 \leq x \leq 1$. This is a boundary value problem which is more difficult than an initial value problem. The exact solution is known as in all the problems considered in this Chapter. The reduced problem obtained for $\varepsilon = 0$ is

$$L_0\, y_0 = \frac{dy_0}{dx} - a = 0 \,,$$

with the solution given by

$$y_0 = ax + A \,.$$

Here A is a constant that must be determined with two boundary conditions. In general, it is not possible to satisfy both conditions. This feature is characteristic of certain singular problems. When $\varepsilon = 0$, the order of the reduced equation becomes lower than the order of the initial equation.

If the boundary condition at $x = 0$ is enforced, the solution becomes

$$y_0 = ax \,.$$

This approximation cannot be uniformly valid since $y_0(1) = a$. Similarly, by enforcing the boundary condition at $x = 1$, the solution becomes

$$y_0 = ax + 1 - a ,\qquad(2.18)$$

which is such that $y_0(0) = 1 - a$. The boundary condition at the origin is not satisfied which indicates necessarily a domain of non-uniformity.

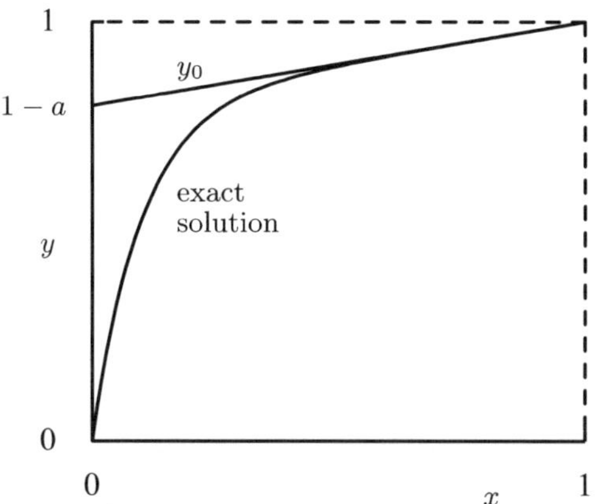

Fig. 2.4. Approximation of the solution of (2.17a, 2.17b) given by (2.18). The exact solution y is given by (2.19)

The exact solution to (2.17a) is

$$y(x, \varepsilon) = ax + (1 - a)\frac{1 - \exp\left(-x/\varepsilon\right)}{1 - \exp\left(-1/\varepsilon\right)} .\qquad(2.19)$$

For $x > 0$, as $\varepsilon \to 0$, it is seen that a good approximation of the exact solution is given by (Fig. 2.4)

$$y = ax + 1 - a + \cdots .$$

This shows that the reduced problem should satisfy the boundary condition at $x = 1$. The domain of non-uniformity is located in the neighbourhood of the origin.

How to answer these questions without knowing the exact solution? The first hints are presented in the next sections.

2.2 Approximation Methods for Singular Perturbation Problems

Many methods have been proposed to solve singular perturbation problems [6, 38, 42, 43, 72, 108, 112]. The most popular methods are briefly described below.

2.2.1 Method of Matched Asymptotic Expansions

The *method of matched asymptotic expansions*, MMAE, has been the subject
of many in-depth mathematical studies and has been used in many practical problems. The underlying ideas have been developed after 1950 when
Friedrichs proposed his model. Afterwards, these ideas have been worked
out and applied to viscous flow equations. Kaplun [45], Lagerstrom [47, 48],
Cole [17] and Van Dyke [107] are among the most important names associated
with the development of MMAE. The most thorough study on the foundations of the method is due to Eckhaus [33, 34]. Despite all the valuable work
devoted to MMAE, it is not possible to formulate a general mathematical
theory of the method. Heuristic rules are available and the applications to
problems of mathematical physics, especially in fluid mechanics, have been
remarkably fruitful.

Considering again Friedrichs' model (2.17a), the examination of the exact
solution shows that

$$\lim_{\varepsilon \to 0} y(x, \varepsilon) = y_0(x) = ax + 1 - a \; ,$$

except in the neighbourhood of the origin where the boundary condition
requires $y|_{x=0} = 0$, whereas $y_0|_{x=0} = 1 - a$.

Two comments play an essential role subsequently.

1. If the limit process expansion is performed with the variable $X = x/\varepsilon$
instead of the variable x, it is obtained

$$\lim_{\varepsilon \to 0} y(x, \varepsilon) = Y_0(X) = (1 - a)\left(1 - e^{-X}\right) \; .$$

This procedure is suggested by the desire to take into account the exponential term. A better approximation is expected because the variable X covers a domain closer to the origin than the variable x. Indeed, the condition
$Y_0|_{x=0} = 0$ is satisfied. However, the condition at $x = 1$ is no longer satisfied
since

$$Y_0|_{x=1} = (1 - a)\left(1 - e^{-1/\varepsilon}\right) \; .$$

This result can be surprising but it must be noted that X belongs to a very
wide domain

$$0 \leq X \leq \frac{1}{\varepsilon} \; ,$$

and terms which are neglected when X is bounded can be non negliglible in
the whole domain.

2. The second comment is the basis of the idea leading to the *asymptotic
matching*. At this point of the discusion, it suffices to note the following
remarkable result

$$\lim_{X \to \infty} Y_0(X) = \lim_{x \to 0} y_0(x) = 1 - a \; .$$

These two comments have been done from the behaviour of the exact solution. Suppose now that the exact solution is not known. How can we conceive a heuristic method to construct an approximation of the solution?

First step. The reduced problem yields

$$y_0 = ax + A \ .$$

In order to determine the constant A, one of the two boundary conditions can be used but the question is to know which boundary condition must be taken into account. It is not even certain that A is determined from a boundary condition. The answer to these questions is discussed for the type of differential equations studied in Chaps. 3 and 6. For equations modelling a physical problem, the answer is guided by the physics of phenomena. For example, in the study of the viscous flow past a flat plate at high Reynolds number, the reduced problem obtained from the Navier-Stokes equations comprises the Euler equations for which the no-slip condition at the wall cannot be applied.

We assume that this problem is solved a priori and that the domain of non-uniformity is known. In the case of Friedrichs' model, this means that the solution of the reduced problem is

$$y_0(x) = ax + 1 - a \ .$$

Second step. The boundary condition at the origin is not satisfied because, in the first step, the solution is related to a region too far from the origin. In order to restore the behaviour of the solution near the origin, it is required to magnify the neighbourhood of the origin by introducing a change of variable

$$X_\alpha = \frac{x}{\varepsilon^\alpha} \ ,$$

where α is a strictly positive number. Then, when x is small, X_α must remain bounded. By setting

$$Y_\alpha(X, \varepsilon) \equiv y(x, \varepsilon) \ ,$$

the governing equation (2.17a) becomes

$$\varepsilon^{1-2\alpha} \frac{\mathrm{d}^2 Y_\alpha}{\mathrm{d}X_\alpha^2} + \varepsilon^{-\alpha} \frac{\mathrm{d}Y_\alpha}{\mathrm{d}X_\alpha} = a \ .$$

Now, the value of α must be tuned for the best. If $\alpha < 1$ or if $\alpha > 1$, the resulting reduced problem leads to a solution unable to reproduce the sharp variation near the origin. Noting that the second derivative must be kept, another choice is the next dominating term be the first derivative. From inspection, it is clear that the optimal choice is $\alpha = 1$.

By setting

$$X_1 = X \quad \text{and} \quad Y_1 = Y \,,$$

the equation becomes

$$\frac{\mathrm{d}^2 Y}{\mathrm{d}X^2} + \frac{\mathrm{d}Y}{\mathrm{d}X} = \varepsilon a \,.$$

Third step. For the above equation, the reduced problem is

$$\frac{\mathrm{d}^2 Y_0}{\mathrm{d}X^2} + \frac{\mathrm{d}Y_0}{\mathrm{d}X} = 0 \,.$$

The general solution is

$$Y_0(X) = A + B e^{-X} \,,$$

where A and B are two undetermined constants. It is natural to satisfy the condition at the origin, which yields

$$Y_0(X) = A \left(1 - e^{-X} \right) \,.$$

It seems now that the second boundary condition at $x = 1$ can also be satisfied. However, the result is false because the domain covered by X is very wide. In reality, in the same manner as $y_0(x)$ is not an approximation valid near the origin, $Y_0(X)$ can not be valid when X is not bounded, especially in the neighbourhood of $x = 1$.

Fourth step. In order to find the missing condition, we assume that an *overlap domain* must exist in which the behaviour of y_0 for small x identifies with the behaviour of $Y_0(X)$ for large X. This can be formulated as the search for an *intermediate domain* formalized with the variable $X_\beta = x/\varepsilon^\beta$. For $0 < \beta < 1$, we obtain

$$y_0(x) = 1 - a + a \varepsilon^\beta X_\beta = 1 - a + \cdots \,,$$
$$Y_0(X) = A \left(1 - e^{-X_\beta / \varepsilon^{1-\beta}} \right) = A + \cdots \,.$$

It is seen that if X_β is kept fixed and if $\varepsilon \to 0$, we obtain $A = 1 - a$.

In this manner, the approximation valid near the origin is found by a technique of matching called later *intermediate matching*

$$Y_0(X) = (1 - a) \left(1 - e^{-X} \right) \,.$$

A more straightforward method consists of taking the limit. The so-called principle of *asymptotic matching*

$$\lim_{X \to \infty} Y_0(X) = \lim_{x \to 0} y_0(x) \,,$$

gives the same value of A because the limits exist. In Chapter 5, it will be seen that if such a principle is easier to implement, the above formulation is too much straightforward.

Fifth step. A *uniformly valid approximation*, UVA, is tentatively constructed by adding the two approximations obtained in their respective domain of validity and by subtracting the common part,

$$y_{\text{app}} = y_0(x) + Y_0(X) - (1 - a) \,,$$

so that

$$y_{\text{app}} = ax + (1 - a)\left(1 - e^{-X}\right) \,. \tag{2.20}$$

It can be checked that y_{app} reduces to $y_0(x)$ and $Y_0(X)$ in their respective domain of validity.

In Fig. 2.5, the exact solution is compared to the composite solution given by (2.20) for $a = 0.2$ and $\varepsilon = 0.25$. We see that the approximation is very good even if the value of ε is not really small. For smaller values of ε, the approximation becomes better. In fact, the smallness of ε is always difficult to estimate.

The ideas described above constitute the basis on which MMAE is constructed.

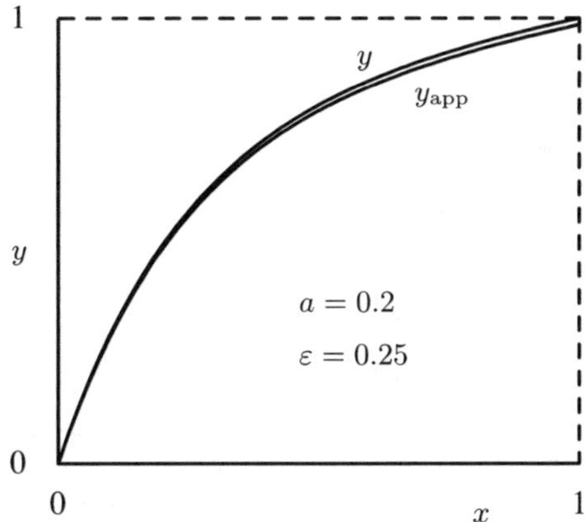

Fig. 2.5. Study of (2.17a, 2.17b). The composite solution y_{app} is given by (2.20). The exact solution y is given by (2.19)

2.2.2 Successive Complementary Expansion Method

As proposed earlier [26, 75], a method consists, at once, of seeking a UVA of the solution by assuming that $y_0(x)$ is known. The approximation has the form

$$y_{\text{a}1} = y_0(x) + Y_0^*(X) \,.$$

For Friedrichs' equation (2.17a), we have

$$
\mathrm{L}_\varepsilon\, y_{\mathrm{a}1} = \varepsilon\frac{\mathrm{d}^2 y_0}{\mathrm{d}x^2} + \frac{\mathrm{d}y_0}{\mathrm{d}x} - a + \frac{1}{\varepsilon}\left[\frac{\mathrm{d}^2 Y_0^*}{\mathrm{d}X^2} + \frac{\mathrm{d}Y_0^*}{\mathrm{d}X}\right] = \frac{1}{\varepsilon}\left[\frac{\mathrm{d}^2 Y_0^*}{\mathrm{d}X^2} + \frac{\mathrm{d}Y_0^*}{\mathrm{d}X}\right] .
$$

This case is very particular because $\dfrac{\mathrm{d}^2 y_0}{\mathrm{d}x^2}$ is zero. If the right hand side is set to zero and if the boundary conditions are satisfied, the solution is

$$
Y_0^*(X) = A + Be^{-X} ,
$$

subject to boundary conditions

$$
Y_0^*(0) = a - 1 \quad \text{and} \quad Y_0^*\left(\frac{1}{\varepsilon}\right) = 0 ,
$$

which yields

$$
Y_0^*(X) = (1 - a)\frac{e^{-1/\varepsilon} - e^{-X}}{1 - e^{-1/\varepsilon}} ,
$$

and, adding to $y_0(x)$, the exact solution is obtained.

The above approach is not usual since Y_0^* depends not only on X *but also on ε*. In the asymptotic method developed in the following Chapters, the functions dependent on ε are clearly separated from the functions independent of ε. Accepting their dependence in ε leads us to a new method called the *successive complementary expansion method*, SCEM.

The method proposed earlier insists on having an independence of Y_0^* with respect to ε. This requirement is achieved by neglecting the terms in $e^{-1/\varepsilon}$ which are very small

$$
Y_0^*(X) = (a - 1)e^{-X} ,
$$

and, adding to y_0, the same approximation (2.20) as in MMAE is recovered

$$
y_{\mathrm{app}} = ax + (1 - a)\left(1 - e^{-X}\right) .
$$

In Chapter 5, it is shown that if the independence with respect to ε is required, SCEM is equivalent to MMAE. As MMAE is relatively easier to implement, the earlier form of SCEM is not used very much. Nevertheless, it must be noted that the asymptotic matching principle is equivalent to the assumed form of the UVA.

2.2.3 Multiple Scale Method

The underlying idea of the method, due to Mahony [62], is based on the search for a UVA. In Friedrichs' model for example, it is known that a UVA cannot be described by a single variable x; another variable X is required. In contrast with SCEM, the structure of the solution is not assumed but

$$
y(x,\varepsilon) \equiv Y(x, X, \varepsilon) \quad \text{with} \quad X = \frac{x}{\varepsilon} \tag{2.21}
$$

is set with the two variables x and X considered *independent*.

The initial equation (2.17a) becomes a partial differential equation

$$\frac{\partial^2 Y}{\partial X^2} + \frac{\partial Y}{\partial X} + \varepsilon \left(2\frac{\partial^2 Y}{\partial x \partial X} + \frac{\partial Y}{\partial x} \right) + \varepsilon^2 \frac{\partial^2 Y}{\partial x^2} = \varepsilon a \ .$$

The function Y being defined in the rectangle $\left[0 \leq x \leq 1, \quad 0 \leq X \leq \frac{1}{\varepsilon} \right]$, the avalaible boundary conditions are insufficient to determine the solution. However, the goal is not to find the exact solution but an approximate one. An expansion is sought in the form

$$Y(x, X, \varepsilon) = Y_0(x, X) + \varepsilon Y_1(x, X) + \mathrm{O}(\varepsilon^2) \ .$$

Two reduced equations result

1. $\dfrac{\partial^2 Y_0}{\partial X^2} + \dfrac{\partial Y_0}{\partial X} = 0$ with the conditions $Y_0(0,0) = 0$ and $Y_0(1, \infty) = 1$,

2. $\dfrac{\partial^2 Y_1}{\partial X^2} + \dfrac{\partial Y_1}{\partial X} = a - \left(2\dfrac{\partial^2 Y_0}{\partial x \partial X} + \dfrac{\partial Y_0}{\partial x} \right)$.

The general solution of the first equation is

$$Y_0(x, X) = A(x) + B(x)e^{-X} \ .$$

The boundary conditions are insufficient to determine the functions $A(x)$ and $B(x)$ since they give

$$A(0) + B(0) = 0 \ ,$$
$$A(1) = 1 \ .$$

However, the second equation gives

$$\frac{\partial^2 Y_1}{\partial X^2} + \frac{\partial Y_1}{\partial X} = a - \frac{\mathrm{d}A}{\mathrm{d}x} + \frac{\mathrm{d}B}{\mathrm{d}x}e^{-X} \ ,$$

which leads to

$$Y_1(x, X) = C(x) + D(x)e^{-X} + X\left(a - \frac{\mathrm{d}A}{\mathrm{d}x} \right) - \frac{\mathrm{d}B}{\mathrm{d}x}Xe^{-X} \ .$$

Then, as in Poincaré–Lighthill's method of strained coordinates discussed in Subsect. 2.2.4, it is plausible to assume that *higher approximations shall be no more singular than the first*. This means that the ratio Y_1/Y_0 must be bounded, independent of ε in the whole considered domain. Then, we set

$$a - \frac{\mathrm{d}A}{\mathrm{d}x} = 0 \ ,$$
$$\frac{\mathrm{d}B}{\mathrm{d}x} = 0 \ .$$

These differential equations are solved with the help of the boundary conditions and we obtain

$$A(x) = ax + 1 - a \,,$$
$$B(x) = a - 1 \,.$$

Knowing $A(x)$ and $B(x)$, the solution Y_0 is expressed by

$$Y_0(x, X) = ax + (1 - a)\left(1 - e^{-X}\right) \,,$$

which is again the approximation obtained from MMAE or from SCEM.

2.2.4 Poincaré–Lighthill's Method

The roots of this method are older but the applications are more limited. A paper by Poincaré dates back to 1892 and he attributed the basic ideas to Lindstedt. Later, in 1949, Lighthill introduced a more general version of the method and Kuo published two papers where the method was applied to viscous flow problems. Tsien, in a review paper published in 1956, called this method the PLK (Poincaré, Lighthill, Kuo) method. Anglo-Saxon authors prefer to call it Lighthill's method or the *strained coordinates method*. To pay tribute to a great mathematician and to a great fluid mechanicist, the method is called here the PL method.

We consider (2.14) already studied

$$L_\varepsilon\, y = (x + \varepsilon y)\frac{\mathrm{d}y}{\mathrm{d}x} + y = 0 \quad \text{with} \quad y|_{x=1} = 1 \,, \tag{2.22}$$

and we seek its solution in the domain $0 \le x \le 1$.

The exact solution is singular on the line $2x = -\varepsilon y$ (Fig. 2.6) and the search for approximations when ε is small transfers the singularity towards $x = 0$. Instead of improving the situation, the next approximations are more and more singular. The idea is to claim that the approximations given by the straightforward expansion have the good shape but not at the right place. Then, y and x are expanded with respect to ε and with respect to a new variable s which replaces x. In a sense, the variable x is slightly strained in such a way that

$$y(x, \varepsilon) = y_0(s) + \varepsilon y_1(s) + \varepsilon^2 y_2(s) + \cdots \,, \tag{2.23a}$$
$$x(s, \varepsilon) = s + \varepsilon x_1(s) + \varepsilon^2 x_2(s) + \cdots \,. \tag{2.23b}$$

By substituting into the original equation and by equating coefficients of like powers of ε, the first two equations are obtained

$$s\frac{\mathrm{d}y_0}{\mathrm{d}s} + y_0 = 0 \quad \text{with} \quad y_0|_{s=1} = 1 \,,$$
$$s\frac{\mathrm{d}y_1}{\mathrm{d}s} + y_1 = -\frac{\mathrm{d}y_0}{\mathrm{d}s}\left(x_1 + y_0 - s\frac{\mathrm{d}x_1}{\mathrm{d}s}\right) \,.$$

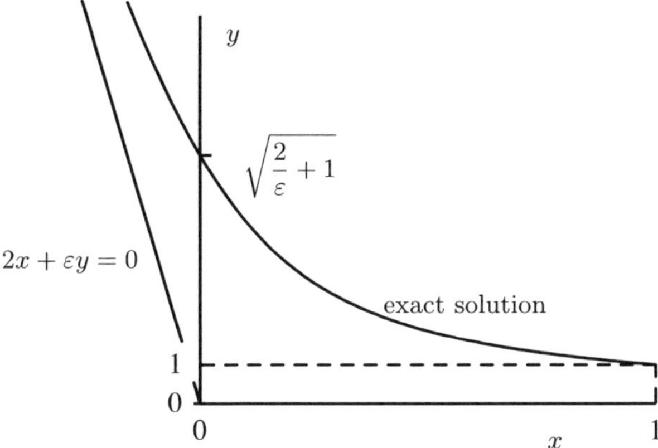

Fig. 2.6. Study of (2.22). The exact solution y is given by (2.16)

The solution of the first equation is

$$y_0(s) = \frac{1}{s} \,,$$

which is identical to the one obtained from the straightforward expansion with x replaced by s.

The second equation, written as

$$\frac{\mathrm{d}}{\mathrm{d}s}(sy_1) = \frac{1}{s^2}\left(x_1 + \frac{1}{s} - s\frac{\mathrm{d}x_1}{\mathrm{d}s}\right) ,$$

has the general solution given by

$$y_1(s) = \frac{A}{s} - \frac{1}{s^2}\left[x_1(s) + \frac{1}{2s}\right] ,$$

where A is an arbitrary constant.

The basic principle formulated by Lighthill is to state that *higher approximations shall be no more singular than the first.*

The unknown function $x_1(s)$ is determined by setting

$$\frac{x_1(s)}{s} + \frac{1}{2s^2} = B(s) ,$$

where $B(s)$ is a bounded function of s. Then, the second order solution yields

$$y_1(s) = \frac{A}{s} - \frac{B(s)}{s} \,.$$

A feature of this method is that the function $x_1(s)$ is not completely determined. Any regular function of s can be chosen for B. For obvious reasons, it is useful that $x_1(s)$ is not zero at $x = 1$. Moreover, simplicity being a good guide, B is taken as a constant. The resulting solution is

$$y(x, \varepsilon) = \frac{1}{s} + \cdots \,,$$

$$x(s, \varepsilon) = s + \frac{\varepsilon}{2}\left(s - \frac{1}{s}\right) + \cdots \,.$$

In this model problem, s can be eliminated and the exact solution is recovered.

Note. The PL method does not apply to Friedrichs' model whereas MMAE does.

2.2.5 Renormalization Group Method

The renormalization group method [15] applies particularly to oscillatory problems. Nevertheless, interesting applications to boundary layer and secular problems have been obtained. The general idea is to give some freedom to the integration constants in order to eliminate the further singularities or to accelerate the convergence of the asymptotic expansion. The contents of the *renormalization group method* is certainly fundamental but its implementation is delicate so that a detailed account is not given.

It suffices here to describe the application to the simplest secular problem

$$L_\varepsilon \, y = \frac{dy}{dt} + \varepsilon y = 0 \,. \tag{2.24}$$

The straightforward solution contains a singularity to the second order when t is large. The "naive" asymptotic exapnsion to this order is

$$y(t, \varepsilon) = A_0 \left[1 - \varepsilon(t - t_0)\right] + \cdots \,,$$

where A_0 and t_0 are two integration constants determined by the initial condition which is not specified. Obviously, this expansion is not uniformly valid when t is large. Taking into account the order of the expansion, we set

$$A_0 = \left[1 + \varepsilon a_1(t_0, \mu)\right] A(\mu) \,. \tag{2.25}$$

In this expression μ is an arbitrary time, A is the renormalized part of A_0 and a_1 is an unknown function. To the considered order, we have

$$y = A(\mu) \left[1 + \varepsilon a_1(t_0, \mu) - \varepsilon(t - \mu) - \varepsilon(\mu - t_0)\right] + \cdots \,.$$

By setting

$$a_1 = \mu - t_0 \,,$$

the divergent part due to t_0 is eliminated and we get

$$y = A(\mu) \left[1 - \varepsilon(t - \mu)\right] + \cdots \,.$$

This form is identical to the naive expression but μ is arbitrary. The renormalization criterion is given by

$$\frac{\partial y}{\partial \mu} = 0$$

for any time t. Then, the differential equation for A is obtained

$$\frac{\mathrm{d}A}{\mathrm{d}\mu} + \varepsilon A = 0 \;,$$

which yields the solution

$$y = A_1 e^{-\varepsilon\mu} \left[1 - \varepsilon(t - \mu) \right] + \cdots \;,$$

where A_1 is a constant.

Setting $\mu = t$, a UVA to the desired order is obtained

$$y(t, \varepsilon) = A_1 e^{-\varepsilon t} + \cdots \;.$$

This approximation is nothing else than the exact solution but the model is very simple.

2.3 Conclusion

The singular perturbation problems are often encountered in physics and many approximation methods have been proposed to solve them. An idea shared in almost all these methods is to correct or to avoid the non-uniformly valid character of a first approximation. The method of matched asymptotic expansion, MMAE, follows this logic. The MMAE consists of seeking approximations in different significant domains and these approximations are matched to render the solution uniformly valid.

The next Chapters are devoted to the construction and to the application of the successive complementary expansion method, SCEM. In its regular form, SCEM leads to the same results as MMAE but without requiring the delicate notion of matching principle. In its non regular form, this method provides definite advantages.

Problems

2-1. Consider the equation

$$x^2 + \varepsilon x - 1 = 0 \;.$$

Solutions are sought when ε is a small parameter.

1. Give the exact solutions and apply a Taylor series expansion near $\varepsilon = 0$ to order ε^2.

2. An iterative method of solution is proposed by writing the equation as

$$x = \pm\sqrt{1 - \varepsilon x} \, .$$

The iterative process is

$$x_n = \pm\sqrt{1 - \varepsilon x_{n-1}} \, .$$

The starting value x_0 is solution of the reduced equation by setting $\varepsilon = 0$. By using partial sums of a Taylor series expansion, give the expansions of solutions obtained by improving the approximation at each step.

3. We assume that the solution has the form

$$x = x_0 + \varepsilon x_1 + \varepsilon^2 x_2 + \cdots \, .$$

Give the values of x_0, x_1, x_2.

4. We set

$$x = x_0 + \delta_1(\varepsilon)x_1 + \delta_2(\varepsilon)x_2 + \cdots \, ,$$

where the sequence δ_1, δ_2 is such that $\delta_2/\delta_1 \to 0$ and $\delta_1 \to 0$ as $\varepsilon \to 0$. Choose δ_1, δ_2 as simply as possible.

2-2. Consider the equation

$$\varepsilon x^2 + x - 1 = 0 \, .$$

The roots are sought when ε is a small parameter.

1. Give the exact solutions and their expansions as $\varepsilon \to 0$.

2. We want to determine the roots by using an iterative process. The reduced equation, obtained by setting $\varepsilon = 0$ has a single root $x = 1$. The other root is lost. The problem facing us is singular. Show that there are two iterative processes, one given by

$$x_n = 1 - \varepsilon x_{n-1}^2 \, ,$$

and the other one given by

$$x_n = -\frac{1}{\varepsilon} + \frac{1}{\varepsilon x_{n-1}} \, ,$$

which enable us to find the results of the previous question.

3. We assume that the roots can be expanded as

$$x^{(1)} = x_0^{(1)} + \varepsilon x_1^{(1)} + \varepsilon^2 x_2^{(1)} + \cdots \, ,$$

$$x^{(2)} = \frac{x_{-1}^{(2)}}{\varepsilon} + x_0^{(2)} + \varepsilon x_1^{(2)} + \cdots \, .$$

Give the coefficients of these expansions.

2-3. Consider the following eigenvalue problem

$$\frac{\mathrm{d}^2 f}{\mathrm{d}x^2} + \lambda^2 f(x) = 0 , \quad \lambda > 0 , \quad \varepsilon \le x \le \pi ,$$

with the boundary conditions

$$f(\varepsilon) = 0 , \quad f(\pi) = 0 .$$

1. Determine the exact solution. In particular, give the set of eigenvalues λ. Give an expansion of λ to order ε.
2. To illustrate the use of a perturbation method, we set

$$f = \varphi_0 + \varepsilon\varphi_1 + \cdots , \quad \lambda = \lambda_0 + \varepsilon\lambda_1 + \cdots .$$

Express the boundary conditions. For the condition at $x = \varepsilon$, an expansion of φ_0 and φ_1 near $x = 0$ will be performed, in such a way that the condition at $x = \varepsilon$ is transferred at $x = 0$.

Determine φ_0, φ_1, λ_0, λ_1. Compare to the exact solution.

2-4. This problem has been proposed by Van Dyke [108]. Consider a two-dimensional, incompressible, inviscid flow. The continuity equation,

$$\frac{\partial u}{\partial x} + \frac{\partial v}{\partial y} = 0 ,$$

leads us to introduce a stream-function ψ such that

$$u = \frac{\partial \psi}{\partial y} , \quad v = -\frac{\partial \psi}{\partial x} .$$

With the stream-function, the continuity equation is automatically satisfied. Moreover, if the flow is steady and inviscid, the curl of the velocity (vorticity) is constant along a streamline. Then, if the freestream (at upstream infinity) is irrotational, the flow is irrotational everywhere. With these conditions, the stream-function satisfies the equation

$$\Delta \psi = 0 .$$

The streamlines are defined by $\psi = \mathrm{cst}$ since the variations of ψ are such that

$$\mathrm{d}\psi = u\mathrm{d}y - v\mathrm{d}x .$$

The above hypotheses are satisfied if a circular cylinder is placed in a uniform, inviscid flow. In polar coordinates, the stream-function is given by

$$\psi = U_\infty(r - \frac{a^2}{r}) \sin\theta ,$$

where $r = 0$ is the center of the circle and a is its radius. The modulus of the freestream velocity is U_∞ and its direction is $\theta = 0$.

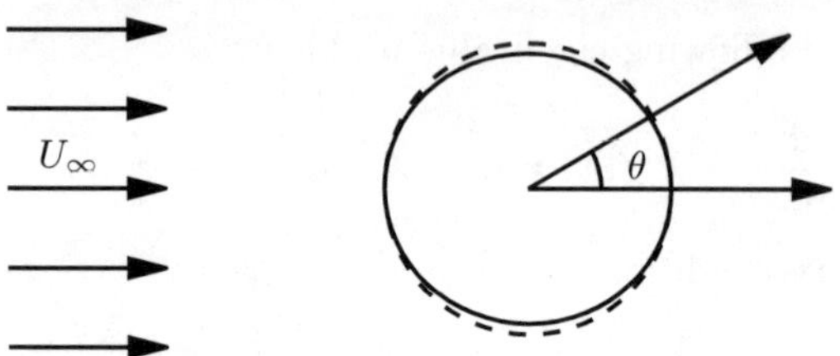

Fig. 2.7. Slightly distorted circular cylinder placed in a uniform flow

The flow around a slightly distorted circular cylinder is studied. The equation of the body is

$$r = a(1 - \varepsilon \sin^2 \theta) \ .$$

This problem is treated with a regular expansion

$$\psi(r, \theta, \varepsilon) = \psi_0(r, \theta) + \varepsilon \psi_1(r, \theta) + \cdots \ .$$

1. Write the equations for ψ_0 and ψ_1. Give the boundary conditions. It is reminded that, with the hypothesis of an inviscid flow, the velocity is tangent to the body wall. The wall is a streamline defined by $\psi = 0$.
2. Give the expession of ψ_1 knowing that the general solution of equation $\Delta \psi = 0$ with the appropriate symmetry conditions is $\sum b_n r^n \sin n\theta$ where n is an integer, positive or negative. It is reminded that

$$\sin^3 \alpha = \frac{1}{4}(3 \sin \alpha - \sin 3\alpha) \ .$$

3. Give the expression of the wall velocity to order ε.
2-5. This problem has been proposed by Van Dyke [108]. Consider a two-dimensional, incompressible, steady, inviscid flow.

We study the flow around a circular cylinder of radius a. The freestream is slightly sheared

$$U^* = U_\infty \left(1 + \varepsilon \frac{y^{*2}}{a^2}\right) \ .$$

The curl of velocity $\omega^* = -\dfrac{\partial u^*}{\partial y^*} + \dfrac{\partial v^*}{\partial x^*}$ is constant along a streamline. Then, ω^* is function only of ψ^*. The equation for ψ^* is

$$\Delta \psi^* = -\omega^* \ ,$$

with $\psi^* = 0$ at $r^* = a$ and $\psi^* \to U_\infty \left(y^* + \dfrac{1}{3}\varepsilon \dfrac{y^{*3}}{a^2}\right)$ when $r^* \to \infty$.

1. Nondimensionalize the problem with the help of U_∞ and a. The dimensionless quantities are denoted without star.

At first, ω is expressed as function of ψ. Using a perturbation method, show that

$$\omega = -2\varepsilon\psi + \frac{2}{3}\varepsilon^2\psi^3 + \cdots .$$

For that, at upstream infinity, the relation $y(\psi)$ will be determined from an iterative method by writing

$$y_n = \psi - \frac{1}{3}\varepsilon y_{n-1}^3 .$$

2. The solution is sought using the following expansion

$$\psi = \psi_0 + \varepsilon\psi_1 + \cdots .$$

Give the equations for ψ_0 and ψ_1 and give the boundary conditions.

Give the solution for ψ_0 and ψ_1. It will be shown that the solution for ψ_1 is

$$\psi_1 = \frac{1}{3}r^3\sin^3\theta - r(\ln r)(\sin\theta) - \frac{1}{4}\frac{1}{r}\sin\theta + \frac{1}{12}\frac{1}{r^3}\sin 3\theta .$$

It is reminded that the expression of a laplacian in polar coordinates is

$$\triangle f = \frac{\partial^2 f}{\partial r^2} + \frac{1}{r}\frac{\partial f}{\partial r} + \frac{1}{r^2}\frac{\partial^2 f}{\partial\theta^2} .$$

Comment the result, in particular as $r \to \infty$.

3 Boundary Layer Structure

The singular perturbation problems considered in this Chapter are based on second order linear differential equations with variable coefficients. We are interested in boundary value problems because not only local existence theorems do not exist but also because these problems are more difficult to treat than initial value problems. Moreover, the solution is not known analytically. Finally if, for a well-posed physical problem, it is relatively easy to localize the domain of non-uniformity, this is not the case here because the problem is purely mathematical. The abstract formulation of the problem does not enable us to determine, a priori, the localization of the boundary layer. The method described in this Chapter to localize the boundary layer is standard and can be used as a good guide for more complex problems.

3.1 Study of a Second Order Differential Equation

In order to illustrate the question of localizing the boundary layer in a singular perturbation problem, we choose to work with a second order differential equation in which the second order derivative is multiplied by a small parameter. The chosen equation does not refer to any physical problem; it is simply a model equation for a singular perturbation problem exhibiting a boundary layer structure. The general form of the equation is

$$\mathrm{L}_\varepsilon\, y = \varepsilon \frac{\mathrm{d}^2 y}{\mathrm{d}x^2} + a\left(x\right)\frac{\mathrm{d}y}{\mathrm{d}x} + b\left(x\right) y = 0 \, , \tag{3.1a}$$

subject to the following boundary conditions

$$y\big|_{x=0} = \alpha \, , \quad y\big|_{x=1} = \beta \, . \tag{3.1b}$$

The function $y\left(x, \varepsilon\right)$ is defined on the domain $x \in [0, 1]$ and ε is a small positive parameter. All quantities are dimensionless.

Below, we seek an approximation of the solution. The functions $a(x)$ and $b(x)$ are defined and continuous on the domain $x \in [0, 1]$. The required additional hypotheses will be stated when necessary. The goal is not to present a complete account of all the features of the problem but the cases considered are sufficient to give the basic ideas useful to treat other cases.

If an approximation is sought as a straightforward expansion, restricted here to the first order, we have

$$y\left(x,\varepsilon\right) = y_0\left(x\right) + \cdots .$$

By substituting into (3.1a), the reduced problem results

$$a\left(x\right)\frac{\mathrm{d}y_0}{\mathrm{d}x} + b\left(x\right)y_0 = 0 . \tag{3.2}$$

The solution is

$$y_0\left(x\right) = C\exp\left[-\int_0^x \frac{b\left(\xi\right)}{a\left(\xi\right)}\,\mathrm{d}\xi\right] , \tag{3.3}$$

where C is a constant yet undetermined. The first additional hypothesis is associated with the existence of the above integral. At certain points, the above integral can be divergent revealing the presence of local singularities. We assume that the integral exists for any value of x. In particular,

$$\lambda = \exp\left[-\int_0^1 \frac{b\left(\xi\right)}{a\left(\xi\right)}\,\mathrm{d}\xi\right] \tag{3.4}$$

is a bounded constant.

The two boundary conditions (3.1b) cannot be satisfied simultaneously except if $\beta = \lambda\alpha$ but this case is not considered here in order to simplify the presentation. Then, it is expected that a domain of fast variation of the function y exists. This domain – the so-called *inner region* – is denoted by D_ε but its localization is not known. We assume that such a domain lies in the neighbourhood of a point x_0 such that $0 \leq x_0 \leq 1$.

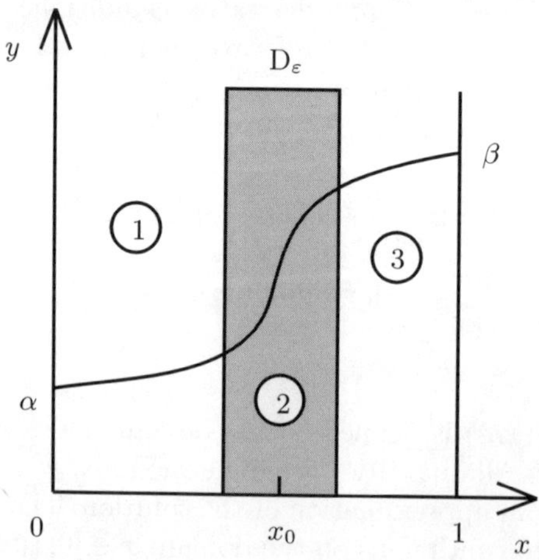

Fig. 3.1. Possible structure of the solution

According to Fig. 3.1, three regions can be identified

Region 1: $x \in [0, x_0[$. This domain is called the *outer region*. The solution is

$$y_0^{(1)}(x) = \alpha \exp\left[-\int_0^x \frac{b(\xi)}{a(\xi)}\,\mathrm{d}\xi\right].$$

Region 3: $x \in]x_0, 1]$. This domain is called the *outer region*. The solution is

$$y_0^{(3)}(x) = \beta \exp\left[-\int_1^x \frac{b(\xi)}{a(\xi)}\,\mathrm{d}\xi\right].$$

Region 2: $x \in D_\varepsilon$. The solution can have very fast variations in this domain which is very small when ε is small. A *boundary layer* forms.

To study the boundary layer, the first step consists of defining the variable adapted to the study of the domain D_ε. The adapted variable is called the *inner variable*. We set

$$X = \frac{x - x_0}{\delta(\varepsilon)}, \tag{3.5}$$

where $\delta(\varepsilon)$ is a strictly positive function, yet undetermined, which tends towards 0 as $\varepsilon \to 0$. This function is a *length scale* of the inner region. The function $\delta(\varepsilon)$ belongs to a class of functions called *order functions* whose properties are presented in Sect. 4.1. Then, the solution is sought as

$$y(x, \varepsilon) \equiv Y(X, \varepsilon).$$

Equation (3.1a) becomes

$$\frac{\varepsilon}{\delta^2}\frac{\mathrm{d}^2 Y}{\mathrm{d}X^2} + \frac{1}{\delta}a(x_0 + \delta X)\frac{\mathrm{d}Y}{\mathrm{d}X} + b(x_0 + \delta X)Y = 0.$$

We assume now that a and b are continuously differentiable functions and that $a(x_0) \neq 0$. After multiplying by δ^2, we obtain

$$\varepsilon\frac{\mathrm{d}^2 Y}{\mathrm{d}X^2} + \delta a(x_0)\frac{\mathrm{d}Y}{\mathrm{d}X} + \mathrm{O}\left(\delta^2\right) = 0. \tag{3.6}$$

The meaning of the symbol $\mathrm{O}\left(\delta^2\right)$ is given in Subsect. 4.1.4. Briefly, it means that the corresponding terms, in D_ε, are as small as δ^2. The possible dominant terms are only the first two terms, so that the magnification of the boundary layer region is tuned by taking $\delta = \varepsilon$.

By letting

$$X = \frac{x - x_0}{\varepsilon} \quad \text{and} \quad y(x, \varepsilon) = Y_0(X) + \cdots,$$

the *inner equation* is obtained

$$\frac{\mathrm{d}^2 Y_0}{\mathrm{d}X^2} + a\left(x_0\right)\frac{\mathrm{d}Y_0}{\mathrm{d}X} = 0 \, .$$

The solution is

$$Y_0\left(X\right) = C\exp\left[-a\left(x_0\right)X\right] + D \, ,$$

where C and D are two constants, yet undetermined. The determination of C and D should come from the asymptotic matching as formulated in Subsect. 2.2.1. We have

- for $x > x_0$

$$\lim_{X \to \infty} Y_0 = \lim_{x \to x_0} y_0^{(3)} = \beta \exp\left[-\int_1^{x_0} \frac{b\left(\xi\right)}{a\left(\xi\right)}\,\mathrm{d}\xi\right] , \qquad (3.7)$$

- for $x < x_0$

$$\lim_{X \to -\infty} Y_0 = \lim_{x \to x_0} y_0^{(1)} = \alpha \exp\left[-\int_0^{x_0} \frac{b\left(\xi\right)}{a\left(\xi\right)}\,\mathrm{d}\xi\right] . \qquad (3.8)$$

These conditions show that Y_0 should have two finite limits as $X \to \pm\infty$, which is impossible. The solution depends on the sign of $a\left(x_0\right)$

- If $a\left(x_0\right) > 0$, then only the limit $X \to +\infty$ is meaningful for Y_0 is not bounded as $X \to -\infty$. It is concluded that $x > x_0$.
- If $a\left(x_0\right) < 0$, then only the limit $X \to -\infty$ is valid. It is concluded that $x < x_0$.

As summarized in Fig. 3.2 several cases are possible:

Case 1. if $a\left(x\right) > 0$, the boundary layer develops in the neighbourhood of $x = 0$.

Case 2. If $a\left(x\right) < 0$, the boundary layer develops in the neighbourhood of $x = 1$.

Case 3. If $a\left(x\right) > 0$ for $x < x_0$ and $a\left(x\right) < 0$ for $x > x_0$, there are two boundary layers, one in the neighbourhood of $x = 0$, the other one in the neighbourhood of $x = 1$.

Case 4. If $a\left(x\right) < 0$ for $x < x_0$ and $a\left(x\right) > 0$ for $x > x_0$, the boundary layer is in the neighbourhood of $x = x_0$. An *inner boundary layer* develops and the outer solution is discontinuous at x_0.

In case 4, it is necessary to reconsider the analysis because it has been assumed that $a\left(x_0\right) \neq 0$. In addition, this restriction on a is sufficient and a study of the case where a could have several zeros is not necessary. Indeed, the qualitative aspect of the solution defined in this way gives the required information on the localization of the boundary layers in a more general case. Figure 3.3 clearly shows that two boundary layers are in the neighbourhood of the end points and two inner boundary layers develop.

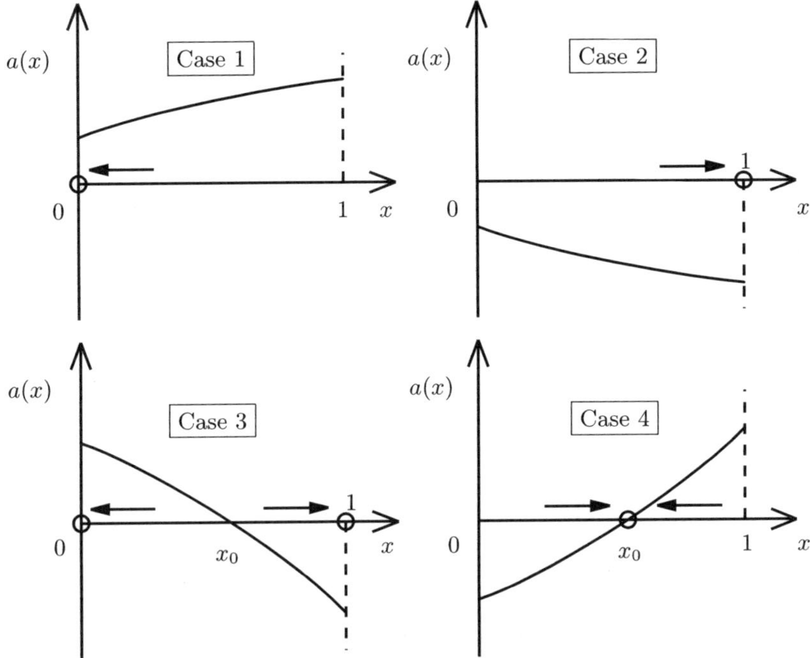

Fig. 3.2. Localization of the bondary layer according to the sign of $a(x)$. The circles indicate the points around which a boundary layer develops

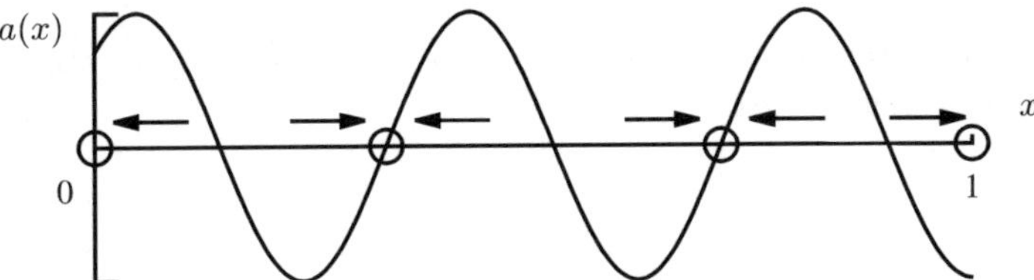

Fig. 3.3. Localization of the boundary layer when $a(x)$ has several zeros. The circles indicate the points around which boundary layers develop

3.2 Analysis of each Case

Case 1: $a\left(x\right) > 0$.

The boundary layer is in the neighbourhood of $x = 0$, region 1 disappears, only regions 2 and 3 remain. We now have

- For the outer region

$$y_0^{(3)}\left(x\right) = y_0\left(x\right) = \beta \exp\left[-\int_1^x \frac{b\left(\xi\right)}{a\left(\xi\right)}\,\mathrm{d}\xi\right]. \tag{3.9}$$

- For the inner region

$$Y_0 \left(X \right) = \left(\alpha - D \right) \exp \left[-a \left(0 \right) X \right] + D \,. \qquad (3.10)$$

In the above equations, we have

$$X = \frac{x}{\varepsilon} \,.$$

The outer approximation $y_0 \left(x \right)$ satisfies $y_0 \left(1 \right) = \beta$ whereas the inner approximation $Y_0 \left(X \right)$ satisfies the condition at the origin $Y_0 \left(0 \right) = \alpha$. The unknown constant D is determined by the asymptotic matching condition

$$\lim_{X \to \infty} Y_0 \left(X \right) = D = \lim_{x \to 0} y_0 \left(x \right) = \frac{\beta}{\lambda} \,,$$

which yields

$$Y_0 \left(X \right) = \left(\alpha - \frac{\beta}{\lambda} \right) \exp \left[-a \left(0 \right) X \right] + \frac{\beta}{\lambda} \,.$$

A UVA can be constructed as

$$y_a \left(x, X \right) = y_0 \left(x \right) + Y_0 \left(X \right) - \frac{\beta}{\lambda} \,,$$

leading to

$$y_a \left(x, X \right) = \left(\alpha - \frac{\beta}{\lambda} \right) \exp \left[-a \left(0 \right) X \right] + \beta \exp \left[- \int_1^x \frac{b \left(\xi \right)}{a \left(\xi \right)} \, \mathrm{d}\xi \right] \,.$$

Above, we assumed that $a(x) > 0$. It is also interesting to study a simple case with $a(x) > 0$ for $x \neq 0$ and $a(0) = 0$ by specifying $a \left(x \right) = x^p$, p being a positive real number. Using the transformation given by (3.5), (3.6) becomes

$$\varepsilon \frac{\mathrm{d}^2 Y}{\mathrm{d}X^2} + \delta^{1+p} X^p \frac{\mathrm{d}Y}{\mathrm{d}X} + \mathrm{O} \left(\delta^2 \right) = 0 \,,$$

with

$$X = \frac{x}{\delta \left(\varepsilon \right)} \,.$$

Here, we assume that $0 \leq p < 1$. It is clear that the boundary layer thickness is such that $\delta \left(\varepsilon \right) = \varepsilon^{1/(1+p)}$ so that the boundary layer variable is

$$X = \frac{x}{\varepsilon^{1/(1+p)}} \,.$$

The inner equation becomes

$$\frac{\mathrm{d}^2 Y_0}{\mathrm{d}X^2} + X^p \frac{\mathrm{d}Y_0}{\mathrm{d}X} = 0 \,.$$

The solution satisfying the condition at the origin is given by

$$Y_0(X) = CG(X) + \alpha ,$$

with

$$G(X) = \int_0^X \exp\left(-\frac{\xi^{1+p}}{1+p}\right) d\xi ,$$

so that the asymptotic matching condition,

$$\lim_{X\to\infty} Y_0(X) = CG(\infty) + \alpha = \lim_{x\to 0} y_0(x) = \frac{\beta}{\lambda} ,$$

leads to the approximation

$$Y_0(X) = \left(\frac{\beta}{\lambda} - \alpha\right)\frac{G(X)}{G(\infty)} + \alpha ,$$

and to the UVA

$$y_a(x, X) = \left(\alpha - \frac{\beta}{\lambda}\right)\left[1 - \frac{G(X)}{G(\infty)}\right] + \beta\exp\left[-\int_1^x \frac{b(\xi)}{a(\xi)} d\xi\right] .$$

Case 3: $a(x) > 0$ for $x < x_0$ and $a(x) < 0$ for $x > x_0$.

Figure 3.4 shows the behaviour of the solution. There are two inner regions $D_\varepsilon^{(1)}$ and $D_\varepsilon^{(3)}$ and an outer region $D^{(2)}$ (see Problems 3-2 and 3-3). The two boundary layers are characterized by the two inner variables

$$X = \frac{x}{\varepsilon} \quad\text{and}\quad X^* = \frac{x-1}{\varepsilon} .$$

In the outer region $D^{(2)}$, according to (3.2) and (3.3), the solution writes

$$y_0(x) = C\exp\left[-\int_0^x \frac{b(\xi)}{a(\xi)} d\xi\right] ,$$

where C is a constant, yet undetermined.

Equation (3.2) written at x_0,

$$a(x_0)\frac{dy_0}{dx} + b(x_0)y_0 = 0 ,$$

indicates that if the derivative of y_0 is bounded at x_0, and if $b(x_0) \neq 0$, then we have $y_0(x_0) = 0$ which implies $C = 0$ and

$$y_0(x) = 0 .$$

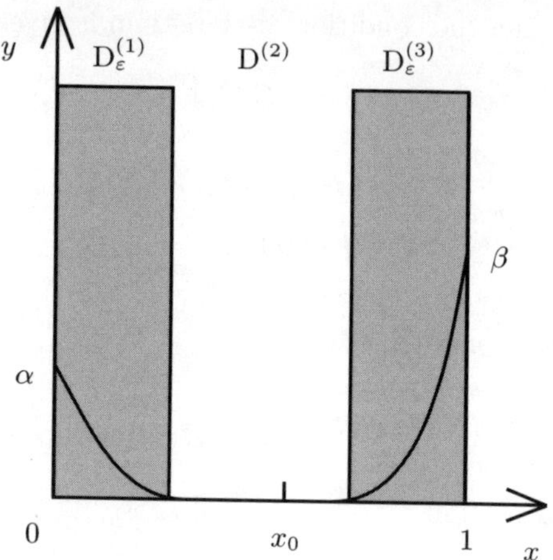

Fig. 3.4. Shape of the solution in case 3

In the inner region, $D_\varepsilon^{(1)}$, the inner equation holds

$$\frac{d^2 Y_0^{(1)}}{dX^2} + a\,(0)\,\frac{dY_0^{(1)}}{dX} = 0 \,,$$

with the solution

$$Y_0^{(1)} = C_1 \exp\left[-a\,(0)\,X\right] + D_1 \,.$$

The two constants C_1 and D_1 are determined from the condition at the origin and from the asymptotic matching principle

$$C_1 + D_1 = \alpha \,,$$

and

$$D_1 = \lim_{X \to \infty} Y_0^{(1)} = \lim_{x \to 0} y_0 = 0 \,.$$

This results

$$Y_0^{(1)} = \alpha \exp\left[-a\,(0)\,X\right] \,.$$

For the inner region $D_\varepsilon^{(3)}$ the inner equation holds

$$\frac{d^2 Y_0^{(3)}}{dX^{*2}} + a\,(1)\,\frac{dY_0^{(3)}}{dX^*} = 0 \,,$$

with the solution

$$Y_0^{(3)} = C_3 \exp\left[-a\,(1)\,X^*\right] + D_3 \,.$$

The two constants C_3 and D_3 are determined from the condition at $x = 1$ and from the asymptotic matching condition

$$C_3 + D_3 = \beta \,,$$

and

$$D_3 = \lim_{X^* \to -\infty} Y_0^{(3)} = \lim_{x \to 1} y_0 = 0 \ .$$

This results

$$Y_0^{(3)} = \beta \exp\left[-a\left(1\right) X^*\right] \ .$$

Finally, the UVA is

$$y_a\left(x, X\right) = \alpha \exp\left[-a\left(0\right) X\right] + \beta \exp\left[-a\left(1\right) X^*\right] \ .$$

Case 4: $a\left(x\right) < 0$ for $x < x_0$ and $a\left(x\right) > 0$ for $x > x_0$.

Figure 3.5 gives the behaviour of the solution. There are two outer regions $D^{(1)}$ and $D^{(3)}$ (see Problem 3-1) and an inner region D_ε.

The boundary layer is characterized by the inner variable

$$X = \frac{x - x_0}{\delta\left(\varepsilon\right)} \ ,$$

where $\delta\left(\varepsilon\right)$ is an order function. To determine $\delta\left(\varepsilon\right)$, we assume that the structure of $a\left(x\right)$ in the neighbourhood of $x = x_0$ is given by

$$a\left(x\right)_{x \to x_0} \cong K^2 \operatorname{sgn}\left(x - x_0\right) \left|x - x_0\right|^p \quad \text{with} \quad 0 < p < 1 \ .$$

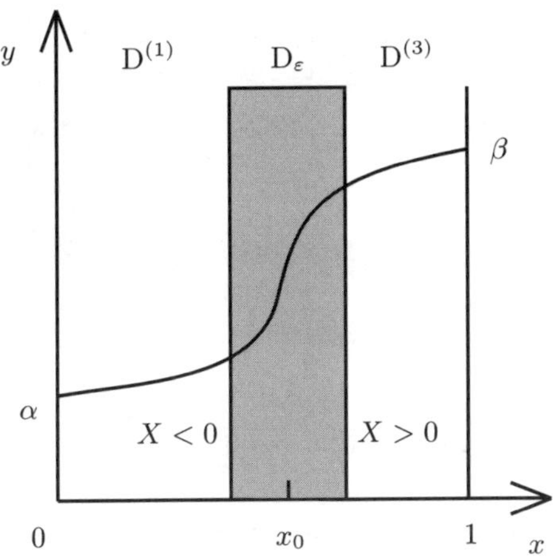

Fig. 3.5. Shape of the solution in case 4

The situation is the same as in case 1 with $a(x) = x^p$. The thickness of the boundary layer is

$$\delta\left(\varepsilon\right) = \varepsilon^{1/(1+p)} \ ,$$

and the inner equation is

$$\frac{\mathrm{d}^2 Y_0}{\mathrm{d}X^2} + K^2 |X|^p \operatorname{sgn}(X)\frac{\mathrm{d}Y_0}{\mathrm{d}X} = 0 \ .$$

The general solution is

$$Y_0 = C_1 \operatorname{sgn} X \int\limits_0^{|X|} \exp\left(-K^2 \frac{\xi^{1+p}}{1+p}\right)\,\mathrm{d}\xi + C_2 \ ,$$

where C_1 and C_2 are two constants determined from the asymptotic matching conditions with the outer approximations.

In the outer regions $\mathrm{D}^{(1)}$ and $\mathrm{D}^{(3)}$, the outer approximations are, respectively,

$$y_0^{(1)}(x) = \alpha \exp\left[-\int_0^x \frac{b(\xi)}{a(\xi)}\,\mathrm{d}\xi\right] \ ,$$

$$y_0^{(3)}(x) = \beta \exp\left[-\int_1^x \frac{b(\xi)}{a(\xi)}\,\mathrm{d}\xi\right] \ .$$

The asymptotic matching yields

$$\lim_{x \to x_0} y_0^{(1)}(x) = \alpha \exp\left[-\int_0^{x_0} \frac{b(\xi)}{a(\xi)}\,\mathrm{d}\xi\right]$$

$$= \lim_{X \to -\infty} Y_0(X) = -C_1 \int_0^\infty \exp\left(-K^2 \frac{\xi^{1+p}}{1+p}\right)\,\mathrm{d}\xi + C_2 \ ,$$

and

$$\lim_{x \to x_0} y_0^{(3)}(x) = \beta \exp\left[-\int_1^{x_0} \frac{b(\xi)}{a(\xi)}\,\mathrm{d}\xi\right] \tag{3.11}$$

$$= \lim_{X \to +\infty} Y_0(X) = C_1 \int_0^\infty \exp\left(-K^2 \frac{\xi^{1+p}}{1+p}\right)\,\mathrm{d}\xi + C_2 \ . \tag{3.12}$$

The above two equations allow the calculation of the two constants C_1 and C_2.

3.3 Conclusion

In this Chapter, a singular perturbation problem has been studied from the analysis of the solution of a second order differential equation. The singularity is introduced by means of a small parameter which multiplies the second order derivative. However, when dealing with ordinary differential equations, it cannot be concluded that a singular perturbation problem is necessarily due to the presence of a small parameter in front of the higher derivative. Many other situations can be encountered.

The topic of this Chapter has been essentially to localize the boundary layer from a method similar to a stability analysis [113]. Using purely mathematical arguments, without being guided by physical considerations, it has been possible to deduce the localization of the boundary layer. This can be useful for more difficult problems, including partial differential equations. In practical problems, the equations model a physical phenomenon. With a good knowledge of the problem, physical arguments can be added to the mathematical ones or even can supplant them to localize the boundary layer.

The discussion of the second order differential equation presented in this Chapter is not exhaustive due to restrictive hypotheses which have been introduced. Moreover, the matching principle has been applied in a very straightforward manner and does not enable us to go further without a more elaborate asymptotic method. All the limits should have a sense which is far from being always the case. The required asymptotics is developed in the next chapters. Differential equations will be used again to study examples whose analysis is not as simple as in this Chapter. In addition, these differential equations will be used to improve the approximations.

Problems

3-1. It is proposed to study an asymptotic approximation of $y(x, \varepsilon)$ such that

$$\mathrm{L}_\varepsilon\, y \equiv \varepsilon \frac{\mathrm{d}^2 y}{\mathrm{d}x^2} + 2\,(x - 1)\,\frac{\mathrm{d}y}{\mathrm{d}x} - 2\,(x - 1)\,y = 0 \,,$$

where

$$0 \leq x \leq 2 \,,$$

with

$$y(0, \varepsilon) = 1 \,, \quad y(2, \varepsilon) = 0 \,.$$

1. Determine the outer region and the corresponding approximation $y_0(x)$.
2. Find the thickness $\delta(\varepsilon)$ of the inner region and determine the general form of the corresponding approximation $Y_0(X)$ where $X = (x - x_0)/\delta$, and x_0 must be determined.
3. Apply the matching principle. Plot the behaviour of the solution.

It is reminded that

$$\int_0^\infty \mathrm{e}^{-s^2}\, \mathrm{d}s = \frac{\sqrt{\pi}}{2} \,.$$

4. Is it possible to give a UVA of $y\,(x,\varepsilon)$ on the domain $0 \le x \le 2$?

3-2. Consider the following problem

$$\varepsilon\frac{\mathrm{d}^2 y}{\mathrm{d}x^2} + (1 + \alpha x)\frac{\mathrm{d}y}{\mathrm{d}x} + \alpha y = 0\,,$$

with

$$y(0,\varepsilon) = 1\,, \quad y(1,\varepsilon) = 1\,.$$

1. Give the general solution $y_0(x)$ outside of any boundary layer.

2. We suppose that $\alpha > -1$. Find $y_0(x)$, the boundary layer solution $Y_0(X)$ and a UVA y_{app}. Show that $X = x/\varepsilon$.

3. We suppose that $\alpha < -1$. Find $y_0(x)$, $Y_0(X)$, $Y_0(X^*)$ and y_{app} with $X^* = (1 - x)/\varepsilon$.

3-3. Consider the following problem

$$\varepsilon\frac{\mathrm{d}^2 y}{\mathrm{d}x^2} + (1 - x)\frac{\mathrm{d}y}{\mathrm{d}x} - y = 0\,, \quad 0 \le x \le 1\,,$$

with

$$y(0,\varepsilon) = 1\,; \quad y(1,\varepsilon) = 1\,.$$

Check that the exact solution has the form

$$y = e^{X^2}\left[A + B\int_0^X e^{-t^2}\,\mathrm{d}t\right]\,,$$

with

$$X = \frac{1 - x}{\sqrt{2\varepsilon}}\,.$$

Determine A and B.

Show that there exists a boundary layer in the neighbourhood of $x = 0$ and another boundary layer in the neighbourhood of $x = 1$.

Give the variable appropriate to each boundary layer.

It is known that as $z \to \infty$

$$\frac{2}{\sqrt{\pi}}\int_0^{z/\sqrt{2}} e^{-t^2}\,\mathrm{d}t = 1 + \frac{2}{\sqrt{\pi}}e^{-z^2/2}\left[-\frac{1}{z} + \frac{1}{z^3} + \cdots\right]\,.$$

4 Asymptotic Expansions

This chapter describes the main tools for asymptotic expansions. In particular, an expansion depends on the use of an asymptotic sequence formed from order functions. Different approaches can be used but, keeping in mind we are interested in practical methods, we use a set of order functions for which a total ordering exists. A more in-depth study can be found in the book by Eckhaus [33].

4.1 Order Functions. Order of a Function

4.1.1 Definition of an Order Function

Definition 4.1. *Let* E *be the set of real functions* $\delta(\varepsilon)$ *of the variable* ε, *strictly positive and continuous in the semiopen interval* $0 < \varepsilon \le \varepsilon_0$ *and such that*

1. $\lim\limits_{\varepsilon \to 0} \delta(\varepsilon)$ *exists (we can have* $\delta(\varepsilon) \xrightarrow[\varepsilon \to 0]{} \infty$*)*,
2. $\forall \delta_1 \in$ E, $\quad \forall \delta_2 \in$ E, $\quad \delta_1 \delta_2 \in$ E.

A function $\delta(\varepsilon) \in$ E *is called an* order function. *According to condition (2), the product of order functions defines an* internal law *on* E.

Note 4.1. If $\delta(\varepsilon)$ *is an order function, then* $1/\delta(\varepsilon)$ *is also an order function.*

Example 4.1.

- $\dfrac{1}{\varepsilon}$, ε, ε^3, $\dfrac{\varepsilon}{1+\varepsilon}$, $\dfrac{1}{\ln(1/\varepsilon)}$, $1+\varepsilon$ are order functions.
- The first condition accepts $1/\varepsilon$, but excludes fast oscillating functions near $\varepsilon = 0$, for example, $1 + \sin^2(1/\varepsilon)$. The second condition excludes any function derived from these functions, for example, $\varepsilon\left[1 + \sin^2(1/\varepsilon)\right]$.

4.1.2 Comparison of Order Functions

A comparison of two order functions, δ_1 and δ_2, requires special notation. *Hardy's notations* are defined as follows:

1. $\delta_1 \preceq \delta_2$, it is said that δ_1 is *asymptotically smaller than or equal to* δ_2

$$\delta_1 \preceq \delta_2 \text{ if } \frac{\delta_1}{\delta_2} \text{ is bounded as } \varepsilon \to 0 \text{ ,}$$

2. $\delta_1 \prec \delta_2$, it is said that δ_1 is *asymptotically smaller* than δ_2

$$\delta_1 \prec \delta_2 \text{ if } \frac{\delta_1}{\delta_2} \to 0 \text{ as } \varepsilon \to 0 \,,$$

3. $\delta_1 \approx \delta_2$, it is said that δ_1 is *asymptotically equal* to δ_2

$$\delta_1 \approx \delta_2 \text{ if } \lim_{\varepsilon \to 0} \frac{\delta_1}{\delta_2} = \lambda \quad (\lambda > 0) \,,$$

where λ is a finite, non zero constant. It is said that δ_1 and δ_2 are *asymptotically identical* when $\lambda = 1$. Sometimes, the following notation is used

$$\delta_1 \cong \delta_2 \text{ if } \lim_{\varepsilon \to 0} \frac{\delta_1}{\delta_2} = 1 \,.$$

Example 4.2. Using the above notation, different order functions are compared:

- $\varepsilon^2 \preceq \varepsilon, \quad 2\varepsilon \preceq \dfrac{\varepsilon}{1+\varepsilon}, \quad \mathrm{e}^{-1/\varepsilon} \preceq \varepsilon^2, \quad \varepsilon^3 \preceq \dfrac{1}{\ln(1/\varepsilon)} \,.$

- $\varepsilon^2 \prec \varepsilon, \quad \mathrm{e}^{-1/\varepsilon} \prec \varepsilon^2, \quad \varepsilon^3 \prec \dfrac{1}{\ln(1/\varepsilon)} \,.$

- $2\varepsilon \approx \dfrac{\varepsilon}{1+\varepsilon} \,.$

- $\varepsilon \cong \dfrac{\varepsilon}{1+\varepsilon} \,.$

4.1.3 Total Ordering

Definition 4.2. *In the set* E, *let* R *be the relation defined by*

$$\mathrm{R}(\delta_1, \delta_2)\colon \delta_1 \approx \delta_2 \text{ or } \delta_1 \prec \delta_2 \,,$$

With this definition, any two elements of E can be compared. The relation is reflexive as $\mathrm{R}(\delta, \delta)$ is always satisfied. It is transitive because, if $\mathrm{R}(\delta_1, \delta_2)$ is satisfied and if $\mathrm{R}(\delta_2, \delta_3)$ is satisfied, then $\mathrm{R}(\delta_1, \delta_3)$ is satisfied. Finally, the relation is skew symmetric because if the two relations $\mathrm{R}(\delta_1, \delta_2)$ and $\mathrm{R}(\delta_2, \delta_1)$ are simultaneously satisfied, then $\delta_1 \approx \delta_2$. The relation R defines a *total ordering* on E.

If the internal law (2) of Subsect. 4.1.1 is not defined on E, the ordering is not total. Then, on such a set, it is possible that two functions δ_1 and δ_2 satisfy neither $\mathrm{R}(\delta_1, \delta_2)$ nor $\mathrm{R}(\delta_2, \delta_1)$. For example, $\delta_1 = \varepsilon$ and $\delta_2 = \varepsilon \left[1 + \sin^2(1/\varepsilon)\right]$ are such that $\delta_1 \preceq \delta_2$ and δ_2/δ_1 has no limit as $\varepsilon \to 0$; it is concluded that the two functions δ_1 and δ_2 cannot be compared with relation R.

Sometimes, on a more general set of order functions that do not satisfy the internal law, the elements of the subset E are called gauge functions. This point of view is not adopted here. It will be seen that the notion of "gauge function" is used here according to another meaning (Subsect. 4.2.3).

4.1.4 Order of a Function

Let $\varphi(x, \varepsilon)$ be a real function of real variables $x = (x_1, x_2, \ldots, x_m)$ and of the parameter ε. Function $\varphi(x, \varepsilon)$ is defined in a domain D of variables $(x_1, x_2, \ldots, x_m)$ and in the interval $0 < \varepsilon \leq \varepsilon_0$. Let $\|\varphi\|$ be a norm of φ in D.

Using *Landau's notation*, we have

1. $\varphi(x, \varepsilon) = O[\delta(\varepsilon)]$ in D if there exists a constant K, independent of ε, such that $\|\varphi\| \leq K\delta$. We say that φ is "big Oh" of δ as $\varepsilon \to 0$.

2. $\varphi(x, \varepsilon) = o[\delta(\varepsilon)]$ in D if $\lim\limits_{\varepsilon \to 0} \dfrac{\|\varphi\|}{\delta} = 0$. We say that φ is "little oh" of δ as $\varepsilon \to 0$.

3. $\varphi(x, \varepsilon) = O_S[\delta(\varepsilon)]$ in D if $\lim\limits_{\varepsilon \to 0} \dfrac{\|\varphi\|}{\delta} = K$ where K is a finite, non zero constant. The symbol O_S means *"is of strict (or sharp) order of"*.

If $\|\varphi\|$ is an order function, Hardy's and Landau's notations are equivalent:

1. $\varphi(x, \varepsilon) = O[\delta(\varepsilon)]$ is equivalent to $\|\varphi\| \preceq \delta$,
2. $\varphi(x, \varepsilon) = o[\delta(\varepsilon)]$ is equivalent to $\|\varphi\| \prec \delta$,
3. $\varphi(x, \varepsilon) = O_S[\delta(\varepsilon)]$ is equivalent to $\|\varphi\| \approx \delta$.

In the rest of this book, the *supremum norm* is used. If a function φ is continuous and bounded in its domain of definition, we have

$$\|\varphi\| = \mathrm{Max}_D\, |\varphi| \; . \tag{4.1}$$

Other norms, for example in L_2, can be used according to the problem being studied. It must be recognized however that the orders of magnitude can be completely different. Considering the function $\varphi(x, \varepsilon) = e^{-x/\varepsilon}$ and $D = [0, 1]$, the supremum norm yields

$$\|\varphi\| = O_S(1) \, .$$

On the other hand, the norm L_2,

$$\|\varphi\| = \left(\int_D \varphi^2 \, dx \right)^{1/2} ,$$

yields

$$\|\varphi\| = O_S\left(\sqrt{\varepsilon}\right) \, .$$

The supremum norm has an essential property for physicists and mechanicists.

Property 4.1. If $\varphi(x, \varepsilon) = \mathrm{O_S}[\delta(\varepsilon)]$, then there exists K such that

$$\forall x \in \mathrm{D}, \quad \forall \varepsilon \in]0, \varepsilon_0], \quad |\varphi| \leq K\delta \, ,$$

where K is a finite, non zero constant independent of ε. This property is not necessarily satisfied when other norms are used. In particular, the hypothesis $\varphi = \mathrm{O_S}(1)$ implies that φ is bounded in its domain of definition.

Note 4.2. Landau's notation is more general than Hardy's notation which can be applied only to order functions. For example, fast oscillating functions can be gauged with Landau's notation

$$\sin\left(\frac{1}{\varepsilon}\right) = \mathrm{O}(1) \, .$$

4.2 Asymptotic Sequence

4.2.1 Definition of an Asymptotic Sequence

Definition 4.3. *A sequence of order functions* δ_n *is called* asymptotic sequence *if*

$$\forall n, \ \delta_{n+1} \prec \delta_n \, .$$

In this definition, n is an integer, positive or zero, in such a way that if ε^n is an asymptotic sequence, ε^{α_n} is not an asymptotic sequence except if

$$\forall n, \ \alpha_{n+1} > \alpha_n \, .$$

4.2.2 Class of Equivalence

In the set E, relation $\delta_1 \approx \delta_2$ is a relation of equivalence **r**. This relation satisfies the following three properties:

a. Reflexivity, $\delta \approx \delta$,
b. Symmetry, $\delta_1 \approx \delta_2$ implies $\delta_2 \approx \delta_1$,
c. Transitivity, $\delta_1 \approx \delta_2$ and $\delta_2 \approx \delta_3$ imply $\delta_1 \approx \delta_3$.

Then, it is possible to define the set $\overline{\mathrm{E}}$ of classes of equivalence, $\overline{\mathrm{E}} = \mathrm{E} \, / \, \mathbf{r}$. When it is necessary to evaluate the order of a function, the choice of a representative of the class of equivalence is a matter of logic but also of intuition or simplicity.

In practice, subsets of E generated by elementary functions are often considered. Obviously, the subset of functions is chosen is such a way that it is possible to define a total ordering.

Example 4.3.

- E_0, the subset generated by ε^n where n is an integer,
- E_1, the subset generated by ε^α where α is rational,
- E_2, the subset generated by $\varepsilon^\alpha \left(\ln \frac{1}{\varepsilon}\right)^\beta$ with $\beta \neq 0$.

4.2.3 Gauge Functions

Definition 4.4. *A* gauge function *is an order function chosen as the* representative of its class of equivalence.

If δ_n and Δ_n are two asymptotic sequences such that

$$\forall n, \ \delta_n \approx \Delta_n \ ,$$

the two sequences are said *asymptotically equivalent.*

The notion of representative of a class of equivalence is present in this definition. For example,

$$\varepsilon^n, \quad \left(\frac{\varepsilon}{1+\varepsilon}\right)^n, \quad (\sin\varepsilon)^n \ ,$$

represent three asymptotically equivalent sequences. Chosing a representative of the class, for example in E_0, has implications on the uniqueness of an asymptotic expansion.

This uniqueness is important in the formal application of an asymptotic matching principle. It is useful to note Du Bois-Reymond's theorem cited by Hardy:

Theorem 4.1. *Given any asymptotic sequence, there exists an infinity of order functions δ^* such that*

$$\forall n, \ \delta^* \prec \delta_n \ .$$

Any function of order $\delta^* (\varepsilon)$ having this property is said asymptotically equivalent to zero with respect to the sequence δ_n.

Then, with the sequence $\delta_n = \varepsilon^n$, any order function of the type $\delta^* (\varepsilon) = e^{-\alpha/\varepsilon}$ with $\alpha > 0$ is asymptotically equivalent to zero. In the same manner, with the sequence $\delta_n (\varepsilon) = [\ln(1/\varepsilon)]^{-n}$, any order function $\delta^* (\varepsilon) = \varepsilon^\alpha$ with $\alpha > 0$ is asymptotically equivalent to zero.

In an asymptotic expansion, order functions which are equivalent to zero are often called transcendentally small terms, TST. Very often, the notation TST is restricted to order functions that are equivalent to zero with respect to the subset E_0 generated by ε^n where n is an integer. The notation EST, exponentially small terms, is well-adapted to the subsets E_0, E_1 and E_2.

4.3 Asymptotic Expansion

4.3.1 Asymptotic Approximation

Two functions $\varphi (x, \varepsilon)$ and $\overline{\varphi}_1 (x, \varepsilon)$ defined in D are said *asymptotically identical* if they have the same order of magnitude and if their difference is negligible

$$\varphi = O_S (\delta_1), \quad \overline{\varphi}_1 = O_S (\delta_1), \quad \varphi - \overline{\varphi}_1 = O_S (\delta_2) \ \text{with} \ \delta_2 \prec \delta_1 \ .$$

The function $\overline{\varphi}_1$ is an *asymptotic approximation* of the function φ. The converse being also true, it must be stressed that the goal of an asymptotic approximation is not to replace a function φ by another function $\overline{\varphi}_1$ which is as complicated as φ or even more complicated, but by a *simpler function*. For example, we can choose $\overline{\varphi}_1 = \varphi$ and, not only $\overline{\varphi}_1$ is an asymptotic approximation of φ, but the accuracy of the approximation is independent of the choice of δ_2 as far as $\lim_{\varepsilon \to 0} \delta_2 = 0$. It is clear that this result has no value. The notion of "simpler function" is the key which enables us to understand why it is interesting to replace the function φ by $\overline{\varphi}_1$.

Since the simplicity, just as the complexity are difficult to define precisely, it is not useful to discuss it further. However, it must be noted that this simplicity comes from the different methods used to construct asymptotic expansions.

In this way, a non trivial approximation $\overline{\varphi}_1$ of φ to order δ_1 is obtained. Subsequently, this notation is important and its origin is that

$$\varphi - \delta_1 \varphi_1 = o(\delta_1) \,,$$

and we set

$$\overline{\varphi}_1 = \delta_1 \varphi_1 \,,$$

with

$$\varphi_1 = O_S(1) \,.$$

If a better approximation is desired, the above process can be repeated. It can be shown that

$$\varphi - \delta_1 \varphi_1 = O_S(\delta_2) \,.$$

If there exists a function $\varphi_2 = O_S(1)$ such that

$$\varphi - \delta_1 \varphi_1 = \delta_2 \varphi_2 + O_S(\delta_3) \ \text{with} \ \delta_3 \prec \delta_2 \,,$$

we can write

$$\varphi = \delta_1 \varphi_1 + \delta_2 \varphi_2 + O_S(\delta_3) \,.$$

The process can be stopped without taking care of the neglected order of magnitude. Then, the following notation is used

$$\varphi = \delta_1 \varphi_1 + \delta_2 \varphi_2 + o(\delta_2) \,.$$

The process can also be continued to the chosen order to obtain

$$\varphi(x, \varepsilon) = \sum_{n=1}^{m} \delta_n(\varepsilon) \varphi_n(x, \varepsilon) + o[\delta_m(\varepsilon)] \,. \tag{4.2}$$

Finally, an m-term *asymptotic expansion* of $\varphi(x, \varepsilon)$ in D has been produced. Taking into account the non-uniqueness of an asymptotic expansion, the number of terms is not a very characteristic feature. It is better to say that an asymptotic expansion to order δ_m is obtained. This expansion can be written more precisely as

$$\varphi\left(x,\varepsilon\right) = \sum_{n=1}^{m} \delta_n\left(\varepsilon\right)\varphi_n\left(x,\varepsilon\right) + \mathrm{O_S}\left[\delta_{m+1}\left(\varepsilon\right)\right] . \qquad (4.3)$$

This expansion is such that

$$\forall n : \varphi_n\left(x,\varepsilon\right) = \mathrm{O_S}\left(1\right) \text{ and } \delta_{n+1} \prec \delta_n .$$

As the supremum norm is used, the definition of sharp order implies that the functions φ_n are *bounded* in their interval of definition.

According to Du Bois-Reymond's theorem, any function $\varphi^*\left(x,\varepsilon\right)$ has the same asymptotic expansion as $\varphi\left(x,\varepsilon\right)$, to the considered order, if

$$\varphi - \varphi^* = \mathrm{O}\left(\delta^*\right) ,$$

where δ^* is asymptotically identical to zero with respect to the asymptotic sequence δ_n of considered order functions. This is one of the reasons of the non-uniqueness of asymptotic expansions.

4.3.2 Regular Functions

If $\varphi\left(x,\varepsilon\right)$ and $\varphi_1\left(x,\varepsilon\right)$ are two continuous functions in a closed and bounded domain D and in the interval $0 < \varepsilon \leq \varepsilon_0$ such that

$$\varphi = \mathrm{O_S}\left(\delta_1\right), \quad \varphi_1 = \mathrm{O_S}\left(1\right), \quad \varphi = \delta_1\varphi_1 + \mathrm{o}\left(\delta_1\right) ,$$

then, necessarily, we have uniformly in D

$$\lim_{\varepsilon \to 0} \left| \frac{\varphi\left(x,\varepsilon\right)}{\delta_1\left(\varepsilon\right)} - \varphi_1\left(x,\varepsilon\right) \right| = 0 .$$

It is precisely in this way that it is checked that the functions φ and $\delta_1\varphi_1$ are asymptotically equivalent. A particularly interesting case of this result [33] occurs when, uniformly in D, we have

$$\lim_{\varepsilon \to 0} \frac{\varphi\left(x,\varepsilon\right)}{\delta_1\left(\varepsilon\right)} = \varphi_1\left(x\right) .$$

An asymptotic approximation of φ can then be written as

$$\varphi\left(x,\varepsilon\right) = \delta_1\left(\varepsilon\right)\varphi_1\left(x\right) + \mathrm{o}\left(\delta_1\right) .$$

A function φ having this property is said *regular*. It must be noted that this property is not necessarily valid to next orders.

Note 4.3. We assumed that $\varphi = \mathrm{O_S}(\delta_1)$ et $\varphi_1 = \mathrm{O_S}(1)$. From the definition of sharp order, this hypothesis implies that the functions φ and φ_1 are *bounded* in their domain of definition.

4.3.3 Regular and Generalized Asymptotic Expansions

The asymptotic approximation of a regular function obtained as above is objectively simpler than the function φ since φ_1 depends only on x. More precisely, if each step of the construction of an asymptotic expansion consists of determining a regular asymptotic approximation, the corresponding asymptotic expansion is said *regular*.

An asymptotic expansion which is not regular is called *non regular*. However, in order to avoid confusion with other concepts, the terminology *generalized expansion* is preferred (see Problem 4-4). Strictly, it is not necessary to add the qualifying adjective "generalized" but, too often, it is thought that an asymptotic expansion is necessarily regular. As, in the next chapters, asymptotic expansions are used in their general framework, the redundancy is not superfluous. An example of *generalized asymptotic expansion* is (4.3)

$$\varphi\left(x,\varepsilon\right) = \sum_{n=1}^{m} \delta_n\left(\varepsilon\right)\varphi_n\left(x,\varepsilon\right) + \mathrm{O}_s\left[\delta_{m+1}\left(\varepsilon\right)\right] . \tag{4.4}$$

Example 4.4. The function $\varphi = \dfrac{1}{1 - \varepsilon x}$ has the following generalized asymptotic expansion

$$\varphi = 1 + \sum_{n=0}^{m} \varepsilon^{2n+1} x^{2n+1}(1 + \varepsilon x) + \mathrm{O}(\varepsilon^{2m+3}) .$$

For an m-term regular asymptotic expansion, the following property holds

$$\forall h < m, \ \lim_{\varepsilon \to 0} \frac{\varphi\left(x,\varepsilon\right) - \sum_{i=1}^{h} \delta_i\left(\varepsilon\right)\varphi_i\left(x\right)}{\delta_{h+1}\left(\varepsilon\right)} = \varphi_{h+1}\left(x\right) .$$

An m-term *regular asymptotic expansion*, also called a Poincaré expansion, takes the form

$$\varphi\left(x,\varepsilon\right) = \sum_{n=1}^{m} \delta_n\left(\varepsilon\right)\varphi_n\left(x\right) + \mathrm{o}\left[\delta_m\left(\varepsilon\right)\right] . \tag{4.5}$$

Example 4.5. The function $\varphi = \dfrac{1}{1 - \varepsilon x}$ has the following regular asymptotic expansion

$$\varphi = \sum_{n=0}^{m} \varepsilon^n x^n + \mathrm{O}(\varepsilon^{m+1}) .$$

An interesting property of regular approximations is: if, for two approximations of the same function, we have

$$\varphi\left(x,\varepsilon\right) = \delta_1\left(\varepsilon\right)\varphi_1\left(x\right) + \mathrm{o}\left(\delta_1\right) \ \text{ and } \ \varphi\left(x,\varepsilon\right) = \overline{\delta}_1\left(\varepsilon\right)\overline{\varphi}_1\left(x\right) + \mathrm{o}\left(\overline{\delta}_1\right) ,$$

then

$$\lim_{\varepsilon \to 0} \frac{\overline{\delta}_1(\varepsilon)}{\delta_1(\varepsilon)} = c \ \text{ and } \ \varphi_1\left(x\right) = c\overline{\varphi}_1(x) ,$$

where c is a finite, non zero constant. The non-uniqueness of asymptotic expansions is also related to such comments.

Obviously, if the asymptotic sequence is chosen in a set of gauge functions and not in a set of order functions, a choice is required. This ensures the uniqueness of the asymptotic expansion.

Example 4.6. Consider the function

$$\varphi(x,\varepsilon) = \left(1 - \frac{\varepsilon}{1+\varepsilon}x\right)^{-1} .$$

Two regular asymptotic expansions are

$$\varphi(x,\varepsilon) = 1 + \sum_{n=1}^{m} \delta_n(\varepsilon)\, x^n + o\left[\delta_m(\varepsilon)\right] \text{ with } \delta_n(\varepsilon) = \left(\frac{\varepsilon}{1+\varepsilon}\right)^n ,$$

$$\varphi(x,\varepsilon) = 1 + \sum_{n=1}^{m} \varepsilon^n x \,(x-1)^{n-1} + o\left[\varepsilon^m\right] .$$

4.3.4 Convergence and Accuracy

A well-known example of an asymptotic expansion of a function $\varphi(\varepsilon)$ can be obtained from its Taylor series expansion when ε is small. For an m times continuously differentiable function near $\varepsilon = 0$, an $(m+1)$-term asymptotic expansion is

$$\varphi(\varepsilon) = \varphi(0) + \varepsilon\varphi'(0) + \cdots + \varepsilon^m \frac{\varphi^{(m)}(0)}{m!} + O_S(\varepsilon^{m+1}) .$$

If m becomes infinite, a series is obtained which can be *convergent* or *divergent*. If the series is convergent, it may not converge to the expanded function. In fact, an asymptotic expansion is different from a series. A series has an infinite number of terms whereas an asymptotic expansion has a finite number of terms. An asymptotic expansion can have an infinite number of terms (in this case we have an asymptotic series) but *the question of the convergence of the series has no connection with the behaviour of the function in the neighbourhood of* $\varepsilon = 0$.

Example 4.7. The Taylor series expansion of the exponential function is

$$e^\varepsilon = 1 + \varepsilon + \frac{\varepsilon^2}{2} + \cdots + \frac{\varepsilon^m}{m!} + \cdots .$$

This asymptotic series converges for any value of ε. The asymptotic expansion

$$f = 1 + \varepsilon + \frac{\varepsilon^2}{2} + \cdots + \frac{\varepsilon^m}{m!} + O_S(\varepsilon^{m+1})$$

is valid *in the neighbourhood of* $\varepsilon = 0$ *but nowhere else.*

Example 4.8. Consider the function $f(x, \varepsilon)$

$$f(x, \varepsilon) = e^{-x/\varepsilon} + e^{-\varepsilon x} \quad \text{for } 2 \leq x \leq 3 . \tag{4.6}$$

An asymptotic expansion of this function is

$$f_{\text{app}} = 1 - \varepsilon x + \varepsilon^2 \frac{x^2}{2} , \tag{4.7}$$

which is obtained by taking the first three terms of the series

$$g(x, \varepsilon) = 1 - \varepsilon x + \varepsilon^2 \frac{x^2}{2} + \cdots + (-1)^m \varepsilon^m \frac{x^m}{m!} + \cdots .$$

Figure 4.1 shows the function $\log \dfrac{|f_{\text{app}} - f|}{f}$ for different values of ε. The relative error due to the approximation f_{app} goes to zero as $\varepsilon \to 0$.

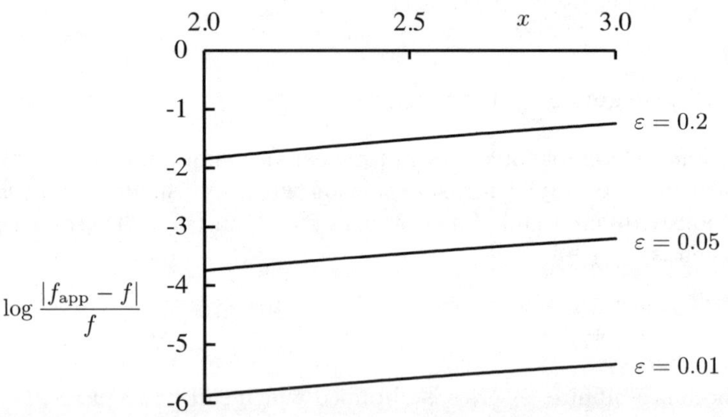

Fig. 4.1. Example of an asymptotic approximation. The curves give the function $\log \dfrac{|f_{\text{app}} - f|}{f}$ for different values of ε. The functions f and f_{app} are given by (4.6) and (4.7)

It is noted that the series $g(x, \varepsilon)$ is convergent for all values of x and ε, but does not converge to $f(x, \varepsilon)$. We have

$$g = e^{-\varepsilon x} .$$

The series g is an asymptotic approximation of f as $\varepsilon \to 0$. The reason is that the term $e^{-x/\varepsilon}$ of the function f is an EST as $\varepsilon \to 0$ and for x being kept fixed, strictly positive.

More generally, it is possible that, for an asymptotic series, the limits $\varepsilon \to 0$ and $m \to \infty$ do not commute (see Problem 4-5). This is an important property of divergent series which can be considered, from the asymptotic point of view, as convergent. It suffices to take ε small enough. If a series diverges, ε must be taken smaller when the number of terms retained is larger. In a certain sense, this leads to a heuristic paradoxical remark: the information contained in the first terms of the expansion is more complete when the series is more divergent. Exaggerating the paradox, it can be said that the divergent series converge more rapidly than the convergent series [4].

Example 4.9. Hinch [42] considers the differential equation

$$f = \frac{1}{x} - \frac{\mathrm{d}f}{\mathrm{d}x} \tag{4.8}$$

whose solution, with $x_0 > 0$, is

$$f = \mathrm{e}^{-1/\varepsilon} \int_{x_0}^{1/\varepsilon} \frac{\mathrm{e}^t}{t}\,\mathrm{d}t\ . \tag{4.9}$$

An asymptotic expansion, as $\varepsilon \to 0$, is

$$f_{\mathrm{app}} = \varepsilon + \varepsilon^2 + 2\varepsilon^3 + \cdots + (m-1)!\,\varepsilon^m\ , \tag{4.10}$$

which is valid for any value of x_0 and corresponds to the first m terms of the series

$$g = \varepsilon + \varepsilon^2 + 2\varepsilon^3 + \cdots + (m-1)!\,\varepsilon^m + \cdots\ .$$

In fact, this series is *divergent* for all values of ε. For a fixed value of ε, $g \to \infty$ as $m \to \infty$.

Figure 4.2 shows a comparison between the exact solution, with $x_0 = 1$ and $\varepsilon = 0.1$, and the asymptotic expansion for different values of m. The approximation is excellent if the number of terms is small. Obviously, the approximation is bad when the number of terms is too large because the series is divergent. The limit number of terms which ensures a good accuracy depends on the value of ε. This number increases when ε decreases. It can also be said that, for a fixed number of terms, the approximation is better when ε is smaller.

There is no contradiction between the different observations. The series is divergent because the limit of g is taken when $m \to \infty$ for a fixed value of ε whereas the asymptotic expansion is valid for a fixed value of m as $\varepsilon \to 0$.

For physical problems, the quality of an asymptotic expansion is not predictable. Sometimes, a good intuition can improve the result. If one writes

$$\sin \varepsilon = \varepsilon - \frac{\varepsilon^3}{6} + \mathrm{O}\left(\varepsilon^5\right)\ ,$$

and

$$\sin \varepsilon = \frac{\varepsilon}{1 + \varepsilon^2/6} + \mathrm{O}\left(\varepsilon^5\right)\ ,$$

it is immediately seen that in the second case, with a single term, the same accuracy can be obtained as in the first case with two terms. This is due to the appropriate choice of the representative in the class of order functions. These convergence improvements are useful in practice.

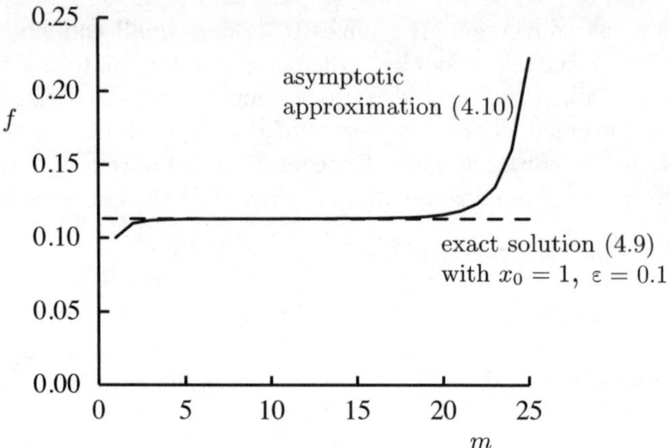

Fig. 4.2. Example of an asymptotic expansion associated with a divergent series

4.3.5 Operations on Asymptotic Expansions

When an approximate solution of a set of partial differential equations is sought, the expansion of the unknown functions is substituted into the equations, and we assume that the elementary operations are valid [26, 108, 114]. Thanks to the properties of the set E of order functions and to the existence of a total ordering on E, it can be shown than the addition, the subtraction, the multiplication or the division of asymptotic expansions are justified if the result is expressed with an asymptotic sequence which is possibly enlarged. The term to term integration with respect to the variables of the problem is also licit. The differentiation, however, can lead to problems. For example, consider the function

$$f(x, \varepsilon) = \sqrt{x + \varepsilon} \,,$$

where ε is a parameter, as small as desired. An asymptotic expansion of this function is

$$f(x, \varepsilon) = \sqrt{x} + \mathrm{o}(1) \,.$$

This expansion is a uniformly valid approximation in the domain $0 \le x \le 1$. The differentiation of f with respect to x yields

$$\frac{\mathrm{d}f}{\mathrm{d}x} = \frac{1}{2\sqrt{x + \varepsilon}} \,,$$

whereas the differentiation of $\sqrt{x}$ gives

$$\frac{\mathrm{d}\sqrt{x}}{\mathrm{d}x} = \frac{1}{2\sqrt{x}} \,.$$

In the domain $0 \leq x \leq 1$, $\dfrac{1}{2\sqrt{x}}$ is not an asymptotic expansion of $\dfrac{\mathrm{d}f}{\mathrm{d}x}$, because $\dfrac{1}{2\sqrt{x}} \to \infty$ as $x \to 0$. By contrast, the term to term integration is possible, for example from $x = 0$

$$\frac{2}{3}(x + \varepsilon)^{3/2} - \frac{2}{3}\varepsilon^{3/2} = \frac{2}{3}x^{3/2} + o(1) \,.$$

The difficulty encountered with the differentiation is due to a singularity near $x = 0$ which appears with a better approximation of function f. The examples considered in Subsects. 6.2.2 and 6.3.2 also show that uniformly valid approximations of a function are not necessary uniformly valid for its derivative.

Finally, Problem 4-7 shows that caution is needed when an asymptotic expansion is substituted into another expansion.

4.4 Conclusion

The construction of an asymptotic expansion is associated with the determination of an asymptotic sequence of order functions. In this book, it is chosen to work with the set E of order functions on which a total ordering is defined. Often, for a given problem, the choice of order functions can be restricted to a subset of E where the total ordering remains defined. Moreover, it can be convenient to use gauge functions which are particular representatives of classes of equivalence of order functions.

It is not always easy to obtain an asymptotic sequence. In certain cases, the sequence appears in a natural manner, but in other cases the sequence is constructed term by term in parallel with the construction of the asymptotic expansion. This difficulty is discussed later when studying different examples.

The successive complementary expansion method, SCEM, studied subsequently for the analysis of singular perturbation problems relies upon the notion of asymptotic expansion. By contrast with a common use, an asymptotic expansion is not necessarily regular. In fact, an important feature of SCEM is to use generalized asymptotic expansions.

Problems

4-1. Consider the following order functions

$$1 \,, \ -\varepsilon \ln \varepsilon \,, \ -\frac{1}{\ln \varepsilon} \,, \ \varepsilon^{\nu} \text{ with } 0 < \nu < 1 \,, \varepsilon \,.$$

Classify these functions according to their order of magnitude by using Hardy's notation as $\varepsilon \to 0$.

4-2. Consider the asymptotic sequence ε^n with n integer, $n \geq 0$. Let φ be the function $\varphi(x, \varepsilon) = \varepsilon \ln(x/\varepsilon)$ with $x > 0$. Compare this function as precisely as possible with the elements of the asymptotic sequence using Landau's notation and the supremum norm. The three following cases will be considered: i) $0 < A_1 \leq x \leq A_2$, ii) $0 < A_1\varepsilon \leq x \leq A_2\varepsilon$, iii) $0 < A_1\varepsilon^2 \leq x \leq A_2\varepsilon^2$. In each case, A_1 and A_2 are strictly positive constants independent of ε.

4-3. Determine if the following approximations are uniformly valid:

1. $e^{\varepsilon x} = 1 + O(\varepsilon)$; $\quad 0 \leq x \leq 1$, $\quad \varepsilon \to 0$.

2. $\dfrac{1}{x + \varepsilon} = O(1)$; $\quad 0 \leq x \leq 1$, $\quad \varepsilon \to 0$.

3. $e^{-x/\varepsilon} = o(\varepsilon^n)$ for any $n > 0$; $\quad 0 \leq x \leq 1$, $\quad \varepsilon \to 0$.

4-4. From the following form of the function φ, give a series expansion of φ as $\varepsilon \to 0$

$$\varphi = \frac{1}{1 + \varepsilon^{\frac{2x-1}{1-x}}} \, .$$

Do we obtain a valid asymptotic expansion in the domain $0 \leq x \leq 1$? (Examine if the successive terms satisfy the definition of an asymptotic expansion in the domain $0 \leq x \leq 1$).

Use the following form of φ,

$$\varphi = \frac{1}{1 + \frac{\varepsilon}{1-x} - 2\varepsilon} \, ,$$

to deduce an asymptotic expansion of φ valid in the whole domain $0 \leq x \leq 1$.

4-5. From successive integrations by parts, show that the function

$$E_1(x) = \int_x^\infty \frac{e^{-t}}{t} \, dt \quad \text{with} \quad x > 0$$

has the following expansion for large values of x

$$E_1(x) \cong \frac{e^{-x}}{x} \left[1 - \frac{1}{x} + \frac{2}{x^2} + (-1)^n \frac{n!}{x^n} + \cdots \right].$$

By giving an upper bound of the neglected term, show that an asymptotic expansion is defined.

Let $R_n(x)$ be the neglected term. Determine the following limits

$$\lim_{n \to \infty, x \text{ fixed}} |R_n(x)| \, ,$$

and

$$\lim_{x \to \infty, n \text{ fixed}} |R_n(x)| \, .$$

Show that the series is divergent.

Taking $x = 3$, show that the asymptotic expansion represents the function $x\,\mathrm{e}^x\,E_1$ with an error wich has a minimum when a certain number of terms is retained.

It is given that $x\,\mathrm{e}^x\,E_1(x) = 0.7862$ for $x = 3$.

4-6. Consider the equation

$$\varepsilon x^2 + x - 1 = 0 ,$$

for which the roots are sought when ε is a small parameter.

The reduced equation, obtained by setting $\varepsilon = 0$, has only one root. To recover the other root, a change of variable is used

$$X = \frac{x}{\delta(\varepsilon)} .$$

We assume that $0 < A_1 \leq |X| \leq A_2$ where A_1 and A_2 are two constants independent of ε. By examining the various possibilities to choose $\delta(\varepsilon)$, show that the second root is recovered by taking $\delta = \varepsilon^{-1}$. The following cases will be considered: i) $\delta \prec 1$, ii) $\delta = 1$, iii) $1 \prec \delta \prec \varepsilon^{-1}$, iv) $\delta \succ \varepsilon^{-1}$, v) $\delta = \varepsilon^{-1}$.

An expansion of the solution is sought as

$$x = \frac{x_0}{\varepsilon} + x_1 + x_2\varepsilon + \cdots .$$

Determine successively x_0, x_1, x_2.

4-7. Let $f(x)$ be the function

$$f(x) = \mathrm{e}^{x^2} ,$$

where x is given by

$$x = \frac{1}{\varepsilon} + \varepsilon .$$

Give an asymptotic expansion of the function $f\left[x(\varepsilon)\right]$. Examine what happens if only the dominant term of x (i.e. $x = 1/\varepsilon$) is kept.

5 Successive Complementary Expansion Method

This chapter is devoted to the asymptotic analysis of singular functions. Two versions of the method of matched asymptotic expansions, MMAE, are reminded with their advantages and drawbacks. The first method, associated with the intermediate matching, is the most popular one since it is apparently the most natural one. The second method, based on Van Dyke's matching principle, VDP, is more puzzling but its application is easier. In both methods, the point is to match two approximations defined in contiguous domains. VDP enables us to understand how a composite approximation, which is taken as a uniformly valid approximation, UVA, can be constructed. This analysis leads us to a modified Van Dyke principle, MVDP, which seems to solve the known counter-examples to VDP, in particular when logarithms are present. This latter method suggests a new approach which is called the successive complementary expansion method, SCEM. The regular form of SCEM is equivalent to MMAE but sets us free of any asymptotic matching. Moreover, the general form of SCEM, with the use of generalized asymptotic expansions, AEs, enables us to treat problems which cannot be solved in an easy manner with regular expansions.

5.1 Method of Matched Asymptotic Expansions

5.1.1 Expansion Operator

Consider a function $\Phi(x, \varepsilon)$ defined in a domain D, for example the interval $[0, 1]$, and assume that a regular AE can be constructed

$$\Phi(x, \varepsilon) = \sum_{i=1}^{n} \delta_0^{(i)}(\varepsilon)\, \Phi_0^{(i)}(x) + \mathrm{o}\left(\delta_0^{(n)}\right), \tag{5.1}$$

where $\delta_0^{(i)}(\varepsilon)$ is an asymptotic sequence of order functions. These expansions are often called Poincaré expansions.

Definition 5.1. *As introduced by Eckhaus [33], an* expansion operator $\mathrm{E}_0^{(n)}$ *is the operator which expresses the asymptotic approximation of Φ to order* $\delta_0^{(n)}$

$$\Phi - \mathrm{E}_0^{(n)}\,\Phi = \mathrm{o}\left(\delta_0^{(n)}\right), \tag{5.2}$$

which is sometimes denoted more imprecisely by

$$\Phi\left(x,\varepsilon\right) \cong \mathrm{E}_0^{(n)}\,\Phi\,, \tag{5.3}$$

with

$$\mathrm{E}_0^{(n)}\,\Phi = \sum_{i=1}^{n} \delta_0^{(i)}\left(\varepsilon\right)\Phi_0^{(i)}\left(x\right)\,.$$

This is an n-term AE but the important point is the order *to which the AE is written.*

If it is desired to obtain the regular AE of Φ for $m \leq n$, it suffices to know $\mathrm{E}_0^{(n)}\,\Phi$. If order functions are used, we have

$$\mathrm{E}_0^{(m)}\,\mathrm{E}_0^{(n)}\,\Phi = \mathrm{E}_0^{(m)}\,\Phi + \mathrm{o}\left(\delta_0^{(m)}\right)\,. \tag{5.4}$$

It is advantegeous to use *gauge functions* $\delta_0^{(i)}\left(\varepsilon\right)$ which answer the question of non-uniqueness of AEs to a large extent. Then, the asymptotic equality (5.4) is replaced by a strict equality

$$\mathrm{E}_0^{(m)}\,\mathrm{E}_0^{(n)}\,\Phi = \mathrm{E}_0^{(m)}\,\Phi\,,$$

which is useful in the application of the matching principles.

5.1.2 Outer Expansion - Inner Expansion

A particularly interesting case occurs when the function Φ is not regular on D, i.e. when the AE of Φ is valid only in a restricted region D_0 of D. For example, D_0 is defined by $0 < A_0 \leq x \leq 1$ where A_0 is a constant independent of ε. The corresponding AE is often called the *outer expansion*. The associated variable x is called the *outer variable*.

For the sake of simplicity, we consider a one-dimensional case in which the singularity occurs in the neighbourhood of $x = 0$. We assume that in the neighbourhood of $x = 0$, there exists at least one AE which can be an asymptotic approximation of Φ obtained by another limit process than the "*outer limit process*" given by $\varepsilon \to 0$ and x being kept fixed, strictly positive.

Before we discuss the method, let us consider the following example

$$\Phi\left(x,\varepsilon\right) = \frac{2}{\sqrt{1-4\varepsilon}}\exp\left(-\frac{x}{2\varepsilon}\right)\mathrm{sh}\left(\frac{\sqrt{1-4\varepsilon}}{2\varepsilon}x\right)\,. \tag{5.5}$$

Assuming $x > 0$, it is easy to construct a 2-term outer AE

$$\Phi\left(x,\varepsilon\right) = \mathrm{e}^{-x} + \varepsilon\,\mathrm{e}^{-x}\left(2-x\right) + \mathrm{o}\left(\varepsilon\right)\,.$$

We set

$$\mathrm{E}_0^{(2)}\,\Phi = \mathrm{e}^{-x} + \varepsilon\,\mathrm{e}^{-x}\left(2 - x\right)\;.$$

This means that a 2-term outer AE has been obtained such that

$$\Phi - \mathrm{E}_0^{(2)}\,\Phi = \mathrm{o}\left(\varepsilon\right)\;.$$

When determining this expansion, we assume that x is *strictly positive* which justifies the expression "outer limit" or "outer expansion". This announces the possibility of a singularity at the origin which is confirmed by the fact that $\Phi\left(0,\varepsilon\right) = 0$ whereas $\mathrm{E}_0^{(1)}\,\Phi = \mathrm{e}^{-x}$ takes the value 1 at the origin. By definition, this singularity cannot be resolved with the second approximation $\mathrm{E}_0^{(2)}\,\Phi$ which takes the value $1 + 2\,\varepsilon$ at the origin. Then, $\mathrm{E}_0^{(1)}\,\Phi$ and $\mathrm{E}_0^{(2)}\,\Phi$ are not asymptotic approximations of Φ in the neighbourhood of the origin. There is a *singular perturbation* and another limit process must be introduced in the neighbourhood of the origin. We say that there is a *boundary layer* behaviour.

The hypothesis $x > 0$ is necessary to neglect a term like $e^{-x/\varepsilon}$, which is not justified if x is very small or equal to zero. The *"inner limit process"* is specified by setting

$$X = \frac{x}{\varepsilon}\;.$$

This new variable X is called the *inner variable* and allows the construction of a regular AE in a neighbourhood of the origin. The so-called *inner expansion* is obtained by applying the limit process in which X is kept fixed and $\varepsilon \to 0$ to the function

$$\Phi^*\left(X,\varepsilon\right) \equiv \Phi\left(\varepsilon X,\varepsilon\right)\;.$$

It is easy to get

$$\Phi\left(x,\varepsilon\right) = \left(1 - \mathrm{e}^{-X}\right) + \varepsilon\left[\left(2 - X\right) - \left(2 + X\right)\mathrm{e}^{-X}\right] + \mathrm{o}\left(\varepsilon\right)\;.$$

We also set

$$\mathrm{E}_1^{(2)}\,\Phi = \left(1 - \mathrm{e}^{-X}\right) + \varepsilon\left[\left(2 - X\right) - \left(2 + X\right)\mathrm{e}^{-X}\right]\;.$$

This means that a 2-term inner AE has been constructed such that

$$\Phi - \mathrm{E}_1^{(2)}\,\Phi = \mathrm{o}\left(\varepsilon\right)\;.$$

5.1.3 Asymptotic Matching

We observe that $\mathrm{E}_0^{(2)}\,\Phi$ is an approximation of Φ in the interval $0 < A_1 \leq x \leq 1$ where A_1 is a constant independent of ε, and $\mathrm{E}_1^{(2)}\,\Phi$ is an approximation of the same function if $0 < B_1 \leq X \leq B_2$ (i.e. $0 < B_1\varepsilon \leq x \leq B_2\varepsilon$) where B_1

et B_2 are two constants independent of ε. Although $\mathrm{E}_0^{(2)}\,\varPhi$ and $\mathrm{E}_1^{(2)}\,\varPhi$ do not have the same structure, there is a link between these two approximations. For example, $\mathrm{E}_0^{(2)}\,\varPhi$ can be regarded as an approximation of $\varPhi$ even closer to the origin. In the same manner, $\mathrm{E}_1^{(2)}\,\varPhi$ can be regarded as an approximation of $\varPhi$ away from the origin. This point of view leads to the notion of *overlap*.

The overlap of $\mathrm{E}_0^{(2)}\,\varPhi$ and $\mathrm{E}_1^{(2)}\,\varPhi$ expresses that these two functions have a common domain of validity, i.e. $\mathrm{E}_0^{(2)}\,\varPhi$ and $\mathrm{E}_1^{(2)}\,\varPhi$ are two simultaneous approximations of $\varPhi$ when x is in a domain between the outer domain and the inner domain. This will be further discussed in Sect. 5.3.

Too close to the origin, $\mathrm{E}_0^{(2)}\,\varPhi$ cannot be an approximation of $\varPhi$, since the perturbation is singular. Similarly, $\mathrm{E}_1^{(2)}\,\varPhi$ is not an approximation of $\varPhi$ too far from the origin. Nevertheless, as suggested by Friedrichs' model (Subsect. 2.2.1), the two limits can be identified. This simplistic idea of asymptotic matching allows us to write

$$\lim_{x\to 0} \mathrm{E}_0^{(2)}\,\varPhi = \lim_{X\to\infty} \mathrm{E}_1^{(2)}\,\varPhi\;.$$

With the example given by (5.5), it is seen that

$$\lim_{x\to 0} \mathrm{E}_0^{(2)}\,\varPhi = 1 + 2\varepsilon\;,$$

whereas $\lim_{X\to\infty} \mathrm{E}_1^{(2)}\,\varPhi$ is not bounded.

To improve the process, it is better to work with the *behaviours* rather than with the limits. Then, by using asymptotic expansions, $\mathrm{E}_0^{(2)}\,\varPhi$ is a function of x and of ε whose behaviour can be determined from the inner process limit. To the same order $\mathrm{O}\left(\varepsilon\right)$, we obtain

$$\mathrm{E}_1^{(2)}\,\mathrm{E}_0^{(2)}\,\varPhi = 1 + \varepsilon\left(2 - X\right)\;,$$

which gives again the previous limit by setting here $X = 0$.

Similarly, we have

$$\mathrm{E}_0^{(2)}\,\mathrm{E}_1^{(2)}\,\varPhi = 1 - x + 2\varepsilon\;.$$

The limit does not exist because now we have $x\to\infty$. However, the two expansions yield the same result

$$\mathrm{E}_1^{(2)}\,\mathrm{E}_0^{(2)}\,\varPhi \equiv \mathrm{E}_0^{(2)}\,\mathrm{E}_1^{(2)}\,\varPhi\;.$$

With a formulation to be discussed later (Sect. 5.4), this method, initially due to Van Dyke [107], is very convenient to produce relations between the outer and inner expansions. It must be noted that the two expansions are based on gauge functions. Otherwise, it would be necessary to replace the sign of identity $\equiv$ by the sign of asymptotic identity $\cong$. It will be seen that, as formulated by Van Dyke, the matching principle cannot always be applied. It is why another method, initiated by the work of Kaplun [45] and Lagerstrom [48], uses the notion of *intermediate matching* [11, 17, 42] with the underlying hypothesis that there exists an overlap domain for the two asymptotic expansions under consideration. The idea is attractive, but it will be seen that the method can be complex to implement and that the method is less efficient than the modified Van Dyke principle.

Let us examine how to apply the rule of intermediate matching to the example given by (5.5). An *"intermediate limit process"* is first introduced with x_δ being kept fixed and $\varepsilon \to 0$, where x_δ is the *intermediate variable* defined by

$$x_\delta = \frac{x}{\delta\left(\varepsilon\right)} \, ,$$

with

$$\varepsilon \prec \delta\left(\varepsilon\right) \prec 1 \, .$$

It is reminded that $\varepsilon \prec \delta$ reads "ε is asymptotically smaller than δ" (see Subsect. 4.1.2).

The following results are obtained

$$\mathrm{E}_\delta \, \Phi = 1 - \delta x_\delta + 2\varepsilon + \mathrm{O}\left(\delta^2\right) + \mathrm{o}\left(\varepsilon\right) \, ,$$

$$\mathrm{E}_\delta \, \mathrm{E}_0^{(2)} \, \Phi = 1 - \delta x_\delta + 2\varepsilon + \mathrm{O}\left(\delta^2\right) + \mathrm{o}\left(\varepsilon\right) \, ,$$

$$\mathrm{E}_\delta \, \mathrm{E}_1^{(2)} \, \Phi = 1 - \delta x_\delta + 2\varepsilon + \mathrm{O}\left(\varepsilon^n\right) \quad \text{for any } n \in R \, .$$

where E_δ is the operator used to expand Φ in the domain where x_δ is such that $0 < C_1 \le x_\delta \le C_2$, C_1 and C_2 being two constants independent of ε.

To order ε, the following results are obtained:

- If $\sqrt{\varepsilon} \preceq \delta\left(\varepsilon\right)$,   then $\mathrm{E}_\delta \, \Phi = \mathrm{E}_\delta \, \mathrm{E}_0^{(2)} \, \Phi$ but $\mathrm{E}_\delta \, \mathrm{E}_0^{(2)} \, \Phi \neq \mathrm{E}_\delta \, \mathrm{E}_1^{(2)} \, \Phi$,
- If $\delta\left(\varepsilon\right) \prec \sqrt{\varepsilon}$,   then $\mathrm{E}_\delta \, \Phi = \mathrm{E}_\delta \, \mathrm{E}_0^{(2)} \, \Phi = \mathrm{E}_\delta \, \mathrm{E}_1^{(2)} \, \Phi = 1 - \delta x_\delta + 2\varepsilon$.

In the first case above, the intermediate matching is not possible. Generally, it cannot be predicted whether the intermediate matching is possible or not.

Figure 5.1 helps to understand the problems encountered. For the sake of simplicity, examine the function

$$\Phi = \mathrm{e}^{-x} \, ,$$

which is the first term of the outer expansion of function in (5.5).

On this figure, each horizontal line is associated with a value of δ. From top to bottom, the first horizontal line corresponds to $\delta = 1$, i.e. to the outer expansion. The second line corresponds to a value of δ such that $\varepsilon \prec \delta \prec 1$, i.e. to an intermediate expansion. The last line corresponds to $\delta = \varepsilon$, i.e. to the inner expansion. Then, along the vertical lines, the variation of δ is indicated from $\delta = 1$ (top) to $\delta = \varepsilon$ (bottom). The orders of magnitude of the

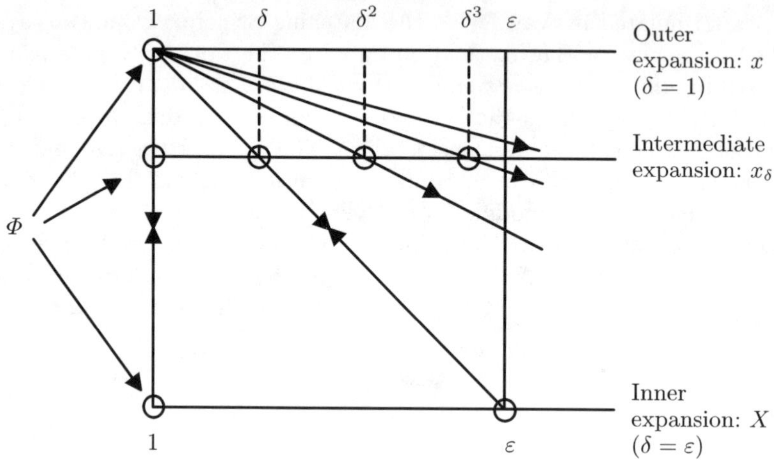

Fig. 5.1. Diagram of orders of magnitude

terms of AEs are indicated along the horizontal lines. The order is limited to
ε on the figure. The oblique lines are related to the behaviour of $\mathrm{E}_0^{(1)}\,\Phi$, (here
$\mathrm{E}_0^{(1)}\,\Phi$ is the function Φ itself), when $x = \delta x_\delta$,

$$\mathrm{e}^{-x} = 1 - \delta x_\delta + \delta^2 x_\delta^2 - \delta^3 x_\delta^3 + \cdots .$$

The small open circles indicate the presence of a term in the expansion.
The figure shows that the intermediate expansion contains four terms whose
order of magnitude is asymptotically larger than ε: the value of δ is such that
$\delta^3 \succ \varepsilon \succ \delta^4$. The outer expansion contains only one term: here, in this simple
case, this term is the function e^{-x} itself. The inner expansion contains two
terms to order ε since

$$\mathrm{e}^{-x} = \mathrm{e}^{-\varepsilon X} = 1 - \varepsilon X + \cdots .$$

The double arrows indicate the sense in which the matching is possible
to this order. For example, the term $-\delta x_\delta$ of the intermediate expansion
matches with the term $-\varepsilon X$ of the inner expansion, but the matching of the
term $\delta^2 x_\delta^2$ requires a term of order ε^2 of the inner expansion which is outside
the figure.

When δ is closer to 1, i.e. when δ is closer to the outer approximation, the
number of terms which must be considered is larger if one wants to reach the
required order $\mathrm{O}(\varepsilon)$. In order to have an intermediate matching in this zone,
a larger and larger number of terms is required in the inner expansion. This
remark underlines the practical difficulties in the application of the technique
of intermediate matching. It will be seen later that there are more complicated
examples where it is not even possible to match with this method.

5.2 Boundary Layer

5.2.1 Expansion Operator to a Given Order

As a preliminary, the notion of expansion operator is completed in this Subsection. A thorough analysis is given by Eckhaus [33].

If $\Phi(x, \varepsilon)$ is defined in the domain D such that $0 \leq x \leq B_0$ and $0 < \varepsilon \leq \varepsilon_0$ where B_0 and ε_0 are positive constants independent of ε, let us assume that the regular AE given by (5.1) can be written as

$$\Phi(x, \varepsilon) = \sum_{i=1}^{n} \delta_0^{(i)}(\varepsilon) \, \Phi_0^{(i)}(x) + \mathrm{o}\left(\delta_0^{(n)}\right) . \tag{5.6}$$

This expansion is called *outer expansion*.

If the function is singular at the origin, this expansion is, a priori, an approximation of Φ to the given order only in the domain D_0 such that $A_0 \leq x \leq B_0$ where A_0 is a constant independent of ε, as small as desired. Nevertheless, it will be seen with Kaplun's extension theorem (Subsect. 5.3.1) that this approximation can be extended in the neighbourhood of $x = 0$ if a lower accuracy is accepted.

Local variables are introduced to study the neighbourhood of the origin

$$x_\nu = \frac{x}{\delta_\nu(\varepsilon)} ,$$

with

$$\delta_\nu(\varepsilon) = \mathrm{o}(1) ,$$

except for $\nu = 0$ where we have

$$\delta_0(\varepsilon) = 1 .$$

The notation $x_0 = x$ is used.

Asymptotic domains D_ν are defined such that $A_\nu \leq x_\nu \leq B_\nu$ where A_ν et B_ν are positive constants independent of ε.

The sequence δ_ν is an asymptotic sequence for which

$$\nu_1 < \nu_2$$

implies

$$\delta_{\nu_2} \prec \delta_{\nu_1} .$$

Then, as ν increases, the corresponding domain is closer and closer to the origin.

We assume that, at each step, a regular AE of $\Phi(x, \varepsilon)$ can be constructed

$$\Phi(x, \varepsilon) = \sum_{i=1}^{n} \delta_\nu^{(i)}(\varepsilon) \, \Phi_\nu^{(i)}(x_\nu) + \mathrm{o}\left(\delta_\nu^{(n)}\right) .$$

The number of terms of the AE depends on ν and it is better to denote it by n_ν rather than by n.

Using the notation

$$\mathrm{E}_\nu^{(n)} \, \Phi = \sum_{i=1}^{n} \delta_\nu^{(i)} \left(\varepsilon\right) \Phi_\nu^{(i)} \left(x_\nu\right) \,, \tag{5.7}$$

we have

$$\Phi - \mathrm{E}_\nu^{(n)} \, \Phi = \mathrm{o}\left(\delta_\nu^{(n)}\right) \,.$$

The operator $\mathrm{E}_\nu^{(n)}$ is called *expansion operator*. Let us specify this notion to order δ. If, for a given order δ, we have

$$\Phi - \mathrm{E}_\nu^{(n)} \, \Phi = \mathrm{o}\left(\delta\right) \,, \tag{5.8}$$

it is said that $\mathrm{E}_\nu^{(n)}$ is an *expansion operator to order* δ. As the number of terms is not an important feature, the exponent (n) is suppressed. Indeed, it is very convenient as n depends in principle on ν.

Then, given an approximation $\mathrm{E}_\nu \, \Phi$ of Φ *to order* δ in the domain D_ν, it is written

$$\Phi - \mathrm{E}_\nu \, \Phi = \mathrm{o}\left(\delta\right) \,.$$

Subsequently, when an approximation is written in this way, it is implicitly understood that the order of the approximation is fixed.

5.2.2 Significant Approximations

At first, we define the meaning of the sentence: *an approximation is contained in another approximation.*

Definition 5.2. *Given two asymptotic approximations defined in different domains* D_μ *and* D_ν, *it is said that* $\mathrm{E}_\nu \, \Phi$ *contains* $\mathrm{E}_\mu \, \Phi$ *if*

$$\mathrm{E}_\mu \, \mathrm{E}_\nu \, \Phi = \mathrm{E}_\mu \, \Phi \,. \tag{5.9}$$

For a *regular function*, it can be said that the outer expansion contains any intermediate expansion. Mathematically, using gauge functions, for any $\delta_\mu \prec 1$, we have

$$\mathrm{E}_\mu \, \mathrm{E}_0 \, \Phi = \mathrm{E}_\mu \, \Phi \,.$$

For *a singular function*, there exists a value of ν, for example $\nu = 1$, such that

$$\mathrm{E}_1 \, \mathrm{E}_0 \, \Phi \neq \mathrm{E}_1 \, \Phi \,.$$

The inner expansion $\mathrm{E}_1 \, \Phi$, *defined to order* δ, *is not contained in the outer expansion.* Similarly, as for the outer expansion, it is said that the approximation given by the inner expansion is *significant* and corresponds to a *distinguished* limit. The corresponding variable x_1 which is often denoted by X is called *boundary layer variable*.

A necessary but not sufficient condition for a regular approximation $\mathrm{E}_\nu\,\Phi$ to be significant is that the approximation is not contained in any other regular approximation to the same order. It will be seen that the presence of logarithms can invalidate the fact that this statement is sufficient.

The links between the various expansions, $\mathrm{E}_0\,\Phi$, $\mathrm{E}_1\,\Phi$ and $\mathrm{E}_\nu\,\Phi$ for $0 < \nu < 1$ are the rules or heuristic principles which define the *asymptotic matching*. The matching has an utmost importance when the function Φ is a solution of an integro-differential equation with initial and boundary conditions. The matching enables us to find the conditions required to solve the reduced inner and outer approximations. In addition, the question to find a UVA and its order will become meaningful later.

5.3 Intermediate Matching

5.3.1 Kaplun's Extension Theorem

Consider a singular function Φ and an outer approximation $\mathrm{E}_0\,\Phi$ to order 1 in a domain $A_0 \leq x \leq B_0$, $(B_0 = 1)$. Kaplun's extension theorem [45, 46] states that, in a certain sense, the validity of the approximation can be extended.

Theorem 5.1. *More precisely, we have*

$$\lim_{\varepsilon \to 0} \left[\Phi - \mathrm{E}_0\,\Phi\right] = 0$$

in the interval $\delta\left(\varepsilon\right) \leq x \leq 1$, *where* $\delta\left(\varepsilon\right)$ *is an order function such that* $\delta\left(\varepsilon\right) \prec 1$.

It must be noted that the accuracy of the approximation is not specified and becomes necessarily lower when x is closer to the origin.

The extension theorem is generalized as:

Theorem 5.2. *Given an approximation* $\mathrm{E}_\nu\,\Phi$ *of* Φ, *defined in a domain* D_ν, *the domain of uniform convergence of* $\mathrm{E}_\nu\,\Phi$ *can be extended.*

More precisely, there exists an order function $\delta_\mu \neq \delta_\nu$ *and therefore a domain* D_μ, *such that* $\mathrm{E}_\nu\,\Phi$ *contains* $\mathrm{E}_\mu\,\Phi$

$$\mathrm{E}_\mu\,\mathrm{E}_\nu\,\Phi = \mathrm{E}_\mu\,\Phi\,.$$

5.3.2 Study of Examples

An extreme care must be taken when applying the extension theorem. To demonstrate the difficulty, we consider two examples.

Example 1. Consider the function

$$\Phi\left(x, \varepsilon\right) = 1 + x + \mathrm{e}^{-x/\varepsilon}\,. \tag{5.10}$$

As it is standard to choose the inner scales by setting

$$\delta_\nu\left(\varepsilon\right) = \varepsilon^\nu ,$$

we obtain to order 1

$$\mathrm{E}_0\,\Phi = 1 + x \quad \text{for} \quad \nu = 0 ,$$
$$\mathrm{E}_\nu\,\Phi = 1 \quad \text{for} \quad 0 < \nu < 1 ,$$
$$\mathrm{E}_1\,\Phi = 1 + e^{-x_1} \quad \text{for} \quad \nu = 1 .$$

What does the extension theorem tell us in the interval $0 < \mu < 1$?

1. If $\nu = 0$, there exists $\mu > 0$ such that

$$1 = \mathrm{E}_\mu\,\mathrm{E}_0\,\Phi = \mathrm{E}_\mu\,\Phi = 1 .$$

Then, $\mathrm{E}_0\,\Phi$ contains $\mathrm{E}_\mu\,\Phi$, and the domain of validity of $\mathrm{E}_\mu\,\Phi$ extends the domain of validity of $\mathrm{E}_0\,\Phi$.

2. If $\nu = 1$, there exists $\mu < 1$ such that

$$1 = \mathrm{E}_\mu\,\mathrm{E}_1\,\Phi = \mathrm{E}_\mu\,\Phi = 1 .$$

Here, $\mathrm{E}_1\,\Phi$ contains $\mathrm{E}_\mu\,\Phi$ and the domain of validity of $\mathrm{E}_\mu\,\Phi$ extends the domain of validity of $\mathrm{E}_1\,\Phi$. The same conclusion is reached when $\mu > 1$ since

$$2 = \mathrm{E}_\mu\,\mathrm{E}_1\,\Phi = \mathrm{E}_\mu\,\Phi = 2 .$$

Finally, the extension theorem is applied without any difficulty, which is a point in favour of the use of intermediate matching.

Example 2. This example illustrates the limits of application of the extension theorem and shows the difficulties encountered in the presence of logarithms. Consider the function

$$\Phi\left(x,\varepsilon\right) = \frac{1}{\ln x} + \frac{e^{-x/\varepsilon}}{\ln \varepsilon} . \tag{5.11}$$

To order $-\dfrac{1}{\ln \varepsilon}$, we obtain

$$\mathrm{E}_0\,\Phi = \frac{1}{\ln x} \quad \text{for} \quad \nu = 0 ,$$
$$\mathrm{E}_\nu\,\Phi = \frac{1}{\nu \ln \varepsilon} \quad \text{for} \quad 0 < \nu < 1 ,$$
$$\mathrm{E}_1\,\Phi = \frac{1 + e^{-x_1}}{\ln \varepsilon} \quad \text{for} \quad \nu = 1 .$$

The theorem states:

1. If $\nu = 0$, there exists $\mu > 0$ such that

$$\frac{1}{\mu \ln \varepsilon} = \mathrm{E}_\mu \, \mathrm{E}_0 \, \Phi = \mathrm{E}_\mu \, \Phi = \frac{1}{\mu \ln \varepsilon} \, .$$

It is concluded that $\mathrm{E}_0 \, \Phi$ contains $\mathrm{E}_\mu \, \Phi$, and the domain of validity of $\mathrm{E}_\mu \, \Phi$ extends the domain of validity of $\mathrm{E}_0 \, \Phi$.

2. If $\nu = 1$, there exists $\mu < 1$ such that

$$\mathrm{E}_\mu \, \mathrm{E}_1 \, \Phi = \mathrm{E}_\mu \, \Phi \, ,$$

but

$$\mathrm{E}_\mu \, \mathrm{E}_1 \, \Phi = \frac{1}{\ln \varepsilon} \quad \text{and} \quad \mathrm{E}_\mu \, \Phi = \frac{1}{\mu \ln \varepsilon} \, .$$

With the scales ε^μ, $\mathrm{E}_1 \, \Phi$ *does not contain* $\mathrm{E}_\mu \, \Phi$.

In this latter case, there is no possibility to satisfy the extension theorem and, consequently, to perform an intermediate matching.

In fact, it is not a counter-example because the set of scales ε^μ is not dense enough to produce all the possibilities. In other words, the extension of the domain corresponding to $\nu = 1$ is too small to be measured with the scale ε^μ. With $\delta_\mu = -\varepsilon \ln \varepsilon$ for example, the extension theorem is satisfied. It is noted that $-\varepsilon \ln \varepsilon \prec \varepsilon^\nu$ for $\nu < 1$ and, to order $-\dfrac{1}{\ln \varepsilon}$, we have

$$\frac{1}{\ln \delta_\mu} = \frac{1}{\ln \varepsilon} \, .$$

We obtain

$$\mathrm{E}_\mu \, \Phi = \mathrm{E}_\mu \, \mathrm{E}_1 \, \Phi = \frac{1}{\ln \varepsilon} \, .$$

With the scale $\delta_\mu = -\varepsilon \ln \varepsilon$, $\mathrm{E}_1 \, \Phi$ contains $\mathrm{E}_\mu \, \Phi$, and the domain of validity of $\mathrm{E}_\mu \, \Phi$ extends the domain of validity of $\mathrm{E}_1 \, \Phi$ (see Problem 5-2).

However, in this example where logarithms are present, it is seen how much the method of intermediate matching can be difficult to apply.

Although the extension theorem is practically useless in this situation, it is observed that, to the considered order

$$\mathrm{E}_1 \, \mathrm{E}_0 \, \Phi = \mathrm{E}_0 \, \mathrm{E}_1 \, \Phi \, .$$

The discussion of this remark is given later (Subsect. 5.4.2).

5.3.3 Rule of Intermediate Matching

Overlap Hypothesis

At first, the hypothesis of an *overlap domain* seems natural and is formulated below. The regular outer expansion of Φ in the domain $0 < A_0 \le x \le 1$ is given by $\mathrm{E}_0^{(n)} \, \Phi$, i.e.

$$\Phi - \mathrm{E}_0^{(n)} \, \Phi = \mathrm{o} \left(\delta_0^{(n)}(\varepsilon) \right) \, .$$

This domain can be extended to the domain $x \in (\tilde{\delta}, 1)$ with $\tilde{\delta} = o(1)$ so that

$$\Phi - \mathrm{E}_0^{(n)}\,\Phi = o(\delta^*) \quad \text{with} \quad \delta_0^{(n)} = \mathrm{O}(\delta^*) , \tag{5.12}$$

which expresses that the extension of the domain of validity is associated with a loss of accuracy of the approximation.

The inner expansion $\mathrm{E}_1^{(m)}\,\Phi$, valid in the domain $0 \le x_1 \le B_1$ where $x_1 = x/\delta_1(\varepsilon)$ is the inner variable, is such that

$$\Phi - \mathrm{E}_1^{(m)}\,\Phi = o\left(\delta_1^{(m)}(\varepsilon)\right) .$$

Its domain of validity can be extended to the domain $x \in (0, \bar{\delta})$ so that

$$\Phi - \mathrm{E}_1^{(m)}\,\Phi = o(\delta^*) \quad \text{with} \quad \delta_1^{(m)} = \mathrm{O}(\delta^*) . \tag{5.13}$$

Note 5.1. In expressions given by (5.12) and (5.13), δ^* is the same quantity. In fact, the choice of δ^* has been made intentionally to be so and it is possible that the accuracy of one or the other extension is better.

Definition 5.3. *There is overlap if*

$$\tilde{\delta} = o(\bar{\delta}) .$$

This definition of overlap can be expressed with the help of expansion operators.

The extensions of the domains of validity of $\mathrm{E}_0^{(n)}$ and $\mathrm{E}_1^{(m)}$ being determined as said above, the overlap of these domains expresses that, for any δ_ν such that

$$\bar{\delta} \succeq \delta_\nu \succeq \tilde{\delta} ,$$

we have, to order δ^*

$$\mathrm{E}_\nu\,\Phi = \mathrm{E}_\nu\,\mathrm{E}_0^{(n)}\,\Phi = \mathrm{E}_\nu\,\mathrm{E}_1^{(m)}\,\Phi . \tag{5.14}$$

In this equation, E_ν is the intermediate expansion operator to order δ^* defined with the intermediate variable $x_\nu = x/\delta_\nu$.

Eckhaus' Rule

In this case, the heuristic rule proposed by Eckhaus is:

If we assume that the extended domains of validity of $\mathrm{E}_0^{(n)}\,\Phi$ and of $\mathrm{E}_1^{(m)}\,\Phi$ overlap and that these continuous, therefore bounded, functions have an intermediate matching, then, for any k, there exist δ_ν, n and m such that

$$\mathrm{E}_\nu^{(k)}\,\Phi = \mathrm{E}_\nu^{(k)}\,\mathrm{E}_0^{(n)}\,\Phi = \mathrm{E}_\nu^{(k)}\,\mathrm{E}_1^{(m)}\,\Phi . \tag{5.15}$$

Practically, it is not always easy to implement such a rule. In effect, the situation is logically the inverse of what was expected. Generally, n and m

are given and it is required to find k in order to be able to write the equality in a certain domain D_ν.

Function given by (5.11) shows the limitations of the application of the intermediate matching. For $k = 1$ and any n, with $\delta_\nu = \varepsilon^\nu$, the following equality holds

$$\mathrm{E}_\nu^{(1)} \, \mathrm{E}_0^{(n)} \, \varPhi = \mathrm{E}_\nu^{(1)} \, \varPhi = \frac{1}{\nu \ln \varepsilon} \text{ for } 0 < \nu < 1 \, ,$$

but there is no value of m leading to the equality $\mathrm{E}_\nu^{(1)} \, \mathrm{E}_1^{(m)} \, \varPhi = \mathrm{E}_\nu^{(1)} \, \varPhi$. In effect, it can be shown that for $\delta_\nu = \varepsilon^\nu$

$$\mathrm{E}_\nu^{(1)} \, \mathrm{E}_1^{(m)} \, \varPhi = \frac{1}{\ln \varepsilon} \left[1 + \sum_{p=1}^{m-1} (1 - \nu)^p \right] \, .$$

In this particular case, there is no overlapping domain with the considered scales ε^ν. It is necessary to take for example the scale $\delta_\nu = -\varepsilon \ln \varepsilon$ in order to achieve the overlap of the extensions of $\mathrm{E}_0^{(n)} \, \varPhi$ and $\mathrm{E}_1^{(m)} \, \varPhi$ and to satisfy Eckhaus' rule.

5.4 Asymptotic Matching Principle

In this section, two asymptotic matching principles are given. The first one is due to Van Dyke [107] and the second one is a modification proposed by Mauss [64].

5.4.1 Van Dyke's Principle

VDP [107] is based on the interpretation of the ideas developed by Kaplun. When the principle works, its application is very simple. Its statement is:
Given n terms of the outer expansion and m terms of the inner expansion, we have

$$\mathrm{E}_1^{(m)} \, \mathrm{E}_0^{(n)} \, \varPhi = \mathrm{E}_0^{(n)} \, \mathrm{E}_1^{(m)} \, \varPhi \, . \tag{5.16}$$

Moreover, assuming that one of the goals of singular asymptotic analysis is to construct an asymptotic expansion of the solution to a given order, a composite approximation has the form [108]

$$\varPhi_{\mathrm{app}} = \mathrm{E}_0^{(n)} \, \varPhi + \mathrm{E}_1^{(m)} \, \varPhi - \mathrm{E}_0^{(n)} \, \mathrm{E}_1^{(m)} \, \varPhi \, . \tag{5.17}$$

Obviously, this form is valid when there is only one boundary layer in the domain where the function is studied, but the result can be generalized to more than two significant domains.

5.4.2 Modified Van Dyke's Principle

With (5.17), the accuracy of the UVA is the accuracy of the least accurate term. It would be better if each term has the same accuracy. Then, it is desirable to use *expansions defined to the same order*, whence the modified principle:

Let $\mathrm{E}_0\,\Phi$ and $\mathrm{E}_1\,\Phi$ be the outer and inner expansions of Φ, to a given order δ. These expansions are defined with an asymptotic sequence of gauge functions. The modified matching principle is

$$\mathrm{E}_0\,\mathrm{E}_1\,\Phi \equiv \mathrm{E}_1\,\mathrm{E}_0\,\Phi \,. \tag{5.18}$$

In addition, in the case of a single boundary layer, a UVA Φ_{app} in D is obtained, to the same order, as the composite expansion

$$\Phi_{\mathrm{app}} = \mathrm{E}_0\,\Phi + \mathrm{E}_1\,\Phi - \mathrm{E}_0\,\mathrm{E}_1\,\Phi \,. \tag{5.19}$$

This principle, proposed by Mauss [64], is called the "modified Van Dyke principle", MVDP. The reason for which this principle has not been stated by Van Dyke is probably that, in applications, the outer and inner expansions are generally organized according to a hierarchy. For example, the knowledge of the first term of $\mathrm{E}_0\,\Phi$ is used to find the order of magnitude of the first term of $\mathrm{E}_1\,\Phi$ which is not necessarily of the same order.

It must be stressed that, with gauge functions, (5.18) can be written as an equality. In order to avoid any ambiguity, we can write

$$\mathrm{E}_1\,\mathrm{E}_0\,\mathrm{E}_1\,\Phi = \mathrm{E}_1\,\mathrm{E}_0\,\Phi \tag{5.20}$$

to specify the use of the variable X.

We can also write

$$\mathrm{E}_0\,\mathrm{E}_1\,\Phi = \mathrm{E}_0\,\mathrm{E}_1\,\mathrm{E}_0\,\Phi \tag{5.21}$$

to specify the use of the variable x. However, in the following examples, or later in SCEM, the distinction between the inner and outer variables has no reason to be a source of problem.

5.5 Examples and Counter-Examples

The various principles and rules discussed above are now applied to different examples.

5.5.1 Example 1

First, consider example given by (5.5), to order ε. We have

$$\mathrm{E}_0\,\Phi = \mathrm{e}^{-x} + \varepsilon\,\mathrm{e}^{-x}\,(2 - x) \,,$$

$$\mathrm{E}_\nu^* \, \varPhi = 1 + 2\varepsilon - \varepsilon^\nu x_\nu + \mathrm{O}\left(\varepsilon^{2\nu}\right) + \mathrm{o}\left(\varepsilon\right) \, ,$$

$$\mathrm{E}_1 \, \varPhi = \left(1 - \mathrm{e}^{-X}\right) + \varepsilon \left[(2 - X) - (2 + X)\,\mathrm{e}^{-X}\right] \, ,$$

where E^* is the asymptotic expansion not written to the given order because the number of terms depends on the value of ν. We have

$$1 + \varepsilon(2 - X) = \mathrm{E}_1\,\mathrm{E}_0\,\varPhi \equiv \mathrm{E}_0\,\mathrm{E}_1\,\varPhi = 1 - x + 2\varepsilon \, .$$

Note 5.2. The presence of a term like εx is impossible. Indeed, with the variable X, this term would be $\varepsilon^2 X$ and, with the use of gauge functions, this term will not appear to order ε considered here.

It is easily seen that, for any $0 \leq \nu \leq 1$, we have

$$\mathrm{E}_\nu\,\varPhi_{\mathrm{app}} = \mathrm{E}_\nu\,\varPhi \, ,$$

with

$$\varPhi_{\mathrm{app}} = \mathrm{e}^{-x} - \mathrm{e}^{-X} + \varepsilon \left[(2 - x)\,\mathrm{e}^{-x} - (2 + X)\,\mathrm{e}^{-X}\right] \, .$$

5.5.2 Example 2

Even if in the preceding example there is a possibility of an intermediate matching, let us examine more precisely the example given by (5.11)

$$\varPhi\left(x, \varepsilon\right) = \frac{1}{\ln x} + \frac{\mathrm{e}^{-x/\varepsilon}}{\ln \varepsilon} \, .$$

To the order $\mathrm{O}\left(\dfrac{1}{(\ln \varepsilon)^2}\right)$, we obtain the following expansions

$$\mathrm{E}_0\,\varPhi = \frac{1}{\ln x} \quad \text{for} \quad \nu = 0 \, ,$$

$$\mathrm{E}_\nu\,\varPhi = \frac{1}{\nu \ln \varepsilon} - \frac{\ln x_\nu}{\nu^2 \, (\ln \varepsilon)^2} \quad \text{for} \quad 0 < \nu < 1 \, ,$$

$$\mathrm{E}_1\,\varPhi = \frac{1 + \mathrm{e}^{-x_1}}{\ln \varepsilon} - \frac{\ln x_1}{(\ln \varepsilon)^2} \quad \text{for} \quad \nu = 1 \, .$$

Here, the outer expansion contains the intermediate expansion

$$\mathrm{E}_\nu\,\varPhi = \mathrm{E}_\nu\,\mathrm{E}_0\,\varPhi = \frac{1}{\nu \ln \varepsilon} - \frac{\ln x_\nu}{\nu^2 \, (\ln \varepsilon)^2} \, .$$

By contrast, the inner expansion does not contain the intermediate expansion because

$$\mathrm{E}_\nu\,\mathrm{E}_1\,\varPhi = \frac{2 - \nu}{\ln \varepsilon} - \frac{\ln x_\nu}{(\ln \varepsilon)^2} \, ,$$

and then $\mathrm{E}_\nu\,\varPhi \neq \mathrm{E}_\nu\,\mathrm{E}_1\,\varPhi$.

With the chosen scales ε^ν, there is no overlap and then no possibility of intermediate matching. Also, Van Dyke's principle fails in certain cases. For example, we have

$$E_1^{(2)}\, E_0^{(1)}\, \Phi = \frac{1}{\ln\varepsilon} - \frac{\ln x_1}{(\ln\varepsilon)^2}\,,$$

but

$$E_0^{(1)}\, E_1^{(2)}\, \Phi = \frac{2}{\ln\varepsilon}\,,$$

and yet, the MVDP applies without any difficulty

$$\frac{1}{\ln\varepsilon} - \frac{\ln x_1}{(\ln\varepsilon)^2} = E_1\, E_0\, \Phi = E_0\, E_1\, \Phi = \frac{2}{\ln\varepsilon} - \frac{\ln x}{(\ln\varepsilon)^2}\,.$$

Then, using the UVA, it can be shown that for any ν

$$E_\nu\, \Phi_{\mathrm{app}} = E_\nu\, \Phi\,.$$

This shows that, to order $\mathrm{O}\left(1/(\ln\varepsilon)^2\right)$ considered here, the same asymptotic expansions are obtained by taking Φ or Φ_{app}, for any value of ν.

This example shows that Van Dyke's principle does not work for any value of the number of terms considered (see Problem 5-3). By contrast, the MVDP indicates exactly the terms which must taken into account in order to ensure the matching and also the possibility of constructing a UVA.

5.5.3 Example 3

Consider the function

$$\Phi\left(x,\varepsilon\right) = \frac{1}{\ln x - \ln\varepsilon + 1}\,. \tag{5.22}$$

To order $\mathrm{O}\left(1/(\ln\varepsilon)^2\right)$, we obtain the following asymptotic expansions

$$E_0\, \Phi = -\frac{1}{\ln\varepsilon} - \frac{1+\ln x}{(\ln\varepsilon)^2} \quad \text{for} \quad \nu = 0\,,$$

$$E_\nu\, \Phi = \frac{1}{(\nu-1)\ln\varepsilon} - \frac{1+\ln x_\nu}{(\nu-1)^2\,(\ln\varepsilon)^2} \quad \text{for} \quad 0 < \nu < 1\,,$$

$$E_1\, \Phi = \frac{1}{\ln x_1 + 1} \quad \text{for} \quad \nu = 1\,.$$

It can be shown that

$$\mathrm{E}_\nu\,\mathrm{E}_0\,\Phi = -\frac{1+\nu}{\ln\varepsilon} - \frac{1+\ln x_\nu}{(\ln\varepsilon)^2}\,,$$

$$\mathrm{E}_\nu\,\mathrm{E}_1\,\Phi = \frac{1}{(\nu-1)\ln\varepsilon} - \frac{1+\ln x_\nu}{(\nu-1)^2\,(\ln\varepsilon)^2}\,.$$

By contrast with the preceding example, the inner expansion contains the intermediate expansion,

$$\mathrm{E}_\nu\,\mathrm{E}_1\,\Phi = \mathrm{E}_\nu\,\Phi\,.$$

On the other hand, the outer expansion does not contain the intermediate expansion,

$$\mathrm{E}_\nu\,\mathrm{E}_0\,\Phi \neq \mathrm{E}_\nu\,\Phi\,.$$

Thus, there is no intermediate matching since there is no overlap domain to this order with the chosen scales ε^ν. In fact, the extension of the domain of validity of $\mathrm{E}_0\,\Phi$ is too small to be measured with the scale ε^ν. The MVDP, however, applies again without any difficulty

$$\mathrm{E}_1\,\mathrm{E}_0\,\Phi = \mathrm{E}_0\,\mathrm{E}_1\,\Phi = -\frac{1}{\ln\varepsilon} - \frac{1+\ln x}{(\ln\varepsilon)^2}\,,$$

and, with the UVA, it can be shown that, for any ν,

$$\mathrm{E}_\nu\,\Phi_{\mathrm{app}} = \mathrm{E}_\nu\,\Phi\,,$$

since $\Phi_{\mathrm{app}} = \Phi$.

5.5.4 Example 4

We combine the functions

$$\Phi_1\,(x,\varepsilon) = \frac{1}{\ln x} + \frac{e^{-x/\varepsilon}}{\ln\varepsilon} \quad \text{and} \quad \Phi_2\,(x,\varepsilon) = \frac{1}{\ln x - \ln\varepsilon + 1}$$

to obtain

$$\Phi = \Phi_1 + \Phi_2\,. \tag{5.23}$$

There is no overlap with the scales ε^ν to order $\mathrm{O}\left(1/\left(\ln\varepsilon\right)^2\right)$ since, by using the results of Subsects. 5.5.2 and 5.5.3, we have

$$\mathrm{E}_\nu\,\mathrm{E}_1\,\Phi \neq \mathrm{E}_\nu\,\Phi\,,$$
$$\mathrm{E}_\nu\,\mathrm{E}_0\,\Phi \neq \mathrm{E}_\nu\,\Phi\,.$$

The extensions of the domains of validity of $\mathrm{E}_0\,\Phi$ and of $\mathrm{E}_1\,\Phi$ are too small, each on their side, to be measured with the scales ε^ν. These extensions are even so small that, *for any scale*, an overlap does not exist and yet, the MVDP works

$$\mathrm{E}_1\,\mathrm{E}_0\,\Phi = \mathrm{E}_0\,\mathrm{E}_1\,\Phi\,.$$

Again, it can be shown that for any ν,

$$\mathrm{E}_\nu\,\Phi_{\mathrm{app}} = \mathrm{E}_\nu\,\Phi\,.$$

In the example of Subsect. 5.5.2, the information on the intermediate expansion is entirely contained in the outer expansion whereas in the example of Subsect. 5.5.3, it is entirely contained in the inner expansion. In the example given by (5.23), the information is contained partly in the outer expansion and partly in the inner expansion.

The intermediate matching requires an overlap domain, i.e. that all the information on the intermediate expansion is contained in the outer expansion as well as in the inner expansion.

5.6 Discussion of the Matching Principle

It appears that the MVDP is able to solve all the known counter-examples to the VDP. However, let us consider the following example

$$\Phi = 1 + \mathrm{e}^{-x/\varepsilon} + \varepsilon\ln\frac{x}{\varepsilon}\,, \tag{5.24}$$

where Φ is defined in the interval $\varepsilon \leq x \leq 1$. To order $\mathrm{O}(\varepsilon)$, we have

$$\mathrm{E}_0\,\Phi = 1 - \varepsilon\ln\varepsilon + \varepsilon\ln x\,, \tag{5.25a}$$

$$\mathrm{E}_1\,\Phi = 1 + \mathrm{e}^{-X} + \varepsilon\ln X \quad\text{with}\quad X = \frac{x}{\varepsilon}\,. \tag{5.25b}$$

Then, to order $\mathrm{O}(-\varepsilon\ln\varepsilon)$, there is no matching

$$\mathrm{E}_0\,\mathrm{E}_1\,\Phi = 1\,,$$

$$\mathrm{E}_1\,\mathrm{E}_0\,\Phi = 1 - \varepsilon\ln\varepsilon\,,$$

but, to this order, it is not possible to construct a UVA. More precisely, the UVA which can be constructed to order $\mathrm{O}(-\varepsilon\ln\varepsilon)$ is the same as the UVA to order $\mathrm{O}(1)$. Therefore, we are led to apply the MVDP, to a certain order, *in association with the construction of a UVA* to the same order. We will see in Subsect. 5.7.2 that the MVDP is a by-product of the regular form of SCEM whose objective is to construct a UVA.

Moreover, the MVDP applies whereas the rule of intermediate matching fails when the outer and inner expansions are defined to a given order, which is always the case in practice. In this respect, it is useful to mention the idea of Van Dyke *"Fortunately, since the two expansions have a common region of validity, it is easy to construct from them a single uniformly valid expansion"*.

In view of the example of Subsect. 5.5.4 where, to the considered order, there is no common domain of validity and where a UVA can be constructed, it appears that the search for a UVA should be the starting point of the asymptotic analysis. We will go back to this question later (Sect. 5.7) but first we shall discuss the applications in physics in which the small parameter ε is not necessary as small as required formally by the preceding mathematics and the application of the idea of overlap becomes unrealistic even in simple cases. Let us consider for example Friedrichs' model of Subsect. 2.1.3. To order 1, we have

$$\begin{aligned}
\mathrm{E}_0\, y = y_0(x) &= ax + 1 - a \,, & \text{(5.26a)} \\
\mathrm{E}_1\, y = Y_0(X) &= (1 - a)(1 - \mathrm{e}^{-X}) \,. & \text{(5.26b)}
\end{aligned}$$

The plot of $y_0(x)$ and $Y_0(X)$ (Fig. 5.2) shows that the intuitive idea of overlap does not apply in reality. Overlap is a mathematical concept valid only as $\varepsilon \to 0$ and subject to limits as discussed before.

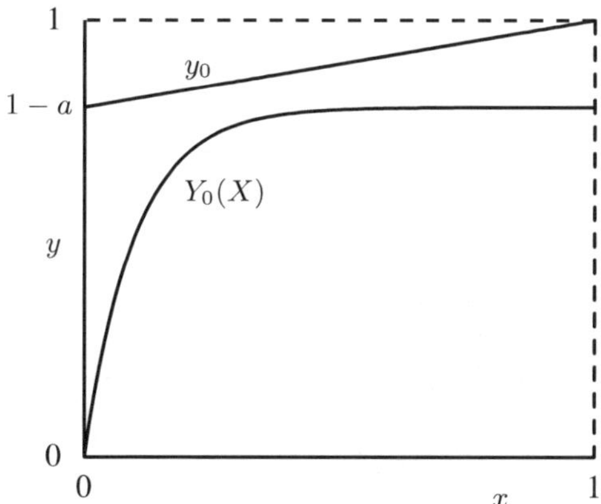

Fig. 5.2. Approximations $y_0(x)$ and $Y_0(X)$ for Friedrichs' problem

Before proceeding further, a few instructive results obtained by Eckhaus [34] are reviewed in a simplified and slightly modified presentation.

5.6.1 Corrective Boundary Layer

Consider the outer expansion $\mathrm{E}_0\, \Phi$. Again, when the number of terms is not specified, this means that the expansion is constructed, with gauge functions, to a prescribed order, for example $\delta\,(\varepsilon)$. A restrictive hypothesis is that $\mathrm{E}_0\, \Phi$ is a continuous, and therefore bounded function in the whole domain D. We define the function $\widetilde{\Phi}$ in D

$$\widetilde{\Phi} = \Phi - \mathrm{E}_0\, \Phi \,. \tag{5.27}$$

Taking into account that $E_0 E_0 \Phi = E_0 \Phi$, we have

$$E_0 \widetilde{\Phi} = 0 \ .$$

We assume that, for any integer k, it is possible to find an order of magnitude $\delta(\varepsilon)$ such that

$$E_0^{(k)} E_1 \widetilde{\Phi} = 0 \ .$$

Then, we can state that the inner expansion $E_1 \widetilde{\Phi}$ contains the outer expansion. For the sake of simplicity, the study is limited to the case where

$$E_0 E_1 \widetilde{\Phi} = 0 \ .$$

It is clear that $E_1 \widetilde{\Phi}$ can be identified as a boundary layer, the adjective "corrective" being added in a natural manner from the definition of $\widetilde{\Phi}$. This property is well verified on the example of Subsect. 5.5.4 for which there is no overlap for Φ.

Then, it can be shown that

$$\Phi = E_0 \Phi + E_1 \Phi - E_0 E_1 \Phi + o(\delta) \ , \tag{5.28}$$

with

$$E_0 E_1 \Phi = E_1 E_0 \Phi \ .$$

From this result, it was deduced by Lagerstrom [48] that there is an overlap domain for $\widetilde{\Phi}$ (Sect. 5.7). An interpretation is that there exists δ_ν such that

$$E_\nu E_1 \widetilde{\Phi} = E_\nu E_0 \widetilde{\Phi} = E_\nu \widetilde{\Phi} = 0 \ .$$

In fact, Lagerstrom used an example given by Fraenkel for which there is no overlap for Φ but the matching principle applies. Again, this idea, sometimes disputable, that the rule of intermediate matching must prevail over the matching principle is underlying. Moreover, if $E_0 \Phi$ is not a bounded function in D, adding a few hypotheses specified below, we can obtain (5.28).

The corrective boundary layer is a key to understand the advantage of the MVDP and of the UVA. Consider again Friedrichs' model (Subsect. 2.1.3). If we set

$$\tilde{y} = y - y_0(x) \ , \tag{5.29}$$

to the specified order, we have

$$E_0 \tilde{y} = 0 \ , \tag{5.30a}$$
$$E_1 \tilde{y} = E_1(y - E_0 y) \ , \tag{5.30b}$$

yielding

$$E_1 \tilde{y} = -(1 - a) e^{-X} \ . \tag{5.31}$$

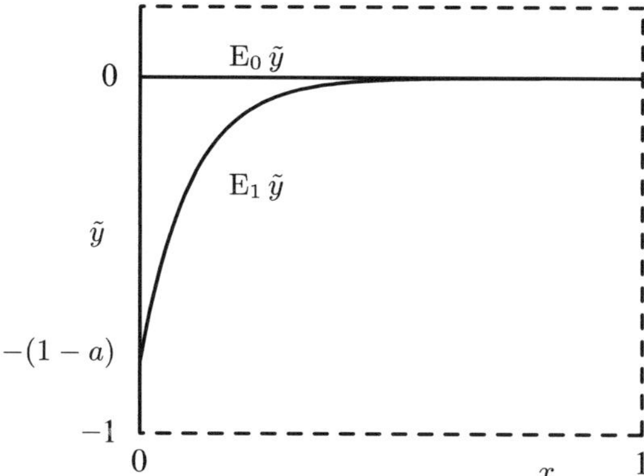

Fig. 5.3. Corrective boundary layer for Friedrichs' problem

As $E_0\,\tilde{y}$ is zero, $E_1\,\tilde{y}$ is in fact a UVA of $\tilde{y}$. It is why now the notion of overlap takes its full meaning (Fig. 5.3). In addition, from the definition of $\tilde{y}$, we obtain

$$E_0\,E_1\,\tilde{y} = E_1\,E_0\,\tilde{y} = 0 \ .$$

As for order 1, we have

$$\tilde{y} = E_1\,\tilde{y} + o(1) \ ,$$

and we obtain the result

$$y = E_0\,y + E_1\,y - E_1\,E_0\,y + o(1) \ . \tag{5.32}$$

Note 5.3. The idea of a defect boundary layer method is exactly along the same lines as the corrective boundary layer. A defect boundary layer formulation has been proposed by Le Balleur [55] to take advantage of the hypothesis that "the calculation domains of inviscid and viscous flows overlap and occupy the whole space". East [32] worked on a similar idea and proposed the use of integral equations based on the difference between the viscous flow and an equivalent inviscid flow to calculate the viscous-inviscid interaction around an airfoil. A defect boundary layer method has also been developed and implemented for hypersonic flows where the characteristics of the inviscid flow vary in a significant manner within the thickness of the boundary layer [5, 7, 8].

5.6.2 The MVDP from the Overlap Hypothesis

The preceding result can be obtained from the hypothesis of overlap. The theorem due to Eckhaus [33] is formulated here in a form adapted to the MVDP. A few conditions are required but they are not very restrictive for applications. The existence of regular expansions $E_0\,\Phi$, $E_1\,\Phi$ and $E_\nu\,\Phi$ to order δ is assumed and, in addition we have

$$\delta = \varepsilon^{m-\gamma} \ ,$$

where m is an integer and γ is a positive number as small as desired. In this way, the possibility to cut an AE between logarithms is excluded. For example, terms of order $\varepsilon \ln \varepsilon$ and ε must be taken together and should not be separated to avoid any problem as discussed by Fraenkel [35] and by Van Dyke [108]. In fact, as discussed before, the MVDP, being associated with the existence of UVAs as given by (5.19), resolves all the known counter-examples to the VDP. Indeed, the question of the cut between logarithms no longer arises. Another important condition is associated with the behaviour of the outer expansion as $x \to 0$ and with the one of the inner expansion as $X \to \infty$.

Condition of behaviour C. *We set*

$$\Phi_0 = \mathrm{E}_0\, \Phi = \sum_{i=1}^{m} \delta_i \left(\varepsilon\right) \varphi_i \left(x\right) \ , \qquad (5.33a)$$

$$\Phi_1 = \mathrm{E}_1\, \Phi = \sum_{i=1}^{m} \delta_i \left(\varepsilon\right) \psi_i \left(X\right) \ , \qquad (5.33b)$$

where, by definition, E_0 and E_1 are operators to order δ_m. The local variable is

$$X = \frac{x}{\nu \left(\varepsilon\right)} \quad \textit{with} \quad \nu \prec 1 \ .$$

Condition C assumes that as $x \to 0$, the behaviour of each function $\varphi_i \left(x\right)$ is

$$\varphi_i \left(x\right) = \sum_{j=1}^{m_i} a_{ij} \Delta_{ij} \left(x\right) + \mathrm{o}\left[\Delta_{im_i} \left(x\right)\right] \ , \qquad (5.34)$$

where a_{ij} is a series of constants and Δ_{ij} is a sequence of gauge functions whose properties are specified below.

We also assume that the functions $\psi_i(X)$ have a behaviour of the same type as $X \to \infty$.

We assume that the gauge functions $\Delta_{ij}(\varepsilon)$ and $\nu\left(\varepsilon\right)$ are elementary functions such as ε^p or $\left(\ln 1/\varepsilon\right)^q$ or products of these functions; p, q are real numbers. To any order δ^* such as $\delta^* \preceq 1$, these functions have the property (Appendix V)

$$\mathrm{E}_0^*\, \mathrm{E}_1^*\, \Delta_{ij}(x) = \mathrm{E}_1^*\, \mathrm{E}_0^*\, \Delta_{ij}(x) \ ,$$

where E_0^* and E_1^* are operators to order δ^*.

It is shown in Appendix V that

$$\mathrm{E}_0\, \mathrm{E}_1\, \Phi_0 = \mathrm{E}_1\, \mathrm{E}_0\, \Phi_0 = \mathrm{E}_1\, \Phi_0 \ , \qquad (5.35)$$

where E_0 and E_1 are operators to order δ_m.

A similar analysis of the behaviour of Φ_1 as $X \to \infty$ yields

$$\mathrm{E}_1\, \mathrm{E}_0\, \Phi_1 = \mathrm{E}_0\, \mathrm{E}_1\, \Phi_1 = \mathrm{E}_0\, \Phi_1 \ . \qquad (5.36)$$

Theorem 5.3. *Eckhaus' theorem adapted to the MVDP, i.e. by working with expansion operators to a given order, states that, if the preceding conditions are satisfied and if there exists an overlap domain such that, to order δ with $\delta \succeq \delta_m$, we have*

$$\mathrm{E}_\nu\, \mathrm{E}_0\, \Phi = \mathrm{E}_\nu\, \mathrm{E}_1\, \Phi = \mathrm{E}_\nu\, \Phi \ ,$$

then

$$\mathrm{E}_0\, \mathrm{E}_1\, \Phi \equiv \mathrm{E}_1\, \mathrm{E}_0\, \Phi \ , \tag{5.37}$$

and

$$\Phi = \mathrm{E}_0\, \Phi + \mathrm{E}_1\, \Phi - \mathrm{E}_0\, \mathrm{E}_1\, \Phi + \mathrm{o}\left(\delta\right) \ . \tag{5.38}$$

Finally, the only important theorem on the asymptotic matching tells us that if there is an overlap domain to a given order, an approximation to the same order is obtained with the MVDP. Nevertheless, the most popular idea is that if the VDP applies, then there is an overlap. Van Dyke has already been cited on this subject [107], Hinch can also be cited: [42] "Van Dyke's matching rule does not always work. Moreover, the rule does not show that the inner and outer expansions are identical in an overlap region".

5.7 Successive Complementary Expansion Method

5.7.1 Principle

SCEM is based upon the idea that the reasoning used in MMAE must be inverted. *At first, a structure of the UVA must be assumed and then the method to construct the UVA is deduced [66].*

The same point of view is adopted in other methods. In the WKB method, a UVA is sought as [42]

$$\Phi_{\mathrm{an}} = \mathrm{e}^X \sum_{i=1}^{n} \delta_i\left(\varepsilon\right) \varphi_i\left(x\right) \quad \text{with} \quad X = \frac{q\left(x,\varepsilon\right)}{\delta\left(\varepsilon\right)} \ ,$$

with q denoting a regular function. This method is well adapted to particular problems as the turning point problem but its usefulness is restricted.

The multiple scale method is another well-known method (Subsect. 2.2.3). The UVA has the form

$$\Phi_{\mathrm{an}} = \sum_{i=1}^{n} \delta_i\left(\varepsilon\right) \varphi_i\left(x, X\right) \ .$$

This method, due to Mahony [62], has been extensively used in many problems. The limitations come from the fact that the order of the equations increases.

SCEM does not require a matching rule or principle. Moreover, in separating the variables x and X, the complexity of the multiple scale method is avoided. The method belongs to a class of so-called multi-variable expansions [75].

In this method, we represent the function Φ by the sum

$$\Phi_{\mathrm{a}}\left(x, X, \varepsilon\right) = \sum_{i=1}^{n} \bar{\delta}_i(\varepsilon) \left[\bar{\varphi}_i(x, \varepsilon) + \bar{\psi}_i(X, \varepsilon)\right] , \tag{5.39}$$

and assume that the above expression is a UVA of Φ defined to order $\bar{\delta}_n$

$$\Phi = \Phi_{\mathrm{a}} + \mathrm{o}(\bar{\delta}_n) . \tag{5.40}$$

The UVA is said *generalized*. It is constructed so that it can be written in the form

$$\Phi_{\mathrm{a}} = \Phi_{\mathrm{ar}} + \mathrm{o}\left(\delta_m\right) , \tag{5.41}$$

where Φ_{ar} is a *regular* UVA such that $\bar{\delta}_n = \mathrm{O}\left(\delta_m\right)$

$$\Phi_{\mathrm{ar}}\left(x, X, \varepsilon\right) = \sum_{i=1}^{m} \delta_i\left(\varepsilon\right) \left[\varphi_i\left(x\right) + \psi_i\left(X\right)\right] . \tag{5.42}$$

The sequence of order functions $\bar{\delta}_i$ may or may not be the same as the sequence δ_i. In addition, the functions δ_i are *gauge functions*.

The transformation of the generalized UVA given by (5.39) into the regular form given by (5.42) is well-suited to boundary layer problems and not to problems involving fast oscillating functions.

We note that the functions on the right hand side of (5.39) or (5.42) are not necessarily bounded separately. Only the sum within square brackets is supposed to be bounded in the domain D.

SCEM has already been used in its regular form [26, 75, 112]. Under the name of method of composite expansions, Nayfeh [72] discusses related techniques developed earlier. The generalized form proposed here is very well adapted to strong interaction problems leading logically to the models of interactive boundary layer (Chap. 8).

With not too strong hypotheses, regular SCEM is equivalent to the MVDP (Subsect. 5.7.2). As already seen, this point is fundamental. The MVDP given by (5.18) takes a precise meaning when associated with a UVA such as the one given by (5.19).

As the application of the MVDP is relatively straightforward, the generalized form given by (5.39) of SCEM is the most interesting. Obviously, a definite advantage should be found in the application of SCEM. Whereas the regular form given by (5.42) does not raise any questioning, the writing of the generalized form given by (5.39) is ambiguous. Indeed, the functions $\bar{\varphi}_i(x, \varepsilon)$ can formally be rewritten as functions of X and ε. Similarly, the functions $\bar{\psi}_i(X, \varepsilon)$ can be rewritten as functions of x and ε. If SCEM appears as essential, guidelines are necessary to construct the successive approximations.

In which cases is SCEM imperative and how to apply it raise two basic questions. In fact, the knowledge of the physics of the problem allows us to answer these questions efficiently. Nevertheless, a few general guidelines can be given. SCEM is recommended in two cases:

1. When the local approximation of the solution presents a complex structure in the neighbourhood of the domain of non-uniformity of the regular approximation. It is the case for a function of the type

$$\Phi(x, \varepsilon) = 1 + \frac{\varepsilon^2}{x + \varepsilon^2}\, e^{-x/\varepsilon}$$

   in the neighbourhood of the origin. The example is intentionally made complicated and the analysis with MMAE indicates the presence of two boundary layer thicknesses. This question is considered in the solution of Problem 5-1 while discussing the solution of a differential equation.
2. When terms occur in AEs whose order of magnitude is not suggested or dictated by the boundary conditions or by the equations as is the case in the example of Subsect. 5.8.3.

How can one be guided to implement SCEM? There is a simple observation when one knows an approximation which is supposed to be uniformly valid Φ_{a}, given by (5.39), which comes from the solution of integro-differential equations. With the corresponding operator formally denoted by L_ε, we have

$$\mathrm{L}_\varepsilon \Phi_{\mathrm{a}} = R_n\,(x, \varepsilon)\ .$$

For the exact solution Φ, we have

$$\mathrm{L}_\varepsilon \Phi = 0\ .$$

Since we assumed that the boundary conditions are exactly satisfied for Φ_{a}, the right hand side R_n should be small. The fact that Φ_{a} is a UVA of Φ when R_n is small in a certain sense is ascertained by estimate theorems which can be obtained, in general, only for linear problems. This is not our purpose here.

Starting with a known UVA,

$$\Phi_{an} = \sum_{i=1}^{n} \bar{\delta}_i(\varepsilon) \left[\bar{\varphi}_i(x, \varepsilon) + \bar{\psi}_i(X, \varepsilon) \right]\ ,$$

the next step is to improve the approximation, at least in the outer region D_0, by adding a term

$$\Phi^*_{a(n+1)} = \Phi_{an} + \bar{\delta}_{n+1}(\varepsilon)\bar{\varphi}_{n+1}(x, \varepsilon)\ .$$

In D_0, $\Phi^*_{a(n+1)}$ is a better approximation than Φ_{an} and, as a result,

$$\left\| \Phi - \Phi^*_{a(n+1)} \right\|_{D_0} = o\left(\bar{\delta}_{n+1}\right) \ .$$

The boundary conditions at $x = 1$ being exactly satisfied, the error at $x = 0$ is $O_S\left(\bar{\delta}_{n+1}\right)$. In certain cases, it can be shown that the additional term is not bounded in the inner region.

Consequently, a UVA is sought as

$$\Phi^*_{a(n+1)} = \Phi_{an} + \bar{\delta}_{n+1}(\varepsilon)\left[\bar{\varphi}_{n+1}(x,\varepsilon) + \bar{\psi}_{n+1}(X,\varepsilon)\right] \ ,$$

so that, in the whole domain D we have

$$\left\| \Phi - \Phi^*_{a(n+1)} \right\|_{D} = o\left(\bar{\delta}_{n+1}\right) \ .$$

This procedure can be repeated and must be initialized by an approximation in the outer region D_0

$$\Phi^*_1 = \bar{\delta}_1(\varepsilon)\bar{\varphi}_1(x,\varepsilon) \ .$$

The minimization of the remainder R_n, "in a certain sense" is one of the keys of SCEM. This will be more precisely discussed when the operators are known. Here, this minimization has only a heuristic value since the estimate theorems are excluded from the presentation.

5.7.2 Equivalence of MVDP and of Regular SCEM

Starting from (5.42), we write

$$\Phi_0\left(x,\varepsilon\right) = \sum_{i=1}^{m} \delta_i\left(\varepsilon\right)\varphi_i\left(x\right) \ ,$$

$$\Phi_1\left(X,\varepsilon\right) = \sum_{i=1}^{m} \delta_i\left(\varepsilon\right)\psi_i\left(X\right) \ .$$

We have

$$\Phi_{ar} = \Phi_0 + \Phi_1 \ ,$$

and, by definition, to order δ_m,

$$\Phi_0 = E_0\,\Phi_0 \ ,$$
$$\Phi_1 = E_1\,\Phi_1 \ ,$$

so that we obtain

$$E_0\,\Phi_{ar} = \Phi_0 + E_0\,\Phi_1 \ ,$$
$$E_1\,\Phi_{ar} = E_1\,\Phi_0 + \Phi_1 \ .$$

This leads to

$$\Phi_{\mathrm{ar}} = \mathrm{E}_0\,\Phi_{\mathrm{ar}} + \mathrm{E}_1\,\Phi_{\mathrm{ar}} - (\mathrm{E}_0\,\Phi_1 + \mathrm{E}_1\,\Phi_0)\ ,$$

and to

$$\mathrm{E}_0\,\Phi_1 + \mathrm{E}_1\,\Phi_0 = \mathrm{E}_0\,\mathrm{E}_1\,\Phi_{\mathrm{ar}} = \mathrm{E}_1\,\mathrm{E}_0\,\Phi_{\mathrm{ar}}\ , \qquad (5.43)$$

and also to the UVA

$$\Phi_{\mathrm{ar}} = \mathrm{E}_0\,\Phi_{\mathrm{ar}} + \mathrm{E}_1\,\Phi_{\mathrm{ar}} - \mathrm{E}_0\,\mathrm{E}_1\,\Phi_{\mathrm{ar}} = \mathrm{E}_0\,\Phi_{\mathrm{ar}} + \mathrm{E}_1\,\Phi_{\mathrm{ar}} - \mathrm{E}_1\,\mathrm{E}_0\,\Phi_{\mathrm{ar}}\ . \quad (5.44)$$

This shows that the MVDP *is contained* in SCEM. In addition, the structure of the asymptotic matching is explicitly given by (5.43).

Proof. Applying E_0 to $\mathrm{E}_1\,\Phi_{\mathrm{ar}}$ and E_1 to $\mathrm{E}_0\,\Phi_{\mathrm{ar}}$, we obtain

$$\mathrm{E}_0\,\mathrm{E}_1\,\Phi_{\mathrm{ar}} = \mathrm{E}_0\,\Phi_1 + \mathrm{E}_0\,\mathrm{E}_1\,\Phi_0\ ,$$
$$\mathrm{E}_1\,\mathrm{E}_0\,\Phi_{\mathrm{ar}} = \mathrm{E}_1\,\Phi_0 + \mathrm{E}_1\,\mathrm{E}_0\,\Phi_1\ ,$$

which yields

$$\mathrm{E}_0\,\Phi_1 + \mathrm{E}_1\,\Phi_0 = \mathrm{E}_0\,\mathrm{E}_1\,\Phi_{\mathrm{ar}} + \mathrm{E}_1\,\mathrm{E}_0\,\Phi_{\mathrm{ar}} - (\mathrm{E}_0\,\mathrm{E}_1\,\Phi_0 + \mathrm{E}_1\,\mathrm{E}_0\,\Phi_1)\ ,$$

or

$$\mathrm{E}_0\,\Phi_1 + \mathrm{E}_1\,\Phi_0 = \mathrm{E}_0\,\mathrm{E}_1\,\Phi_{\mathrm{ar}} + \mathrm{E}_1\,\mathrm{E}_0\,\Phi_0 - \mathrm{E}_0\,\mathrm{E}_1\,\Phi_0\ ,$$

or

$$\mathrm{E}_0\,\Phi_1 + \mathrm{E}_1\,\Phi_0 = \mathrm{E}_1\,\mathrm{E}_0\,\Phi_{\mathrm{ar}} + \mathrm{E}_0\,\mathrm{E}_1\,\Phi_1 - \mathrm{E}_1\,\mathrm{E}_0\,\Phi_1\ .$$

If condition C of Subsect. 5.6.2 is valid, (5.35) and (5.36) imply

$$\mathrm{E}_0\,\Phi_1 = \mathrm{E}_0\,\mathrm{E}_1\,\Phi_1 \equiv \mathrm{E}_1\,\mathrm{E}_0\,\Phi_1\ ,$$
$$\mathrm{E}_1\,\Phi_0 = \mathrm{E}_1\,\mathrm{E}_0\,\Phi_0 \equiv \mathrm{E}_0\,\mathrm{E}_1\,\Phi_0\ ,$$

yielding the result already mentioned (see (5.43) and (5.44)). $\square$

These conclusions enable us to adapt certain results of Lagerstrom [48]. Even if the overlap of $\mathrm{E}_0\,\Phi$ and $\mathrm{E}_1\,\Phi$ does not exist, there is an overlap on the function $\widetilde{\Phi}$ describing the corrective boundary layer defined in Subsect. 5.6.1.

Proof. Taking into account (5.40) and (5.41), we have demonstrated that to order δ_m

$$\mathrm{E}_0\,\mathrm{E}_1\,\Phi = \mathrm{E}_1\,\mathrm{E}_0\,\Phi\ , \qquad (5.45a)$$
$$\Phi_{\mathrm{ar}} = \mathrm{E}_0\,\Phi + \mathrm{E}_1\,\Phi - \mathrm{E}_1\,\mathrm{E}_0\,\Phi\ . \qquad (5.45b)$$

Therefore, in the domain D, we have

$$\Phi - \Phi_{\mathrm{ar}} = \mathrm{o}(\delta_m)\ .$$

We define function $\widetilde{\Phi}$ by

$$\widetilde{\Phi} = \Phi - \mathrm{E}_0\,\Phi\,.$$

It can be shown that, in D

$$\widetilde{\Phi} - \mathrm{E}_1\,\widetilde{\Phi} = \Phi - \mathrm{E}_0\,\Phi - \mathrm{E}_1\,\Phi + \mathrm{E}_1\,\mathrm{E}_0\,\Phi = \mathrm{o}(\delta_m)\,.$$

It is concluded that $\mathrm{E}_1\,\widetilde{\Phi}$ is an approximation of $\widetilde{\Phi}$ to order δ_m in the whole domain D. On the other hand, with $\mathrm{E}_0\,\widetilde{\Phi} = \mathrm{E}_0\,\mathrm{E}_0\,\widetilde{\Phi}$, we obtain

$$\mathrm{E}_0\,\widetilde{\Phi} = \mathrm{E}_0\,\widetilde{\Phi} - \mathrm{E}_0\,\mathrm{E}_0\,\widetilde{\Phi} = 0\,,$$

and

$$\widetilde{\Phi} - \mathrm{E}_0\,\widetilde{\Phi} = \Phi - \mathrm{E}_0\,\Phi\,.$$

Now, $\mathrm{E}_0\,\Phi$ is an approximation of Φ to order δ_m in a certain domain included in D. In the same manner, it is deduced that $\mathrm{E}_0\,\widetilde{\Phi}$ is an approximation of $\widetilde{\Phi}$ to order δ_m.

Finally, $\mathrm{E}_0\,\widetilde{\Phi}$ and $\mathrm{E}_1\,\widetilde{\Phi}$ are two approximations of $\widetilde{\Phi}$ which have a common domain of validity. Therefore, there is an overlap for $\widetilde{\Phi}$ whereas for $\mathrm{E}_0\,\Phi$ and $\mathrm{E}_1\,\Phi$ nothing tells us that there is an overlap. $\square$

Again, the importance of a UVA is stressed. The notion of corrective boundary layer, which is nothing else than a UVA, makes the link between the intuitive notion of overlap and the asymptotic matching.

5.8 Applications of SCEM

In this section, we examine the application of SCEM to two analytical functions (5.5) and (5.23) already studied; a differential equation is also considered in Subsect. 5.8.3.

5.8.1 Example 1

We consider again the function given by (5.5),

$$\Phi\left(x,\varepsilon\right) = \frac{2}{\sqrt{1-4\varepsilon}}\exp\left(-\frac{x}{2\varepsilon}\right)\mathrm{sh}\left(\frac{\sqrt{1-4\varepsilon}}{2\varepsilon}x\right)\,, \tag{5.46}$$

for which a UVA is sought in the domain $x \geq 0$.

The significant approximations to order ε are given by

$$\mathrm{E}_0\,\Phi = \mathrm{e}^{-x} + \varepsilon\,\mathrm{e}^{-x}\left(2-x\right)\,,$$
$$\mathrm{E}_1\,\Phi = \left(1-\mathrm{e}^{-X}\right) + \varepsilon\left[\left(2-X\right)-\left(2+X\right)\mathrm{e}^{-X}\right]\,,$$

with the boundary layer variable

$$X = \frac{x}{\varepsilon} \ .$$

The procedure leading to a UVA according to SCEM is described below.
Step 1. The first regular outer approximation is

$$\varphi_1 (x) = \mathrm{e}^{-x} \ ,$$

but, since

$$\varphi_1 (0) = 1 \ ,$$

the boundary condition at $x = 0$ is not satisfied. We seek a function $\psi_1 (X)$ such that

$$\Phi_{\mathrm{a1}} (x, X) = \varphi_1 (x) + \psi_1 (X) \ ,$$

which is supposed to be a UVA to order 1 of $\Phi(x, \varepsilon)$.

To the same order, applying the expansion operator E_1, with the equality

$$\psi_1 = \mathrm{E}_1 (\Phi - \varphi_1) \ ,$$

it can be shown that

$$\psi_1 (X) = - \mathrm{e}^{-X} \ .$$

Since $\psi_1 (0) = -1$, the condition at $x = 0$ is satisfied.
Step 2. Since $\psi_1 (X)$ is an EST for large X and, since the condition at $x = 0$ is exactly satisfied, a second approximation is sought by simple iteration. To order ε, and taking into account

$$\mathrm{E}_0 \, \Phi = \varphi_1 (x) + \varepsilon\varphi_2 (x) \ ,$$

we have

$$\varphi_2 (x) = (2 - x) \, \mathrm{e}^{-x} \ .$$

From this expression, we observe that $\varphi_2 (0) = 2$. Since the condition at $x = 0$ is not satisfied, we seek a function $\psi_2 (X)$ such that

$$\Phi_{\mathrm{a2}} (x, X, \varepsilon) = \Phi_{\mathrm{a1}} (x, X) + \varepsilon \left(\varphi_2 (x) + \psi_2 (X) \right) \ ,$$

which is supposed to be a UVA to order ε of $\Phi(x, \varepsilon)$.

Using the inner expansion operator, we obtain

$$\varepsilon\psi_2 = \mathrm{E}_1 (\Phi - \Phi_{\mathrm{a1}} - \varepsilon\varphi_2) \ .$$

It can again be shown that

$$\psi_2 (X) = - (2 + X) \, \mathrm{e}^{-X} \ .$$

Since $\psi_2 (0) = -2$, the boundary condition at $x = 0$ is satisfied. At this stage, a UVA to order ε has been constructed

$$\Phi_{\mathrm{a2}} (x, X, \varepsilon) = \mathrm{e}^{-x} - \mathrm{e}^{-X} + \varepsilon \left[(2 - x) \, \mathrm{e}^{-x} - (2 + X) \, \mathrm{e}^{-X} \right] \ .$$

Then, to order ε, we have

$$\Phi_{a2}\left(x, X, \varepsilon\right) = E_0\,\Phi + E_1\,\Phi - \left[1 + \varepsilon\left(2 - X\right)\right] \ .$$

In agreement with results of Sect. 5.7, the MVDP

$$E_0\,E_1\,\Phi = E_1\,E_0\,\Phi = 1 + \varepsilon\left(2 - X\right)$$

is a *consequence of the assumed structure of the UVA and not the converse.*

In addition, taking into account (5.43), the variable X is appropriate in the application of MVDP because

$$\Phi_1 = -\,e^{-X} - \varepsilon(2 + X)\,e^{-X} \ ,$$

and therefore $E_0\,\Phi_1 = 0$.

Note 5.4. SCEM has been used here in its regular form.

5.8.2 Example 2

Consider the function given by (5.23)

$$\Phi\left(x, \varepsilon\right) = \frac{1}{\ln x} + \frac{e^{-x/\varepsilon}}{\ln \varepsilon} + \frac{1}{\ln x - \ln \varepsilon + 1} \ . \tag{5.47}$$

A UVA is sought in the domain $x \geq \varepsilon$. We observe that

$$\Phi\left(\varepsilon, \varepsilon\right) = 1 + \frac{1 + e^{-1}}{\ln \varepsilon} \ .$$

In addition, to order $O\left(-1/\ln \varepsilon\right)$, we have

$$E_0\,\Phi = \frac{1}{\ln x} - \frac{1}{\ln \varepsilon} = \varphi_1\left(x, \varepsilon\right) \ .$$

The boundary condition at $x = \varepsilon$ is not satisfied for $\varphi_1\left(\varepsilon, \varepsilon\right) = 0$. Then, a correction $\psi_1\left(X, \varepsilon\right)$ is sought to obtain a UVA to the same order

$$\Phi_{a1}\left(x, X, \varepsilon\right) = \varphi_1\left(x, \varepsilon\right) + \psi_1\left(X, \varepsilon\right) \quad \text{with} \quad X = \frac{x}{\varepsilon} \ .$$

By using the expansion operator E_1, we note that

$$\psi_1 = E_1\left(\Phi - \varphi_1\right) \ .$$

It is easy to check that

$$E_1\,\varphi_1 = 0 \ ,$$

and to show

$$\psi_1 = \mathrm{E}_1 \,\Phi = \frac{1}{\ln X + 1} + \frac{1 + \mathrm{e}^{-X}}{\ln \varepsilon} \ .$$

In particular, we have

$$\psi_1 \left(1, \varepsilon\right) = 1 + \frac{1 + \mathrm{e}^{-1}}{\ln \varepsilon} \ .$$

Then, it is noted that the UVA yields the exact solution

$$\Phi_{\mathrm{a1}} \left(x, X, \varepsilon\right) = \frac{1}{\ln x} - \frac{1}{\ln \varepsilon} + \frac{1}{\ln X + 1} + \frac{1 + \mathrm{e}^{-X}}{\ln \varepsilon} \ .$$

As for the first example, we note that

$$\Phi_{\mathrm{a1}} \left(x, X, \varepsilon\right) = \mathrm{E}_0 \,\Phi + \mathrm{E}_1 \,\Phi \ ,$$

with

$$\mathrm{E}_0 \,\mathrm{E}_1 \,\Phi = \mathrm{E}_1 \,\mathrm{E}_0 \,\Phi = 0 \ ,$$

showing again that *the MVDP is a consequence of the assumed structure of the UVA*.

5.8.3 Example 3

This example is concerned with an ordinary differential equation introduced by Eckhaus [33]. The analysis with regular expansions as given by Lagerstrom [48] is somewhat involved.

The problem consists of solving the equation

$$\mathrm{L}_\varepsilon \,\Phi = (\varepsilon + x)\frac{\mathrm{d}^2 \Phi}{\mathrm{d}x^2} + \frac{\mathrm{d}\Phi}{\mathrm{d}x} - 1 = 0 \tag{5.48a}$$

in the domain $0 \leq x \leq 1$, subject to the boundary conditions

$$\Phi(0) = 0 \ , \quad \Phi(1) = 2 \ . \tag{5.48b}$$

The reduced equation, obtained by setting $\varepsilon = 0$, is still of second order but is singular at $x = 0$ because the function multiplying the second derivative is zero at this point.

An outer expansion φ_1 is obtained from the equation

$$\mathrm{L}_0 \,\varphi_1 = x\frac{\mathrm{d}^2 \varphi_1}{\mathrm{d}x^2} + \frac{\mathrm{d}\varphi_1}{\mathrm{d}x} - 1 = 0 \tag{5.49}$$

the solution of which is

$$\varphi_1 = 1 + x + A_1 \ln x \ ,$$

where the condition $\varphi_{1(x=1)} = 2$ has been taken into account. The condition at the origin cannot be satisfied and a UVA is sought as

$$\Phi_{\mathrm{a}} = \varphi_1 + \psi_1(X, \varepsilon) \quad \text{with} \quad X = \frac{x}{\varepsilon} \; .$$

Substituting into the original equation and taking into account (5.49), we have

$$\mathrm{L}_\varepsilon \, \Phi_{\mathrm{a}} = -\varepsilon \frac{A_1}{x^2} + \frac{1}{\varepsilon} \left[(1 + X) \frac{\mathrm{d}^2 \psi_1}{\mathrm{d}X^2} + \frac{\mathrm{d}\psi_1}{\mathrm{d}X} \right] \; .$$

It is clear that if one wants to have $\mathrm{L}_\varepsilon \, \Phi_{\mathrm{a}} = 0$ and to satisfy exactly the boundary conditions, then Φ_{a} is the exact solution. Generally, this is not the case. Here, the exact solution is effectively obtained by taking

$$A_1 = 0 \quad \text{and} \quad (1 + X) \frac{\mathrm{d}^2 \psi_1}{\mathrm{d}X^2} + \frac{\mathrm{d}\psi_1}{\mathrm{d}X} = 0 \; ,$$

with the boundary conditions for ψ_1

$$\psi_{1(X=0)} = -1 \; , \quad \psi_{1(X=1/\varepsilon)} = 0 \; .$$

The solution for ψ_1 is

$$\psi_1 = B_1 \ln(1 + X) + B_2 \; , \quad B_1 = \frac{1}{\ln\left(1 + 1/\varepsilon\right)} \; , \quad B_2 = -1 \; ,$$

whence the exact solution

$$\Phi = \Phi_{\mathrm{a}} = x + \frac{\ln\left(1 + x/\varepsilon\right)}{\ln\left(1 + 1/\varepsilon\right)} \; . \tag{5.50}$$

The generalized expansions offer a decisive advantage in the determination of the solution $\psi_1(X, \varepsilon)$.

5.9 Conclusion

The method of matched asymptotic expansions, MMAE, is extensively used to analyze singular perturbation problems, particularly when a boundary layer is present. Aerodynamics constitute a very wide field of application of MMAE [38, 41, 108, 116].

The principle is first to seek significant approximations in domains associated with the scales of the problem and then to find the link between the approximations. The answer is provided by the notion of matching. The most popular techniques are based either on the intermediate variables with the idea of overlap or on Van Dyke's matching principle, VDP.

The overlap which postulates the intermediate matching is a hypothesis which remains illusive for values of parameters which are small but fixed. By contrast, the notion of corrective boundary layer provides a sense to both the overlap and the modified Van Dyke principle, MVDP, thanks to the idea of uniformly valid approximation, UVA. Again, the MVDP, to a given order, must be used in conjunction with the hypothesis of the existence of a UVA to the same order. When this is the case, MVDP is more convenient and has a wider range of application than the rule of intermediate matching and allows the solution of the known counter-examples in the application of VDP.

The fact that the existence of a UVA is in the heart of the MVDP leads naturally to the successive complementary expansion method, SCEM.

Compared to MMAE, the point of view of SCEM is reversed. In this latter method, we first assume the form of UVA which is sought and the method of construction results. Consequently, it is not necessary to call for a matching principle. In certain applications, the UVA is more important than the asymptotic expansion, AE, in the boundary layer. In addition, the use of generalized expansions enables us to solve problems which are difficult or even impossible with regular asymptotic expansions.

Problems

5-1. Consider the equation

$$\mathrm{L}_\varepsilon\, y \equiv (x + \varepsilon)\frac{\mathrm{d}y}{\mathrm{d}x} + (1 + \varepsilon)y + xy = 0\ ,$$

with the boundary condition

$$y(0, \varepsilon) = 1\ .$$

1. Find a UVA to order ε. The MVDP will be used.
2. Use SCEM to keep the term $x + \varepsilon$ which is at the origin of the singularity.

5-2. To order $\mathrm{O}\left\{1/\left[\ln(1/\varepsilon)\right]^2\right\}$, a function $\varPhi(x, \varepsilon)$ has the following outer and inner expansions

$$\varPhi = \frac{1}{\ln(1/\varepsilon)} - \frac{1 + \ln x}{[\ln(1/\varepsilon)]^2}\ ,$$

$$\varPhi = \frac{1}{\ln X + 1}\ ,$$

with

$$X = \frac{x}{\varepsilon}\ .$$

We want to check the rule of intermediate matching. We set

$$\eta = \varepsilon^\alpha X = \frac{x}{\varepsilon^{1-\alpha}} \quad \text{with} \quad 0 < \alpha < 1\ .$$

According to this method, the outer expansion is written with the variable η and the behaviour of the result is studied as $\varepsilon \to 0$ assuming that η is kept fixed. On the other hand, the inner expansion is written with the variable η and the behaviour of the result is studied as $\varepsilon \to 0$ assuming that η is kept fixed. Draw the conclusion. Try to check the rule of intermediate matching with

$$\eta = x \ln \frac{1}{\varepsilon} = X\varepsilon \ln \frac{1}{\varepsilon} \ .$$

5-3. Consider the equation

$$\frac{\mathrm{d}^2 \Phi}{\mathrm{d}x^2} + \frac{1}{x}\frac{\mathrm{d}\Phi}{\mathrm{d}x} + \Phi \frac{\mathrm{d}\Phi}{\mathrm{d}x} = 0 \ ,$$

with the boundary conditions

$$x = \varepsilon : \quad \Phi = 0 \ ; \quad x \to \infty : \quad \Phi = 1 \ .$$

Hinch [42] proposes the solution in which the outer and inner expansions are

$$\Phi = 1 + \frac{g_1(x)}{\ln(1/\varepsilon)} + \frac{g_2(x)}{[\ln(1/\varepsilon)]^2} + \cdots \ ,$$

$$\Phi = A_1 \frac{\ln X}{\ln(1/\varepsilon)} + A_2 \frac{\ln X}{[\ln(1/\varepsilon)]^2} + \cdots \ ,$$

with

$$X = \frac{x}{\varepsilon} \ .$$

We have

$$g_1(x) = B_1 \int_x^\infty \frac{\mathrm{e}^{-t}}{t} \, \mathrm{d}t = B_1 E_1(x) \ ,$$

$$g_2(x) = B_2 E_1(x) + B_1^2 \left[2E_1(2x) - \mathrm{e}^{-x} E_1(x) \right] \ .$$

As $x \to 0$, we have the following behaviours

$$E_1(x) \cong - \ln x - \gamma + x \ ; \quad \gamma = 0.57722 \ldots \ ,$$

$$2E_1(2x) - \mathrm{e}^{-x} E_1(x) \cong - \ln x - \gamma - \ln 4 - x \ln x + (3 - \gamma)x \ .$$

1. The matching between the outer and inner expansions is performed with the method of intermediate expansion. We set

$$\eta = \varepsilon^\alpha X = \frac{x}{\varepsilon^{1-\alpha}} \quad \text{with} \quad 0 < \alpha < 1 \ .$$

According to this method, the outer expansion is written with the variable η and the behaviour of the result is studied as $\varepsilon \to 0$ assuming that η is kept fixed. On the other hand, the inner expansion is written with the variable η and the behaviour of the result is studied as $\varepsilon \to 0$ assuming that η is kept fixed. From the comparison between the two expressions to order $\dfrac{1}{\ln \frac{1}{\varepsilon}}$, determine the constants A_1, B_1, A_2, B_2.

2. Examine the application of the VDP

$$\mathrm{E}_0^{(m)} \, \mathrm{E}_1^{(n)} \, \varPhi = \mathrm{E}_1^{(n)} \, \mathrm{E}_0^{(m)} \, \varPhi \, ,$$

where $\mathrm{E}_0^{(m)}$ denotes the outer expansion in which m terms are retained and $\mathrm{E}_1^{(n)}$ denotes the inner expansion in which n are retained. Express

$$\mathrm{E}_0^{(1)} \, \mathrm{E}_1^{(1)} \, \varPhi \text{ and } \mathrm{E}_1^{(1)} \, \mathrm{E}_0^{(1)} \, \varPhi \, ,$$
$$\mathrm{E}_0^{(2)} \, \mathrm{E}_1^{(1)} \, \varPhi \text{ and } \mathrm{E}_1^{(1)} \, \mathrm{E}_0^{(2)} \, \varPhi \, .$$

In each case, it will be concluded if the matching principle is satisfied or not with the constants determined above.

5-4. A function $y(x, \varepsilon)$ is given by its outer and inner expansions

$$y = \mathrm{e}^{1-x} \left[1 + \varepsilon(1 - x) \right] + \mathrm{O}(\varepsilon^2) \, ,$$
$$y = A_0 \left(1 - \mathrm{e}^{-X} \right) + \varepsilon \left[(A_1 - A_0 X) - (A_1 + A_0 X) \, \mathrm{e}^{-X} \right] + \mathrm{O}(\varepsilon^2) \, ,$$

with

$$X = \frac{x}{\varepsilon} \, .$$

Write the matching of these two expansions by using the expansion operators E_0 and E_1, firstly to order 1, secondly to order ε. Give the value of the constants A_0 and A_1.

Give a composite approximation, uniformly valid to order ε.

5-5. A spaceship is in the gravitational field of the Earth (mass M_T) and of the Moon (mass M_L). The following notations are used: r is the distance between the spaceship and the Earth, d is the distance between the Earth and the Moon, G is the universal gravity constant.

1. We set

$$x = \frac{r}{d} \, , \qquad \varepsilon = \frac{M_L}{M_T + M_L} \, ,$$

where ε is the reduced mass of the Moon. Show that the convenient characteristic time is

$$T = \frac{d^{3/2}}{\sqrt{(M_L + M_T)G}} \, ,$$

which yields the mathematical model

$$\frac{\mathrm{d}^2 x}{\mathrm{d}t^2} = -\frac{1 - \varepsilon}{x^2} + \frac{\varepsilon}{(1 - x)^2} \, .$$

2. With a good approximation, the model reduces to

$$\frac{1}{2} \left(\frac{\mathrm{d}x}{\mathrm{d}t} \right)^2 = \frac{1 - \varepsilon}{x} + \frac{\varepsilon}{1 - x} \qquad \text{with} \quad 0 \le x \le 1 \, .$$

The initial condition on function $x(t, \varepsilon)$ is given by

$$x(0, \varepsilon) = 0 .$$

In particular, it is assumed that the energy of the spaceship is zero when x is large.

An approximation is sought with MMAE.

Express the equation as

$$\frac{\mathrm{d}t}{\mathrm{d}x} = G(x) .$$

An outer approximation is sought as

$$t = t_0(x) + \varepsilon t_1(x) .$$

Determine $t_0(x)$ and $t_1(x)$. Give the initial conditions.

3. Justify the choice of the boundary layer variable $X = (1 - x)/\varepsilon$.

An inner expansion is sought as

$$t = \varepsilon T_0(X) + \varepsilon^2 T_1(X) .$$

Give T_0 and T_1. Give the integration constants by using the matching principle.

Deduce a UVA t_{app} of t in the domain $0 \le x \le 1$.

We have

$$\int \sqrt{\frac{X}{1 + X}}\, \mathrm{d}X = \sqrt{X(1 + X)} - \ln\left[\sqrt{X} + \sqrt{1 + X}\right] + \mathrm{cst} .$$

4. The regular form of SCEM is used. The first approximation is

$$t_{\mathrm{a1}} = f_0(x) .$$

In fact, t_{a1} is a UVA to order 1. Give f_0. Determine the remainder of the equation

$$\mathrm{L}_\varepsilon(t_{\mathrm{a1}}) = \frac{\mathrm{d}t_{\mathrm{a1}}}{\mathrm{d}x} - G(x) ,$$

on the one hand when $0 < A_1 \le x \le A_2 < 1$ where A_1 et A_2 are constants independent of ε and, on the other hand, when $0 < B_1 \le X \le B_2$ where B_1 and B_2 are two constants independent of ε and X is the inner variable

$$X = \frac{1 - x}{\varepsilon} .$$

The UVA to order ε has the form

$$t_{\mathrm{a2}} = f_0(x) + \varepsilon\left[f_1(x) + F_1(X)\right] .$$

Give $f_1(x)$ and $F_1(X)$. Compare to the composite approximation obtained from MMAE. At each step of the construction of the approximation, take care to examine the order of magnitude of terms present in the equations when $0 < A_1 \leq x \leq A_2 < 1$ or $0 < B_1 \leq X \leq B_2 < 1$.

Determine the remainder of the equation

$$L_\varepsilon(t_{a2}) = \frac{dt_{a2}}{dx} - G(x)$$

when $0 < A_1 \leq x \leq A_2 < 1$ and when $0 < B_1 \leq X \leq B_2 < 1$.

5. The generalized form of SCEM is applied. We seek an approximation in the form

$$t_{a1} = y_0(x, \varepsilon)$$

which satisfies the initial condition and which is such that the remainder,

$$L_\varepsilon(t_{a1}) = \frac{dt_{a1}}{dx} - G(x) \, ,$$

is $O(\varepsilon)$ when $0 < A_1 \leq x \leq A_2 < 1$ and when $0 < B_1 \leq X \leq B_2 < 1$. Determine the equation for y_0.

The approximation is complemented as

$$t_{a2} = y_0(x, \varepsilon) + \varepsilon y_1(x, \varepsilon) \, ,$$

which satisfies the initial condition and which is such that the remainder

$$L_\varepsilon(t_{a2}) = \frac{dt_{a2}}{dx} - G(x)$$

is $O(\varepsilon^2)$ when $0 < A_1 \leq x \leq A_2 < 1$ and when $0 < B_1 \leq X \leq B_2 < 1$. Determine the equation for y_1.

6. Calculate the value of t obtained at $x = 1$ i) from the numerical solution of the exact equation, ii) from regular SCEM (approximation t_{a2}), and iii) from generalized SCEM (approximation t_{a2}).

5-6. This problem is treated in the book of Cole [17]. The deflection w^* of an elastic beam with a constant tension τ that supports a load distribution $p^*(x^*)$ per unit length is given by equation

$$EI\frac{d^4 w^*}{dx^{*4}} - \tau\frac{d^2 w^*}{dx^{*2}} = p^*(x^*) \quad \text{with} \quad 0 \leq x^* \leq L \, ,$$

where E is Young's modulus and I is the moment of inertia which is constant in a plane cross section.

The singular perturbation problem occurs when the two effects due to Young's modulus are relatively small compared to the tension effects. Two boundary layers form in the neighbourhood of the beam ends.

We consider the case where the slope and the deflection are zero at both ends

$$w^* = 0 \, ; \quad \frac{dw^*}{dx^*} = 0 \quad \text{at} \quad x^* = 0 \quad \text{and} \quad x^* = L \, .$$

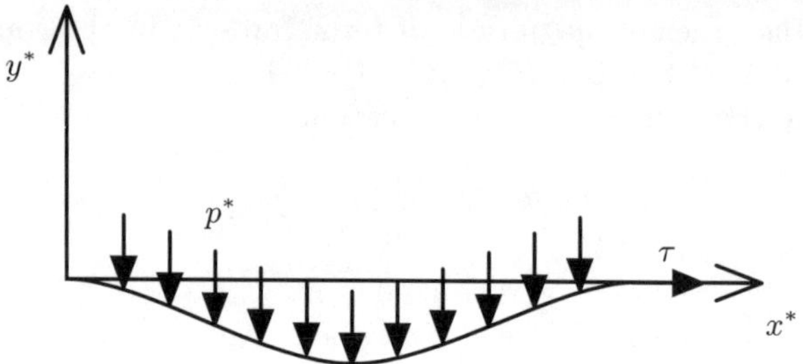

Fig. 5.4. Deflection of a beam

The equation is nondimensionalized with

$$x = \frac{x^*}{L} \ , \quad p = \frac{p^*}{\mathcal{P}} \ , \quad w = \frac{\tau}{\mathcal{P}L^2}w^* \ .$$

We obtain

$$\varepsilon\frac{\mathrm{d}^4w}{\mathrm{d}x^4} - \frac{\mathrm{d}^2w}{\mathrm{d}x^2} = p(x) \ , \quad 0 \le x \le 1 \ ,$$

with the conditions

$$w(0) = 0 \ , \quad w(1) = 0 \ , \quad \frac{\mathrm{d}w}{\mathrm{d}x}(0) = 0 \ , \quad \frac{\mathrm{d}w}{\mathrm{d}x}(1) = 0 \ .$$

The small parameter ε is introduced

$$\varepsilon = \frac{EI}{\tau L^2} \ .$$

Note that $\mathcal{P}$ is arbitrary. We can take $\mathcal{P} = \tau/L$ so that $w = w^*/L$.

The problem is studied with MMAE.

1. We seek an outer expansion in the form

$$w = w_0(x) + \nu_1(\varepsilon)w_1(x) + \nu_2(\varepsilon)w_2(x) + \cdots \ ,$$

where $\nu_i(\varepsilon)$ forms an asymptotic sequence. Write the equation for $w_0(x)$. Show that the solution has the form

$$w_0 = B_0 + A_0x - \int_0^x p(\lambda)(x - \lambda) \, \mathrm{d}\lambda \ .$$

Study the form of w_0 as $x \to 0$ and as $x \to 1$. It will be shown that

$$w_0(x) = B_0 + A_0x - p(0)\frac{x^2}{2} - p'(0)\frac{x^3}{3!} + \mathrm{O}(x^4) \text{ as } x \to 0 \ ,$$

$$w_0(x) = B_0 + A_0 - \int_0^1 p(\lambda)(1 - \lambda) \, d\lambda + \left[A_0 - \int_0^1 p(\lambda) \, d\lambda\right](x - 1)$$

$$-p(1)\frac{(x - 1)^2}{2} - p'(1)\frac{(x - 1)^3}{3!} + \mathrm{O}\left[(x - 1)^4\right] \text{ as } x \to 1 \ .$$

2. The inner expansion is studied in the neighbourhood of $x = 0$. The inner variable is

$$X = \frac{x}{\delta(\varepsilon)} \ .$$

The inner expansion has the form

$$w = \mu_0(\varepsilon)W_0(X) + \mu_1(\varepsilon)W_1(X) + \cdots \ .$$

Determine δ.

The matching with the outer expansion leads us to choose

$$\mu_0 = \varepsilon^{1/2} \ .$$

Determine the equation for W_0 and show that the solution has the form

$$W_0 = C_0(X + e^{-X} - 1) \ .$$

3. Write the matching between the inner expansion in the neighbourhood of $x = 0$ and the outer expansion by using the expansion operators E_0 and E_1 to order $\varepsilon^{1/2}$ assuming that $\nu_1 = \varepsilon^{1/2}$. Deduce that the first term of the outer expansion is

$$w_0 = C_0 x - \int_0^x p(\lambda)(x - \lambda) \, d\lambda \ .$$

Write the equation for w_1. Show that the solution is

$$w_1 = -C_0 + A_1 x \ .$$

4. We study the inner expansion in the neighbourhood of $x = 1$. The inner variable is

$$X^+ = \frac{x - 1}{\delta^+(\varepsilon)} \ .$$

Determine $\delta^+(\varepsilon)$.

The outer expansion has the form

$$w_0 = \mu_0^+ W_0^+ + \mu_1^+ W_1^+ \ .$$

Determine μ_0^+. Write the equation for W_0^+. Show that the solution has the form

$$W_0^+ = C_0^+(X^+ + 1 - e^{X^+}) \ .$$

Write the matching with the outer expansion by using the operators E_0 and E_1^+ to order $\varepsilon^{1/2}$.

The following notations are used

$$k = \int_0^1 p(\lambda) \, d\lambda \text{ where } k \text{ is the total load,}$$

$$M^{(0)} = \int_0^1 p(\lambda)\lambda \, d\lambda$$ where $M^{(0)}$ is the moment of the load about the point $x = 0$,

$$M^{(1)} = -\int_0^1 p(\lambda)(1 - \lambda) \, d\lambda$$ where $M^{(1)}$ is the moment of the load about the point $x = 1$.

Calculate C_0, C_0^+ and A_1 as function of k, $M^{(0)}$ and $M^{(1)}$.

6 Ordinary Differential Equations

The practical implementation of the successive complementary expansion method, SCEM, is presented in this Chapter with applications to functions implicitly defined as solutions of ordinary differential equations. In all the cases, the solution depends on a small parameter leading to a singular perturbation problem. More precisely, the structure of the solution exhibits a boundary layer. The boundary layer is always located in the neighbourhood of a singular point, but this point is not necessarily a part of the boundary of the considered domain. Among the variety of examples found in the literature [17, 42, 48, 72], a selection has been made to enlighten the various aspects of SCEM and its advantages. This is a first step toward the treatment of physical problems governed by partial differential equations.

Compared to the method of matched asymptotic expansions, MMAE, an essential difference is the objective. Indeed, with SCEM, the main point is to form a uniformly valid approximation, UVA, from an assumed structure of the UVA. No matching principle is necessary to obtain the result. By contrast, the principle of MMAE is first to seek approximations in significant subsets of the considered domain. A matching principle is absolutely necessary to make the link between the so obtained approximations. A UVA is constructed at the end.

6.1 Example 1

We consider the equation

$$\mathrm{L}_\varepsilon\,\Phi \equiv \varepsilon\frac{\mathrm{d}^2\Phi}{\mathrm{d}x^2} + a\,(x)\,\frac{\mathrm{d}\Phi}{\mathrm{d}x} + b\,(x)\,\Phi = 0 \ , \ a(x) > 0 \ , \tag{6.1a}$$

where $\Phi\,(x,\varepsilon)$ is defined in the domain D $(0 \leq x \leq 1)$ and the functions a and b are given. The boundary conditions are

$$\Phi\,(0,\varepsilon) = \alpha \ , \quad \Phi\,(1,\varepsilon) = \beta \ . \tag{6.1b}$$

6.1.1 Application of MMAE

The first approximation $\varphi_1(x)$ is obtained by setting $\varepsilon = 0$ in (6.1a)

$$a(x)\frac{\mathrm{d}\varphi_1}{\mathrm{d}x} + b(x)\varphi_1 = 0 . \qquad (6.2)$$

The solution is

$$\varphi_1(x) = C \exp\left(-\int_0^x \frac{b(\xi)}{a(\xi)}\,\mathrm{d}\xi\right) .$$

All the sufficient hypotheses on a and b are supposed to be satisfied if required. In particular, we assume that the above integral exists.

This approximation cannot cope with the two boundary conditions (6.1b) if $\beta \neq \lambda\alpha$ where

$$\lambda = \exp\left[-\int_0^1 \frac{b(\xi)}{a(\xi)}\,\mathrm{d}\xi\right] .$$

Now, with $a(x) > 0$, it was shown in Chap. 3 that there exists a *boundary layer* in the neighbourhood of $x = 0$. It is then natural to impose the condition at $x = 1$ to determine the value of the constant C so that the first outer approximation of $\Phi(x, \varepsilon)$ is the function

$$\varphi_1(x) = \beta \exp\left(+\int_x^1 \frac{b(\xi)}{a(\xi)}\,\mathrm{d}\xi\right) . \qquad (6.3)$$

In the neighbourhood of $x = 0$, a *significant degeneracy* of the equation is obtained with the local variable $X = x/\varepsilon$ since the original equation becomes

$$\frac{\mathrm{d}^2\Phi}{\mathrm{d}X^2} + a(\varepsilon X)\frac{\mathrm{d}\Phi}{\mathrm{d}X} + \varepsilon b(\varepsilon X)\Phi = 0 .$$

If $a(x)$ is a sufficiently regular function, a *first regular inner approximation* of Φ satisfies the equation

$$\frac{\mathrm{d}^2\psi_1}{\mathrm{d}X^2} + a(0)\frac{\mathrm{d}\psi_1}{\mathrm{d}X} = 0 .$$

With $\psi_1(0) = \alpha$, the solution is

$$\psi_1(X) = (\alpha - A)\,e^{-a(0)X} + A .$$

The constant A can be calculated by applying the *modified Van Dyke matching principle*, MVDP, to order 1

$$\mathrm{E}_0\Phi = \varphi_1 , \quad \mathrm{E}_1\Phi = \psi_1 ,$$

whence

$$A = \mathrm{E}_0\mathrm{E}_1\Phi = \mathrm{E}_1\mathrm{E}_0\Phi = \frac{\beta}{\lambda} \ .$$

The solution becomes

$$\psi_1\left(X\right) = \left(\alpha - \frac{\beta}{\lambda}\right) e^{-a(0)X} + \frac{\beta}{\lambda} \ ,$$

and, according to (5.19), the following *UVA* is obtained

$$\Phi_{\mathrm{a}1}\left(x, X\right) = \beta \exp\left(+\int_x^1 \frac{b\left(\xi\right)}{a\left(\xi\right)}\, \mathrm{d}\xi\right) + \left(\alpha - \frac{\beta}{\lambda}\right) e^{-a(0)X} \ . \tag{6.4}$$

Since the boundary conditions are satisfied to order ε^n for any positive integer n, a better approximation can be obtained by simple iteration on the equations. A *regular outer expansion* to order ε is

$$\mathrm{E}_0\Phi = \varphi_1\left(x\right) + \varepsilon\varphi_2\left(x\right) \ ,$$

with

$$a\left(x\right)\frac{\mathrm{d}\varphi_2}{\mathrm{d}x} + b\left(x\right)\varphi_2 = -\frac{\mathrm{d}^2\varphi_1}{\mathrm{d}x^2} \ .$$

The boundary condition at $x = 1$ being satisfied by φ_1, we take

$$\varphi_2(1) = 0 \ .$$

To the same order, a *regular inner expansion* is

$$\mathrm{E}_1\Phi = \psi_1\left(X\right) + \varepsilon\psi_2\left(X\right) \ ,$$

with the equation

$$\frac{\mathrm{d}^2\psi_2}{\mathrm{d}X^2} + a(0)\frac{\mathrm{d}\psi_2}{\mathrm{d}X} = -Xa'(0)\frac{\mathrm{d}\psi_1}{\mathrm{d}X} - b(0)\psi_1 \ , \quad a'\left(x\right) = \frac{\mathrm{d}a(x)}{\mathrm{d}x} \ .$$

The boundary condition at $X = 0$ being satisfied by ψ_1, we take

$$\psi_2(0) = 0 \ .$$

The solutions are easily calculated since the missing boundary conditions are obtained from the application of the MVDP to order ε

$$\mathrm{E}_0\mathrm{E}_1\Phi = \mathrm{E}_1\mathrm{E}_0\Phi \ .$$

Note 6.1. It is reminded that, in the multiple scale method, the essential argument stated by Lighthill, implies that ψ_2 should be no more singular than ψ_1. Clearly, this is not the case here in view of the right hand side of the equation for ψ_2.

6.1.2 Application of SCEM

The construction of a UVA starts as with MMAE. The first approximation
is obtained in the form

$$\Phi = \hat{\varphi}_1(x, \varepsilon) + \cdots ,$$

where $\hat{\varphi}_1$ can be a function of x *and* of ε since *generalized expansions* are
accepted with SCEM. Substituting this expansion into (6.1a) and neglecting
$O(\varepsilon)$ terms, (6.2) is recovered. With the same arguments as those already
presented in Subsect. 6.1.1, the application of the boundary condition at
$x = 1$ again yields the solution given by (6.3), that is

$$\hat{\varphi}_1(x, \varepsilon) = \varphi_1(x) .$$

The principle of SCEM consists of *complementing this approximation* to
get a first UVA in the form

$$\hat{\Phi}_{\mathrm{a1}}(x, X, \varepsilon) = \varphi_1(x) + \hat{\psi}_1(X, \varepsilon) . \tag{6.5}$$

Equation (6.1a) yields

$$L_\varepsilon \hat{\Phi}_{\mathrm{a1}} \equiv \frac{1}{\varepsilon}\left(\frac{\mathrm{d}^2\hat{\psi}_1}{\mathrm{d}X^2} + a(x)\frac{\mathrm{d}\hat{\psi}_1}{\mathrm{d}X}\right) + b(x)\hat{\psi}_1 + \varepsilon\frac{\mathrm{d}^2\varphi_1}{\mathrm{d}x^2} = R_1 .$$

The approximation $\hat{\Phi}_{\mathrm{a1}}$ is close to the solution Φ if the remainder R_1 is
small *in a certain sense*. This remainder has two parts

$$R_{11} = \frac{1}{\varepsilon}\left(\frac{\mathrm{d}^2\hat{\psi}_1}{\mathrm{d}X^2} + a(x)\frac{\mathrm{d}\hat{\psi}_1}{\mathrm{d}X}\right) + b(x)\hat{\psi}_1 ,$$

$$R_{12} = \varepsilon\frac{\mathrm{d}^2\varphi_1}{\mathrm{d}x^2} .$$

As φ_1 is a UVA in D_0 – domain which is outside the boundary layer – $\hat{\psi}_1$ is
negligible in D_0. In other words, $\hat{\psi}_1$ is of order 1 only in the boundary layer
domain D_1 of extension ε. This is why R_{11} cannot be viewed in the same
way as R_{12}. Without commenting more, it is said that: *the estimates giving
an upper bound of* $\left|\Phi - \hat{\Phi}_{\mathrm{a1}}\right|$ *in* D *are of integral nature* [63, 75]. However,
the conclusion is drawn that, if R_{12} is $O(\varepsilon)$ in D, it is sufficient that R_{11}
is $O(1)$ in the same domain in order that the two parts have an equivalent
contribution to the result

$$\left|\Phi - \hat{\Phi}_{\mathrm{a1}}\right| < K_1\varepsilon ,$$

where K_1 is a positive constant independent of ε. The solution of Problem 6-1
gives details on this point. As also discussed in [17, 26], under certain conditions, a UVA of the solution of the considered problem is the sum of an outer
approximation and of a boundary layer term.

The requirement that R_{11} should be "as small as possible" can be satisfied
in different ways. Two methods are examined below.

Method a

Function $\hat{\psi}_1$ is sought as the solution of

$$\frac{\mathrm{d}^2\hat{\psi}_1}{\mathrm{d}X^2} + a\left(0\right)\frac{\mathrm{d}\hat{\psi}_1}{\mathrm{d}X} = 0 \, .$$

Forcing $\hat{\varPhi}_{\mathrm{a}1}$ to satisfy the boundary conditions *exactly*, the following boundary conditions on $\hat{\psi}_1$ are obtained,

$$\hat{\psi}_1\left(0,\varepsilon\right) = \alpha - \frac{\beta}{\lambda} \, , \; \hat{\psi}_1\left(\frac{1}{\varepsilon},\varepsilon\right) = 0 \, .$$

The solution is

$$\hat{\psi}_1\left(X,\varepsilon\right) = \left(\alpha - \frac{\beta}{\lambda}\right)\frac{e^{-a(0)X} - e^{-a(0)/\varepsilon}}{1 - e^{-a(0)/\varepsilon}} \, .$$

In the construction of the UVA, it is important to note that the corrective term is an EST for large values of X. Clearly, the second boundary condition on $\hat{\psi}_1$ can be replaced by $\hat{\psi}_1 \to 0$ as $X \to \infty$. In fact, the EST can be eliminated by taking

$$\hat{\psi}_1\left(X\right) = \left(\alpha - \frac{\beta}{\lambda}\right)e^{-a(0)X} \quad \text{with} \quad \lim_{X\to\infty}\hat{\psi}_1 = 0 \, .$$

With this result, the UVA given by (6.5) is identical to the composite solution (6.4) deduced from MMAE *but the asymptotic matching principle is now a result*, in agreement with the conclusions of Sect. 5.7.

In order to improve the accuracy of the approximation, we keep ESTs if ε is not really small. In this way, in particular in the neighbourhood of $x = 1$, the UVA $\hat{\varPhi}_{\mathrm{a}1}$ remains close to the exact solution.

Since the boundary conditions are satisfied to order ε^n for any positive integer n, a better approximation is obtained by a simple iteration on the equation. Another procedure is to examine *the remainder of the equation*

$$\mathrm{L}_\varepsilon\,\hat{\varPhi}_{\mathrm{a}1} \equiv \frac{a\left(x\right) - a\left(0\right)}{\varepsilon}\frac{\mathrm{d}\hat{\psi}_1}{\mathrm{d}X} + b\left(x\right)\hat{\psi}_1 + \varepsilon\frac{\mathrm{d}^2\varphi_1}{\mathrm{d}x^2} \, . \tag{6.6}$$

At this stage, there is a significant difference between MMAE and SCEM because $\hat{\psi}_1$ is a boundary layer term in the outer domain. Then, in the remainder, the first and the second terms are ESTs in the outer domain and are of order 1 in the inner domain whereas the third term is uniformly of order ε in the whole domain. Using the estimate theorem given in [63] yields

$$\left|\varPhi - \hat{\varPhi}_{\mathrm{a}1}\right| < K_1\varepsilon \, ,$$

under the assumption that all functions and their derivatives in (6.6) are bounded.

The construction of a better approximation can be obtained by writing

$$\Phi = \hat{\Phi}_{a1} + \varepsilon\hat{\varphi}_2(x,\varepsilon) + \cdots . \qquad (6.7)$$

Equation (6.1a) becomes

$$L_\varepsilon\,\Phi = L_\varepsilon\,\hat{\Phi}_{a1} + \varepsilon\,L_\varepsilon\,\hat{\varphi}_2(x,\varepsilon) + \cdots ,$$

where

$$L_\varepsilon\,\hat{\varphi}_2(x,\varepsilon) = \varepsilon\frac{\mathrm{d}^2\hat{\varphi}_2}{\mathrm{d}x^2} + a\,(x)\,\frac{\mathrm{d}\hat{\varphi}_2}{\mathrm{d}x} + b(x)\hat{\varphi}_2 .$$

As already mentioned, when x is in D_0, $\hat{\psi}_1$ and $\dfrac{\mathrm{d}\hat{\psi}_1}{\mathrm{d}X}$ are ESTs. Neglecting $O(\varepsilon^2)$ terms in D_0 yields the equation for $\hat{\varphi}_2$

$$a\,(x)\,\frac{\mathrm{d}\hat{\varphi}_2}{\mathrm{d}x} + b\,(x)\,\hat{\varphi}_2 = -\frac{\mathrm{d}^2\hat{\varphi}_1}{\mathrm{d}x^2} .$$

Since the boundary conditions are satisfied by the first UVA, we simply take

$$\hat{\varphi}_2\,(1,\varepsilon) = 0 .$$

The solution shows that $\hat{\varphi}_2$ is a function of x only and we have

$$\hat{\varphi}_2(x,\varepsilon) = \varphi_2(x) ,$$

where $\varphi_2(x)$ is the same function as in MMAE.

Clearly, the approximation given by (6.7) must be corrected in the neighbourhood of $x = 0$ and this can be done by seeking a second UVA in the form

$$\hat{\Phi}_{a2}\,(x,X,\varepsilon) = \hat{\Phi}_{a1} + \varepsilon\left[\varphi_2\,(x) + \hat{\psi}_2\,(X,\varepsilon)\right] . \qquad (6.8)$$

Substituting (6.8) into (6.1a) yields

$$L_\varepsilon\,\hat{\Phi}_{a2} = \frac{a\,(x) - a\,(0)}{\varepsilon}\frac{\mathrm{d}\hat{\psi}_1}{\mathrm{d}X} + b\,(x)\,\hat{\psi}_1 + \varepsilon^2\frac{\mathrm{d}^2\varphi_2}{\mathrm{d}x^2} + \frac{\mathrm{d}^2\hat{\psi}_2}{\mathrm{d}X^2} + a\,(x)\,\frac{\mathrm{d}\hat{\psi}_2}{\mathrm{d}X} + \varepsilon b\,(x)\,\hat{\psi}_2 .$$

We assume that φ_2 is uniformly of order 1 in the studied domain. By neglecting $O(\varepsilon)$ terms in the domain D_1, that is, when $0 < A_1 \leq X \leq A_2$ where A_1 and A_2 are constants independent of ε, we obtain the following equation for $\hat{\psi}_2$,

$$\frac{\mathrm{d}^2\hat{\psi}_2}{\mathrm{d}X^2} + a(0)\frac{\mathrm{d}\hat{\psi}_2}{\mathrm{d}X} = -\frac{a\,(x) - a\,(0)}{\varepsilon}\frac{\mathrm{d}\hat{\psi}_1}{\mathrm{d}X} - b\,(x)\,\hat{\psi}_1 .$$

The boundary conditions are exactly satisfied with

$$\hat{\psi}_2\,(0,\varepsilon) = -\varphi_2\,(0) , \quad \hat{\psi}_2\left(\frac{1}{\varepsilon},\varepsilon\right) = 0 .$$

Finally, with a similar analysis as for $\mathrm{L}_\varepsilon \, \hat{\Phi}_{\mathrm{a}1}$, the expression of $\mathrm{L}_\varepsilon \, \hat{\Phi}_{\mathrm{a}2}$ shows that the accuracy associated with the remainder of $\hat{\Phi}_{\mathrm{a}2}$ is better than for $\hat{\Phi}_{\mathrm{a}1}$ (see (6.6))

$$\mathrm{L}_\varepsilon \, \hat{\Phi}_{\mathrm{a}2} \equiv [a\,(x) - a\,(0)] \frac{\mathrm{d}\hat{\psi}_2}{\mathrm{d}X} + \varepsilon b\,(x)\,\hat{\psi}_2 + \varepsilon^2 \frac{\mathrm{d}^2\varphi_2}{\mathrm{d}x^2} \ .$$

On the basis of the preceding remarks, there exists a positive constant $\widehat{K}_2$, independent of ε such that

$$\left| \Phi - \hat{\Phi}_{\mathrm{a}2} \right| < \widehat{K}_2 \varepsilon^2 \ .$$

Method b

The first UVA is sought in the form

$$\bar{\Phi}_{\mathrm{a}1} = \varphi_1 + \bar{\psi}_1(X, \varepsilon) \ .$$

With this expression, we can write (6.1a) as

$$\mathrm{L}_\varepsilon \, \bar{\Phi}_{\mathrm{a}1} = \frac{1}{\varepsilon} \left(\frac{\mathrm{d}^2\bar{\psi}_1}{\mathrm{d}X^2} + a\,(x) \frac{\mathrm{d}\bar{\psi}_1}{\mathrm{d}X} \right) + b\,(x)\,\bar{\psi}_1 + \varepsilon \frac{\mathrm{d}^2\varphi_1}{\mathrm{d}x^2} \ .$$

To improve the accuracy of the approximation, the idea is to *include more information in the first UVA*. Practically, in D_1, instead of neglecting $\mathrm{O}(1)$ terms, $\mathrm{O}(\varepsilon)$ terms are neglected and we obtain

$$\frac{1}{\varepsilon} \left(\frac{\mathrm{d}^2\bar{\psi}_1}{\mathrm{d}X^2} + a\,(x) \frac{\mathrm{d}\bar{\psi}_1}{\mathrm{d}X} \right) + b\,(0)\,\bar{\psi}_1 = 0 \ .$$

This provides us with a better UVA which contains the second order of the boundary layer expansion. The remainder for $\bar{\Phi}_{\mathrm{a}1}$ is

$$\mathrm{L}_\varepsilon \, \bar{\Phi}_{\mathrm{a}1} \equiv [b(x) - b(0)]\,\bar{\psi}_1 + \varepsilon \frac{\mathrm{d}^2\varphi_1}{\mathrm{d}x^2} \ . \tag{6.9}$$

In agreement with the preceding heuristic remarks, the contribution of the term $[b(x) - b(0)]\,\bar{\psi}_1$ is smaller than the contribution of the term $\varepsilon \dfrac{\mathrm{d}^2\varphi_1}{\mathrm{d}x^2}$.

Then, the next UVA is sought in the form

$$\bar{\Phi}_{\mathrm{a}2} = \bar{\Phi}_{\mathrm{a}1} + \varepsilon\bar{\varphi}_2\,(x, \varepsilon) \ .$$

Substituting into (6.1a) yields

$$\mathrm{L}_\varepsilon \, \bar{\Phi}_{\mathrm{a}2} = [b(x) - b(0)]\,\bar{\psi}_1 + \varepsilon \frac{\mathrm{d}^2\varphi_1}{\mathrm{d}x^2} + \varepsilon^2 \frac{\mathrm{d}^2\bar{\varphi}_2}{\mathrm{d}x^2} + a\,(x)\,\varepsilon \frac{\mathrm{d}\bar{\varphi}_2}{\mathrm{d}x} + b(x)\varepsilon\bar{\varphi}_2 \ .$$

Neglecting $O(\varepsilon^2)$ terms in D_0 gives the following equation for $\bar{\varphi}_2$

$$a\left(x\right)\frac{d\bar{\varphi}_2}{dx} + b\left(x\right)\bar{\varphi}_2 = -\frac{d^2\varphi_1}{dx^2} \,.$$

The *exact boundary conditions* are satisfied with

$$\bar{\varphi}_2\left(1,\varepsilon\right) = 0 \,, \quad \bar{\psi}_1\left(0,\varepsilon\right) = \alpha - \frac{\beta}{\lambda} - \varepsilon\bar{\varphi}_2\left(0,\varepsilon\right) \,, \quad \bar{\psi}_1\left(\frac{1}{\varepsilon},\varepsilon\right) = 0 \,.$$

We have again

$$\bar{\varphi}_2(x,\varepsilon) = \varphi_2(x) \,,$$

where $\varphi_2(x)$ is the same function as in MMAE.

For confirmation, if this UVA is complemented as

$$\bar{\bar{\Phi}}_{a2} = \bar{\bar{\Phi}}_{a1} + \varepsilon\varphi_2(x) + \varepsilon\bar{\psi}_2(X,\varepsilon) \,,$$

it is shown that $\bar{\psi}_2 = 0$. In effect, (6.1a) becomes

$$L_\varepsilon\,\bar{\bar{\Phi}}_{a2} = [b(x) - b(0)]\,\bar{\psi}_1 + \varepsilon^2\frac{d^2\varphi_2}{dx^2} + \frac{d^2\bar{\psi}_2}{dX^2} + a(x)\frac{d\bar{\psi}_2}{dX} + b(x)\varepsilon\bar{\psi}_2 \,.$$

Neglecting $O(\varepsilon)$ terms in D_1 yields

$$\frac{d^2\bar{\psi}_2}{dX^2} + a(0)\frac{d\bar{\psi}_2}{dX} = 0 \,,$$

with

$$\bar{\psi}_2\left(0,\varepsilon\right) = 0 \,, \quad \bar{\psi}_2\left(\frac{1}{\varepsilon},\varepsilon\right) = 0 \,.$$

The solution is $\bar{\psi}_2 = 0$.

Finally, the remainder writes

$$L_\varepsilon\,\bar{\bar{\Phi}}_{a2} \equiv [b\left(x\right) - b\left(0\right)]\,\bar{\psi}_1 + \varepsilon^2\frac{d^2\varphi_2}{dx^2} \,,$$

which leads to

$$\left|\Phi - \bar{\bar{\Phi}}_{a2}\right| < \overline{K}_2\varepsilon^2 \,.$$

The question is to evaluate if this problem is simpler to solve than the initial problem.

Until now, the implicit hypotheses require that the functions introduced in the calculations and their derivatives are bounded in the whole domain D. As shown below, this is not always necessary.

6.2 Example 2

We consider the equation

$$\mathrm{L}_\varepsilon \, \varPhi \equiv \varepsilon \frac{\mathrm{d}^2\varPhi}{\mathrm{d}x^2} + x^{1/4}\frac{\mathrm{d}\varPhi}{\mathrm{d}x} - \varPhi = 0 \, , \qquad (6.10\mathrm{a})$$

where the function $\varPhi$ is defined in the domain $0 \le x \le 1$ with the boundary conditions

$$\varPhi(0) = 0 \, , \quad \varPhi(1) = \mathrm{e}^{4/3} \, . \qquad (6.10\mathrm{b})$$

6.2.1 Application of MMAE

A straightforward outer expansion,

$$\mathrm{E}_0\varPhi = \varphi_1(x) + \varepsilon\varphi_2(x) + \cdots \, ,$$

leads to equations

$$x^{1/4}\frac{\mathrm{d}\varphi_1}{\mathrm{d}x} - \varphi_1 = 0 \, ,$$

$$x^{1/4}\frac{\mathrm{d}\varphi_2}{\mathrm{d}x} - \varphi_2 = -\frac{\mathrm{d}^2\varphi_1}{\mathrm{d}x^2} \, ,$$

with

$$\varphi_1(1) = \mathrm{e}^{4/3} \, , \quad \varphi_2(1) = 0 \, .$$

The solutions

$$\varphi_1 = \exp\left(\frac{4}{3}x^{3/4}\right) \, ,$$

$$\varphi_2 = -\left(\frac{1}{2}x^{-1/2} + 4x^{1/4} - \frac{9}{2}\right)\exp\left(\frac{4}{3}x^{3/4}\right) \, ,$$

produce *singular terms in the neighbourhood of the origin*. The approximation is not uniformly valid and another one must be sought near $x = 0$.

A significant degeneracy of the initial equation is obtained with the local variable defined by

$$X = \frac{x}{\varepsilon^{4/5}} \, , \qquad (6.11)$$

and (6.10a) becomes

$$\frac{\mathrm{d}^2\varPhi}{\mathrm{d}X^2} + X^{1/4}\frac{\mathrm{d}\varPhi}{\mathrm{d}X} - \varepsilon^{3/5}\varPhi = 0 \, . \qquad (6.12)$$

The inner expansion starts with a term $\psi_1(X)$ governed by the equation

$$\frac{d^2\psi_1}{dX^2} + X^{1/4}\frac{d\psi_1}{dX} = 0 \ .$$

The solution can satisfy only one boundary condition $\psi_1(0) = 0$ and a matching condition $\psi_1(\infty) = 1$

$$\psi_1 = \frac{G_{5/4}(X)}{G_{5/4}(\infty)} \quad \text{with} \quad G_{5/4}(X) = \int_0^X \exp\left(-\frac{4}{5}t^{5/4}\right) dt \ .$$

The condition $\psi_1(\infty) = 1$ results from the application of the MVDP to order 1

$$1 = E_1 E_0 \Phi = E_0 E_1 \Phi = \psi_1(\infty) \ .$$

With the outer expansion and the inner equation, the inner expansion is

$$E_1 \Phi = \psi_1(X) + \varepsilon^{3/5}\psi_2(X) + \varepsilon\psi_3(X) + O(\varepsilon^{6/5}) \ . \tag{6.13}$$

The equation for ψ_2 is

$$\frac{d^2\psi_2}{dX^2} + X^{1/4}\frac{d\psi_2}{dX} = \psi_1 \ ,$$

with $\psi_2(0) = 0$. Taking into account that $\psi_1 \cong 1 + \text{EST}$ as $X \to \infty$, the behaviour of ψ_2 as $X \to \infty$ is given by

$$\psi_2 \cong \frac{4}{3}X^{3/4} + A - \frac{1}{2}X^{-1/2} + O(X^{-7/4}) \ .$$

The equation for ψ_3 is

$$\frac{d^2\psi_3}{dX^2} + X^{1/4}\frac{d\psi_3}{dX} = 0 \ ,$$

with $\psi_3(0) = 0$. The solution is

$$\psi_3(X) = BG_{5/4}(X) \ .$$

The MVDP is used and gives

- to order $\varepsilon^{3/5}$,

$$E_0 E_1 \Phi = 1 + \frac{4}{3}x^{3/4} + A\varepsilon^{3/5} \ ,$$

$$E_1 E_0 \Phi = 1 + \frac{4}{3}\varepsilon^{3/5}X^{3/4} \ ,$$

which yields $A = 0$. Then, we obtain a composite expansion supposed to be a UVA to order $\varepsilon^{3/5}$

$$\Phi_{a2} = \Phi_{a1}(x, \varepsilon) + \varepsilon^{3/5} \left[\psi_2(X) - \frac{4}{3} X^{3/4} \right] , \tag{6.14}$$

where Φ_{a1} is a UVA to order 1,

$$\Phi_{a1}(x, \varepsilon) = \exp\left(\frac{4}{3} x^{3/4} \right) + \frac{G_{5/4}(X)}{G_{5/4}(\infty)} - 1 , \tag{6.15}$$

- to order ε ,

$$\mathrm{E}_0 \mathrm{E}_1 \Phi = 1 + \frac{4}{3} x^{3/4} - \frac{1}{2} \varepsilon x^{-1/2} + B\varepsilon G_{5/4}(\infty) ,$$

$$\mathrm{E}_1 \mathrm{E}_0 \Phi = 1 + \varepsilon^{3/5} \left[\frac{4}{3} X^{3/4} - \frac{1}{2} X^{-1/2} \right] + \frac{9}{2} \varepsilon ,$$

which yields $B = \dfrac{9}{2 G_{5/4}(\infty)}$, whence a composite expansion which is supposed to be a UVA to order ε

$$\Phi_{a3} = \Phi_{a2}(x, \varepsilon) + \varepsilon \left[\frac{9}{2} \frac{G_{5/4}(X)}{G_{5/4}(\infty)} + \varphi_2(x) + \frac{1}{2} x^{-1/2} - \frac{9}{2} \right] . \tag{6.16}$$

6.2.2 Application of SCEM

A first approximation $\bar{\varphi}_1(x)$ satisfies the equation

$$x^{1/4} \frac{\mathrm{d}\bar{\varphi}_1}{\mathrm{d}x} - \bar{\varphi}_1 = 0 .$$

With the boundary condition $\bar{\varphi}_1(1) = \mathrm{e}^{4/3}$, the solution is

$$\bar{\varphi}_1 = \exp\left(\frac{4}{3} x^{3/4} \right) .$$

For later use, we give the expression of $\varepsilon \dfrac{\mathrm{d}^2 \bar{\varphi}_1}{\mathrm{d}x^2}$ written with the variable X as $\varepsilon \to 0$

$$\varepsilon \frac{\mathrm{d}^2 \bar{\varphi}_1}{\mathrm{d}x^2} = -\frac{1}{4} X^{-5/4} + \frac{2}{3} \varepsilon^{3/5} X^{-1/2} + \frac{10}{9} \varepsilon^{6/5} X^{1/4} + \cdots .$$

A UVA is defined by complementing the first approximation

$$\bar{\Phi}_{a1} = \bar{\varphi}_1 + \bar{\psi}_1(X, \varepsilon) , \tag{6.17}$$

where the function $\bar{\psi}_1$ satisfies the equation

$$\frac{\mathrm{d}^2 \bar{\psi}_1}{\mathrm{d}X^2} + X^{1/4} \frac{\mathrm{d}\bar{\psi}_1}{\mathrm{d}X} = 0$$

subject to the boundary conditions

$$\bar{\psi}_1(0,\varepsilon) = -1 \ , \quad \bar{\psi}_1\left(\varepsilon^{-4/5},\varepsilon\right) = 0 \ .$$

The solution is

$$\bar{\psi}_1 = \frac{G_{5/4}(X)}{G_{5/4}\left(\varepsilon^{-4/5}\right)} - 1 \ .$$

It is noted that $\bar{\varPhi}_{\mathrm{a1}}$ is a UVA of the function $\varPhi$ but not for its first and second derivatives because $\dfrac{\mathrm{d}\bar{\varphi}_1}{\mathrm{d}x}$ and $\dfrac{\mathrm{d}^2\bar{\varphi}_1}{\mathrm{d}x^2}$ tend towards infinity as $x \to 0$. In particular, we observe that

$$\mathrm{L}_\varepsilon\,\bar{\varPhi}_{\mathrm{a1}} = \varepsilon\frac{\mathrm{d}^2\bar{\varphi}_1}{\mathrm{d}x^2} - \bar{\psi}_1 \ .$$

The right hand side cannot be integrated in the interval $0 \le x \le 1$. SCEM, in its generalized form, enables us to answer these questions. In effect, the use of generalized AEs allows us to consider the following approximation

$$\bar{\varPhi}_{\mathrm{a2}} = \bar{\varphi}_1 + \bar{\psi}_1 + \varepsilon^{3/5}\left(\bar{\varphi}_2 + \bar{\psi}_2\right) \ , \tag{6.18}$$

where $\bar{\varphi}_2$ is solution of equation

$$x^{1/4}\frac{\mathrm{d}\bar{\varphi}_2}{\mathrm{d}x} - \bar{\varphi}_2 = 0 \ ,$$

with the condition $\bar{\varphi}_2(1,\varepsilon) = 0$. The solution is $\bar{\varphi}_2 = 0$.

The original equation becomes

$$\mathrm{L}_\varepsilon\,\bar{\varPhi}_{\mathrm{a2}} = \frac{\mathrm{d}^2\bar{\psi}_2}{\mathrm{d}X^2} + X^{1/4}\frac{\mathrm{d}\bar{\psi}_2}{\mathrm{d}X} - \bar{\psi}_1 - \varepsilon^{3/5}\bar{\psi}_2 + \varepsilon\frac{\mathrm{d}^2\bar{\varphi}_1}{\mathrm{d}x^2} \ .$$

We note that the term $\varepsilon\dfrac{\mathrm{d}^2\bar{\varphi}_1}{\mathrm{d}x^2}$ behaves like $X^{-5/4}$ as $\varepsilon \to 0$. Then, this term is kept in the equation for $\bar{\psi}_2$

$$\frac{\mathrm{d}^2\bar{\psi}_2}{\mathrm{d}X^2} + X^{1/4}\frac{\mathrm{d}\bar{\psi}_2}{\mathrm{d}X} = \bar{\psi}_1 - \varepsilon\frac{\mathrm{d}^2\bar{\varphi}_1}{\mathrm{d}x^2} \ .$$

The boundary conditions are

$$\bar{\psi}_2(0,\varepsilon) = 0 \ ; \quad \bar{\psi}_2\left(\varepsilon^{-4/5},\varepsilon\right) = 0 \ .$$

We observe that $\bar{\varPhi}_{\mathrm{a2}}$ is a UVA not only for $\varPhi$ but also for its first and second derivatives. In effect, by definition

$$\bar{\psi}_2 = \varepsilon^{-3/5}\left(\bar{\varPhi}_{\mathrm{a2}} - \bar{\varphi}_1 - \bar{\psi}_1\right) \ ,$$

and it is possible to form an equation for $\bar{\bar{\Phi}}_{a2}$

$$\varepsilon\frac{\mathrm{d}^2\bar{\bar{\Phi}}_{a2}}{\mathrm{d}x^2} + x^{1/4}\frac{\mathrm{d}\bar{\bar{\Phi}}_{a2}}{\mathrm{d}x} = \bar{\psi}_1 + \bar{\varphi}_1 \,, \tag{6.19}$$

with the boundary conditions

$$\bar{\bar{\Phi}}_{a2}(x=0) = 0 \,, \quad \bar{\bar{\Phi}}_{a2}(x=1) = \mathrm{e}^{4/3} \,.$$

No singularity is expected for $\bar{\bar{\Phi}}_{a2}$ and for its first and second derivatives. In particular, we have

$$\mathrm{L}_\varepsilon\,\bar{\bar{\Phi}}_{a2} = -\varepsilon^{3/5}\bar{\psi}_2 \,,$$

which clearly shows that $\bar{\bar{\Phi}}_{a2}$ is a UVA.

The next UVA is

$$\bar{\bar{\Phi}}_{a3} = \bar{\bar{\Phi}}_{a2} + \varepsilon(\bar{\varphi}_3 + \bar{\psi}_3) \,. \tag{6.20}$$

Taking into account the expression for $\bar{\psi}_2$ expressed with the variable x as $\varepsilon \to 0$, the equation for $\bar{\varphi}_3$ is

$$x^{1/4}\frac{\mathrm{d}\bar{\varphi}_3}{\mathrm{d}x} - \bar{\varphi}_3 = \varepsilon^{-2/5}\bar{\psi}_2 \,,$$

with $\bar{\varphi}_3(1,\varepsilon) = 0$. The solution is

$$\bar{\varphi}_3 = \varepsilon^{-2/5}\left[\int_1^x t^{-1/4}\exp\left(-\frac{4}{3}t^{3/4}\right)\bar{\psi}_2\,\mathrm{d}t\right]\exp\left(\frac{4}{3}x^{3/4}\right) \,.$$

The equation for $\bar{\psi}_3$ is

$$\frac{\mathrm{d}^2\bar{\psi}_3}{\mathrm{d}X^2} + X^{1/4}\frac{\mathrm{d}\bar{\psi}_3}{\mathrm{d}X} = 0 \,,$$

subject to the boundary conditions

$$\bar{\psi}_3(0,\varepsilon) = -\bar{\varphi}_3(0,\varepsilon) \,, \quad \bar{\psi}_3(\varepsilon^{-4/5},\varepsilon) = 0 \,.$$

The solution is

$$\bar{\psi}_3 = \bar{\varphi}_3(0,\varepsilon)\left[\frac{G_{5/4}(X)}{G_{5/4}(\varepsilon^{-4/5})} - 1\right] \,.$$

6.2.3 Identification with MMAE Results

As shown above, a UVA deduced from the application of SCEM is

$$\bar{\bar{\Phi}}_{a3} = \bar{\varphi}_1 + \bar{\psi}_1 + \varepsilon^{3/5}\bar{\psi}_2 + \varepsilon(\bar{\varphi}_3 + \bar{\psi}_3) \,.$$

The functions $\bar{\varphi}_1$, $\bar{\psi}_1$, $\bar{\psi}_2$, $\bar{\varphi}_3$, $\bar{\psi}_3$ being approximated by regular expansions, the expression for $\bar{\bar{\Phi}}_{a3}$ becomes

$$\bar{\bar{\Phi}}_{a3} = \hat{f}_1(x) + \hat{F}_1(X) + \varepsilon^{3/5}\hat{F}_2(X) + \varepsilon\left[\hat{f}_3(x) + \hat{F}_3(X)\right] + \mathrm{o}(\varepsilon) \,.$$

It is shown (Problem 6-6) that this expansion is identical to the composite expansion given by (6.16) calculated from MMAE. *Therefore, SCEM results contain MMAE results and the matching principle is a consequence.* Obviously, it is simply a confirmation of the general result given in Sect. 5.7.

Note 6.2. The regular expansions can be obtained directly by applying the regular form of SCEM, but then it is not possible to show that the generalized expansions contain the regular expansions.

6.2.4 Numerical Results

Numerical results showing MMAE and SCEM approximations are plotted in Fig. 6.1. Comparison with the numerical solution of the exact equation is also given. Within the numerical approximation, this solution can be considered as a reference solution. For all the values of ε, SCEM results are a very good approximation of the numerical solution. With MMAE, we observe that the approximation Φ_{a1} is better than the approximation Φ_{a2}. Moreover, when ε is not small enough, the results of MMAE are markedly different from the exact solution.

6.3 Example 3

Consider the equation

$$\mathrm{L}_\varepsilon\, \Phi \equiv \varepsilon \frac{\mathrm{d}^2\Phi}{\mathrm{d}x^2} + x^{1/3}\frac{\mathrm{d}\Phi}{\mathrm{d}x} - \Phi = 0 \,, \tag{6.21a}$$

where the function Φ is defined in the domain $0 \le x \le 1$ with the boundary conditions

$$\Phi(0) = 0 \,, \quad \Phi(1) = \mathrm{e}^{3/2} \,. \tag{6.21b}$$

6.3.1 Application of MMAE

A straightforward outer expansion begins with

$$\mathrm{E}_0\Phi(x,\varepsilon) = \varphi_1(x) + \cdots \,,$$

which yields the equation

$$x^{1/3}\frac{\mathrm{d}\varphi_1}{\mathrm{d}x} - \varphi_1 = 0 \,,$$

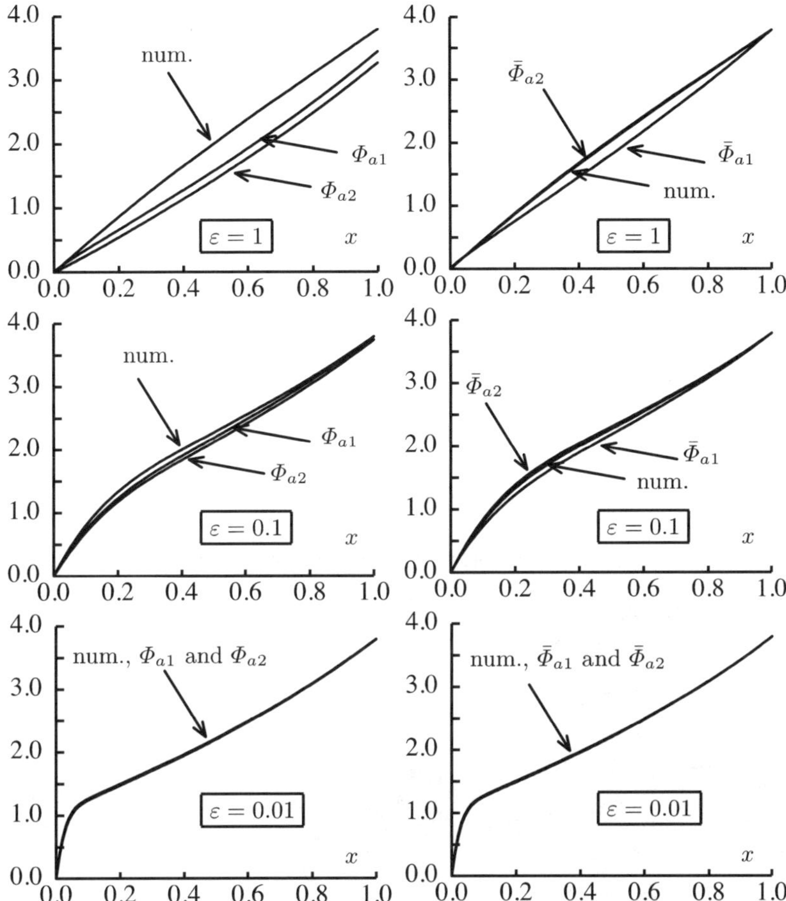

Fig. 6.1. Study of problem given by (6.10a). On the left: comparison of the numerical solution with MMAE approximations: Φ_{a1} (6.15), Φ_{a2} (6.14). On the right: comparison of the numerical solution with SCEM approximations: $\bar{\bar{\Phi}}_{a1}$ (6.17), $\bar{\bar{\Phi}}_{a2}$ (6.19)

with

$$\varphi_1(1) = \mathrm{e}^{3/2} \ .$$

The solution,

$$\varphi_1 = \exp\left(\frac{3}{2}x^{2/3}\right) \ ,$$

contains singular terms in the neighbourhood of the origin. Therefore, it is necessary to seek an appropriate inner approximation. A significant degeneracy of the initial equation is obtained with the local variable

$$X = \frac{x}{\varepsilon^{3/4}} \tag{6.22}$$

because (6.21a) becomes

$$L_\varepsilon \Phi = \varepsilon^{-1/2} \left(\frac{\mathrm{d}^2 \Phi}{\mathrm{d}X^2} + X^{1/3} \frac{\mathrm{d}\Phi}{\mathrm{d}X} \right) - \Phi = 0 \ . \tag{6.23}$$

The inner expansion begins with a term $\psi_1(X)$ governed by

$$\frac{\mathrm{d}^2 \psi_1}{\mathrm{d}X^2} + X^{1/3} \frac{\mathrm{d}\psi_1}{\mathrm{d}X} = 0 \ .$$

The solution satisfies the boundary condition $\psi_1(0) = 0$ and the matching condition $\psi_1(\infty) = 1$,

$$\psi_1 = \frac{G_{4/3}(X)}{G_{4/3}(\infty)} \quad \text{with} \quad G_{4/3}(X) = \int_0^X \exp\left(-\frac{3}{4} t^{4/3} \right) \mathrm{d}t \ .$$

The condition $\psi_1(\infty) = 1$ results from the application of the MVDP to order 1

$$1 = \mathrm{E}_1 \mathrm{E}_0 \Phi = \mathrm{E}_0 \mathrm{E}_1 \Phi = \psi_1(\infty) \ .$$

Taking into account the outer expansion and the inner equation, the inner expansion is improved as

$$\mathrm{E}_1 \Phi(x, \varepsilon) = \psi_1(X) + \varepsilon^{1/2} \psi_2(X) + \cdots \ , \tag{6.24}$$

and the equation for ψ_2 becomes

$$\frac{\mathrm{d}^2 \psi_2}{\mathrm{d}X^2} + X^{1/3} \frac{\mathrm{d}\psi_2}{\mathrm{d}X} = \psi_1 \ ,$$

with

$$\psi_2(0) = 0 \ .$$

Knowing that $\psi_1 \cong 1 + \mathrm{EST}$ as $X \to \infty$, the behaviour of ψ_2 as $X \to \infty$ is given by

$$\psi_2 \cong \frac{3}{2} X^{2/3} + A - \frac{1}{2} X^{-2/3} + \cdots \ .$$

On the other hand, the next term of the outer expansion is necessarily of order ε because all the terms whose order is between 1 and ε are zero

$$\mathrm{E}_0 \Phi(x, \varepsilon) = \varphi_1(x) + \varepsilon \varphi_2(x) + \cdots \ . \tag{6.25}$$

The equation for φ_2 is

$$x^{1/3} \frac{\mathrm{d}\varphi_2}{\mathrm{d}x} - \varphi_2 = -\frac{\mathrm{d}^2 \varphi_1}{\mathrm{d}x^2} \ ,$$

with

$$\varphi_2(1) = 0 \ .$$

The solution is

$$\varphi_2 = - \left(\frac{1}{2} x^{-2/3} + \ln x - \frac{1}{2} \right) \exp \left(\frac{3}{2} x^{2/3} \right) \ .$$

Applying the MVDP to order $\varepsilon^{1/2}$ yields $A = 0$ because we have

$$1 + \frac{3}{2} \varepsilon^{1/2} X^{2/3} = \mathrm{E}_1 \mathrm{E}_0 \varPhi = \mathrm{E}_1 \mathrm{E}_0 \mathrm{E}_1 \varPhi = 1 + A \varepsilon^{1/2} + \frac{3}{2} \varepsilon^{1/2} X^{2/3} \ .$$

The outer expansion given by (6.25) enables us to write, to order ε,

$$\mathrm{E}_1 \mathrm{E}_0 \varPhi = 1 + \varepsilon^{1/2} \left(\frac{3}{2} X^{2/3} - \frac{1}{2} X^{-2/3} \right)$$
$$- \frac{3}{4} \varepsilon \ln \varepsilon - \varepsilon \left(-\frac{9}{8} X^{4/3} + \ln X + \frac{1}{4} \right) \ . \tag{6.26}$$

Note 6.3. With the outer expansion given by (6.25) and the expressions for φ_1 and φ_2, the determination of $\mathrm{E}_1 \mathrm{E}_0 \varPhi$ is instructive only to order 1, to order $\varepsilon^{1/2}$ or to order ε. The result obtained to any order strictly between 1 and $\varepsilon^{1/2}$ is identical to the result obtained to order 1. Similarly, the result obtained to any order strictly between $\varepsilon^{1/2}$ and ε is identical to the result obtained to order $\varepsilon^{1/2}$. In particular, the expression for $\mathrm{E}_1 \mathrm{E}_0 \varPhi$ to order $-\varepsilon \ln \varepsilon$ is identical to the expression for $\mathrm{E}_1 \mathrm{E}_0 \varPhi$ to order $\varepsilon^{1/2}$.

Equation (6.26) suggests the next terms of the inner expansion

$$\mathrm{E}_1 \varPhi(x, \varepsilon) = \psi_1(X) + \varepsilon^{1/2} \psi_2(X) - \varepsilon \ln \varepsilon \, \psi_3^*(X) + \varepsilon \psi_3(X) + \cdots \ . \tag{6.27}$$

The equations for ψ_3^* and ψ_3 are

$$\frac{\mathrm{d}^2 \psi_3^*}{\mathrm{d}X^2} + X^{1/3} \frac{\mathrm{d}\psi_3^*}{\mathrm{d}X} = 0 \ ,$$
$$\frac{\mathrm{d}^2 \psi_3}{\mathrm{d}X^2} + X^{1/3} \frac{\mathrm{d}\psi_3}{\mathrm{d}X} = \psi_2 \ ,$$

with

$$\psi_3^*(0) = 0 \ ,$$
$$\psi_3(0) = 0 \ .$$

We obtain

$$\psi_3^* = B \frac{G_{4/3}(X)}{G_{4/3}(\infty)} \ ,$$

and, from the behaviour of ψ_2 as $X \to \infty$, we deduce the behaviour of ψ_3

$$\psi_3 \cong \frac{9}{8} X^{4/3} + C - \ln X + \cdots \ .$$

Then, to order ε, we have

$$\mathrm{E}_1\mathrm{E}_0\mathrm{E}_1\varPhi = 1 + \varepsilon^{1/2}\left(\frac{3}{2}X^{2/3} - \frac{1}{2}X^{-2/3}\right)$$

$$-B\varepsilon\ln\varepsilon - \varepsilon\left(-\frac{9}{8}X^{4/3} + \ln X - C\right). \tag{6.28}$$

Thus, the MVDP written to order ε,

$$\mathrm{E}_1\mathrm{E}_0\varPhi \equiv \mathrm{E}_1\mathrm{E}_0\mathrm{E}_1\varPhi,$$

together with (6.26), gives $B = 3/4$ and $C = -1/4$.

Note 6.4. To determine the constant B, the MVDP must be written to order ε and not to order $-\varepsilon\ln\varepsilon$ because the expression of $\mathrm{E}_1\mathrm{E}_0\varPhi$ to order $-\varepsilon\ln\varepsilon$ does not contain any term of order $-\varepsilon\ln\varepsilon$.

Finally, the following UVAs are obtained:
- to order 1,

$$\varPhi_{\mathrm{a}1} = \varphi_1(x) + \psi_1(X) - 1, \tag{6.29}$$

- to order $\varepsilon^{1/2}$,

$$\varPhi_{\mathrm{a}2}(x, X, \varepsilon) = \varPhi_{\mathrm{a}1}(x, X) + \varepsilon^{1/2}\left[\psi_2(X) - \frac{3}{2}X^{2/3}\right], \tag{6.30}$$

- to order ε,

$$\varPhi_{\mathrm{a}3}(x, X, \varepsilon) = \varPhi_{\mathrm{a}2}(x, X, \varepsilon) - \varepsilon\ln\varepsilon\left[\psi_3^*(X) - \frac{3}{4}\right]$$

$$+\varepsilon\left[\varphi_2(x) + \psi_3(X) + \frac{1}{2}x^{-2/3} - \frac{9}{8}X^{4/3} + \ln X + \frac{1}{4}\right]. \tag{6.31}$$

6.3.2 Application of SCEM

With generalized expansions, the UVAs given below are obtained. To order 1, we have

$$\bar{\varPhi}_{\mathrm{a}1} = \bar{\varphi}_1 + \bar{\psi}_1(X, \varepsilon), \tag{6.32}$$

with

$$\bar{\varphi}_1 = \exp\left(\frac{3}{2}x^{2/3}\right), \quad \bar{\psi}_1 = \frac{G_{4/3}(X)}{G_{4/3}(\varepsilon^{-3/4})} - 1.$$

Here again, the remainder $\mathrm{L}_\varepsilon\,\bar{\varPhi}_{\mathrm{a}1}$,

$$\mathrm{L}_\varepsilon\,\bar{\varPhi}_{\mathrm{a}1} = \varepsilon\frac{\mathrm{d}^2\bar{\varphi}_1}{\mathrm{d}x^2} - \bar{\psi}_1,$$

cannot be integrated on the domain of definition. To order $\varepsilon^{1/2}$, the UVA is

$$\bar{\varPhi}_{\mathrm{a}2} = \bar{\varPhi}_{\mathrm{a}1} + \varepsilon^{1/2}\bar{\psi}_2. \tag{6.33}$$

Taking into account the expression for $\varepsilon \dfrac{\mathrm{d}^2 \bar{\varphi}_1}{\mathrm{d}x^2}$, written with the variable X, as $\varepsilon \to 0$,

$$\varepsilon \frac{\mathrm{d}^2 \bar{\varphi}_1}{\mathrm{d}x^2} = -\frac{1}{3} X^{-4/3} + \frac{1}{2} \varepsilon^{1/2} X^{-2/3} + \frac{9}{8}\varepsilon + \cdots ,$$

the equation for $\bar{\psi}_2$ is

$$\frac{\mathrm{d}^2 \bar{\psi}_2}{\mathrm{d}X^2} + X^{1/3} \frac{\mathrm{d}\bar{\psi}_2}{\mathrm{d}X} = -\varepsilon \frac{\mathrm{d}^2 \bar{\varphi}_1}{\mathrm{d}x^2} + \bar{\psi}_1 , \tag{6.34}$$

with the boundary conditions

$$\bar{\psi}_2(0,\varepsilon) = 0 , \quad \bar{\psi}_2(\varepsilon^{-3/4},\varepsilon) = 0 .$$

It can be checked that

$$\mathrm{L}_\varepsilon \, \bar{\varPhi}_{a2} = -\varepsilon^{1/2} \bar{\psi}_2 ,$$

which shows that $\bar{\varPhi}_{a2}$ is a UVA. To proceed to order ε, we set

$$\bar{\varPhi}_{a3} = \bar{\varPhi}_{a2} + \varepsilon(\bar{\varphi}_3 + \bar{\psi}_3) . \tag{6.35}$$

From the equation for $\bar{\psi}_2$, it may be shown that the behaviour of $\bar{\psi}_2$, as $X \to \infty$, has the form

$$\bar{\psi}_2 = \varepsilon^{1/2} f(x) + \cdots .$$

Then, the equation for $\bar{\varphi}_3$ is

$$x^{1/3} \frac{\mathrm{d}\bar{\varphi}_3}{\mathrm{d}x} - \bar{\varphi}_3 = \varepsilon^{-1/2} \bar{\psi}_2 , \tag{6.36}$$

with $\bar{\varphi}_3(1,\varepsilon) = 0$.

On the other hand, the behaviour of $\bar{\varphi}_3$, as $x \to 0$, is given by

$$\varepsilon^{3/2} \frac{\mathrm{d}^2 \bar{\varphi}_3}{\mathrm{d}x^2} = F(X) + \cdots ,$$

hence the equation for $\bar{\psi}_3$

$$\frac{\mathrm{d}^2 \bar{\psi}_3}{\mathrm{d}X^2} + X^{1/3} \frac{\mathrm{d}\bar{\psi}_3}{\mathrm{d}X} = -\varepsilon^{3/2} \frac{\mathrm{d}^2 \bar{\varphi}_3}{\mathrm{d}x^2} , \tag{6.37}$$

with the boundary conditions

$$\bar{\psi}_3(0,\varepsilon) = -\bar{\varphi}_3(0,\varepsilon) , \quad \bar{\psi}_3(\varepsilon^{-3/4},\varepsilon) = 0 .$$

The remainder is

$$\mathrm{L}_\varepsilon \, \bar{\varPhi}_{a3} = -\varepsilon \bar{\psi}_3 ,$$

which is satisfactory.

Note 6.5. We observe that the generalized AE deduced from SCEM does not contain any term of order $\varepsilon \ln \varepsilon$.

6.3.3 Identification with MMAE Results

As shown above, a UVA deduced from the application of SCEM is

$$\bar{\Phi}_{a3} = \bar{\varphi}_1 + \bar{\psi}_1 + \varepsilon^{1/2}\bar{\psi}_2 + \varepsilon(\bar{\varphi}_3 + \bar{\psi}_3) \ .$$

The functions $\bar{\varphi}_1$, $\bar{\psi}_1$, $\bar{\psi}_2$, $\bar{\varphi}_3$, $\bar{\psi}_3$ being approximated by regular expansions, the expression for $\bar{\Phi}_{a3}$ becomes

$$\bar{\Phi}_{a3} = \hat{f}_1(x) + \hat{F}_1(X) + \varepsilon^{1/2}\hat{F}_2(X) + \varepsilon \ln \varepsilon \hat{F}_3(X) + \varepsilon \left[\hat{f}_4(x) + \hat{F}_4(X) \right] + \mathrm{o}(\varepsilon) \ .$$

It is shown (Problem 6-7) that this expansion is identical to the composite expansion given by (6.31) calculated from SCEM. *Therefore, SCEM results contain MMAE results and the matching principle is a consequence.* Obviously, it is simply a confirmation of the general result given in Sect. 5.7.

Note 6.6. The SCEM generalized expansions do not contain any term of order $-\varepsilon \ln \varepsilon$. These terms appear when the approximation is based on regular expansions. Such terms are also present in MMAE expansions. Now, logarithms are sources of difficulty in MMAE. For this reason, SCEM proves to be very advantageous.

6.4 Stokes-Oseen's Flow Model

We consider the equation

$$\mathrm{L}_\varepsilon\, \Phi \equiv \frac{\mathrm{d}^2\Phi}{\mathrm{d}x^2} + \frac{1}{x}\frac{\mathrm{d}\Phi}{\mathrm{d}x} + \Phi\frac{\mathrm{d}\Phi}{\mathrm{d}x} = 0 \ , \qquad (6.38\mathrm{a})$$

where the function Φ is defined in the domain $x \geq \varepsilon$ with the boundary conditions

$$x = \varepsilon : \quad \Phi = 0 \quad \text{and} \quad x \to \infty : \quad \Phi = 1 \ . \qquad (6.38\mathrm{b})$$

This problem, proposed by Lagerstrom [48], simulates the difficulties encountered in the analysis of Stokes-Oseen's flow, i.e. the two-dimensional flow at low Reynolds number around a circular cylinder.

6.4.1 Application of SCEM

A singular perturbation problem arises in the neighbourhood of $x = \varepsilon$. Elsewhere, the solution tends to $\Phi = 1$ as $\varepsilon \to 0$. Thus, we seek a first approximation in the form

$$\Phi = 1 + \bar{\delta}_1(\varepsilon)\bar{\varphi}_1(x, \varepsilon) + \cdots \ , \qquad (6.39)$$

where $\bar{\delta}_1$ is an order function such that $\bar{\delta}_1 \to 0$ as $\varepsilon \to 0$.

Equation (6.38a) becomes

$$\frac{\mathrm{d}^2 \bar{\varphi}_1}{\mathrm{d}x^2} + \frac{1}{x}\frac{\mathrm{d}\bar{\varphi}_1}{\mathrm{d}x} + \frac{\mathrm{d}\bar{\varphi}_1}{\mathrm{d}x} = 0 \ . \tag{6.40}$$

The exact boundary conditions on Φ are satisfied with

$$x = \varepsilon : \quad \bar{\delta}_1 \bar{\varphi}_1 = -1 \ , \quad x \to \infty : \quad \bar{\varphi}_1 = 0 \ .$$

The gauge $\bar{\delta}_1(\varepsilon)$ is determined to within a multiplicative constant simultaneously with the solution by applying the boundary conditions. The solution is

$$\bar{\varphi}_1 = -E_1(x) \ ,$$

with

$$E_1(x) = \int_x^\infty \frac{\mathrm{e}^{-t}}{t}\, \mathrm{d}t \ ,$$

and

$$\bar{\delta}_1(\varepsilon) = \frac{1}{E_1(\varepsilon)} \ .$$

Note the behaviour of $E_1(\varepsilon)$ as $\varepsilon \to 0$

$$E_1(\varepsilon) \cong -\ln \varepsilon - \gamma + \varepsilon + \cdots \ ,$$

where γ is the Euler constant $\gamma = 0.577215$.

We seek a UVA by complementing the preceding approximation

$$\bar{\Phi}_{\mathrm{a}1} = 1 + \bar{\delta}_1 \bar{\varphi}_1 + \bar{\delta}_1 \bar{\psi}_1(X, \varepsilon) \quad \text{with} \quad X = \frac{x}{\varepsilon} \ .$$

In fact, it is shown that $\bar{\psi}_1 = 0$. Therefore, the UVA is

$$\bar{\Phi}_{\mathrm{a}1} = 1 - \frac{E_1(x)}{E_1(\varepsilon)} \ . \tag{6.41}$$

Note 6.7. Chen et al. [15] obtained the same result by using a method based on the renormalization group.

It is interesting to examine the remainder $\mathrm{L}_\varepsilon \, \bar{\Phi}_{\mathrm{a}1}$,

$$\mathrm{L}_\varepsilon \, \bar{\Phi}_{\mathrm{a}1} = \frac{E_1(x)}{E_1^2(\varepsilon)}\frac{\mathrm{d}E_1(x)}{\mathrm{d}x} = -\frac{E_1(x)}{E_1^2(\varepsilon)}\frac{\mathrm{e}^{-x}}{x} \ .$$

We note that, x being kept fixed, $\mathrm{L}_\varepsilon \, \bar{\Phi}_{\mathrm{a}1} \to 0$ as $\varepsilon \to 0$, but if $x = \varepsilon$ then $\mathrm{L}_\varepsilon \, \bar{\Phi}_{\mathrm{a}1} \to -\infty$ as $\varepsilon \to 0$. Nevertheless, the integral of the remainder remains finite, non zero, as $\varepsilon \to 0$. In fact, this integral is independent of ε

$$\int_\varepsilon^\infty \mathrm{L}_\varepsilon \, \bar{\Phi}_{\mathrm{a}1} \, \mathrm{d}x = \frac{1}{2 E_1^2(\varepsilon)} \left[E_1^2(x) \right]_\varepsilon^\infty = -\frac{1}{2} \ .$$

We seek a better approximation in the form

$$\Phi = 1 - \frac{E_1(x)}{E_1(\varepsilon)} + \bar\delta_2(\varepsilon)\bar\varphi_2(x,\varepsilon) + \cdots .$$

The equation for $\bar\varphi_2$ is

$$\frac{\mathrm{d}^2\bar\varphi_2}{\mathrm{d}x^2} + \frac{1}{x}\frac{\mathrm{d}\bar\varphi_2}{\mathrm{d}x} + \frac{\mathrm{d}\bar\varphi_2}{\mathrm{d}x} + \frac{1}{\bar\delta_2 E_1^2(\varepsilon)}E_1(x)\frac{\mathrm{d}E_1(x)}{\mathrm{d}x} = 0 . \qquad (6.42)$$

As the boundary conditions are satisfied by the approximation $\bar\Phi_{\mathrm{a}1}$, we take

$$x = \varepsilon : \quad \bar\varphi_2 = 0 \quad \text{and} \quad x \to \infty : \quad \bar\varphi_2 = 0 .$$

The gauge $\bar\delta_2(\varepsilon)$ is determined to within a multiplicative constant simultaneously with the solution by applying the boundary conditions. The solution is

$$\bar\delta_2\bar\varphi_2 = \frac{F_1(x)}{E_1^2(\varepsilon)} - \frac{F_1(\varepsilon)E_1(x)}{E_1^3(\varepsilon)} ,$$

with

$$F_1(x) = 2E_1(2x) - \mathrm{e}^{-x}E_1(x) ,$$

and

$$\bar\delta_2 = \frac{F_1(\varepsilon)}{E_1^3(\varepsilon)} ,$$

Note that the behaviour of $F_1(\varepsilon)$ as $\varepsilon \to 0$ is

$$F_1(\varepsilon) \cong -\ln\varepsilon - \gamma - 2\ln 2 - \varepsilon\ln\varepsilon + (3 - \gamma)\varepsilon + \cdots .$$

We seek a UVA in the form

$$\bar\Phi_{\mathrm{a}2} = 1 - \frac{E_1(x)}{E_1(\varepsilon)} + \bar\delta_2(\varepsilon)\bar\varphi_2(x,\varepsilon) + \bar\delta_2\bar\psi_2(X,\varepsilon) .$$

It is shown that $\bar\psi_2 = 0$ so that the UVA $\bar\Phi_{\mathrm{a}2}$ is

$$\bar\Phi_{\mathrm{a}2} = 1 - \frac{E_1(x)}{E_1(\varepsilon)} + \frac{F_1(x)}{E_1^2(\varepsilon)} - \frac{F_1(\varepsilon)E_1(x)}{E_1^3(\varepsilon)} . \qquad (6.43)$$

6.4.2 Numerical Results

The numerical results in Fig. 6.2 show that $\bar\Phi_{\mathrm{a}1}$ and $\bar\Phi_{\mathrm{a}2}$ are excellent approximations of the numerical solution of the complete equation even for large values of ε.

It is noted that the order functions $\bar\delta_1$ and $\bar\delta_2$ are determined to within a multiplicative constant by applying the exact boundary conditions. On the contrary, with MMAE, many choices are possible and the range of values of

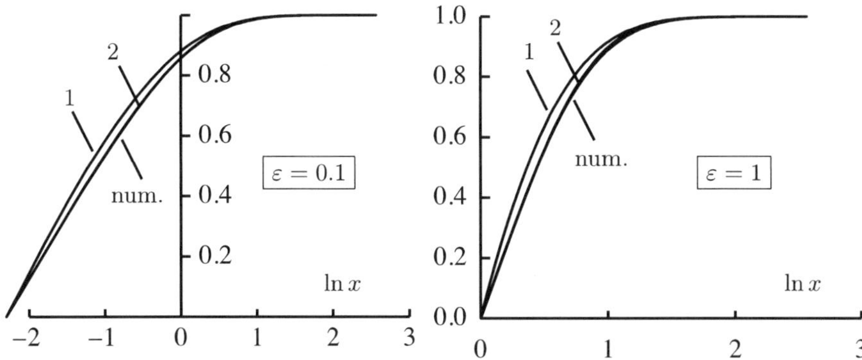

Fig. 6.2. Stokes-Oseen's flow model. Comparison of the numerical solution with SCEM approximations: 1 = first order UVA (6.41), 2 = second order UVA (6.43)

ε over which the numerical accuracy of the MMAE approximation is good depends strongly on this choice [42, 48]. Often, the order functions are taken as $-1/\ln \varepsilon$ and $1/(\ln \varepsilon)^2$. With this choice, it is clear that the approximation cannot be valid if $\varepsilon = 1$, and the accuracy of the approximation is correct only for very small values of ε. This question is discussed in detail by Lagerstrom [48] who also proposes the use of gauges $\bar{\delta}_1$ and $\bar{\delta}_2$ as determined here.

With this example, it is again observed that the difficulty associated with the presence of logarithms is completely solved with SCEM.

6.5 Terrible Problem

Consider the equation

$$L_\varepsilon \, \Phi \equiv \frac{\mathrm{d}^2 \Phi}{\mathrm{d}x^2} + \frac{1}{x}\frac{\mathrm{d}\Phi}{\mathrm{d}x} + \left(\frac{\mathrm{d}\Phi}{\mathrm{d}x}\right)^2 + \Phi \frac{\mathrm{d}\Phi}{\mathrm{d}x} = 0 \;, \tag{6.44a}$$

where the function Φ is defined in the domain $x \geq \varepsilon$ with the boundary conditions

$$x = \varepsilon : \quad \Phi = 0 \quad \text{and} \quad x \to \infty : \quad \Phi = 1 \;. \tag{6.44b}$$

This problem has also been proposed by Lagerstrom [48]. With MMAE, the difficulties which are raised are such that the problem has been termed "terrible problem" by Hinch [42]. The solution requires the knowledge of an infinite number of terms, which is not really along the lines of the method.

The solution can be obtained with the change of function e^Φ but to follow the general procedure, this transformation is not used here.

6.5.1 Application of SCEM

We seek a first approximation in the form

$$\Phi = 1 + \bar{\delta}_1(\varepsilon)\bar{\varphi}_1(x,\varepsilon) + \cdots . \tag{6.45}$$

Susbstituting into (6.44a) yields

$$\bar{\delta}_1 \frac{\mathrm{d}^2 \bar{\varphi}_1}{\mathrm{d}x^2} + \frac{\bar{\delta}_1}{x}\frac{\mathrm{d}\bar{\varphi}_1}{\mathrm{d}x} + \bar{\delta}_1^2\left(\frac{\mathrm{d}\bar{\varphi}_1}{\mathrm{d}x}\right)^2 + \bar{\delta}_1\frac{\mathrm{d}\bar{\varphi}_1}{\mathrm{d}x} + \bar{\delta}_1^2\bar{\varphi}_1\frac{\mathrm{d}\bar{\varphi}_1}{\mathrm{d}x} + \cdots = 0 . \tag{6.46}$$

Neglecting $\mathrm{O}(\bar{\delta}_1^2)$ terms, the equation for $\bar{\varphi}_1$ is

$$\bar{\delta}_1 \frac{\mathrm{d}^2 \bar{\varphi}_1}{\mathrm{d}x^2} + \frac{\bar{\delta}_1}{x}\frac{\mathrm{d}\bar{\varphi}_1}{\mathrm{d}x} + \bar{\delta}_1\frac{\mathrm{d}\bar{\varphi}_1}{\mathrm{d}x} = 0 . \tag{6.47}$$

With the exact boundary conditions

$$x = \varepsilon : \quad \bar{\delta}_1\bar{\varphi}_1 = -1 \quad \text{and} \quad x \to \infty : \quad \bar{\varphi}_1 = 0 ,$$

the solution of (6.47) is

$$\bar{\varphi}_1 = -E_1(x) ,$$

with

$$E_1(x) = \int_x^\infty \frac{\mathrm{e}^{-t}}{t}\,\mathrm{d}t ,$$

and

$$\bar{\delta}_1(\varepsilon) = \frac{1}{E_1(\varepsilon)} .$$

Then, we seek a UVA in the form

$$\bar{\Phi}_{\mathrm{a1}} = 1 + \bar{\delta}_1\bar{\varphi}_1 + \bar{\delta}_1\bar{\psi}_1(X,\varepsilon) \quad \text{with} \quad X = \frac{x}{\varepsilon} ,$$

but we find that $\bar{\psi}_1 = 0$. Therefore, we have

$$\bar{\Phi}_{\mathrm{a1}} = 1 + \bar{\delta}_1\bar{\varphi}_1 .$$

Let us examine the remainder $\mathrm{L}_\varepsilon\,\bar{\Phi}_{\mathrm{a1}}$

$$\begin{aligned}
\mathrm{L}_\varepsilon\,\bar{\Phi}_{\mathrm{a1}} &= \frac{1}{E_1^2(\varepsilon)}\left(\frac{\mathrm{d}E_1(x)}{\mathrm{d}x}\right)^2 + \frac{E_1(x)}{E_1^2(\varepsilon)}\frac{\mathrm{d}E_1(x)}{\mathrm{d}x} \\
&= \frac{\mathrm{e}^{-2x}}{x^2 E_1^2(\varepsilon)} - \frac{E_1(x)}{E_1^2(\varepsilon)}\frac{\mathrm{e}^{-x}}{x} .
\end{aligned}$$

We deduce that, x being kept fixed, $\mathrm{L}_\varepsilon\,\bar{\Phi}_{\mathrm{a1}} \to 0$ as $\varepsilon \to 0$. If $x = \varepsilon$, then $\mathrm{L}_\varepsilon\,\bar{\Phi}_{\mathrm{a1}} \to \infty$ as $\varepsilon \to 0$, but the integral of the remainder is such that

$$\int_{\varepsilon}^{\infty} \mathrm{L}_{\varepsilon}\,\bar{\varPhi}_{\mathrm{a}1}\,\mathrm{d}x \xrightarrow[\varepsilon\to 0]{} \infty \ .$$

As shown below, a difficulty should be expected. We seek a better approximation in the form

$$\varPhi = 1 + \bar{\delta}_1(\varepsilon)\bar{\varphi}_1(x,\varepsilon) + \bar{\delta}_2(\varepsilon)\bar{\varphi}_2(x,\varepsilon) + \cdots \ , \tag{6.48}$$

which leads to

$$\bar{\delta}_2\frac{\mathrm{d}^2\bar{\varphi}_2}{\mathrm{d}x^2} + \frac{\bar{\delta}_2}{x}\frac{\mathrm{d}\bar{\varphi}_2}{\mathrm{d}x} + \bar{\delta}_2\frac{\mathrm{d}\bar{\varphi}_2}{\mathrm{d}x} = -\frac{\bar{\delta}_1^2}{\bar{\delta}_2}\left[\left(\frac{\mathrm{d}\bar{\varphi}_1}{\mathrm{d}x}\right)^2 + \bar{\varphi}_1\frac{\mathrm{d}\bar{\varphi}_1}{\mathrm{d}x}\right] \ . \tag{6.49}$$

With the exact boundary conditions

$$x = \varepsilon : \quad \bar{\varphi}_2 = 0 \quad \text{and} \quad x \to \infty : \quad \bar{\varphi}_2 = 0 \ ,$$

the solution is

$$\bar{\delta}_2\bar{\varphi}_2 = -\frac{F_1(\varepsilon) - \frac{1}{2}E_1^2(\varepsilon)}{E_1^3(\varepsilon)}E_1(x) + \frac{F_1(x) - \frac{1}{2}E_1^2(x)}{E_1^2(\varepsilon)} \ , \tag{6.50}$$

with

$$F_1(x) = 2E_1(2x) - \mathrm{e}^{-x}E_1(x) \ .$$

Equation (6.50) contains the term $\dfrac{1}{2}\dfrac{E_1(x)}{E_1(\varepsilon)}$, which is of order $-1/\ln\varepsilon$. Now, to this order, the UVA $\bar{\varPhi}_{\mathrm{a}1}$ should be complete. It is concluded that *the expansion is not asymptotic.* The source of difficulty lies in the term $\bar{\delta}_1^2\left(\dfrac{\mathrm{d}\bar{\varphi}_1}{\mathrm{d}x}\right)^2$ which has been neglected in (6.46) in the derivation of the equation for $\bar{\varphi}_1$. It is concluded that this term must be kept and the equation for $\bar{\varphi}_1$ becomes

$$\bar{\delta}_1\frac{\mathrm{d}^2\bar{\varphi}_1}{\mathrm{d}x^2} + \frac{\bar{\delta}_1}{x}\frac{\mathrm{d}\bar{\varphi}_1}{\mathrm{d}x} + \delta_1^2\left(\frac{\mathrm{d}\bar{\varphi}_1}{\mathrm{d}x}\right)^2 + \bar{\delta}_1\frac{\mathrm{d}\bar{\varphi}_1}{\mathrm{d}x} = 0 \ . \tag{6.51}$$

It is surprising to keep a $O(\bar{\delta}_1^2)$ term but it happens that the term $\bar{\delta}_1^2\left(\dfrac{\mathrm{d}\bar{\varphi}_1}{\mathrm{d}x}\right)^2$ disappears when combined with the term $\bar{\delta}_1\dfrac{\mathrm{d}^2\bar{\varphi}_1}{\mathrm{d}x^2}$ if the solution is a logarithmic function. In this way, a term in $-1/\ln\varepsilon$ is no longer present in the next order term of the expansion.

Indeed, setting

$$\bar{\delta}_1\bar{\varphi}_1 = \ln\left(1 + \bar{\delta}_1\bar{f}_1\right) \ , \tag{6.52}$$

we have

$$\bar{\delta}_1 \frac{\mathrm{d}^2 \bar{\varphi}_1}{\mathrm{d}x^2} = \frac{\bar{\delta}_1 \dfrac{\mathrm{d}^2 \bar{f}_1}{\mathrm{d}x^2}(1 + \bar{\delta}_1 \bar{f}_1) - \bar{\delta}_1^2 \left(\dfrac{\mathrm{d}\bar{f}_1}{\mathrm{d}x}\right)^2}{\left(1 + \bar{\delta}_1 \bar{f}_1\right)^2} \ ,$$

$$\bar{\delta}_1^2 \left(\frac{\mathrm{d}\bar{\varphi}_1}{\mathrm{d}x}\right)^2 = \frac{\bar{\delta}_1^2 \left(\dfrac{\mathrm{d}\bar{f}_1}{\mathrm{d}x}\right)^2}{\left(1 + \bar{\delta}_1 \bar{f}_1\right)^2} \ ,$$

and, neglecting $\mathrm{O}(\bar{\delta}_1^2)$ terms, (6.51) becomes

$$\frac{\mathrm{d}^2 \bar{f}_1}{\mathrm{d}x^2} + \frac{1}{x}\frac{\mathrm{d}\bar{f}_1}{\mathrm{d}x} + \frac{\mathrm{d}\bar{f}_1}{\mathrm{d}x} = 0 \ .$$

Imposing the exact boundary conditions on Φ, we take

$$x = \varepsilon : \quad 1 + \bar{\delta}_1 \bar{f}_1 = \frac{1}{\mathrm{e}} \quad \text{and} \quad x \to \infty : \quad \bar{f}_1 = 0 \ .$$

Thus, the solution is

$$\bar{\delta}_1 \bar{\varphi}_1 = \ln\left[1 + \left(\frac{1}{\mathrm{e}} - 1\right)\frac{E_1(x)}{E_1(\varepsilon)}\right] \ ,$$

with

$$\bar{\delta}_1 = \frac{1}{E_1(\varepsilon)} \quad \text{and} \quad \bar{f}_1 = \left(\frac{1}{\mathrm{e}} - 1\right)E_1(x) \ .$$

The next step is to seek a UVA in the form

$$\bar{\Phi}_{a1} = 1 + \ln\left[1 + \bar{\delta}_1 \bar{f}_1 + \bar{\delta}_1 \bar{g}_1(X, \varepsilon)\right] \ ,$$

and it is shown that $\bar{g}_1 = 0$. Therefore, the first UVA is

$$\bar{\Phi}_{a1} = 1 + \ln\left[1 + \left(\frac{1}{\mathrm{e}} - 1\right)\frac{E_1(x)}{E_1(\varepsilon)}\right] \ . \tag{6.53}$$

Let us examine the remainder $\mathrm{L}_\varepsilon \bar{\Phi}_{a1}$,

$$\mathrm{L}_\varepsilon \bar{\Phi}_{a1} = \left(\frac{1}{\mathrm{e}} - 1\right)\frac{1}{E_1(\varepsilon)}\frac{\mathrm{d}E_1(x)}{\mathrm{d}x}\frac{\ln\left[1 + \left(\frac{1}{\mathrm{e}} - 1\right)\frac{E_1(x)}{E_1(\varepsilon)}\right]}{1 + \left(\frac{1}{\mathrm{e}} - 1\right)\frac{E_1(x)}{E_1(\varepsilon)}}$$

$$= -\left(\frac{1}{\mathrm{e}} - 1\right)\frac{1}{E_1(\varepsilon)}\frac{\mathrm{e}^{-x}}{x}\frac{\ln\left[1 + \left(\frac{1}{\mathrm{e}} - 1\right)\frac{E_1(x)}{E_1(\varepsilon)}\right]}{1 + \left(\frac{1}{\mathrm{e}} - 1\right)\frac{E_1(x)}{E_1(\varepsilon)}} \ .$$

Note that, x being kept fixed, $\mathrm{L}_\varepsilon \bar{\Phi}_{a1} \to 0$ as $\varepsilon \to 0$. If $x = \varepsilon$, then $\mathrm{L}_\varepsilon \bar{\Phi}_{a1} \to -\infty$ as $\varepsilon \to 0$ but now the integral of the remainder remains finite

$$\int_\varepsilon^\infty \mathrm{L}_\varepsilon \, \bar{\Phi}_{\mathrm{a}1} \, \mathrm{d}x = \left[\frac{1}{2} \ln^2 \left\{ 1 + \left(\frac{1}{\mathrm{e}} - 1 \right) \frac{E_1(x)}{E_1(\varepsilon)} \right\} \right]_\varepsilon^\infty = -\frac{1}{2} \, .$$

According to SCEM, we seek a better approximation in the form

$$\Phi = 1 + \bar{\delta}_1 \bar{\varphi}_1 + \bar{\delta}_2 \bar{\varphi}_2 + \cdots \, .$$

For the same reasons as before, the term $\bar{\delta}_2^2 \left(\dfrac{\mathrm{d}\bar{\varphi}_2}{\mathrm{d}x} \right)^2$ is kept in the equation for $\bar{\varphi}_2$. The following UVA is obtained

$$\bar{\Phi}_{\mathrm{a}2} = \bar{\Phi}_{\mathrm{a}1} + \ln \left[1 + \left(\frac{1}{\mathrm{e}} - 1 \right)^2 \left(\frac{F_1(x)}{E_1^2(\varepsilon)} - \frac{F_1(\varepsilon)}{E_1^3(\varepsilon)} E_1(x) \right) \right] \, . \qquad (6.54)$$

It is more straightforward to seek an approximation in the form

$$\Phi = 1 + \ln \left[1 + \bar{\delta}_1 \bar{f}_1 + \bar{\delta}_2 \bar{f}_2 + \cdots \right] \, .$$

The equation for $\bar{f}_2$ is

$$\frac{\mathrm{d}^2 \bar{f}_2}{\mathrm{d}x^2} + \frac{1}{x} \frac{\mathrm{d}\bar{f}_2}{\mathrm{d}x} + \frac{\mathrm{d}\bar{f}_2}{\mathrm{d}x} = \frac{1}{\bar{\delta}_2} \left(\frac{1}{\mathrm{e}} - 1 \right)^2 \frac{E_1(x)}{E_1^2(\varepsilon)} \, .$$

Note that the next term $\bar{\delta}_2 \bar{g}_2(X, \varepsilon)$ is zero and, finally, the second UVA becomes

$$\begin{aligned}
\bar{\Phi}_{\mathrm{a}2} = 1 + \ln \Bigg[&1 + \left(\frac{1}{\mathrm{e}} - 1 \right) \frac{E_1(x)}{E_1(\varepsilon)} \\
&+ \left(\frac{1}{\mathrm{e}} - 1 \right)^2 \left(\frac{F_1(x)}{E_1^2(\varepsilon)} - \frac{F_1(\varepsilon)}{E_1^3(\varepsilon)} E_1(x) \right) \Bigg] \, .
\end{aligned} \qquad (6.55)$$

6.5.2 Numerical Results

Figure 6.3 shows a good agreement between the UVAs constructed with SCEM and the numerical solution of (6.44a), even for values of ε which are not small compared to 1. The prescription of the exact boundary conditions has enabled us, as with the Stokes-Oseen's flow model, to choose well-appropriate gauges. With respect to the Stokes-Oseen's flow model, the additional difficulty is the presence of a non linear term which is eliminated with a logarithmic term.

6.6 Conclusion

The application of the successive complementary expansion method, SCEM, to singular perturbation problems occurring in the solution of ordinary differential equations has shown that the method is flexible and efficient. In its

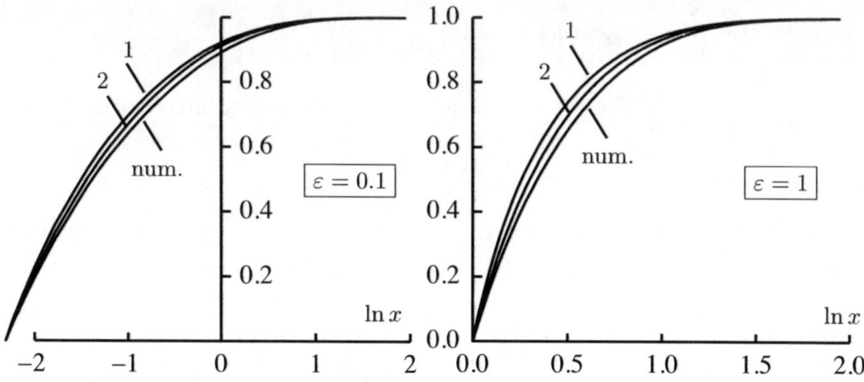

Fig. 6.3. Terrible problem. Comparison of the numerical solution with SCEM approximations: 1 = first order UVA (6.53), 2 = second order UVA (6.55)

simpler form, with regular expansions, SCEM leads to the same results as the method of matched asymptotic expansion, MMAE. The construction, however, is different. With SCEM, we seek a uniformly valid approximation, UVA, starting from an assumed structure of the UVA. A matching principle is not required. Even, the matching principle is a by-product of the method. With MMAE, first we seek approximations in different significant domains. A matching principle is necessary to ensure the consistency of the approximations and, finally, a UVA is formed.

By accepting generalized expansions, SCEM enables us to go further. For instance, more information can be included in the first terms of the expansion in order to improve the accuracy. This feature is especially important because the asymptotic series are often divergent. Moreover, the boundary conditions can be imposed, not asymptotically, but exactly from the beginning of the construction. In certain cases, it can be shown that the idea to impose the exact boundary conditions leads to a very appropriate choice of gauge functions forming the asymptotic sequence. These properties take all their value if it is recognized that the achievement of asymptotic methods is to obtain results which are valid even if the small parameter is not really small compared to unity.

With generalized expansions, SCEM results contain MMAE results. This means that a SCEM approximation is richer than the corresponding MMAE approximation.

Another advantage of SCEM is the removal of the problem of logarithms with the associated difficulty of the application of a matching principle. In all the examples studied in this Chapter, the logarithms appear only when one goes back to regular expansions.

The counterpart of these assets is the necessity of a deeper analysis of the properties of the solution. The improvement of the accuracy of the UVA to a given step implies generally an increase in the effort devoted to obtain the solution. The question is to know if it is worth the trouble. In comparison, the application of MMAE is more systematic.

Problems

6-1. Consider the equation

$$L_\varepsilon\, y \equiv \varepsilon \frac{\mathrm{d}^2 y}{\mathrm{d}x^2} + \frac{\mathrm{d}y}{\mathrm{d}x} + y = 0 \,.$$

The function $y(x)$ is defined in the domain $0 \leq x \leq 1$ and the boundary conditions are

$$y(0) = a \,, \quad y(1) = b \,.$$

The solution is studied with MMAE. A boundary layer forms in the neighbourhood of the point $x = 0$. The variables adapted to the outer region and to the inner region are respectively x and $X = x/\varepsilon$. The associated expansions are

$$y(x,\varepsilon) = y_1(x) + \varepsilon y_2(x) + \cdots + \varepsilon^n y_n(x) + \mathrm{O}(\varepsilon^{n+1}) \,,$$
$$y(x,\varepsilon) = Y_1(X) + \varepsilon Y_2(X) + \cdots + \varepsilon^n Y_n(X) + \mathrm{O}(\varepsilon^{n+1}) \,.$$

1. Express the functions $y_1(x)$, $y_2(x)$, $Y_1(X)$ and $Y_2(X)$. Give also the composite approximations y_{a1} and y_{a2} which are supposed to be UVAs to order 1 and to order ε .
2. Give the order of magnitude of the remainders $L_\varepsilon\, y_{a1}$ and $L_\varepsilon\, y_{a2}$, on the one hand in the domain $0 \leq x \leq 1$ and on the other hand in the domain $0 < A_0 \leq x \leq 1$ where A_0 is a constant independent of ε .

6-2. Consider the problem

$$\varepsilon \frac{\mathrm{d}^2 y}{\mathrm{d}x^2} + \frac{\mathrm{d}y}{\mathrm{d}x} + y = 0 \,, \quad 0 \leq x \leq 1 \,,$$

with

$$y(0) = e \,, \quad y(1) = 1 \,.$$

An exact solution is easily obtained but the problem is studied here with SCEM. The exact boundary conditions will be imposed.

Find the location of the boundary layer.

The first approximation has the form

$$y = y_0(x) \,.$$

Give $y_0(x)$.

We seek a UVA in the form

$$y_{a1} = y_0(x) + Y_0(X, \varepsilon) , \quad X = \frac{x}{\delta(\varepsilon)} .$$

Give δ and Y_0. In the equation for Y_0, only the dominant terms will be kept for $0 < A_1 \leq X \leq A_2$ where A_1 and A_2 are constants independent of ε. It will be shown that $Y_0 = 0$.

The next approximation has the form

$$y = y_0(x) + Y_0(X, \varepsilon) + \nu(\varepsilon) y_1(x, \varepsilon) .$$

Give ν and y_1. In the equation for y_1, only the dominant terms will be kept for $0 < B_1 \leq x \leq 1$ where B_1 is a constant independent of ε.

We seek a UVA in the form

$$y_{a2} = y_0(x) + Y_0(X, \varepsilon) + \nu(\varepsilon) y_1(x, \varepsilon) + \nu(\varepsilon) Y_1(x, \varepsilon) .$$

Give Y_1. In the equation for Y_1, only the dominant terms will be kept for $0 < A_1 \leq X \leq A_2$.

Give the solution by applying SCEM in its regular form.

6-3. Consider the problem

$$\varepsilon \frac{d^2 y}{dx^2} + (1 - x) \frac{dy}{dx} - y = 0 , \quad 0 \leq x \leq 1 ,$$

with

$$y(0) = 1 , \quad y(1) = 1$$

A boundary layer develops near $x = 0$ and another one near $x = 1$.

The regular form of SCEM is applied.

Show that the variables appropriate to the boundary layers are $\xi = x/\varepsilon$ near $x = 0$ and $\zeta = (1 - x)/\varepsilon^{1/2}$ near $x = 1$.

Determine the solution $y_0(x)$ of the reduced equation (obtained by setting $\varepsilon = 0$). The integration constant will not be calculated. The approximation is complemented as

$$y = y_0(x) + Z_0(\zeta) .$$

Show that $y_0(x) = 0$. Check that the solution for Z_0 has the form

$$Z_0 = e^{\zeta^2/2} \left[A + B \int_0^{\zeta/\sqrt{2}} e^{-t^2} \, dt \right] .$$

The solution for Z_0 must satisfy the condition $y(1) = 1$. Deduce a relation for the integration constants of Z_0. The other relation will be obtained later.

Finally, we seek a UVA in the form

$$y_a = Z_0(\zeta) + Y_0(\xi) .$$

The condition at $x = 0$ must be satisfied by the solution $Y_0(\xi)$. Deduce a relation between the integration constants of Y_0.

Apply the boundary conditions at $x = 0$ and $x = 1$ to y_{a}. Deduce the missing relations to determine all the integration constants. Give the solution y_{a}.

It is reminded that

$$\int_0^\infty e^{-t^2}\, \mathrm{d}t = \frac{\sqrt{\pi}}{2}\,.$$

6-4. From the lubrication theory of Reynolds, the pressure $p(x)$ in a highly loaded slider bearing is related to the thickness $h(x)$ of the film of fluid by the dimensionless equation

$$\varepsilon\frac{\mathrm{d}}{\mathrm{d}x}\left[h^3 p\frac{\mathrm{d}p}{\mathrm{d}x}\right] = \frac{\mathrm{d}(ph)}{\mathrm{d}x}\,, \quad 0 \le x \le 1\,,$$

with

$$p(0) = 1\,, \quad p(1) = 1\,.$$

We set

$$h_0 = h(0)\,, \quad h_1 = h(1) = 1\,.$$

A boundary layer develops in the neighbourhood of $x = 1$.

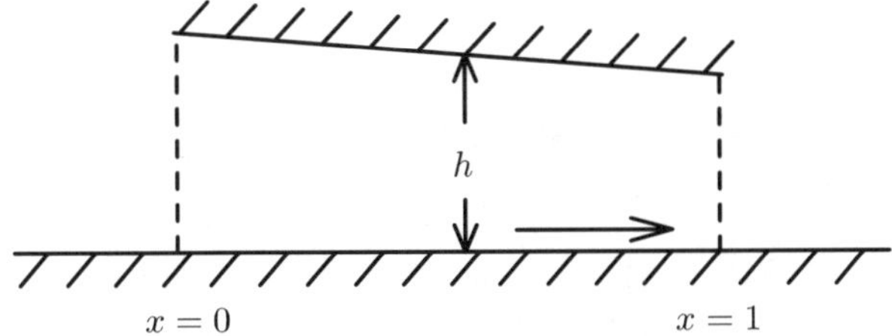

Fig. 6.4. Slider bearing

The regular form of SCEM is applied. The first approximation is

$$p = p_0(x)\,.$$

Give the solution for $p_0(x)$.

Show that the boundary layer variable is

$$X = \frac{1 - x}{\varepsilon}\,.$$

We seek a UVA in the form

$$p = p_0(x) + P_0(X)\,.$$

Give the solution in the form $X = X(P_0)$.

6-5. This problem has been studied by Van Dyke [108] with MMAE. Consider an inviscid, incompressible, two-dimensional, irrotational flow around an elliptic airfoil with a zero angle of attack. The equation of the airfoil is

$$y = \pm \varepsilon T(x) \,,$$

with

$$T = \sqrt{1 - x^2} \quad \text{for} \quad -1 \le x \le 1, \quad \text{otherwise} \quad T = 0 \,.$$

At freestream infinity, we have

$$u = 1 \,.$$

To take into account the singularities which occur near $x = -1$ and $x = 1$, the inner variables are used

$$S_1 = \frac{1 + x}{\varepsilon^2} \,, \quad S_2 = -\frac{1 - x}{\varepsilon^2} \,, \quad Y = \frac{y}{\varepsilon^2} \,.$$

The flow is defined by the potential equation: $\triangle \phi = 0$. A UVA of the potential ϕ is sought in the form

$$\phi = x + \varepsilon \varphi_1(x, y) + \varepsilon^2 \left[\Phi_1(S_1, Y) + \Psi_1(S_2, Y) \right] + \varepsilon^3 \left[\Phi_2(S_1, Y) + \Psi_2(S_2, Y) \right] \,.$$

1. Give the velocity components as function of φ_1, Φ_1, Φ_2, Ψ_1, Ψ_2. It is reminded that

$$u = \frac{\partial \phi}{\partial x} \,, \quad v = \frac{\partial \phi}{\partial y} \,.$$

2. Express the slip condition at the wall. The following identity enables us to solve the singularities at $x = -1$ and $x = 1$

$$\varepsilon T' = \varepsilon T'(x) - \varepsilon f(x) + \varepsilon g(x) + F(S_1) - G(S_2) \,,$$

where T' is the derivative $\dfrac{\mathrm{d}T}{\mathrm{d}x}$ and

$$f = \frac{1}{\sqrt{2(x + 1)}} \quad \text{for} \quad x > -1 \,,$$

$$g = \frac{1}{\sqrt{2(1 - x)}} \quad \text{for} \quad x < 1 \,,$$

$$F = \frac{1}{\sqrt{2S_1}} \quad \text{for} \quad S_1 > 0 \,,$$

$$G = \frac{1}{\sqrt{-2S_2}} \quad \text{for} \quad S_2 < 0 \,.$$

Outside the intervals of definition given above, we have $f = 0$, $g = 0$, $F = 0$, $G = 0$.

3. Give the equations for φ_1, Φ_1, Φ_2, Ψ_1, Ψ_2. It will be noted that the function φ_1 is given by the thin airfoil theory and that $\Phi_1 + S_1$, $\Phi_2 + S_1$, $\Psi_1 + S_2$, $\Psi_2 + S_2$ correspond to the potential around parabolas.

It is reminded that the velocity components of the flow around a parabola given by the equation

$$y = \sqrt{2Rx}$$

are

$$\frac{u}{V_\infty} = 1 - \frac{\sqrt{R}}{2} \frac{\sqrt{r + \frac{R}{2} - x}}{r} \, ,$$

$$\frac{v}{V_\infty} = \frac{\sqrt{R}}{2} \frac{\sqrt{r + x - \frac{R}{2}}}{r} \, ,$$

with

$$r = \sqrt{\left(x - \frac{R}{2}\right)^2 + y^2} \, .$$

It is also reminded that the velocity u on the axis corresponding to the potential φ_1 is given by

$$u = \frac{1}{\pi} \oint_{-\infty}^{+\infty} \frac{\frac{\partial \varphi_1}{\partial y}(x, 0_+)}{x - \xi} \, \mathrm{d}\xi \, .$$

The notation $\dfrac{\partial \varphi_1}{\partial y}(x, 0_+)$ means that the derivative $\dfrac{\partial \varphi_1}{\partial x}$ is evaluated along the upper surface $(y = 0_+)$.

It is found that the velocity u, for $-1 < x < 1$ corresponding to the potential φ_1 is equal to 1 because

$$\frac{1}{\pi} \oint_{-1}^{1} \frac{T'}{x - \xi} \, \mathrm{d}\xi = 1 \quad \text{for} \quad -1 \le x \le 1 \, ,$$

$$\frac{1}{\pi} \oint_{-1}^{\infty} \frac{f}{x - \xi} \, \mathrm{d}\xi = 0 \quad \text{for} \quad x > -1 \, ,$$

$$\frac{1}{\pi} \oint_{-\infty}^{1} \frac{g}{x - \xi} \, \mathrm{d}\xi = 0 \quad \text{for} \quad x < 1 \, .$$

Give the components and the modulus of the velocity at the wall of the ellipse.

The exact solution for the distribution of the velocity at the wall of the ellipse is

$$\frac{q}{V_\infty} = \frac{1 + \varepsilon}{\sqrt{1 + \varepsilon^2 \dfrac{x^2}{1 - x^2}}} \, .$$

Compare the approximate results with the exact solution by plotting the results for $\varepsilon = 0.1$, $\varepsilon = 0.25$ and $\varepsilon = 0.5$.

6-6. Demonstrate the results of Subsect. 6.2.3.

6-7. Demonstrate the results of Subsect. 6.3.3.

7 High Reynolds Number Flows

As an introduction to the second part of this book devoted to the asymptotic analysis of high Reynolds number flows, this Chapter reminds standard results which are useful subsequently for the general understanding of the subject. Then, most of these results are given without detailed proof.

Prandtl's boundary layer theory [78] has been a major step in the understanding of the flow behaviour in aerodynamics and it proved to be an extremely useful and fruitful practical engineering tool. The theory has been formalized much later with the implementation of the method of matched asymptotic expansions, MMAE [47, 107]. Further progress has been made with the second order boundary layer theory [105].

Soon after the introduction of the boundary layer concept, numerical solutions of boundary layer equations ran into difficulties when, in the presence of adverse pressure gradients, the skin-friction decreases and vanishes. This problem has been analyzed by Landau [51] and Goldstein [40]. In a general manner, the question has been raised to determine the solution of the boundary layer equations downstream of a station where the velocity profile is given [39]. Among different results, Goldstein has shown that generally the solution of boundary layer equations is singular if the velocity profile has a zero derivative at the wall (zero wall shear-stress) and that it is not possible to continue the boundary layer calculations downstream of the point of zero wall shear-stress. It is interesting to note in passing that the method used for this analysis is very close to MMAE but, at that time, this method was not formalized. Goldstein suggested also that *inverse methods* could solve the separation singularity. In these inverse methods, instead of the external velocity, the distribution of a quantity associated with the boundary layer is prescribed, for example the distribution of the displacement thickness. Then, the velocity at the edge of the boundary layer becomes an unknown which is determined from the solution of the boundary layer equations. As far as the distribution of the displacement thickness is a regular function, the solution of the boundary layer equations is regular even in the presence of separated flow. These results have been shown numerically by Catherall and Mangler [12].

With Lighthill's analysis of the upstream influence phenomenon in supersonic flow [58], a great leap forward has been taken. The problem is to know how a disturbance in a flat plate boundary layer, for example a small deflection of the wall, affects the boundary layer when the external flow is supersonic. A small perturbation theory has been proposed in which the perturbed flow is structured into three layers. In the region farthest from the wall, the perturbations obey the linearized equations of an inviscid supersonic flow. In the region corresponding to the usual boundary layer, the perturbations obey the small perturbation equations of a parallel, inviscid, compressible flow. Finally, close to the wall, a viscous layer is introduced in order to satisfy the no-slip conditions at the wall. In this latter region, the equations are the Orr-Sommerfeld equations which also provides the evolution of the stability of an incompressible boundary layer. Then, the solution shows the possibility for the perturbations to propagate *upstream*. The order of magnitude of the length of interaction is $LRe^{-3/8}$ where L is the distance between the prescribed disturbance and the flat plate leading edge and Re is the Reynolds number based on L.

This problem addresses the question of the viscous-inviscid interaction, i.e. the interaction between the boundary layer and the inviscid flow region. Lighthill's analysis shed new light on the understanding of this phenomenon. This analysis has been complemented with the triple deck theory. A discussion of the problems associated with separation and with the structure of separated flows, in relation with the triple deck theory in particular, can be found in [86].

The triple deck theory is attributed to Stewartson and Williams and to Neyland [74, 95, 97, 100, 101]. Messiter [69] has also arrived to this theory by analyzing the flow near the trailing edge of a flat plate. Stewartson and Williams consider that their theory is a non linear extension of the theory proposed by Lighthill. In fact, the triple deck theory has been a major advance in fluid mechanics wich led to significant progress in the understanding of many types of flows [41, 116].

Besides these theoretical breakthroughs, practical methods have been devised to solve the interaction between the boundary layer and the inviscid flow, in particular with the goal to calculate separated flows [9, 10, 13, 53, 54, 56, 60, 109]. A justification of the interactive methods has been provided, at least partially, by the triple deck theory [81, 110].

In this Chapter, a simplified analysis of the problems set by boundary layer separation is presented. The discussion is based upon the use of a boundary layer integral method. The method is approximate but reproduces the properties of boundary layer equations fairly well. Moreover, it provides us with a simple way to gain a general understanding of the mathematical and numerical questions set by the viscous-inviscid interaction [23, 24, 53]. To this extent, this is a pedagogical model.

7.1 Boundary Layer Theories

Very comprehensive papers on high Reynolds number flows and their asymptotic structure have been written by different authors [82, 83, 86, 92, 98]. In particular, in these references, the problem of separation is discussed. In this section, we present the main results of standard boundary layer and triple deck theories without mathematical proof.

7.1.1 Prandtl's Boundary Layer

We consider laminar flow past a wall, for example around an airfoil in an unbounded atmosphere (see also Problem 7-2). We assume that the flow is incompressible, two-dimensional, steady and is described by the Navier-Stokes equations (Appendix I).

We use an orthonormal axis system (x, y) with all quantities expressed in dimensionless form. Coordinates x and y are reduced by a reference length L, velocity by a reference velocity V and pressure by ϱV^2. The dimensionless *Navier-Stokes equations* become

$$\frac{\partial \mathcal{U}}{\partial x} + \frac{\partial \mathcal{V}}{\partial y} = 0 \, , \tag{7.1a}$$

$$\mathcal{U}\frac{\partial \mathcal{U}}{\partial x} + \mathcal{V}\frac{\partial \mathcal{U}}{\partial y} = -\frac{\partial \mathcal{P}}{\partial x} + \frac{1}{\mathcal{R}}\frac{\partial^2 \mathcal{U}}{\partial x^2} + \frac{1}{\mathcal{R}}\frac{\partial^2 \mathcal{U}}{\partial y^2} \, , \tag{7.1b}$$

$$\mathcal{U}\frac{\partial \mathcal{V}}{\partial x} + \mathcal{V}\frac{\partial \mathcal{V}}{\partial y} = -\frac{\partial \mathcal{P}}{\partial y} + \frac{1}{\mathcal{R}}\frac{\partial^2 \mathcal{V}}{\partial x^2} + \frac{1}{\mathcal{R}}\frac{\partial^2 \mathcal{V}}{\partial y^2} \, , \tag{7.1c}$$

where $\mathcal{U} = u/V$ and $\mathcal{V} = v/V$ are the velocity components in the x- and y-directions, $\mathcal{P} = p/\varrho V^2$ is the pressure, ϱ the fluid density and μ its dynamic viscosity coefficient. The kinematic viscosity coefficient $\nu = \mu/\varrho$ can also be used. The Reynolds number $\mathcal{R}$ is defined by

$$\mathcal{R} = \frac{\varrho V L}{\mu} \, .$$

The objective is to simplify the Navier-Stokes model when *the Reynolds number of the flow is large compared to unity.*

Two regions are identified: an *inviscid flow* region far from the wall and a *boundary layer* region in its vicinity. In the inviscid flow region, significant variations of velocity occur over distances whose order of magnitude is the same in any direction of space. The length scale L is the chord of the airfoil immersed in the flow. In the boundary layer, *two length scales* are needed. Along the direction parallel to the wall, the length scale is again the chord of the airfoil but the appropriate length scale ℓ in the direction normal to the wall is

$$\ell = L\mathcal{R}^{-1/2} \, .$$

A relation between the scales ℓ and L is obtained by assuming that *the viscosity characteristic time ℓ^2/ν is of the same order as the convection characteristic time L/V*. This relation is fundamental in the boundary layer theory.

The small parameter ε is defined by

$$\varepsilon^2 = \frac{1}{\mathcal{R}}\,. \tag{7.2}$$

In the inviscid flow region, the Navier-Stokes equations reduce to *the Euler equations*. In this region, the flow velocity components and the pressure are expanded as

$$\mathcal{U} = u_1(x,y) + \cdots \,,\ \mathcal{V} = v_1(x,y) + \cdots \,,\ \mathcal{P} = p_1(x,y) + \cdots \,, \tag{7.3}$$

and the Euler equations can be written as

$$\frac{\partial u_1}{\partial x} + \frac{\partial v_1}{\partial y} = 0\,, \tag{7.4a}$$

$$u_1\frac{\partial u_1}{\partial x} + v_1\frac{\partial u_1}{\partial y} = -\frac{\partial p_1}{\partial x}\,, \tag{7.4b}$$

$$u_1\frac{\partial v_1}{\partial x} + v_1\frac{\partial v_1}{\partial y} = -\frac{\partial p_1}{\partial y}\,. \tag{7.4c}$$

Note 7.1. With dimensional variables, the Euler equations are

$$\frac{\partial u}{\partial x} + \frac{\partial v}{\partial y} = 0\,, \tag{7.5a}$$

$$\varrho u\frac{\partial u}{\partial x} + \varrho v\frac{\partial u}{\partial y} = -\frac{\partial p}{\partial x}\,, \tag{7.5b}$$

$$\varrho u\frac{\partial v}{\partial x} + \varrho v\frac{\partial v}{\partial y} = -\frac{\partial p}{\partial y}\,, \tag{7.5c}$$

where u, v, p correspond to u_1, v_1, p_1 respectively. For the coordinates x and y, the same notations as for the dimensionless form are kept.

Since the no-slip conditions at the wall cannot be satisfied, it is necessary to introduce a *boundary layer* structure. The axis system used here is tied to the wall (Fig. 7.1). For convenience, the variables x and y are also used (in the case of the flat plate flow, these variables are identical to those used in the Navier-Stokes equations). The x-axis is taken along the wall and the y-axis is normal to the wall. In the boundary layer region, the velocity components and the pressure are expanded as

$$\mathcal{U} = U(x,Y) + \cdots \,,\ \mathcal{V} = \varepsilon V(x,Y) + \cdots \,,\ \mathcal{P} = P(x,Y) + \cdots \,, \tag{7.6}$$

where Y is the local variable

$$Y = \frac{y}{\varepsilon}\,. \tag{7.7}$$

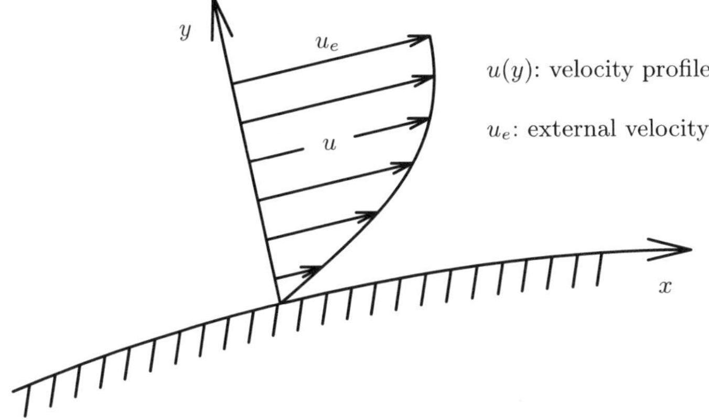

Fig. 7.1. Velocity profile in a boundary layer

The *first order boundary layer equations* are

$$\frac{\partial U}{\partial X} + \frac{\partial V}{\partial Y} = 0 \,, \tag{7.8a}$$

$$U\frac{\partial U}{\partial X} + V\frac{\partial U}{\partial Y} = -\frac{\partial P}{\partial X} + \frac{\partial^2 U}{\partial Y^2} \,, \tag{7.8b}$$

$$0 = -\frac{\partial P}{\partial Y} \,. \tag{7.8c}$$

Note 7.2. With dimensional variables, the boundary layer equations can be written as

$$\frac{\partial u}{\partial x} + \frac{\partial v}{\partial y} = 0 \,, \tag{7.9a}$$

$$\varrho u \frac{\partial u}{\partial x} + \varrho v \frac{\partial u}{\partial y} = -\frac{\partial p}{\partial x} + \mu \frac{\partial^2 u}{\partial y^2} \,, \tag{7.9b}$$

$$0 = \frac{\partial p}{\partial y} \,, \tag{7.9c}$$

where u, v, p correspond to U, εV, P respectively.

Thanks to the presence of the boundary layer, the *no-slip condition* at the wall can be satisfied. Locally, at $Y = 0$, we have

$$U = 0 \,, \quad V = 0 \,.$$

If we denote by E and I the outer and inner expansion operators (which correspond, respectively, to the invisicid region and to the boundary layer), the MVDP becomes

$$\mathrm{I}\,\mathrm{E}\,\mathcal{U} = \mathrm{E}\,\mathrm{I}\,\mathcal{U} \,, \tag{7.10a}$$

$$IE\,\mathcal{V} = EI\,\mathcal{V}\,, \tag{7.10b}$$

$$IE\,\mathcal{P} = EI\,\mathcal{P}\,, \tag{7.10c}$$

and, to order 1, we have

$$\lim_{Y\to\infty} U(x,Y) = u_1(x,0)\,, \tag{7.11a}$$

$$v_1(x,0) = 0\,, \tag{7.11b}$$

$$\lim_{Y\to\infty} P(x,Y) = p_1(x,0)\,. \tag{7.11c}$$

Very often in the literature, the notation U_e is used for the velocity calculated from the Euler equations at the wall of the airfoil. The velocity U_e is related to the static pressure by Bernoulli's equation and, since the pressure is constant along a normal to the wall in the boundary layer, we have

$$\frac{\partial P}{\partial x} = \frac{\mathrm{d}P}{\mathrm{d}x} = -U_e\frac{\mathrm{d}U_e}{\mathrm{d}x}\,. \tag{7.12}$$

Note 7.3. In dimensional form, the previous equation is

$$\frac{\partial p}{\partial x} = \frac{\mathrm{d}p}{\mathrm{d}x} = -\varrho u_e\frac{\mathrm{d}u_e}{\mathrm{d}x}\,. \tag{7.13}$$

We observe that *there is no boundary condition on the velocity component v at the edge of the boundary layer.* The matching with the inviscid flow is performed to next order (see (7.15)).

With conditions of uniform flow at infinity and the wall condition $v_1(x,0) = 0$, the Euler equations can be solved *independently* from the boundary layer equations. Then, an output of the solution of Euler equations is the velocity distribution $U_e(x) = u_1(x,0)$ which is an input for the boundary layer equations.

Therefore, theoretically, the following calculations are performed *sequentially*:

Step 1. The inviscid flow is calculated around the real airfoil by solving the Euler equations with a zero velocity component normal to the wall. This calculation provides us, in particular, with the velocity $U_e(x)$ at the wall.

Step 2. The evolution of the boundary layer is calculated with the velocity distribution $U_e(x)$ as input.

Step 3. The inviscid flow is corrected by solving the linearized form of Euler equations. Indeed, the expansion of the outer flow is

$$\mathcal{U} = u_1(x,y) + \varepsilon u_2(x,y) + \cdots\,, \tag{7.14a}$$

$$\mathcal{V} = v_1(x,y) + \varepsilon v_2(x,y) + \cdots\,, \tag{7.14b}$$

$$\mathcal{P} = p_1(x,y) + \varepsilon p_2(x,y) + \cdots\,, \tag{7.14c}$$

and it is easy to show that u_2, v_2, p_2 satisfy the linearized Euler equations.

The calculation of the inviscid flow perturbed by the presence of the boundary layer introduces the notion of displacement thickness discussed by Lighthill [59], and recovered in MMAE [107]. Equivalently, this condition is obtained by applying the MVDP to the velocity $\mathcal{V}$ to order ε. We get

$$v_2(x,0) = \lim_{Y \to \infty} \left[V - Y \left(\frac{\partial v_1}{\partial y} \right)_{y=0} \right] , \qquad (7.15)$$

or, by using the continuity equation

$$v_2(x,0) = \int_0^\infty \frac{\partial}{\partial x} \left[-U + u_1(x,0) \right] \, \mathrm{d}Y = \frac{\mathrm{d}}{\mathrm{d}x} \left[U_e \Delta_1 \right] . \qquad (7.16)$$

The displacement thickness Δ_1, expressed here in dimensionless form, represents *the effect of the boundary layer on the inviscid flow* and is defined by

$$\Delta_1 = \int_0^\infty \left(1 - \frac{U}{U_e} \right) \, \mathrm{d}Y , \qquad (7.17)$$

or, in dimensional form

$$\delta_1 = \int_0^\infty \left(1 - \frac{u}{u_e} \right) \, \mathrm{d}y . \qquad (7.18)$$

The calculation of the inviscid flow perturbed by the boundary layer can be performed in different ways. One of them consists of calculating the inviscid flow around a modified airfoil: the wall of the airfoil is displaced normal to itself over a distance equal to the displacement thickness. Another convenient way consists of simulating the same effect with the help of a *blowing velocity* v_b distributed along the *real* wall of the airfoil

$$v_b = \frac{\mathrm{d}}{\mathrm{d}x} \left[u_e \delta_1 \right] . \qquad (7.19)$$

In this method, the corrected inviscid flow is calculated by prescribing the velocity v_b at the wall.

According to the above procedure, the boundary layer equations are solved with a prescribed pressure distribution. It is said that the boundary layer equations are solved in the *standard mode* or *direct mode*.

It is observed that the calculations are performed *sequentially*. The process is arranged according to a *hierarchy* in which the inviscid and viscous regions are considered in turn.

In engineering calculation methods, the third step is performed by solving the Euler equations and not their linearized form. Sometimes, the iterative cycle between the inviscid and viscous flows is repeated.

7.1.2 Triple Deck

The singular behaviour of solutions of boundary layer equations at separation in the direct mode [40, 51] (see also Problem 7-4) has been considered as a limitation of the model. This restriction is severe since solutions do not exist downstream of the separation point. For a long time, the validity of the boundary layer model has been blamed because the component of velocity normal to the wall tends towards infinity, which is in contradiction with the boundary layer hypotheses.

The *triple deck theory* has enabled us to better understand the nature of the problem.

The main ideas of this theory are given, to a large extent, in Lighthill's paper [58]. As already said in the introduction of the present Chapter, the problem was to construct a theory to reproduce the phenomenon of upstream influence observed when a flat plate laminar boundary layer developing in a supersonic external flow is subject to a local perturbation. For example, if the wall is slightly deflected, a variation of the wall pressure distribution is recorded upteam of the deflection. It can be thought that this upstream influence is in contradiction with the fact that the boundary layer is governed by a parabolic system of equations (if it is assumed that the pressure is prescribed) and that the external inviscid flow is governed by a hyperbolic system of equations. In addition, the order of magnitude of the distance of upstream influence seems to be incompatible with the boundary layer thickness because it is much greater than this thickness. Finally, in experiments, a strong modification of the flow is observed although the boundary layer is thin. The explanation of this latter point was known; a small perturbation leading to a pressure increase induces a thickening of the boundary layer which, in turn, provokes a larger pressure increase. Lighthill considers the boundary layer *perturbations*. In the basic flow, viscosity plays a crucial role but the perturbations occur on such scales that viscosity cannot influence them. Then, the evolution of boundary layer perturbations is described by the linearized equations of compressible, inviscid flow. This hypothesis is no longer valid very near the wall where the viscous forces are of the same order as the pressure and inertia forces and the no-slip conditions hold at the wall. In the region very close to the wall, Lighthill assumes that the perturbations follow the Orr-Sommerfeld equation established initially for analyzing the linear stability of a laminar boundary layer subject to small perturbations. The qualitative description of the structure is completed by assuming that the perturbations can affect the external flow. The perturbations of the external flow obey the linearized equations of supersonic, inviscid flow. The equations proposed in the different regions are linked together by coupling conditions which ensure the continuity of the different functions characterizing the flow, the pressure and the velocity for example. According to this formulation, Lighthill's theory can be considered as a linearized version of the triple deck theory. Lighthill's theory allows the correct evaluation of the length of upstream influence and many other problems can be solved with the non linear formulation.

We consider a steady, two-dimensional, incompressible, laminar flow on a semi-infinite plate. The velocities, lengths and pressure are nondimensionalized with reference quantities V, L and ϱV^2. The reference velocity is the freestream velocity and the reference length is the length of development of the boundary layer. The streamwise coordinate x and the coordinate normal to the wall y are dimensionless. The Reynolds number is defined by

$$\mathcal{R} = \frac{\varrho V L}{\mu} \; .$$

At distance L from the plate leading edge, the boundary layer is perturbed, for example, by a small hump at the wall. This point corresponds to $x = x_0$ (with the chosen reference length, $x_0 = 1$). The hump can induce boundary layer separation.

The objective is to define a model which is able to avoid the singular behaviour of the boundary layer but which is simpler than the Navier-Stokes model. Therefore, a significant degeneracy of Navier-Stokes equations is sought. It must be underlined that the model describes the *perturbations* of the base flow.

Around the hump, the perturbed flow is structured in *three decks* as shown in Fig. 7.2: *a lower deck, a main deck and an upper deck*. The triple deck structure establishes a link between the unperturbed upstream flow and the downstream flow.

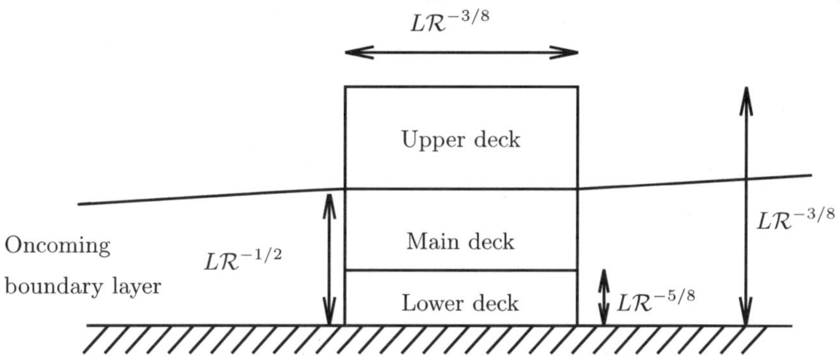

Fig. 7.2. Triple deck structure

The streamwise and transverse length scales of the perturbed region are $L\mathcal{R}^{-3/8}$. Inside the perturbed region, there are three decks. The thickness of the lower deck is $L\mathcal{R}^{-5/8}$; the viscous effects are important in this deck. The main deck is the continuation of the oncoming boundary layer. Its thickness is

$L\mathcal{R}^{-1/2}$, and the viscous effects are negligible (for the perturbations) because the dimensions of the perturbed region are such that the viscosity has no significant effect. In the upper deck, the viscous effects are negligible; its thickness is $L\mathcal{R}^{-3/8}$.

Studies have been performed by specifying the order of magnitude of the hump dimensions [22, 79, 80]; results are discussed briefly in Problem 10-3. Here, the characteristics of the hump are not specificied, but we assume that the dimensions are compatible with the results of the theory.

The triple deck theory describes, for example, the flow around a hump whose height is of order $L\mathcal{R}^{-5/8}$ and length is of order $L\mathcal{R}^{-3/8}$. It is fundamental to have in mind that *the dimensions of the disturbance vary with the Reynolds number* and tend towards zero as the Reynolds number tends towards infinity.

The scales show the *local character* of the theory. As the Reynolds number tends towards infinity, the triple deck domain shrinks to a point. Then, the theory is not able, a priori, to describe the global structure of a flow as the Reynolds number tends towards infinity. In the same way, Prandtl's boundary layer theory describes the flow in the vicinity of the wall as the Reynolds number tends towards infinity.

Below, we recall the main results of the triple deck theory without justification of the scales and of the expansions in the different regions. The arguments leading to the definition of the different gauges are given in Note 7.5, p. 146. More detailed presentations of the construction of the theory are given in [82, 86].

The small parameter of the problem ε, asymptotic dimensionless thickness of the oncoming boundary layer, is related to the Reynolds number by

$$\varepsilon = \mathcal{R}^{-1/2} .$$

In each deck, the following variables are used

$$\text{Upper deck: } X = \varepsilon^{-3/4}(x - x_0), \quad Y^* = \varepsilon^{-3/4}y , \qquad (7.20\text{a})$$

$$\text{Main deck: } \ X = \varepsilon^{-3/4}(x - x_0), \quad Y \ = \ \varepsilon^{-1}y , \qquad (7.20\text{b})$$

$$\text{Lower deck: } X = \varepsilon^{-3/4}(x - x_0), \quad \widetilde{Y} \ = \varepsilon^{-5/4}y , \qquad (7.20\text{c})$$

We denote by $U_0(Y)$ the non perturbed velocity profile of the boundary layer at point $x = x_0$ and its slope at the wall, λ, is defined by

$$\lambda = \left(\frac{\mathrm{d}U_0}{\mathrm{d}Y}\right)_{Y=0} . \qquad (7.21)$$

The appropriate expansions in each deck are given below.
• Upper deck

$$\mathcal{U} = 1 + \varepsilon^{1/2}U_1^*(X, Y^*) + \cdots ,$$

$$\mathcal{V} = \varepsilon^{1/2}V_1^*(X, Y^*) + \cdots ,$$

$$\mathcal{P} = \varepsilon^{1/2}P_1^*(X, Y^*) + \cdots .$$

- Main deck

$$\mathcal{U} = U_0(Y) + \varepsilon^{1/4}U_1(X, Y) + \cdots ,$$

$$\mathcal{V} = \varepsilon^{1/2}V_1(X, Y) + \cdots ,$$

$$\mathcal{P} = \varepsilon^{1/2}P_1(X, Y) + \cdots .$$

- Lower deck

$$\mathcal{U} = \varepsilon^{1/4}\widetilde{U}_1(X, \widetilde{Y}) + \cdots ,$$

$$\mathcal{V} = \varepsilon^{3/4}\widetilde{V}_1(X, \widetilde{Y}) + \cdots ,$$

$$\mathcal{P} = \varepsilon^{1/2}\widetilde{P}_1(X, \widetilde{Y}) + \cdots .$$

All the scales and asymptotic structures are discussed in Nayfeh [73] and Mauss et al. [64, 65].

In the main deck, the first term is $U_0(Y)$ which does not depend on X. This means that, to the considered order, the basic velocity profile does not vary significantly in the perturbed domain. The first term in the upper deck is 1, i.e. the value of the first approximation for the inviscid flow outside the boundary layer. This value matches with the limit of $U_0(Y)$ as $Y \to \infty$.

The equations in the three decks are given below.

- Upper deck

$$\frac{\partial U_1^*}{\partial X} + \frac{\partial V_1^*}{\partial Y^*} = 0 , \tag{7.22a}$$

$$\frac{\partial U_1^*}{\partial X} = -\frac{\partial P_1^*}{\partial X} , \tag{7.22b}$$

$$\frac{\partial V_1^*}{\partial X} = -\frac{\partial P_1^*}{\partial Y^*} . \tag{7.22c}$$

- Main deck

$$\frac{\partial U_1}{\partial X} + \frac{\partial V_1}{\partial Y} = 0 , \tag{7.22d}$$

$$U_0\frac{\partial U_1}{\partial X} + V_1\frac{\partial U_0}{\partial Y} = 0 , \tag{7.22e}$$

$$\frac{\partial P_1}{\partial Y} = 0 . \tag{7.22f}$$

- Lower deck

$$\frac{\partial \widetilde{U}_1}{\partial X} + \frac{\partial \widetilde{V}_1}{\partial \widetilde{Y}} = 0 , \tag{7.22g}$$

$$\widetilde{U}_1 \frac{\partial \widetilde{U}_1}{\partial X} + \widetilde{V}_1 \frac{\partial \widetilde{U}_1}{\partial \widetilde{Y}} = -\frac{\partial \widetilde{P}_1}{\partial X} + \frac{\partial^2 \widetilde{U}_1}{\partial \widetilde{Y}^2} , \tag{7.22h}$$

$$\frac{\partial \widetilde{P}_1}{\partial \widetilde{Y}} = 0 . \tag{7.22i}$$

In the main deck, the solution is

$$U_1 = A(X)U_0'(Y) \quad \text{with} \quad U_0'(Y) = \frac{\mathrm{d}U_0}{\mathrm{d}Y} , \tag{7.23a}$$

$$V_1 = -A'(X)U_0(Y) \quad \text{with} \quad A'(X) = \frac{\mathrm{d}A}{\mathrm{d}X} , \tag{7.23b}$$

where the function $A(X)$ is an unknown which must be determined such that $A \to 0$ as $X \to -\infty$.

Let E, M and I denote the expansion operators in the upper, main and lower decks respectively. The application of the MVDP leads to the results given below.

To order $\varepsilon^{1/4}$ for $\mathcal{U}$, the condition $\mathrm{IM}\mathcal{U} = \mathrm{MI}\mathcal{U}$ gives

$$\lim_{\widetilde{Y} \to \infty} \left(\widetilde{U}_1 - \lambda \widetilde{Y} \right) = \lambda A . \tag{7.24}$$

To order $\varepsilon^{1/2}$ for $\mathcal{V}$, the condition $\mathrm{EM}\mathcal{V} = \mathrm{ME}\mathcal{V}$ gives

$$V_1^*(X,0) = \lim_{Y \to \infty} V_1(X,Y) , \tag{7.25}$$

or, taking into account (7.23b) and the fact that $U_0 \to 1$ as $Y \to \infty$

$$V_1^*(X,0) = -\frac{\mathrm{d}A}{\mathrm{d}X} . \tag{7.26}$$

To order $\varepsilon^{1/2}$ for $\mathcal{P}$, the conditions $\mathrm{IM}\mathcal{P} = \mathrm{MI}\mathcal{P}$ and $\mathrm{EM}\mathcal{P} = \mathrm{ME}\mathcal{P}$ give

$$P_1(X,0) = \lim_{\widetilde{Y} \to \infty} \widetilde{P}_1(X,\widetilde{Y}) , \tag{7.27a}$$

$$P_1^*(X,0) = \lim_{Y \to \infty} P_1(X,Y) . \tag{7.27b}$$

Moreover, as we have $\dfrac{\partial P_1}{\partial Y} = 0$ and $\dfrac{\partial \widetilde{P}_1}{\partial \widetilde{Y}} = 0$, we deduce

$$P_1^*(X,0) = P_1(X) = \widetilde{P}_1(X) . \tag{7.28}$$

Equation (7.24) is one of the conditions necessary to solve (7.22g, 7.22h, 7.22i).

Condition (7.26) on velocity $V_1^*(X,0)$ allows the solution of the upper deck equations (7.22a–7.22c). The velocity $V_1^*(X,0)$ can be identified with the perturbation of blowing velocity v_b given by (7.19) used in the boundary layer studies to simulate viscous effects on the inviscid flow. Function A is called *displacement function*.

Note 7.4. In the main deck, the slope of the streamlines is given by

$$\frac{\varepsilon\,dY}{\varepsilon^{3/4}dX} = \frac{v}{u} = -\frac{\varepsilon^{1/2}A'(X)U_0(Y)}{U_0(Y) + \varepsilon^{1/4}A(X)U_0'(Y)} \,,$$

that is, to first order

$$\frac{dY}{dX} = -\varepsilon^{1/4}A'(X) \,.$$

Thus, the equation of the streamlines is

$$Y = -\varepsilon^{1/4}A(X) + C \,,$$

with the condition $A \to 0$ as $X \to -\infty$; C is a constant which depends on the considered streamline.

With respect to the non perturbed streamlines whose equation is $Y = C$, we note that the streamlines are displaced by the quantity $-\varepsilon^{1/4}A(X)$ which depends only on X and not on Y. In the main deck, all streamlines undergo an identical displacement normally to the wall. Moreover, the velocity along a streamline is

$$\left[\sqrt{(U_0 + \varepsilon^{1/4}A\,U_0')^2 + \varepsilon A'^2 U_0^2}\,\right]_{Y=-\varepsilon^{1/4}A(X)+C} = U_0(Y = C) + \mathrm{O}(\varepsilon^{1/2}) \,.$$

Neglecting terms of order $\varepsilon^{1/2}$, the velocity is constant along the streamlines in the main deck.

From the solution in the upper deck, the pressure and the velocity normal to the wall are related by a Hilbert integral (Appendix III)

$$P_1^*(X,0) = -\frac{1}{\pi}\fint_{-\infty}^{\infty} \frac{V_1^*(X,0)}{X - \xi}\,d\xi \,,$$

or

$$P_1^*(X,0) = \frac{1}{\pi}\fint_{-\infty}^{\infty} \frac{A'(X)}{X - \xi}\,d\xi \,, \tag{7.29}$$

where the sign $\fint$ means that the Cauchy principal part of the integral is taken. We also have

$$P_1^*(X,0) = -U_1^*(X,0) \,.$$

With (7.28), we deduce

$$\widetilde{P}_1(X) = \frac{1}{\pi}\fint_{-\infty}^{\infty} \frac{A'(X)}{X - \xi}\,d\xi \,. \tag{7.30}$$

We note that the main deck and lower deck equations are included in the standard boundary layer equations. The lower deck equations are even exactly the same as the standard equations but the boundary conditions are not the usual ones. At the wall, we have

$$\widetilde{U}_1 = 0 \,,\; \widetilde{V}_1 = 0 \,, \tag{7.31}$$

but the matching between the lower deck and the main deck gives condition (7.24). Moreover, the perturbations must vanish at upstream infinity in order to ensure the matching with the non perturbed boundary layer.

Solving the triple deck equations reduces essentially to solving the lower deck equations (7.22g) and (7.22h), with the interaction law (7.30) and the boundary conditions (7.31), (7.24). Function $A(X)$ is a part of the solution. In practice, it is required to introduce a disturbance. If it is a local deformation of the wall defined by $\widetilde{Y} = F(X)$, this function adds to the displacement function and, everywhere, function A is replaced by $(F + A)$ if Prandtl's transformation is used to reduce the wall to the equation $\widetilde{Y} = 0$ [80].

In contrast with the standard boundary layer theory, the inviscid flow and viscous flow equations form a *strongly coupled system*. The upper deck solution depends on the lower and main deck solution through function A whereas the lower and main deck solution depends on the upper deck solution through the pressure distribution. The upper deck solution cannot be determined independently from the lower and main deck solution; conversely, the lower and main deck solution cannot be determined independently from the upper deck solution. It is said that the inviscid flow and the viscous flow *interact*. The main deck has a passive role which consists of transmitting the pressure and the displacement effect between the boundary layer (lower deck) and the inviscid flow (upper deck).

The matching condition (7.25) on the velocity normal to the wall between the upper deck and the main deck results from the identity of gauges for this velocity component. This is an essential feature which guarantees the *absence of hierarchy between the three decks*. In addition, in the lower deck, the perturbation of streamwise velocity is of order $\varepsilon^{1/4}$, since the expansion in this deck is

$$u = \varepsilon^{1/4}\widetilde{U}_1 + \cdots .$$

Now, in this deck, the base velocity profile – Blasius' profile – is given by

$$U_0 = \lambda Y = \varepsilon^{1/4}\lambda\widetilde{Y} .$$

Thus, in the lower deck, the base profile and the perturbation profile have the same order of magnitude. Then, it is possible that the resultant velocity has negative values. Associated with the interaction between the decks, this property gives access to the calculation of *separated flows*. However, it must be noted that the results given here do no constitute a solution to Goldstein's singularity because, in the triple deck theory, the disturbance which leads to separation tends towards zero as the Reynolds number tends towards infinity.

Note 7.5. The choice of order functions is crucial for the consistency of the results and the success of the model. The constraints used to determine these order functions are given below.

At first, it is observed that the pressure, to first order, matches directly between the different decks. The pressure perturbations are of the same order everywhere.

The main deck is the continuation of the oncoming boundary layer. In this way, the thickness of the main deck is known.

In the upper deck, the dimensions of the perturbed domain are the same in both directions of space.

The two terms of the continuity equation are always of the same order in the different decks so that this equation is never trivial.

In the lower deck, the viscous terms, the inertia terms and the pressure term have the same order in the streamwise momentum equation so that the no-slip condition at the wall applies.

The wall shear-stress is given by the slope of the velocity profile at the wall in the lower deck. The choice of the order function is such that the perturbation of the wall shear-stress has the same order as the wall shear-stress of the oncoming boundary layer. Thus, the resultant wall shear-stress can be negative and the solution can describe a separated boundary layer.

The velocity component normal to the wall has the same order in the upper deck and in the main deck. This essential condition prevents the hierarchy between the different decks from occurring.

Note 7.6. Without any external disturbance, the triple deck model has the trivial solution $A = 0$. There exists, however, an eigensolution such that [98]

$$p = -\alpha(-X)^{1/2} \,, \quad \frac{\mathrm{d}A}{\mathrm{d}X} = 0 \qquad \text{if } X < 0 \,,$$

$$p = 0 \,, \quad \frac{\mathrm{d}A}{\mathrm{d}X} = -\alpha X^{1/2} \text{ if } X > 0 \,,$$

where α is an arbitrary constant. From Sychev [98, 101], an appropriate choice of α makes the solution compatible with downstream conditions, and the singularity at $X = 0$ is smoothed. This model is implemented in association with Kirchhoff's free streamline model in order to analyze the separation problem on a regular surface, for example, on a circular cylinder [102].

Note 7.7. In supersonic flow, the triple deck structure is very close to the incompressible case but the interaction law (7.30) is replaced by Ackeret's law which, with appropriate scales, can be written as

$$p = -\frac{\mathrm{d}A}{\mathrm{d}X} \,.$$

Lighthill's linearized form of the triple deck has an eigensolution [58]

$$p = a_1 \exp \kappa X + \cdots \,,$$

which provides us with the key of the free interaction problem in which perturbations propagates upstream. These solutions can be interpreted as the generation of spontaneous perturbations.

7.2 Analysis of an Integral Method

7.2.1 Integral Method

We consider a laminar, two-dimensional, incompressible, steady boundary layer flow. The boundary layer equations are given by (7.9a–7.9c) and the pressure is related to the external velocity by (7.13).

An integral method [20] is based on the integrated form of the local equations, the integration being performed with respect to y over the whole boundary layer, i.e. between the wall and the boundary layer edge. The integral equations represent global balance equations over the thickness of the boundary layer. The choice of integral equations is practically infinite: we can take, for example, the integrated form of the continuity equation or of the momentum equation, we can take also the integrated form of the kinetic energy equation or of any other moment equation. Here, we choose the *integrated form of the kinetic energy equation and of the momentum equation* [20]

$$\frac{\mathrm{d}\delta_3}{\mathrm{d}x} + 3\frac{\delta_3}{u_e}\frac{\mathrm{d}u_e}{\mathrm{d}x} = 2C_D \, , \qquad (7.32\mathrm{a})$$

$$\frac{\mathrm{d}\theta}{\mathrm{d}x} + \theta\frac{H+2}{u_e}\frac{\mathrm{d}u_e}{\mathrm{d}x} = \frac{C_f}{2} \, . \qquad (7.32\mathrm{b})$$

In these equations, u_e is the velocity at the boundary layer edge; δ_1, δ_3 and θ represent the displacement thickness, the kinetic energy thickness and the momentum thickness respectively; H is the shape factor. We have

$$\delta_1 = \int_0^\infty \left(1 - \frac{u}{u_e}\right) \, \mathrm{d}y \, , \quad \delta_3 = \int_0^\infty \frac{u}{u_e} \left(1 - \frac{u^2}{u_e^2}\right) \, \mathrm{d}y \, ,$$

$$\theta = \int_0^\infty \frac{u}{u_e} \left(1 - \frac{u}{u_e}\right) \, \mathrm{d}y \, , \quad H = \frac{\delta_1}{\theta} \, .$$

The dissipation coefficient C_D and the skin-friction coefficient C_f are defined by

$$C_D = \frac{1}{\varrho u_e^3} \int_0^\infty \mu \left(\frac{\partial u}{\partial y}\right)^2 \, \mathrm{d}y \, , \quad \frac{C_f}{2} = \frac{\tau_w}{\varrho u_e^2} \, ,$$

where τ_w is the wall shear-stress

$$\tau_w = \left(\mu\frac{\partial u}{\partial y}\right)_{y=0} \, .$$

The dissipation coefficient represents, with the boundary layer hypotheses, the integral of the deformation work of viscous stress $\left(\mu\dfrac{\partial u}{\partial y}\right)\dfrac{\partial u}{\partial y}$ which is responsible for the transformation of kinetic energy into heat.

The standard method of solution of boundary layer equations – the direct mode – consists of assuming that the distribution of the external velocity u_e is known and prescribed. An inviscid flow calculation provides us with this distribution. Then, the integral equations (7.32a) and (7.32b) contain five unknown functions $\delta_1(x)$, $\delta_3(x)$, $\theta(x)$, $C_D(x)$ and $C_f(x)$. Function $H(x)$ is not an additional unknown since, by definition, we have $H = \delta_1/\theta$. The set of equations being open, it is necessary to complete the integral equations with closure relationships. They have the form

$$\frac{2C_D\mathcal{R}_\theta}{H_{32}} = F_1(H) , \qquad (7.33\text{a})$$

$$\frac{C_f}{2}\mathcal{R}_\theta = F_2(H) , \qquad (7.33\text{b})$$

$$H_{32} = F_3(H) , \qquad (7.33\text{c})$$

where $\mathcal{R}_\theta$ is the Reynolds number based on the momentum thickness

$$\mathcal{R}_\theta = \frac{\varrho u_e \theta}{\mu} ,$$

and H_{32} is defined by

$$H_{32} = \frac{\delta_3}{\theta} .$$

Functions $F_1(H)$, $F_2(H)$ and $F_3(H)$ are obtained from Falkner-Skan's self-similar solutions [20]. These functions are given on Figs. 7.3, 7.4 and 7.5.

Thus, the boundary layer integral equations (7.32a) and (7.32b) associated with the closure relations (7.33a), (7.33b) and (7.33c) constitute an *integral method* for calculating boundary layers.

For the analyses presented below, it is more convenient to rewrite the equations by taking into account the definition of H_{32}. We have

$$\frac{\mathrm{d}\delta_3}{\mathrm{d}x} = H_{32}\frac{\mathrm{d}\theta}{\mathrm{d}x} + \theta H'_{32}\frac{\mathrm{d}}{\mathrm{d}x}\left(\frac{\delta_1}{\theta}\right)$$

$$= (H_{32} - HH'_{32})\frac{\mathrm{d}\theta}{\mathrm{d}x} + H'_{32}\frac{\mathrm{d}\delta_1}{\mathrm{d}x} ,$$

with

$$H'_{32} = \frac{\mathrm{d}H_{32}}{\mathrm{d}H} .$$

The set of integral equations becomes

$$(H_{32} - HH'_{32})\frac{\mathrm{d}\theta}{\mathrm{d}x} + H'_{32}\frac{\mathrm{d}\delta_1}{\mathrm{d}x} + 3\frac{\delta_3}{u_e}\frac{\mathrm{d}u_e}{\mathrm{d}x} = 2C_D , \qquad (7.34\text{a})$$

$$\frac{\mathrm{d}\theta}{\mathrm{d}x} + \theta\frac{H+2}{u_e}\frac{\mathrm{d}u_e}{\mathrm{d}x} = \frac{C_f}{2} . \qquad (7.34\text{b})$$

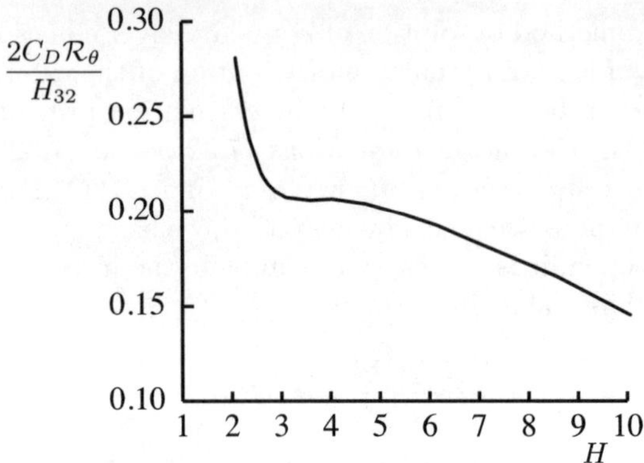

Fig. 7.3. Dissipation function

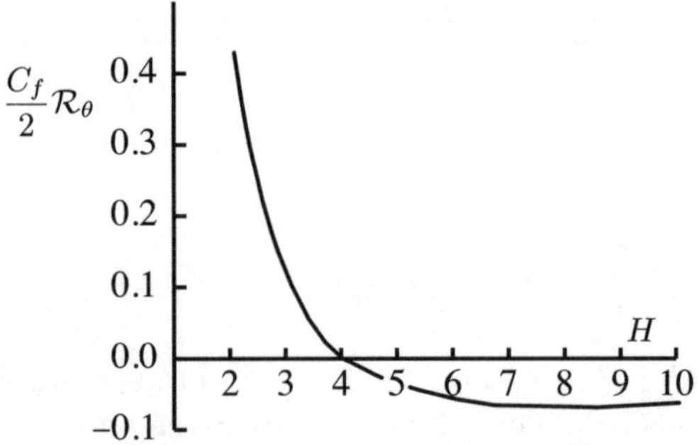

Fig. 7.4. Skin-friction coefficient

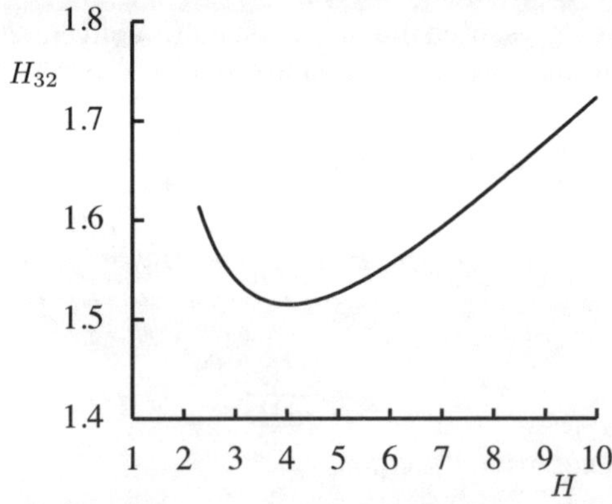

Fig. 7.5. Function $H_{32}(H)$

Three problems are discussed subsequently

Problem 1. The external velocity $u_e(x)$ is known. The boundary layer problem is set under its standard form – *the direct mode*. It is examined if the boundary layer calculation is always possible.

Problem 2. The distribution of the displacement thickness $\delta_1(x)$ is assumed to be known and the distribution of the external velocity becomes an unknown function. The boundary layer is calculated in *the inverse mode*. It is examined if the calculation of the boundary layer is always possible.

Problem 3. We analyze the flow in a diffuser whose geometry is known, and we assume that the velocity distribution in the core of the flow is uniform in a cross-section. The viscous and inviscid flow equations interact and it is examined if the calculation is always possible.

7.2.2 Direct Mode

We assume that the distribution of the external velocity $u_e(x)$ is prescribed. Taking into account the closure relationships (7.33a–7.33c), the main unknowns in (7.34a), (7.34b) are $\delta_1(x)$ and $\theta(x)$. Then, the question is to calculate their derivatives $\dfrac{\mathrm{d}\delta_1}{\mathrm{d}x}$ and $\dfrac{\mathrm{d}\theta}{\mathrm{d}x}$, from the system

$$(H_{32} - HH'_{32})\frac{\mathrm{d}\theta}{\mathrm{d}x} + H'_{32}\frac{\mathrm{d}\delta_1}{\mathrm{d}x} = 2C_D - 3\frac{\delta_3}{u_e}\frac{\mathrm{d}u_e}{\mathrm{d}x} , \qquad (7.35a)$$

$$\frac{\mathrm{d}\theta}{\mathrm{d}x} = \frac{C_f}{2} - \theta\frac{H+2}{u_e}\frac{\mathrm{d}u_e}{\mathrm{d}x} . \qquad (7.35b)$$

The determinant of this system is

$$\Delta_1 = -H'_{32} .$$

We assume that the shape factor H is larger than 2.21 which corresponds to the stagnation point in two-dimensional flow. In this domain, the determinant Δ_1 is zero when $H = 4.029$ (Fig. 7.5), which corresponds to *boundary layer separation*; the skin-friction vanishes for this value of H (Fig. 7.4).

At separation point, the resolution is either impossible or indeterminate. The indetermination occurs if the compatibility relation is satisfied

$$2C_D - 3\frac{\delta_3}{u_e}\frac{\mathrm{d}u_e}{\mathrm{d}x} = H_{32}\left(\frac{C_f}{2} - \theta\frac{H+2}{u_e}\frac{\mathrm{d}u_e}{\mathrm{d}x}\right) ,$$

which implies that the distribution of external velocity follows a particular law which is not verified generally.

Then, in general, the resolution of boundary layer equations is *impossible* at separation. It follows that the derivative $\dfrac{\mathrm{d}\delta_1}{\mathrm{d}x}$ becomes infinite because $\dfrac{\mathrm{d}\theta}{\mathrm{d}x}$ can be calculated from (7.35b) and, substituting in (7.35a), we obtain a finite, non zero value of $H'_{32}\dfrac{\mathrm{d}\delta_1}{\mathrm{d}x}$. As H'_{32} vansishes at separation, it is deduced that the value of the derivative of δ_1 becomes infinite. Moreover, the calculation of the boundary layer downstream of separation is impossible because the derivative $\dfrac{\mathrm{d}H_{32}}{\mathrm{d}x}$ is non zero at separation point if the compatibility relation is not satisfied. Then, the value of H_{32} becomes less than the minimum value given by relation $F_3(H)$.

The separation singularity and the impossibility of calculating the boundary layer downstream of separation when the external velocity (or the pressure) is prescribed are results similar to those obtained by Goldstein [40] (see Problems 7-4 and 7-7).

7.2.3 Inverse Mode

Now, we assume that the distribution of the displacement thickness is known and prescribed. The distribution of the external velocity becomes an unknown and must be calculated from the boundary layer equations. In practice, this problem has a meaning only if the boundary layer equations are associated with the inviscid flow equations and if an algorithm is devised to solve the set of both systems of equations. This question is discussed in Sect. 7.3.

In this mode, putting the main unknowns on the left hand side, the boundary layer equations can be written as

$$(H_{32} - HH'_{32})\frac{\mathrm{d}\theta}{\mathrm{d}x} + 3\frac{\delta_3}{u_e}\frac{\mathrm{d}u_e}{\mathrm{d}x} = 2C_D - H'_{32}\frac{\mathrm{d}\delta_1}{\mathrm{d}x} , \qquad (7.36a)$$

$$\frac{\mathrm{d}\theta}{\mathrm{d}x} + \theta\frac{H+2}{u_e}\frac{\mathrm{d}u_e}{\mathrm{d}x} = \frac{C_f}{2} . \qquad (7.36b)$$

The unknowns are the derivatives $\dfrac{\mathrm{d}\theta}{\mathrm{d}x}$ and $\dfrac{\mathrm{d}u_e}{\mathrm{d}x}$. The determinant of this system is

$$\Delta_2 = \theta\frac{H+2}{u_e}(H_{32} - HH'_{32}) - 3\frac{\delta_3}{u_e} = \frac{\theta}{u_e}\left[(H_{32} - HH'_{32})(H+2) - 3H_{32}\right] .$$

For $H > 2.21$, it is shown that $\Delta_2 \neq 0$. In the inverse mode, we are certain that the resolution of the boundary layer equations presents no singularity if, obviously, the distribution of δ_1 is sufficiently regular. Even at separation, there is no difficulty to solve the boundary layer equations.

This result is similar to the conclusions of Catherall and Mangler [12] who have shown numerically that the solution of the boundary layer equations in inverse mode is regular through a separation point.

An important conclusion is that the separation singularity cannot be attributed completely to the use of boundary layer equations. The way in which these equations are solved plays an important role (see Problem 7-3).

7.2.4 Simultaneous Mode

We consider a flow in a symmetrical plane diffuser whose geometry is given. Function $h(x)$ is known (Fig. 7.6). For the sake of simplicity, we assume that the velocity u_e of the flow in the inviscid region is uniform in a cross-section. We also assume that the evolution of the cross-section is slow. By using the definition of the displacement thickness δ_1, the mass conservation in the diffuser yields

$$u_e(h - \delta_1) = \text{Cst} ,$$

or, after differentiation,

$$u_e \frac{\mathrm{d}h}{\mathrm{d}x} - u_e \frac{\mathrm{d}\delta_1}{\mathrm{d}x} + (h - \delta_1)\frac{\mathrm{d}u_e}{\mathrm{d}x} = 0 .$$

The shape of the diffuser being known, the main unknowns are the displacement thickness $\delta_1(x)$, the momentum thickness $\theta(x)$ and the velocity $u_e(x)$. The equations are

$$(H_{32} - HH'_{32})\frac{\mathrm{d}\theta}{\mathrm{d}x} + H'_{32}\frac{\mathrm{d}\delta_1}{\mathrm{d}x} + 3\frac{\delta_3}{u_e}\frac{\mathrm{d}u_e}{\mathrm{d}x} = 2C_D , \qquad (7.37\mathrm{a})$$

$$\frac{\mathrm{d}\theta}{\mathrm{d}x} + \theta\frac{H+2}{u_e}\frac{\mathrm{d}u_e}{\mathrm{d}x} = \frac{C_f}{2} , \qquad (7.37\mathrm{b})$$

$$- u_e\frac{\mathrm{d}\delta_1}{\mathrm{d}x} + (h - \delta_1)\frac{\mathrm{d}u_e}{\mathrm{d}x} = -u_e\frac{\mathrm{d}h}{\mathrm{d}x} . \qquad (7.37\mathrm{c})$$

The unknowns being the derivatives of θ, δ_1 and u_e, the determinant is

$$\Delta_3 = (H_{32} - HH'_{32})(H + 2)\theta - [H'_{32}(h - \delta_1) + 3H_{32}\theta] .$$

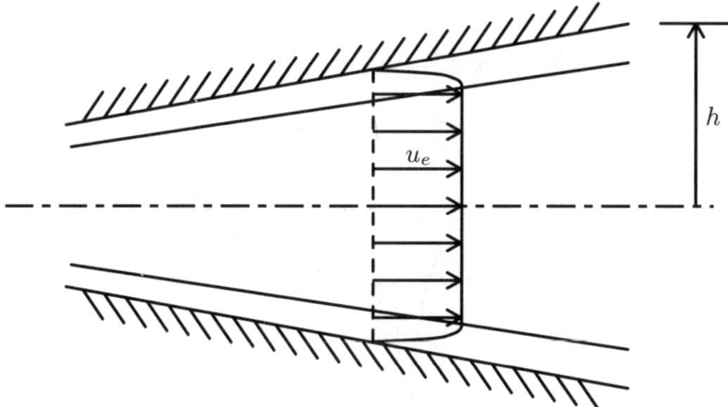

Fig. 7.6. Flow in a diffuser

This determinant vanishes when

$$\frac{h}{\theta} = f(H) \quad \text{with} \quad f(H) = \frac{(H_{32} - HH'_{32})(H+2) + HH'_{32} - 3H_{32}}{H'_{32}} \;.$$

Function $f(H)$ has a local minimum $f = 183.5$ when $H = 6.67$ (Fig. 7.7). This means that the resolution of (7.37a), (7.37b) and (7.37c) is possible when $h/\theta < 183.5$ (by assuming that θ remains positive). In this domain, the simultaneous resolution of viscous and inviscid equations do not raise any difficulty. When the cross-section of the diffuser is too large compared to the boundary layer thickness ($h/\theta > 183.5$), it is possible that the hypothesis of a one-dimensional inviscid flow, uniform in a cross-section, is not sufficient to express the interaction with the boundary layer. It would be necessary to consider a two-dimensional inviscid flow.

We observe that the determinant Δ_1 is a minor of the determinant Δ_3 which occurs when the viscous and inviscid equations are solved separately as in the direct mode. Therefore, this analysis shows that the separation singularity is associated with the technique used to solve the whole flow. However, if the separated zone is too extended, the standard boundary layer hypotheses must be revised because certain hypotheses are restrictive, for example the hypothesis of a constant static pressure along a normal to the wall.

It must be noted that all the conclusions drawn above are obtained with a *finite Reynolds number*. As the Reynolds number tends towards infinity, the interaction between the boundary layer and the inviscid flow becomes inefficient to solve the separation singularity because, for a given diffuser geometry, the momentum thickness θ tends towards zero and $\dfrac{h}{\theta}$ becomes infinite. This result is similar to the one obtained by Stewartson [96] who has established

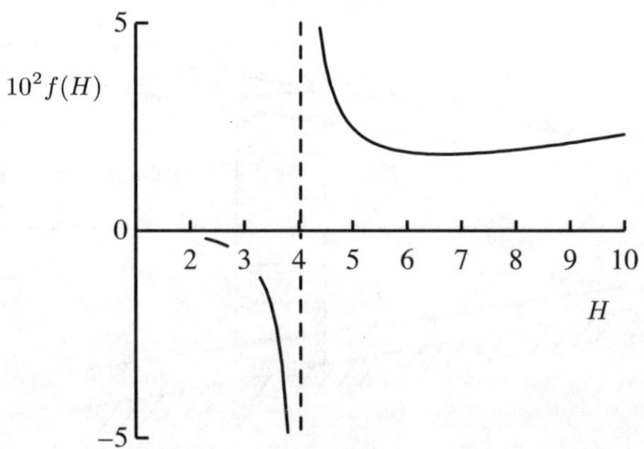

Fig. 7.7. Function $f(H) = \dfrac{(H_{32} - HH'_{32})(H+2) + HH'_{32} - 3H_{32}}{H'_{32}}$

that the triple deck structure is not able to solve the separation singularity when the boundary layer develops in the presence of a pressure gradient *independent* of the Reynolds number. There is no contradiction with the triple deck theory because, in this theory, the dimension of the disturbance and therefore the associated pressure gradient vary with the Reynolds number.

7.3 Viscous-Inviscid Interaction

In aerodynamics, a standard problem is to calculate the flow around an airfoil at least when the flow is not separated but it is also desired to calculate separated flows to know, for example, the value of the maximum lift and the corresponding angle of attack.

When the Reynolds number is large enough, an approximate solution of the Navier-Stokes equations is obtained from the viscous-inviscid interaction. The problem consists of solving the system comprising the inviscid equations, the boundary layer equations and the interaction law which connects the two systems of equations. Historically, these methods have been developed on a purely intuitive basis. Nowadays, a justification, at least partial, has been provided by different theories including the boundary layer theory and the triple deck theory. Other analyses bring additional elements [22, 80]. The study presented in Subsects. 7.2.2–7.2.4 with the integral method justifies also the interaction methods.

An essential difference between the asymptotic theories and the viscous-inviscid interaction methods must be noted. The latter methods are devoted to solve problems at *finite Reynolds numbers*, which is the practical problem, whereas the asymptotic methods study the behaviour of flows as *the Reynolds number tends towards infinity*. The conclusions are not necessarily identical.

According to the standard boundary layer theory, the calculation of the interaction is performed *sequentially*. At first, the inviscid flow is calculated around the real airfoil by applying the condition that the velocity normal to the wall is zero at the wall. Afterwards, the boundary layer is calculated with, as input, the streamwise wall velocity determined by the inviscid flow. Finally, the inviscid flow is corrected by taking into account the displacement effect. The procedure is called *direct-direct*: direct for the inviscid flow and direct for the boundary layer (Fig. 7.8).

In the presence of separation, the procedure is no longer valid because the solution of boundary layer layer equations is singular and it is not possible to calculate the boundary layer downstream of the separation point. To solve this problem, inverse methods can be used (Fig. 7.9). These methods can be associated with inverse methods to calculate the inviscid flow: the input is the pressure calculated from the boundary layer equations and the result is the shape of the body corresponding to the pressure distribution (in fact the real shape modified by the displacement effect). In practice, this type of method,

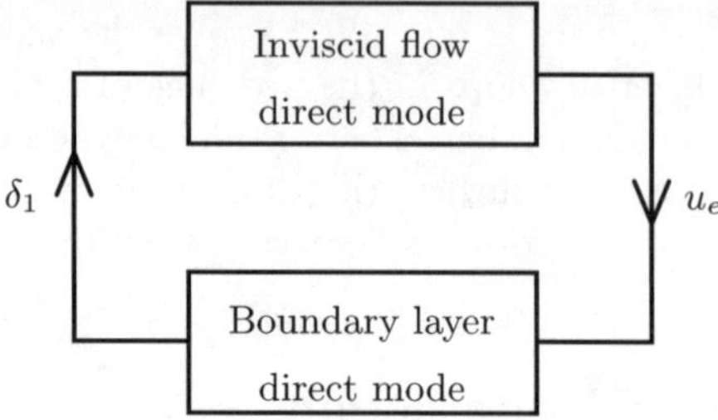

Fig. 7.8. Viscous-inviscid interaction. Direct mode

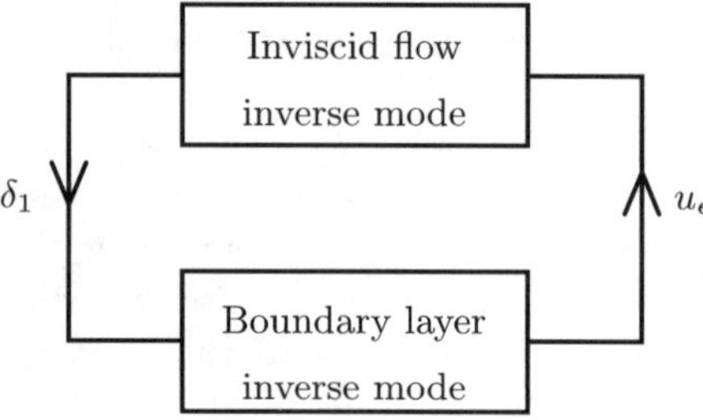

Fig. 7.9. Viscous-inviscid interaction. Inverse mode

called *inverse-inverse*, is not easy to implement and other procedures have been developed [102].

Semi-inverse methods are a very efficient example [9, 10, 53, 54, 56]. These methods consist of solving the boundary layer equations in the inverse mode and the inviscid flow equations in the direct mode (Fig. 7.10). For a given distribution of the displacement thickness, the boundary layer equations yield a distribution of velocity $u_{e\,\mathrm{BL}}(x)$. For the same distribution of the displacement thickness, the inviscid flow equations yield a distribution of the wall velocity $u_{e\,\mathrm{IN}}(x)$. Generally, for any distribution of the displacement thickness, the two velocity distributions are not identical. Iterative procedures have been devised to obtain $u_{e\,\mathrm{BL}}(x) = u_{e\,\mathrm{IN}}(x)$. For example, Carter [9, 10] proposed to determine the new estimate of the displacement thickness at iteration $(n + 1)$ by

$$\delta_1^{n+1}(x) = \delta_1^n(x) \left[1 + \omega \left(\frac{u_{e\,\mathrm{BL}}^n(x)}{u_{e\,\mathrm{IN}}^n(x)} - 1\right)\right] ,$$

where ω is a relaxation factor.

Another approach has been developed by Veldman [109]. In agreement with the triple deck theory, the inviscid flow and the boundary layer are strongly coupled and there is no hierarchy between the systems of equations. In a simultaneous method, the external velocity $u_e(x)$ and the displacement

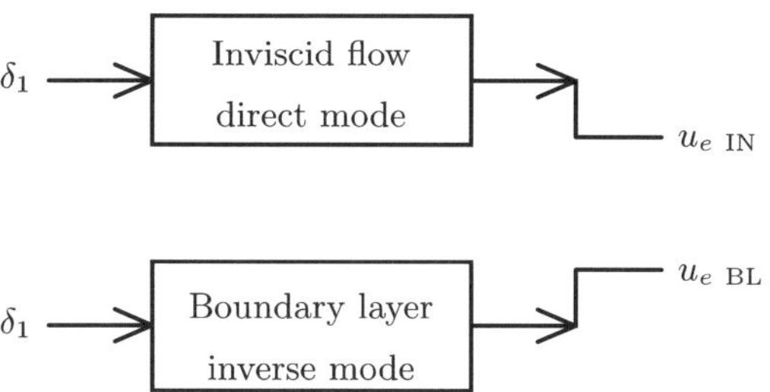

Fig. 7.10. Viscous-inviscid interaction. Semi-inverse mode

thickness $\delta_1(x)$ are calculated simultaneouly from the set of viscous and inviscid equations.

For example, let us consider a flow on a flat plate perturbed by a small local deformation of the wall. The external velocity u_e is given by

$$u_e(x) = u^0 + \delta u_e(x) \,,$$

where u^0 is the velocity induced by the shape of the real wall calculated from the linearized Euler equations and $\delta u_e(x)$ is the perturbation due to the boundary layer. This perturbation is expressed by a Hilbert integral

$$\delta u_e = \frac{1}{\pi} \!\!\!\fint_{x_a}^{x_b} \frac{v_b}{x - \xi} \, \mathrm{d}\xi \,, \tag{7.38}$$

In (7.38) v_b is the blowing velocity given by (7.19)

$$v_b(\xi) = \frac{\mathrm{d}}{\mathrm{d}\xi} \left[u_e(\xi)\delta_1(\xi) \right] \,,$$

which simulates the boundary layer effect in the domain (x_a, x_b). The Hilbert integral and the boundary layer equations are solved simultaneously with an iterative method [109]. An application is presented in Sect. 9.1.

This method has been extended to the calculation of the flow around wings with compressibility effects [13].

7.4 Conclusion

At high Reynolds number, the study of the flow around a streamlined body benefits from the structure comprising an inviscid region and a boundary layer. Historically, the standard boundary layer theory and the triple deck theory participated in the understanding of interaction between the two regions to a great extent. From a practical point of view, the viscous-inviscid methods are very efficient tools. Different numerical techniques, partially inspired by results obtained by the application of MMAE, have been proposed to answer the needs to calculate flows in strong interaction. In Chap. 8, the successive complementary expansion method, SCEM, is applied to flows at high Reynolds number. Thanks to the use of generalized expansions, it will be shown that the theory of interactive boundary layer, IBL, is fully justified.

Problems

7-1. We consider the Navier-Stokes equations describing the steady flow of an incompressible Newtonian fluid around a semi-infinite flat plate deformed by an indentation whose equation is $y = \varepsilon F(x)$ for $x > 0$. The freestream, of velocity V_∞, is uniform and parallel to the plate. The leading edge is at $x = 0$ and $\varepsilon = R_e^{-\frac{1}{2}}$ is a small parameter with R_e denoting the Reynolds number

$$R_e = \frac{V_\infty L}{\nu},$$

where L is the characteristic length of development of the boundary layer.

Below, all the variables are dimensionless.

For the stream-function ψ, the outer expansion is

$$\psi(x, y, \varepsilon) = \psi_0(x, y) + \delta_1(\varepsilon)\psi_1(x, y) + \cdots.$$

The inner expansion writes

$$\psi(x, y, \varepsilon) = \Delta_0(\varepsilon)\phi_0(x, Y) + \cdots,$$

with

$$Y = \frac{y}{\varepsilon}.$$

Give ψ_0, Δ_0, δ_1 and write the equation for ϕ_0.

We seek a solution for ϕ_0 in the form $\phi_0 = \sqrt{2x}f(\eta)$ with $\eta = \overline{Y}/\sqrt{2x}$ and $\overline{Y} = Y - F(x)$. Write the equation for $f(\eta)$.

It is noted that

$$f(\eta) = \eta - \beta_0 + \text{EST} \quad \text{as} \quad \eta \to \infty,$$

and

$$f(\eta) = \frac{1}{2}\alpha_0\eta^2 + O(\eta^5) \quad \text{as} \quad \eta \to 0.$$

Deduce the equation for the streamline $\psi = 0$ to second order. To write this equation, the matching between the outer and inner expansions will be used.

7-2. *Hiemenz's problem.* We study the boundary layer in the vicinity of a stagnation point of a circular cylinder placed in a uniform freestream [37].

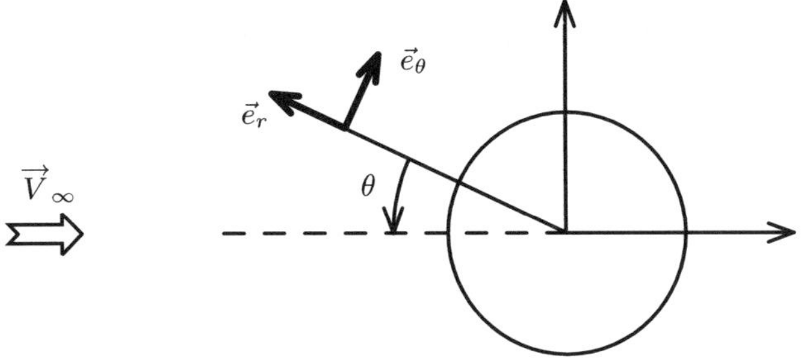

Fig. 7.11. Flow around a circular cylinder

The small parameter ε of the problem is defined by

$$\varepsilon^2 = \frac{1}{Re} \ , \quad Re = \frac{V_\infty a}{\nu} \ ,$$

where a is the radius of the circle and V_∞ is the freestream velocity.

The following dimensionless quantities are used

$$x = \frac{x^*}{a} \ , \quad y = \frac{y^*}{a} \ , \quad u_\theta = \frac{u_\theta^*}{V_\infty} \ , \quad u_r = \frac{u_r^*}{V_\infty} \ , \quad p = \frac{p^*}{\varrho V_\infty^2} \ .$$

In dimensionless form, using polar coordinates (Fig. 7.11), the Navier-Stokes equations write

$$\frac{\partial u_\theta}{\partial \theta} + \frac{\partial}{\partial r}(r u_r) = 0 \ ,$$

$$\frac{u_\theta}{r}\frac{\partial u_\theta}{\partial \theta} + u_r \frac{\partial u_\theta}{\partial r} + \frac{u_\theta u_r}{r} = -\frac{1}{r}\frac{\partial p}{\partial \theta} + \frac{\varepsilon^2}{r}\frac{\partial}{\partial r}\left[\frac{\partial u_r}{\partial \theta} + r\frac{\partial u_\theta}{\partial r} - u_\theta\right]$$

$$+ \frac{\varepsilon^2}{r}\frac{\partial}{\partial \theta}\left[2\left(\frac{1}{r}\frac{\partial u_\theta}{\partial \theta} + \frac{u_r}{r}\right)\right]$$

$$+ \frac{\varepsilon^2}{r}\left(\frac{1}{r}\frac{\partial u_r}{\partial \theta} + \frac{\partial u_\theta}{\partial r} - \frac{u_\theta}{r}\right) \ ,$$

$$\frac{u_\theta}{r}\frac{\partial u_r}{\partial \theta} + u_r \frac{\partial u_r}{\partial r} - \frac{u_\theta^2}{r} = -\frac{\partial p}{\partial r} + \frac{\varepsilon^2}{r}\frac{\partial}{\partial r}\left[2r\frac{\partial u_r}{\partial r}\right]$$

$$+ \frac{\varepsilon^2}{r}\frac{\partial}{\partial \theta}\left[\frac{1}{r}\frac{\partial u_r}{\partial \theta} + \frac{\partial u_\theta}{\partial r} - \frac{u_\theta}{r}\right]$$

$$- 2\frac{\varepsilon^2}{r}\left[\frac{1}{r}\frac{\partial u_\theta}{\partial \theta} + \frac{u_r}{r}\right] \ .$$

MMAE is applied. Check that the first appoximation, solution of the reduced equations, is

$$u_{\theta 1} = \sin\theta \left(1 + \frac{1}{r^2}\right) ,$$

$$u_{r1} = \cos\theta \left(-1 + \frac{1}{r^2}\right) ,$$

$$p_1 = p_\infty + \frac{1}{2}\left[1 - (u_{\theta 1}^2 + u_{r1}^2)\right] .$$

Give the evolution of $u_{\theta 1}$, u_{r1}, p_1 in the neighbourhood of the stagnation point ($\theta = 0$, $r = 1$). Deduce the form of the inner expansion

$$u_\theta = \varepsilon U_{\theta 1}(\Theta, R) + \cdots ,$$
$$u_r = \varepsilon U_{r1}(\Theta, R) + \cdots ,$$
$$p = P_0 + \varepsilon^2 P_1(\Theta, R) + \cdots ,$$

where

$$\Theta = \frac{\theta}{\varepsilon} , \quad R = \frac{r-1}{\varepsilon} .$$

Write the equations for $U_{\theta 1}$, U_{r1}, P_1. Give the boundary conditions and the matching conditions.

We seek the solution in the form

$$U_{\theta 1} = \Theta\varphi'(R) , \quad U_{r1} = -\varphi(R) , \quad P_1 = -2(\Theta^2 + \Phi(R)) .$$

Write the equations for φ and Φ.

Calculate the difference of stagnation pressure between the wall and the boundary layer edge ($R \to \infty$).

7-3. The evolution of a laminar boundary layer is described by the integral method proposed in Subsect. 7.2.1. We assume that the shape factor H of the boundary layer is a known function of x: $H = H(x)$. Write the equations in the form of a system for $\dfrac{\mathrm{d}\theta}{\mathrm{d}x}$ and $\dfrac{\mathrm{d}u_e}{\mathrm{d}x}$. Is the calculation of these derivatives always possible?

7-4. *Goldstein's singularity.* Goldstein [39] studied the structure of the solution of the boundary layer equations downstream of a point x_0 where the velocity profile is prescribed.

All quantities are dimensionless. The reference quantities are ℓ, u_0, ϱ, ν. The Reynolds number is $R = u_0\ell/\nu$. The quantities x, y, u, v and p are dimensionless, the reference quantities being ℓ, $\ell/R^{1/2}$, u_0, $u_0/R^{1/2}$, $\varrho u_0^{1/2}$ respectively.

1. Show that the boundary layer equations write

$$\frac{\partial u}{\partial x} + \frac{\partial v}{\partial y} = 0 ,$$

$$u\frac{\partial u}{\partial x} + v\frac{\partial v}{\partial y} = -\frac{\mathrm{d}p}{\mathrm{d}x} + \frac{\partial^2 u}{\partial y^2} .$$

2. We assume that the velocity profile at x_0 is given by

$$u = a_1 y + a_2 y^2 + a_3 y^3 + \cdots ,$$

where a_1, a_2, ... are functions of x. The profile satisfies the condition $u = 0$ at $y = 0$.

We also assume that the pressure gradient can be written as

$$-\frac{\mathrm{d}p}{\mathrm{d}x} = p_0 + p_1(x - x_0) + p_2(x - x_0)^2 + \cdots ,$$

where p_1, p_2, ... are constants.

From the boundary layer equations, show that the following relations hold

$$2a_2 + p_0 = 0 , \quad a_3 = 0 , \quad a_1\frac{\mathrm{d}a_1}{\mathrm{d}x} - 24a_4 = 0 , \quad \frac{2}{3}a_1\frac{\mathrm{d}a_2}{\mathrm{d}x} - 20a_5 = 0 .$$

3. By differentiating the boundary layer equations with respect to x show that

$$2\frac{\mathrm{d}a_2}{\mathrm{d}x} + p_1 = 0 , \quad \frac{\mathrm{d}a_3}{\mathrm{d}x} = 0 .$$

4. Show that

$$2a_2 + p_0 = 0 , \quad a_3 = 0 , \quad 5!\,a_5 + 2a_1 p_1 = 0 .$$

Coefficients a_1, a_4, ... are free. Note that the pressure gradient and therefore the coefficients p_i are prescribed.

The above conditions are called compatibility conditions. If they are not satisfied, singularities occur when solving the boundary layer equations downstream of point x_0. A particular case occurs if $a_1 = 0$, which corresponds to boundary layer separation. Then, show that the compatibility conditions are

$$2a_2 + p_0 = 0 , \quad a_3 = 0 , \quad a_4 = 0 , \quad a_5 = 0 ,$$

$$6!\,a_6 = 2p_0 p_1 , \quad a_7 = 0 .$$

Coefficients a_8, a_{12}, a_{16}, a_{20}, ... are free.

In general, all the compatibility conditions are not satisfied. Goldstein assumes that the condition $2a_2 + p_0 = 0$ is satisfied but not the other conditions. Show that

$$a_1 = \sqrt{48a_4(x - x_0)} ,$$

and deduce that the derivative $\dfrac{\mathrm{d}a_1}{\mathrm{d}x}$ is infinite at point $x = x_0$. If the solution exists upstream of point x_0, coefficient a_4 must be negative. Then, the solution is impossible downstream of point x_0. This behaviour is known as Goldstein's singularity. Goldstein confirmed this conclusion by studying in detail the structure of the solution in the neighbourhood of the separation point.

7-5. *Goldstein's wake.* Goldstein [39] studied the structure of the solution of boundary layer equations, with a prescribed pressure distribution, when the velocity profile is given at point $x = x_0$.

The following dimensionless quantities are used

$$u = \frac{u^*}{u_0} \, , \quad v = \frac{v^*}{u_0} R^{1/2} \, , \quad x = \frac{x^*}{\ell} \, , \quad y = \frac{y^*}{\ell} R^{1/2} \, , \quad p = \frac{p^*}{\varrho u_0^2} \, ,$$

and the Reynolds number is

$$R = \frac{u_0 \ell}{\nu} \, .$$

In these relations, u_0 is a reference velocity and ℓ is a reference length.

In dimensionless form, the boundary layer equations are

$$\frac{\partial u}{\partial x} + \frac{\partial v}{\partial y} = 0 \, ,$$

$$u \frac{\partial u}{\partial x} + v \frac{\partial u}{\partial y} = -\frac{\mathrm{d}p}{\mathrm{d}x} + \frac{\partial^2 u}{\partial y^2} \, .$$

At point of abscissa x_0, the velocity profile is given by

$$u(x_0, y) = a_0 + a_1 y + a_2 y^2 + a_3 y^3 + \cdots \, .$$

The case $a_0 = 0$, $a_1 \neq 0$ is studied here and we consider the formation of a symmetric wake downstream of a flat plate. The singularity is due to a change of boundary conditions at $y = 0$ because, downstream of point x_0, we must have

$$y = 0 : \quad v = 0 \, , \quad \frac{\partial u}{\partial y} = 0 \, .$$

The pressure gradient is prescribed as

$$-\frac{\mathrm{d}p}{\mathrm{d}x} = p_0 + p_1(x - x_0) + p_2(x - x_0)^2 + \cdots \, .$$

MMAE is applied. The proposed structure comprises two layers: an outer layer in which the appropriate variables are

$$\xi = (x - x_0)^{1/n} \, , \quad y \, ,$$

and an inner layer in which the appropriate variables are

$$\xi = (x - x_0)^{1/n} \, , \quad \eta = \frac{y}{n(x - x_0)^{1/n}} \, .$$

The solution is studied in the downstream neighbourhood of point x_0. The small parameter of the problem is ξ.

The case $n = 1$, leading to a regular solution, is discarded. We assume that $n > 1$.

We seek an outer expansion in the form

$$u = F_0'(y) + \xi F_1'(y) + \xi^2 F_2'(y) + \cdots \, .$$

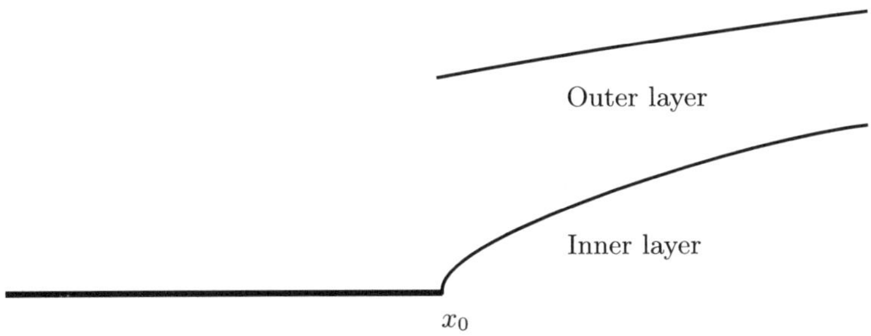

Fig. 7.12. Formation of a wake

At $\xi = 0$, we must have $u = u(x_0, y)$, hence

$$F_0' = a_0 + a_1 y + a_2 y^2 + \cdots .$$

Express v from the continuity equation.

From the momentum equation, give the equation for F_1. Show that the solution has the form

$$F_1 = k F_0' .$$

The impossibility to satisfy all the boundary conditions, due to the absence of viscous term, leads us to introduce the inner layer. We seek the solution in the form

$$u = f_0'(\eta) + \xi f_1'(\eta) + \xi^2 f_2'(\eta) + \cdots .$$

Express v from the continuity equation.

From the momentum equation, give the value of n. Give the equations for f_0' et f_1' and give the boundary conditions. Write the matching on the velocity u between the outer and inner layers. Show that $f_0 = 0$ and $F_1 = 0$. The following result will be used

$$f_1 \underset{\eta \to \infty}{\cong} \alpha \eta^2 + \text{EST} .$$

Note 7.8. In a very small neighbourhood of the trailing edge, it is necessary to refine Goldstein's solution by considering a triple deck structure (see Subsect. 9.2.1). Goldstein's solution is correct outside this neighbourhood and is used as a boundary condition for the structure closer to the trailing edge.

7-6. *Flat plate leading edge.* Goldstein [39] studied the structure of the solution of boundary layer equations, with a prescribed pressure distribution, when the velocity profile is given at point $x = x_0$.

The following dimensionless quantities are used

$$u = \frac{u^*}{u_0} , \quad v = \frac{v^*}{u_0} R^{1/2} , \quad x = \frac{x^*}{\ell} , \quad y = \frac{y^*}{\ell} R^{1/2} , \quad p = \frac{p^*}{\varrho u_0^2} ,$$

and the Reynolds number is

$$R = \frac{u_0 \ell}{\nu} \ .$$

In these relations, u_0 is a reference velocity and ℓ is a reference length.

In dimensionless form, the boundary layer equations are

$$\frac{\partial u}{\partial x} + \frac{\partial v}{\partial y} = 0 \ ,$$

$$u\frac{\partial u}{\partial x} + v\frac{\partial u}{\partial y} = -\frac{\mathrm{d}p}{\mathrm{d}x} + \frac{\partial^2 u}{\partial y^2} \ .$$

At point of abscissa x_0, the velocity profile is given by

$$u(x_0, y) = a_0 + a_1 y + a_2 y^2 + a_3 y^3 + \cdots \ .$$

The case $a_0 \neq 0$ is studied here. This represents, for example, the formation of the boundary layer in the neighbourhood of a flat plate leading edge. The singularity is due to the change of boundary conditions at $y = 0$ because, downstream of point x_0, we must have

$$y = 0 : \quad u = 0 \ , \quad v = 0 \ .$$

The pressure gradient is prescribed as

$$-\frac{\mathrm{d}p}{\mathrm{d}x} = p_0 + p_1(x - x_0) + p_2(x - x_0)^2 + \cdots \ .$$

MMAE is applied. The proposed structure comprises two layer: an outer layer in which the appropriate variables are

$$\xi = (x - x_0)^{1/n} \ , \quad y \ ,$$

and an inner layer in which the appropriate variables are

$$\xi = (x - x_0)^{1/n} \ , \quad \eta = \frac{y}{n(x - x_0)^{1/n}} \ .$$

The solution is studied in the downstream neighbourhood of point x_0. The small parameter of the problem is ξ.

The case $n = 1$, leading to a regular solution, is discarded. We assume that $n > 1$.

We seek an outer expansion in the form

$$u = F_0'(y) + \xi F_1'(y) + \xi^2 F_2'(y) + \cdots \ .$$

At $\xi = 0$, we must have $u = u(x_0, y)$, hence

$$F_0' = a_0 + a_1 y + a_2 y^2 + \cdots \ .$$

Express v from the continuity equation.

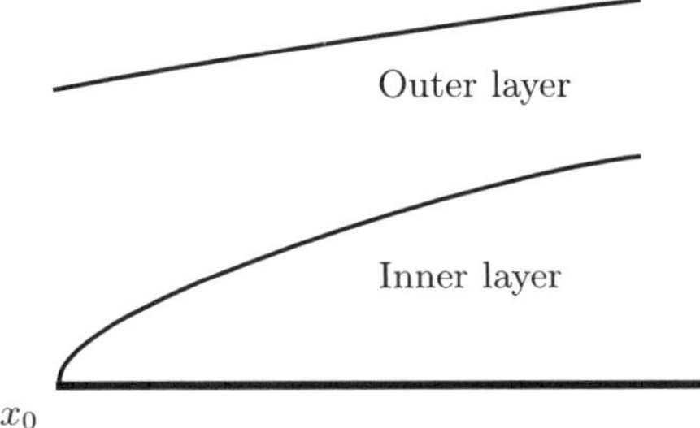

Fig. 7.13. Formation of a boundary layer

From the momentum equation, give the equation for F_1. Show that the solution has the form

$$F_1 = kF_0' \, .$$

The impossibility to satisfy all the boundary conditions, due to the absence of viscous term, leads us to introduce the inner layer. We seek the solution in the form

$$u = f_0'(\eta) + \xi f_1'(\eta) + \xi^2 f_2'(\eta) + \cdots \, .$$

Express v from the continuity equation.

From the momentum equation, give the value of n. Give the equations for f_0' and f_1' and give the boundary conditions. Do not seek to solve analytically the equations for f_0' and f_1'. It is known that as $\eta \to \infty$, we have

$$f_0 \underset{\eta \to \infty}{\cong} A_0 \eta + B_0 + \mathrm{EST} \, ,$$

with $B_0 = -0.86 A_0^{1/2}$. Show that the behaviour of f_1 is

$$f_1 \underset{\eta \to \infty}{\cong} A_1 \eta^2 + B_1 \eta + C_1 + \cdots \, .$$

Express A_1 and B_1.

Write the matching on the velocity u between the outer and inner layers up to order ξ. Express A_0 and k as function of a_0.

Note 7.9. In a very small neighbourhood of the leading edge, it is necessary to refine Goldstein's solution by considering the Navier-Stokes equations.

7-7. *Neighbourhood of separation.* Goldstein [40] studied the structure of the solution of boundary layer equations, with a prescribed pressure, when we give the velocity profile at point $x = x_0$ where the boundary layer separates. The solution is studied here upstream of the separation point.

The following dimensionless variables are defined by

$$x = \frac{x_0^* - x^*}{\ell} \, , \quad y = \frac{y^*}{\ell} R^{1/2} \, , \quad u = \frac{u^*}{u_{e0}^*} \, , \quad v = \frac{v^*}{u_{e0}^*} R^{1/2} \, ,$$

$$p = \frac{p^*}{\varrho u_{e0}^2} \ , \quad \psi = \frac{\psi^*}{u_{e0}^* \ell} R^{1/2} \ .$$

The velocity u_{e0}^* is the velocity at the edge of the boundary layer at point x_0^*. The stream-function ψ is such that

$$u = \frac{\partial \psi}{\partial y} \ , \quad v = \frac{\partial \psi}{\partial x} \ .$$

The reference length ℓ and the Reynolds number are defined by

$$\ell = -\frac{u_{e0}^*}{\left(\dfrac{\mathrm{d}u_e^*}{\mathrm{d}x^*}\right)_{x_0^*}} \ , \quad R = \frac{u_{e0}^* \ell}{\nu} \ .$$

The boundary layer momentum equation becomes

$$-u\frac{\partial u}{\partial x} + v\frac{\partial u}{\partial y} = \frac{\mathrm{d}p}{\mathrm{d}x} + \frac{\partial^2 u}{\partial y^2} \ .$$

The pressure gradient is prescribed as

$$\frac{\mathrm{d}p}{\mathrm{d}x} = -(1 + p_1 x + p_2 x^2 + \cdots) \ .$$

At $x = 0$, the velocity profile is given by

$$u(0, y) = a_2 y^2 + a_3 y^3 + \cdots \ .$$

MMAE is applied. Two regions are identified. In the inner region, the variables are

$$\xi = x^{1/n} \ , \quad \eta = \frac{y}{2^{1/2} x^{1/n}} \ .$$

In the outer region, the variables are

$$\xi = x^{1/n} \ , \quad y \ .$$

We assume that the outer and inner expansions are respectively

$$u = 2(f_0'(\eta) + \xi f_1'(\eta) + \xi^2 f_2'(\eta) + \cdots) \ ,$$
$$u = \chi_0'(y) + \xi \chi_1'(y) + \xi^2 \chi_2'(y) + \cdots \ .$$

The small parameter of the problem is ξ. We assume that χ_i' can be expanded in Taylor series in the neighbourhood of $y = 0$. Show that

$$\lim_{\eta \to \infty} \frac{f_r'}{\eta^r} = \frac{a_r}{2} 2^{r/2} \ .$$

We study the inner expansion. The stream-function has the expansion

$$\psi = 2^{3/2}(\xi f_0 + \xi^2 f_1 + \xi^3 f_2 + \xi^4 f_3 + \xi^5 f_4 + \cdots) \,.$$

With $a_0 = 0$ and $a_1 = 0$, the previous conditions imply that $f_0 = 0$ and $f_1 = 0$. Show that the balance between viscous and convection terms leads us to take $n = 4$. Give the equations for f_2, f_3 et f_4. Give the boundary conditions at $y = 0$. Check that the solutions for f_2, f_3 and f_4 are

$$f_2 = \frac{\eta^3}{6} \,,$$
$$f_3 = \alpha_1 \eta^2 \,,$$
$$f_4 = \alpha_2 \eta^2 - \frac{\alpha_1^2}{15}\eta^5 \,,$$

where α_1 and α_2 are constants yet undetermined.

Show that we must have $a_2 = 1/2$ and $a_3 = 0$.

Show that α_1 is given by

$$a_4 = -\frac{\alpha_1^2}{6} \,.$$

Now, we study the outer expansion. The stream-function has the expansion

$$\psi = \chi_0 + \xi\chi_1 + \xi^2\chi_2 + \xi^3\chi_3 + \cdots \,.$$

Show that

$$\chi_1 = 0 \,,$$
$$\chi_2 = 2^{3/2}\alpha_1\chi_0' \,,$$
$$\chi_3 = 2^{3/2}\alpha_2\chi_0' \,,$$

with

$$\chi_0' = \frac{y^2}{2} + a_4 y^4 + \cdots \,.$$

From the inner expansion, calculate $\left(\dfrac{\partial u}{\partial y}\right)_{y=0}$. From the outer expansion, show that $v \to \infty$ and $\dfrac{\partial u}{\partial x} \to \infty$ as $\xi \to 0$.

Note 7.10. The study of the solution downstream of point x_0^* show that a_4 has the form $a_4 = \beta_1^2/6$. This solution is compatible with the solution upstream of point x_0^* only if $a_4 = 0$. For a given distribution of the external velocity, this condition is not satisfied in general. This condition can be satisfied only for a particular distribution of the external velocity. Catherall and Mangler [12] proposed to use an inverse method to produce a regular solution.

8 Interactive Boundary Layer

The method of matched asymptotic expansions, MMAE, has been extensively used in fluid mechanics and contributed to the remarkable advances in the description of flows [107]. The study of high Reynolds number flows past streamlined bodies is one of the most famous example.

The boundary layer theory emerged at first from the very inspired ideas of Prandtl. Much later, a sound mathematical basis has been introduced with the formalism of MMAE. Thanks to this mathematical tool, Van Dyke [105] also proposed an improvement with a second-order boundary layer theory which takes into account various effects, for example the influence of the streamwise or transverse wall curvature, the influence of external vorticity or the influence of a gradient of stagnation enthalpy. A few years later, a breakthrough in the understanding of interactions between the inviscid flow and the boundary layer occurred with the triple deck theory whose consistency is fully based on the use of MMAE [69, 74, 100].

The viscous-inviscid interaction at high Reynolds number is analyzed here by using the successive complementary expansion method, SCEM. As for ordinary differential equations, the principle of SCEM is to seek a UVA, an approximation which is uniformly valid in the whole flow field. Moreover, the introduction of *generalized expansions* proved to be very fruitful. The first step consists of the inviscid flow approximation which applies far from the walls. Obviously, this approximation must be improved near the walls by adding a correction which takes into account the effects of viscosity. Thanks to generalized expansions, a strong coupling occurs between the viscous and inviscid regions. The hierarchy obtained with MMAE is broken, which is a major difference between MMAE results and generalized SCEM results. This notion is called "interactive boundary layer", IBL. This means that the effect of the boundary layer on the inviscid flow and the reciprocal effect are considered simultaneously. The construction of UVAs does not require any matching principle, only the boundary conditions of the problem are applied.

The principle of a strong interaction between the inviscid flow and the boundary layer is known for a long time and has been implemented in coupling – or interactive – methods [9, 13, 53, 109, 111]. However, Sychev et al. [102], in commenting these methods, noted that: "No rational mathematical arguments (based, say, on asymptotic analysis of the Navier-Stokes equations) have been given to support the model approach". The objective of this chapter is precisely to lay the theoretical foundations of the IBL methods thanks to SCEM.

8.1 Application of SCEM

We consider a flow at high Reynolds number past a flat wall. The flow is laminar, incompressible, two-dimensional and steady. The dimensionless Navier-Stokes equations are (Appendix I)

$$\frac{\partial \mathcal{U}}{\partial x} + \frac{\partial \mathcal{V}}{\partial y} = 0 \,, \tag{8.1a}$$

$$\mathcal{U}\frac{\partial \mathcal{U}}{\partial x} + \mathcal{V}\frac{\partial \mathcal{U}}{\partial y} = -\frac{\partial \mathcal{P}}{\partial x} + \varepsilon^2 \left(\frac{\partial^2 \mathcal{U}}{\partial x^2} + \frac{\partial^2 \mathcal{U}}{\partial y^2} \right) \,, \tag{8.1b}$$

$$\mathcal{U}\frac{\partial \mathcal{V}}{\partial x} + \mathcal{V}\frac{\partial \mathcal{V}}{\partial y} = -\frac{\partial \mathcal{P}}{\partial y} + \varepsilon^2 \left(\frac{\partial^2 \mathcal{V}}{\partial x^2} + \frac{\partial^2 \mathcal{V}}{\partial y^2} \right) \,, \tag{8.1c}$$

with

$$\varepsilon^2 = \frac{1}{\mathcal{R}} = \frac{\mu}{\varrho V L} \,, \tag{8.2}$$

where the Reynolds number $\mathcal{R}$ is based on the reference velocity V and on the reference length L. All the variables are reduced with these reference quantities. The coordinate along the wall is x and the coordinate normal to the wall is y; the velocity components along x and y are $\mathcal{U}$ and $\mathcal{V}$, respectively; the pressure is $\mathcal{P}$.

For the needs of SCEM, the momentum equations are written by putting all the terms on the left hand side and, symbolically, (8.1b) and (8.1c) become

$$\mathrm{L}_\varepsilon \,\mathcal{U} = 0 \,,$$
$$\mathrm{L}_\varepsilon \,\mathcal{V} = 0 \,.$$

It is understood that, for any UVA, the continuity equation is identically satisfied.

8.1.1 Outer Approximation

At first, we seek an outer approximation with a *generalized expansion* beginning with the terms

$$\mathcal{U} = u_1(x, y, \varepsilon) + \cdots \,,$$
$$\mathcal{V} = v_1(x, y, \varepsilon) + \cdots \,,$$
$$\mathcal{P} = p_1(x, y, \varepsilon) + \cdots \,.$$

By neglecting $O(\varepsilon^2)$ terms, the Navier-Stokes equations reduce to the Euler equations,

$$\frac{\partial u_1}{\partial x} + \frac{\partial v_1}{\partial y} = 0 \, , \tag{8.3a}$$

$$u_1 \frac{\partial u_1}{\partial x} + v_1 \frac{\partial u_1}{\partial y} = -\frac{\partial p_1}{\partial x} \, , \tag{8.3b}$$

$$u_1 \frac{\partial v_1}{\partial x} + v_1 \frac{\partial v_1}{\partial y} = -\frac{\partial p_1}{\partial y} \, . \tag{8.3c}$$

The solution of these equations requires boundary conditions. At infinity, the most common condition is to prescribe a uniform flow. If the oncoming flow is rotational, the conditions must be examined specifically for each particular case. Along the walls, boundary conditions are also required but, at this stage of the discussion, it is not possible to specify them. It is known only that the no-slip conditions cannot be imposed and the approximation already obtained needs refinement.

8.1.2 Determination of a Uniformly Valid Approximation

The application of SCEM consists of adding a *correction* to the outer approximation (Fig. 8.1)

$$\mathcal{U} = u_1(x, y, \varepsilon) + U_1(x, Y, \varepsilon) + \cdots \, , \tag{8.4a}$$

$$\mathcal{V} = v_1(x, y, \varepsilon) + \varepsilon V_1(x, Y, \varepsilon) + \cdots \, , \tag{8.4b}$$

$$\mathcal{P} = p_1(x, y, \varepsilon) + \Delta(\varepsilon) P_1(x, Y, \varepsilon) + \cdots \, , \tag{8.4c}$$

where Δ is a gauge function yet undetermined and Y is the boundary layer variable

$$Y = \frac{y}{\varepsilon} \, . \tag{8.5}$$

The term εV_1 in the expansion of $\mathcal{V}$ is justified by the continuity equation which must be non trivial, and then, the terms involving derivatives with respect to the streamwise and transverse variables must be of the same order of magnitude. The form of the expansion for pressure $\mathcal{P}$ is discussed later.

Note 8.1. The idea to add a correction to the outer approximation meets the concepts of corrective boundary layer and of defect formulation mentioned on page 79.

With expansions given by (8.4a–8.4c) and using (8.3a–8.3c), the Navier-Stokes equations become

$$\frac{\partial U_1}{\partial x} + \frac{\partial \varepsilon V_1}{\partial y} = 0 \, , \tag{8.6a}$$

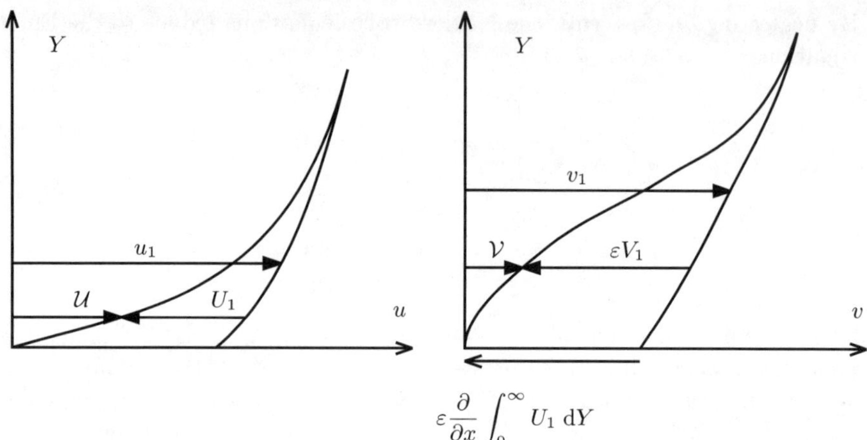

$$\varepsilon \frac{\partial}{\partial x} \int_0^\infty U_1 \, \mathrm{d}Y$$

Fig. 8.1. Velocity components in the boundary layer

$$U_1 \frac{\partial}{\partial x} (u_1 + U_1) + u_1 \frac{\partial U_1}{\partial x} + \varepsilon V_1 \frac{\partial}{\partial y} (u_1 + U_1) + v_1 \frac{\partial U_1}{\partial y}$$

$$= -\frac{\partial \Delta P_1}{\partial x} + \varepsilon^2 \frac{\partial^2}{\partial x^2} (u_1 + U_1) + \varepsilon^2 \frac{\partial^2}{\partial y^2} (u_1 + U_1) \ , \qquad (8.6\mathrm{b})$$

$$U_1 \frac{\partial}{\partial x} (v_1 + \varepsilon V_1) + u_1 \frac{\partial \varepsilon V_1}{\partial x} + \varepsilon V_1 \frac{\partial}{\partial y} (v_1 + \varepsilon V_1) + v_1 \frac{\partial \varepsilon V_1}{\partial y}$$

$$= -\frac{\partial \Delta P_1}{\partial y} + \varepsilon^2 \frac{\partial^2}{\partial x^2} (v_1 + \varepsilon V_1) + \varepsilon^2 \frac{\partial^2}{\partial y^2} (v_1 + \varepsilon V_1) \ . \qquad (8.6\mathrm{c})$$

The simplification of these equations leads to the *first and second order IBL models* discussed below [25]. To achieve this goal, the derivatives with respect to y must be evaluated carefully. For example, the diffusion terms in the y-direction become

$$\varepsilon^2 \frac{\partial^2}{\partial y^2} (u_1 + U_1) = \varepsilon^2 \frac{\partial^2 u_1}{\partial y^2} + \frac{\partial^2 U_1}{\partial Y^2} \ ,$$

$$\varepsilon^2 \frac{\partial^2}{\partial y^2} (v_1 + \varepsilon V_1) = \varepsilon^2 \frac{\partial^2 v_1}{\partial y^2} + \varepsilon \frac{\partial^2 V_1}{\partial Y^2} \ ,$$

and the derivative of the pressure is

$$\frac{\partial \Delta P_1}{\partial y} = \frac{\Delta}{\varepsilon} \frac{\partial P_1}{\partial Y} \ .$$

Thus, (8.6a–8.6c) become

$$\frac{\partial U_1}{\partial x} + \frac{\partial V_1}{\partial Y} = 0 \ , \qquad (8.7\mathrm{a})$$

$$U_1\frac{\partial u_1}{\partial x} + U_1\frac{\partial U_1}{\partial x} + u_1\frac{\partial U_1}{\partial x} + \varepsilon V_1\frac{\partial u_1}{\partial y} + V_1\frac{\partial U_1}{\partial Y} + \frac{v_1}{\varepsilon}\frac{\partial U_1}{\partial Y}$$

$$= -\Delta\frac{\partial P_1}{\partial x} + \varepsilon^2\frac{\partial^2 u_1}{\partial x^2} + \varepsilon^2\frac{\partial^2 U_1}{\partial x^2} + \varepsilon^2\frac{\partial^2 u_1}{\partial y^2} + \frac{\partial^2 U_1}{\partial Y^2}\,, \qquad (8.7\text{b})$$

$$U_1\frac{\partial v_1}{\partial x} + \varepsilon U_1\frac{\partial V_1}{\partial x} + \varepsilon u_1\frac{\partial V_1}{\partial x} + \varepsilon V_1\frac{\partial v_1}{\partial y} + \varepsilon V_1\frac{\partial V_1}{\partial Y} + v_1\frac{\partial V_1}{\partial Y}$$

$$= -\frac{\Delta}{\varepsilon}\frac{\partial P_1}{\partial Y} + \varepsilon^2\frac{\partial^2 v_1}{\partial x^2} + \varepsilon^3\frac{\partial^2 V_1}{\partial x^2} + \varepsilon^2\frac{\partial^2 v_1}{\partial y^2} + \varepsilon\frac{\partial^2 V_1}{\partial Y^2}\,. \qquad (8.7\text{c})$$

8.1.3 Gauge for the Pressure

At the wall, v_1 is equal to $-\varepsilon V_1$, so that, in the boundary layer, v_1 can be considered formally as being of order ε (Fig. 8.1). By neglecting $\mathrm{O}(\varepsilon^2)$ terms, the y-momentum equation (8.7c) becomes

$$U_1\frac{\partial v_1}{\partial x} + \varepsilon U_1\frac{\partial V_1}{\partial x} + \varepsilon u_1\frac{\partial V_1}{\partial x} + \varepsilon V_1\frac{\partial V_1}{\partial Y} + v_1\frac{\partial V_1}{\partial Y} = -\frac{\Delta}{\varepsilon}\frac{\partial P_1}{\partial Y} + \varepsilon\frac{\partial^2 V_1}{\partial Y^2}\,. \quad (8.8)$$

This equation shows that we must take $\Delta = \mathrm{O_S}(\varepsilon^2)$. Indeed, if we take $\Delta \succ \varepsilon^2$, we have a result without interest $\dfrac{\partial P_1}{\partial Y} = 0$; if we take $\Delta \prec \varepsilon^2$, the resulting equation can not be verified because this equation is formed with terms coming from an independent set of equations. Thus, we choose $\Delta = \varepsilon^2$.

8.2 First Order Interactive Boundary Layer

8.2.1 Generalized Boundary Layer Equations

In the x-momentum equation (8.7b), $\mathrm{O}(\varepsilon)$ terms are neglected. Then, to first order, the generalized boundary layer equations are

$$\frac{\partial U_1}{\partial x} + \frac{\partial V_1}{\partial Y} = 0\,, \qquad (8.9\text{a})$$

$$U_1\frac{\partial u_1}{\partial x} + U_1\frac{\partial U_1}{\partial x} + u_1\frac{\partial U_1}{\partial x} + V_1\frac{\partial U_1}{\partial Y} + \frac{v_1}{\varepsilon}\frac{\partial U_1}{\partial Y} = \frac{\partial^2 U_1}{\partial Y^2}\,, \qquad (8.9\text{b})$$

and the y-momentum equation enables us to calculate $\dfrac{\partial P_1}{\partial Y}$ from

$$\frac{1}{\varepsilon}U_1\frac{\partial v_1}{\partial x} + U_1\frac{\partial V_1}{\partial x} + u_1\frac{\partial V_1}{\partial x} + V_1\frac{\partial V_1}{\partial Y} + \frac{v_1}{\varepsilon}\frac{\partial V_1}{\partial Y} = -\frac{\partial P_1}{\partial Y} + \frac{\partial^2 V_1}{\partial Y^2}\,. \quad (8.9\text{c})$$

The above equations are rewritten in a form which is closer to the usual form by setting

$$u = u_1 + U_1 \,, \tag{8.10a}$$

$$v = v_1 + \varepsilon V_1 \,, \tag{8.10b}$$

$$p = p_1 + \varepsilon^2 P_1 \quad \text{or} \quad \frac{\partial p}{\partial y} = \frac{\partial p_1}{\partial y} + \varepsilon \frac{\partial P_1}{\partial Y} \,. \tag{8.10c}$$

Using the Euler equations for u_1, v_1 and p_1, (8.9a, 8.9b) become

$$\left. \begin{aligned} & \frac{\partial u}{\partial x} + \frac{\partial v}{\partial y} = 0 \\ & u\frac{\partial u}{\partial x} + v\frac{\partial u}{\partial y} - v\frac{\partial u_1}{\partial y} = u_1\frac{\partial u_1}{\partial x} + \frac{1}{\mathcal{R}}\frac{\partial^2 (u - u_1)}{\partial y^2} \end{aligned} \right\} \,, \tag{8.11}$$

and the y-momentum equation (8.9c) becomes

$$u\frac{\partial v}{\partial x} + v\frac{\partial v}{\partial y} + (v_1 - v)\frac{\partial v_1}{\partial y} = -\frac{\partial p}{\partial y} + \frac{1}{\mathcal{R}}\frac{\partial^2 (v - v_1)}{\partial y^2} \,. \tag{8.12}$$

Equations (8.11) must be solved in association with the Euler equations for u_1, v_1 and p_1. The solution gives a *UVA over the whole domain* and not only in the boundary layer

8.2.2 Boundary Conditions

The boundary conditions are

$$\begin{aligned} \text{at the wall}: &\quad U_1 + u_1 = 0 \,, \quad \varepsilon V_1 + v_1 = 0 \,, \\ \text{at infinity}: &\quad U_1 = 0 \qquad\quad , \quad V_1 = 0 \,, \end{aligned} \tag{8.13}$$

or

$$\begin{aligned} \text{at the wall}: &\quad u = 0 \qquad\quad , \quad v = 0 \,, \\ y \to \infty \quad : &\quad u - u_1 \to 0 \,, \quad v - v_1 \to 0 \,. \end{aligned} \tag{8.14}$$

Conditions at infinity are also imposed on the Euler equations.

The condition $v - v_1 \to 0$ as $y \to \infty$ requires that the generalized boundary layer equations (8.11) and the Euler equations (8.3a–8.3c) must be solved *simultaneously*. It is not possible to solve the Euler equations independently from the boundary layer equations because the condition that the normal velocity is zero at the wall does not apply to the Euler equations. *The two systems of equations interact; one system influences the other system and vice versa.*

In the triple deck theory, the coupling between the different decks also originates from the conditions imposed to the velocity normal to the wall. In particular, the identity of gauges for this velocity component between the upper deck and the main deck is essential to prevent the hierarchy between the decks from occurring (Subsect. 7.1.2). This property establishes a close relationship between the two theories.

The idea of IBL is not new and different forms have been used [9, 13, 53, 109, 111]. Until now, the justifications rested on the analysis of the inverse mode for example (Subsect. 7.2.3) or on the triple deck theory. Here, the IBL concept is fully justified thanks to the use of *generalized expansions* in SCEM.

8.2.3 Estimate of the Remainders of Equations

The remainders of the Navier-Stokes equations are

$$
\mathrm{L}_\varepsilon\, u = \varepsilon V_1 \frac{\partial u_1}{\partial y} - \varepsilon^2 \left[-\frac{\partial P_1}{\partial x} + \frac{\partial^2 u_1}{\partial x^2} + \frac{\partial^2 u_1}{\partial y^2} + \frac{\partial^2 U_1}{\partial x^2} \right],
$$

$$
\mathrm{L}_\varepsilon\, v = \varepsilon V_1 \frac{\partial v_1}{\partial y} - \varepsilon^2 \left[\frac{\partial^2 v_1}{\partial x^2} + \frac{\partial^2 v_1}{\partial y^2} + \varepsilon \frac{\partial^2 V_1}{\partial x^2} \right].
$$

Taking into account that the boundary conditions are exactly satisfied, if these remainders were zero, we would have the exact solution. Obviously, the remainders are not zero but uniformly small.

8.3 Second Order Interactive Boundary Layer

8.3.1 Generalized Boundary Layer Equations

In order to construct a second order model, $\mathrm{O}(\varepsilon^2)$ terms are neglected in the x-momentum equation (8.7b). The second order generalized boundary layer equations are

$$
\frac{\partial U_1}{\partial x} + \frac{\partial V_1}{\partial Y} = 0, \qquad (8.15\mathrm{a})
$$

$$
U_1 \frac{\partial u_1}{\partial x} + U_1 \frac{\partial U_1}{\partial x} + u_1 \frac{\partial U_1}{\partial x} + \varepsilon V_1 \frac{\partial u_1}{\partial y} + V_1 \frac{\partial U_1}{\partial Y} + \frac{v_1}{\varepsilon} \frac{\partial U_1}{\partial Y} = \frac{\partial^2 U_1}{\partial Y^2}, \quad (8.15\mathrm{b})
$$

and, neglecting $\mathrm{O}(\varepsilon^2)$ terms in the y-momentum equation (8.7c), we obtain

$$
\frac{1}{\varepsilon} U_1 \frac{\partial v_1}{\partial x} + U_1 \frac{\partial V_1}{\partial x} + u_1 \frac{\partial V_1}{\partial x} + V_1 \frac{\partial v_1}{\partial y} + V_1 \frac{\partial V_1}{\partial Y} + \frac{v_1}{\varepsilon} \frac{\partial V_1}{\partial Y}
$$

$$
= -\frac{\partial P_1}{\partial Y} + \frac{\partial^2 V_1}{\partial Y^2}. \qquad (8.15\mathrm{c})
$$

Note 8.2. In (8.15b), the pressure P_1 is absent. To the considered order, the corresponding term is negligible so that everything is as if, in the boundary layer, the pressure were equal to the pressure p_1 solution of Euler equations. The same remark applies to the first order IBL model.

If we set

$$u = u_1 + U_1 \,, \tag{8.16a}$$

$$v = v_1 + \varepsilon V_1 \,, \tag{8.16b}$$

$$p = p_1 + \varepsilon^2 P_1 \quad \text{or} \quad \frac{\partial p}{\partial y} = \frac{\partial p_1}{\partial y} + \varepsilon \frac{\partial P_1}{\partial Y} \,, \tag{8.16c}$$

the boundary layer equations (8.15a, 8.15b) can also be written as

$$\left. \begin{aligned} &\frac{\partial u}{\partial x} + \frac{\partial v}{\partial y} = 0 \\ &u\frac{\partial u}{\partial x} + v\frac{\partial u}{\partial y} = u_1\frac{\partial u_1}{\partial x} + v_1\frac{\partial u_1}{\partial y} + \frac{1}{\mathcal{R}}\frac{\partial^2 (u - u_1)}{\partial y^2} \end{aligned} \right\} \,, \tag{8.17}$$

and the y-momentum equation (8.15c) becomes

$$u\frac{\partial v}{\partial x} + v\frac{\partial v}{\partial y} = -\frac{\partial p}{\partial y} + \frac{1}{\mathcal{R}}\frac{\partial^2 (v - v_1)}{\partial y^2} \,. \tag{8.18}$$

Equations (8.17) are coupled to the Euler equations. It is not possible to solve one system independently from the other. As in the first order model, the solution gives a *UVA in the whole flow field*.

Note 8.3. Equations (8.17), proposed by DeJarnette and Radcliffe [29] (see also [31]) on a heuristic basis, are fully justified here. Again, let us note that this model results from the use of generalized expansions.

8.3.2 Boundary Conditions

The boundary conditions are the same as for the first order model

$$\left. \begin{aligned} &\text{at the wall}: &u = 0 &\quad, &v = 0 \\ &y \to \infty &: \quad u - u_1 \to 0 \,, &v - v_1 \to 0 \end{aligned} \right\} \,. \tag{8.19}$$

Conditions at infinity are also imposed on the field described by the Euler equations.

We note that the conditions $u - u_1 \to 0$ and $v - v_1 \to 0$ as $y \to \infty$ enable the x-momentum equation to be identically satisfied beyond the boundary layer edge.

8.3.3 Estimate of the Remainders of Equations

Obviously, the remainders in the Navier-Stokes equations are smaller than in the first order model. We have

$$\mathrm{L}_\varepsilon\, u = -\varepsilon^2 \left[-\frac{\partial P_1}{\partial x} + \frac{\partial^2 u_1}{\partial x^2} + \frac{\partial^2 u_1}{\partial y^2} + \frac{\partial^2 U_1}{\partial x^2} \right] \,,$$

$$\mathrm{L}_\varepsilon\, v = -\varepsilon^2 \left[\frac{\partial^2 v_1}{\partial x^2} + \frac{\partial^2 v_1}{\partial y^2} + \varepsilon \frac{\partial^2 V_1}{\partial x^2} \right] \,.$$

If the external flow is irrotational, we observe that a better accuracy is expected since

$$\mathrm{L}_\varepsilon\, u = -\varepsilon^2 \left[-\frac{\partial P_1}{\partial x} + \frac{\partial^2 U_1}{\partial x^2} \right] \,,$$

$$\mathrm{L}_\varepsilon\, v = -\varepsilon^3 \frac{\partial^2 V_1}{\partial x^2} \,.$$

We note that the remainders comprise only boundary layer terms.

This point is discussed in Chap. 9 with applications of IBL models (Subsect. 9.3.4).

8.4 Displacement Effect

In the first and second order IBL models, the viscous-inviscid interaction results to a large extent from the condition,

$$\lim_{y \to \infty} (v - v_1) = 0 \,. \tag{8.20}$$

This condition can be interpreted in terms of displacement effect. From the continuity equation, we have

$$v = - \int_0^y \frac{\partial u}{\partial x} \, \mathrm{d}y' \,,$$

$$v_1 = v_{10} - \int_0^y \frac{\partial u_1}{\partial x} \, \mathrm{d}y' \,,$$

where y' denotes the integration variable with respect to y and v_{10} is the value of v_1 at the wall. We deduce

$$v - v_1 = \int_0^y \left(\frac{\partial u_1}{\partial x} - \frac{\partial u}{\partial x} \right) \, \mathrm{d}y' - v_{10} \,.$$

Condition (8.20) becomes

$$v_{10} = \frac{\mathrm{d}}{\mathrm{d}x} \left[\int_0^\infty (u_1 - u) \, \mathrm{d}y \right] \,. \tag{8.21}$$

Thus, the viscous-inviscid interaction is represented by a blowing velocity v_{10} at the wall, which implies a displacement effect of the inviscid flow with respect to a fictitious flow without boundary layer. This displacement is expressed by the integral $\int_0^\infty (u_1 - u) \, \mathrm{d}y$.

8.5 Reduced Model for an Irrotational External Flow

We consider the flow around a streamlined body in an unbounded atmosphere. The freestream is *irrotational*. Even, in aerodynamics, the freestream is often uniform. Then, the inviscid flow defined in IBL model is irrotational.

The UVA associated with the first order IBL model is given by (8.10a–8.10c). The thickness of the boundary layer being of order ε, we have $y \ll 1$ in the boundary layer. If u_1, v_1 and p_1 are expandable in the neighbourhood of the wall, we can write in the boundary layer

$$u = u_{10} + U_1 + \cdots ,$$

$$v = v_{10} - yu_{1x0} + \varepsilon V_1 + \cdots ,$$

$$\frac{\partial p}{\partial y} = p_{1y0} + yp_{1yy0} + \varepsilon\frac{\partial P_1}{\partial Y} + \cdots ,$$

where the continuity equation (8.3a) is used and we defined

$$u_{1x} = \frac{\partial u_1}{\partial x}, \quad p_{1y} = \frac{\partial p_1}{\partial y} , \quad p_{1yy} = \frac{\partial^2 p_1}{\partial y^2} .$$

The index "0" denotes a value at the wall.

Then, we set

$$U(x, Y, \varepsilon) = u_{10} + U_1 ,$$

$$V(x, Y, \varepsilon) = V_1 + \frac{1}{\varepsilon}(v_{10} - yu_{1x0}) ,$$

$$\frac{\partial P}{\partial Y}(x, Y, \varepsilon) = \frac{\partial P_1}{\partial Y} + \frac{1}{\varepsilon}(p_{1y0} + yp_{1yy0}) ,$$

and the UVA becomes

$$u = U + u_1 - u_{10} , \tag{8.22a}$$

$$v = \varepsilon V + v_1 - v_{10} + yu_{1x0} , \tag{8.22b}$$

$$\frac{\partial p}{\partial y} = \varepsilon\frac{\partial P}{\partial Y} + p_{1y} - p_{1y0} - yp_{1yy0} . \tag{8.22c}$$

The boundary conditions given by (8.14) yield

$$\text{at the wall :} \quad U = 0 , \quad V = 0 , \tag{8.23a}$$

$$\lim_{Y \to \infty} U = u_{10} , \quad \lim_{Y \to \infty} (V + Yu_{1x0}) = \frac{v_{10}}{\varepsilon} . \tag{8.23b}$$

While the generalized boundary layer equations (8.9a, 8.9b) or (8.11) are valid everywhere, they can be simplified if restricted to the boundary layer. Indeed, in this region, we have $y \ll 1$. Then, the characteristics of the external flow are expanded in Taylor series in the neighbourhood of $y = 0$

$$u_1 = u_{10} + y \left(\frac{\partial u_1}{\partial y} \right)_{y=0} + \cdots ,$$

$$\frac{\partial u_1}{\partial x} = u_{1x0} + y \left(\frac{\partial^2 u_1}{\partial x \partial y} \right)_{y=0} + \cdots .$$

If the *inviscid flow is irrotational*, by neglecting the wall curvature effects and by observing that v_1 is $\mathrm{O}(\varepsilon)$ in the boundary layer, we find that $\dfrac{\partial u_1}{\partial y}$ is $\mathrm{O}(\varepsilon)$ in the boundary layer. Morevover, $\dfrac{\partial^2 v_1}{\partial y^2}$ is also $\mathrm{O}(\varepsilon)$ in the boundary layer. Therefore, in this region, we have

$$u_1 = u_{10} + \mathrm{O}(\varepsilon^2) ,$$

$$\frac{\partial u_1}{\partial x} = u_{1x0} + \mathrm{O}(\varepsilon^2) ,$$

$$v_1 = v_{10} - y u_{1x0} + \mathrm{O}(\varepsilon^3) ,$$

$$\frac{\partial^2 v_1}{\partial y^2} = \mathrm{O}(\varepsilon) .$$

In addition, in the boundary layer and from (8.22a–8.22c), we have

$$u = U + \mathrm{O}(\varepsilon^2) ,$$

$$v = \varepsilon V + \mathrm{O}(\varepsilon^3) ,$$

$$\frac{\partial p}{\partial y} = \varepsilon \frac{\partial P}{\partial Y} + \mathrm{O}(\varepsilon^2) .$$

By neglecting $\mathrm{O}(\varepsilon^2)$ terms, the generalized boundary layer equations restricted to the boundary layer become

$$\left. \begin{aligned} \frac{\partial U}{\partial x} + \frac{\partial V}{\partial Y} &= 0 \\[2mm] U \frac{\partial U}{\partial x} + V \frac{\partial U}{\partial Y} &= u_{10} u_{1x0} + \frac{\partial^2 U}{\partial Y^2} \end{aligned} \right\} . \tag{8.24}$$

With a change of notation – u_{10} is often replaced by u_e – these equations are exactly Prandtl's equations (Subsect. 7.1.1), but as discussed hereafter, the boundary conditions are not the usual conditions.

The y-momentum equation (8.12) restricted to the boundary layer becomes

$$U \frac{\partial V}{\partial x} + V \frac{\partial V}{\partial Y} = - \frac{\partial P}{\partial Y} + \frac{\partial^2 V}{\partial Y^2} . \tag{8.25}$$

At the wall, the boundary conditions are

$$U(x, 0, \varepsilon) = 0 , \tag{8.26a}$$

$$V(x, 0, \varepsilon) = 0 , \tag{8.26b}$$

$$\lim_{Y \to \infty} U = u_{10} , \tag{8.26c}$$

$$\lim_{Y \to \infty} (V + Y u_{1x0}) = \frac{v_{10}}{\varepsilon} . \tag{8.26d}$$

We note that condition (8.26c) is identical to the condition used in Prandtl's theory but condition (8.26d) brings a new element. Due to this condition, calculations of separated flows are not excluded.

Thus, for an irrotational *external flow, the first order IBL model reduces to the standard boundary layer equations (8.24) strongly coupled to the Euler equations.* The strong coupling is due to the boundary conditions (8.26c), (8.26d).

With the same hypotheses, the second order IBL model leads to the same reduced model, given by (8.24–8.25), and to the same boundary conditions (8.26a–8.26d).

Note 8.4. From the continuity equation, we have

$$v = -\int_0^y \frac{\partial u}{\partial x}\, \mathrm{d}y' \, ,$$

where y' denotes the integration variable with respect to y. Then, condition (8.26d) for v_{10} becomes

$$v_{10} = \int_0^\infty \left(\frac{\mathrm{d}u_{10}}{\mathrm{d}x} - \frac{\partial u}{\partial x} \right) \mathrm{d}y$$

or

$$v_{10} = \frac{\mathrm{d}(u_{10}\delta_1)}{\mathrm{d}x} \quad \text{with} \quad \delta_1 = \int_0^\infty \left(1 - \frac{u}{u_{10}} \right) \mathrm{d}y \, . \tag{8.27}$$

The viscous-inviscid interaction is expressed through a blowing velocity v_{10} related to the displacement thickness δ_1 by (8.27).

8.6 Conclusion

Different approximations of Navier-Stokes equations for the study of high Reynolds number flows past streamlined bodies have been obtained by applying SCEM.

The procedure used to get these approximations is close to the one followed for an ordinary differential equation. As a first approximation, the flow is described by the Euler equations. Obviously, this model is not valid near the wall. Thanks to the use of *generalized* expansions, the search for a UVA over the whole flow field leads to the *interactive boundary layer*, IBL, model, to first and second orders.

UVAs obtained with MMAE and SCEM are different because MMAE is based on *regular expansions*. A major consequence is that the wall boundary condition for the first order inviscid flow approximation in MMAE is a zero normal velocity. It results that a *hierarchy* is established between the inviscid and viscous sets of equations which are solved *sequentially*. At first, the inviscid flow equations are solved *independently* from the boundary layer equations. In the second step, the boundary layer equations are solved using results obtained from the previous calculation. In the third step, the solution of the second order inviscid flow equations takes into account the boundary layer effects and provides a correction to the first estimate. Finally, the second order boundary layer can be calculated. With IBL, the hierarchy between the inviscid flow equations and the boundary layer is broken. The slip condition at the walls for the inviscid flow no longer exists. This condition is replaced by the condition that the normal velocity must tend towards the viscous flow normal velocity far away from the wall. Consequently, *the inviscid flow equations and the boundary layer equations interact.* One system of equations influences the other and vice versa. The two sets of equations must be solved simultaneously. With appropriate numerical techniques, IBL model lets us calculate separated flows. An example is presented in Chap. 9.

The triple deck theory also contains, in its remarkable achievements, the treatment of separated flows thanks to the absence of hierarchy between the decks. In this regard, the identity of gauges for the normal velocity in the upper and main decks is essential. In fact, this property establishes a close parallel between the triple deck and IBL theories. However, it must be kept in mind that the triple deck theory is very local since the interaction region shrinks to a single point as the Reynolds number tends towards infinity whereas IBL theory remains global through its streamwise extent.

In Chapter 10, it is shown that the second order IBL model *contains* the triple deck model and also Van Dyke's second order boundary layer model. These two models are obtained by applying *regular expansions* to IBL model as the Reynolds number tends towards infinity.

The idea of IBL is not new since this notion has been extensively applied to calculate flows around airfoils or wings. Rational arguments to support this approach were missing until now. Here, this lack is filled with a full justification thanks to the application of SCEM in its generalized form. Moreover, the boundary layer equations are a generalized form of Prandtl's equations.

For irrotational external flows, the generalized boundary layer equations simplify if their validity is restricted to the boundary layer region. Then, the standard boundary layer equations apply while remaining strongly coupled to the inviscid flow equations. The first or second order IBL models can be interesting if the characteristics of the inviscid flow vary significantly within the boundary layer thickness.

Problems

8-1. We analyze the laminar, incompressible flow at high Reynolds number past a semi-infinite flat plate of zero thickness. The freestream is parallel to the plate and the leading edge of the plate is orthogonal to the freestream. The freestream is uniform with a velocity V_∞.

1. Write the Navier-Stokes equations in dimensionless form. The dimensionless quantities are denoted by x, y, $\mathcal{U}$, $\mathcal{V}$, $\mathcal{P}$. The dimensionless equations contain the Reynolds number Re

$$Re = \frac{V_\infty L}{\nu},$$

where L is a reference length which represents the length of development of the boundary layer. We suppose that $Re \gg 1$ and we introduce the small parameter ε

$$\varepsilon = Re^{-1/2} .$$

2. We study the flow with SCEM in its regular form. We seek a first (outer) approximation in the form

$$\mathcal{U} = u_1(x, y) + \cdots ,$$
$$\mathcal{V} = v_1(x, y) + \cdots ,$$
$$\mathcal{P} = p_1(x, y) + \cdots .$$

Write the equations for u_1, v_1, p_1.
3. We seek a UVA in the form

$$\mathcal{U} = u_1(x, y) + U_1(x, Y) + \cdots ,$$
$$\mathcal{V} = v_1(x, y) + \varepsilon V_1(x, Y) + \cdots ,$$
$$\mathcal{P} = p_1(x, y) + \Delta(\varepsilon) P_1(x, Y) + \cdots ,$$

with

$$Y = \frac{y}{\varepsilon} ,$$

and Δ is a gauge yet undetermined.

Write the equations for U_1, V_1, P_1. Give the gauge Δ.

Give the boundary conditions by noting that we seek a regular expansion, i.e. an expansion for which we have for example $u_1 = u_1(x, y)$ and $U_1 = U_1(x, Y)$.

Give the solution for u_1, v_1, p_1.
4. We set

$$U = u_1 + U_1 ,$$
$$V = v_1 + \varepsilon V_1 .$$

Write the equations for U and V. Give the boundary conditions. Identify with Prandtl's formulation.

For the flat plate, the solution of these equations is Blasius' solution obtained by seeking a self-similarity solution. For $x > 0$, the solution has the form

$$U = f'(\eta) \quad \text{with} \quad \eta = \frac{Y}{\sqrt{2x}} ,$$

and Blasius' solution is

$$f''' + f f'' = 0 \quad \text{with} \quad f(\eta) = \int_0^\eta f'(\zeta)\, \mathrm{d}\zeta ,$$

with the boundary conditions

$$f(0) = 0 \,, \quad f'(0) = 0 \,, \quad f' \xrightarrow[\eta\to\infty]{} 1 \,.$$

We deduce the following asymptotic behaviours

$$f \underset{\eta\to\infty}{\cong} \eta - \beta_0 + \mathrm{EST} \,,$$

$$f \underset{\eta\to 0}{\cong} \frac{\alpha_0}{2}\eta^2 + \mathrm{O}(\eta^5) \,.$$

The numerical calculation gives the values of constants

$$\alpha_0 = 0.469600 \,,$$

$$\beta_0 = 1.21678 \,.$$

Deduce the behaviour of V_1 and P_1 as $Y \to \infty$.

5. Since we have neglected certain terms in the equations, we make an error as $Y \to \infty$. Indeed, we have

$$V_1 \xrightarrow[Y\to\infty]{} \frac{\beta_0}{\sqrt{2x}} \,,$$

which does not enable us, with the approximation obtained until now, to satisfy the condition that the velocity V tends towards zero at infinity.

If we want to have a better approximation, taking into account the previous results, we must write

$$\mathcal{U} = 1 + U_1(x, Y) + \varepsilon u_2(x, y) + \cdots \,,$$
$$\mathcal{V} = \varepsilon \left[V_1(x, Y) + v_2(x, y) \right] + \cdots \,,$$
$$\mathcal{P} = \varepsilon p_2(x, y) + \varepsilon^2 P_1(x, Y) + \cdots \,.$$

Write the equations for u_2, v_2, p_2 and the required boundary conditions.

Show that it is interesting to perform the following change of functions

$$u_2^* = u_2 \,,$$

$$v_2^* = v_2 + \frac{\beta_0}{\sqrt{2x}} \,,$$

$$p_2^* = p_2 + \frac{\beta_0}{2\sqrt{2}} x^{-3/2} y \,.$$

Then, the solution is

$$u_2^* = -\frac{\beta_0}{2} \frac{y}{\sqrt{x^2 + y^2}\sqrt{x + \sqrt{x^2 + y^2}}} \,,$$

$$v_2^* = \frac{\beta_0}{2} \frac{\sqrt{x + \sqrt{x^2 + y^2}}}{\sqrt{x^2 + y^2}} \,.$$

We note the very particular feature of the solution for u_2 which vanishes at $y = 0$ except at the origin where the solution is singular.

9 Applications of Interactive Boundary Layer Models

The successive complementary expansion method, SCEM, in its generalized form, fully justifies the notion of IBL. A UVA is obtained by solving two sets of equations – the generalized boundary layer equations and the inviscid flow equations. These two systems are strongly coupled. In the framework of the standard boundary layer theory, a hierarchy is established between the two systems: first, we can solve the inviscid flow equations and then the boundary layer equations. With IBL, this hierarchy is broken; the two systems interact, one system influences the other and vice versa.

For an irrotational freestream, the boundary layer equations reduce, in the boundary layer region, to standard Prandtl's equations. However, the interactive character is still present because the viscous and inviscid equations remain strongly coupled. This feature is essential for flows including separated zones. An example is presented in Sect. 9.1.

For a rotational freestream, this reduced model does not apply. The validation of IBL for such flows is therefore necessary. Thus, in Sect. 9.3, the IBL model is applied to several examples in which the external flow is rotational. The results are compared to the numerical solutions of Navier-Stokes equations and to Van Dyke's model [105].

The objective of this chapter is not to give an exhaustive account of applications of IBL models but, simply, to show a few illustrative examples. We discuss applications of IBL methods to aerodynamic flows but we insist on the influence of external vorticity whose study is not very common. The IBL methods have been developed by different authors [9, 10, 30, 53, 54, 55, 109, 110] and applied extensively in aerodynamics [1, 2]; it has been shown that these methods are very efficient [13, 52, 56, 57]. The reader can find in the literature detailed comparisons between the applications of the triple deck theory, IBL and numerical solutions of the Navier-Stokes equations in external flow [81] and in internal flow [49, 50]. Internal flows are also discussed in Chap. 12.

9.1 Calculation of a Flow with Separation

9.1.1 Definition of the Flow

We consider a flow on a flat plate deformed by a small hump (Fig. 9.1). The wall geometry is defined by

$$\frac{y}{L} = \pm \frac{0.03}{\cosh\left[4\left(\frac{x}{L} - 2.5\right)\right]} \, , \tag{9.1}$$

where L is a reference length. The freestream is uniform and its velocity is V_∞. Therefore, the inviscid flow is *irrotational*. The Reynolds number based on velocity V_∞ and on length L is equal to 8×10^4.

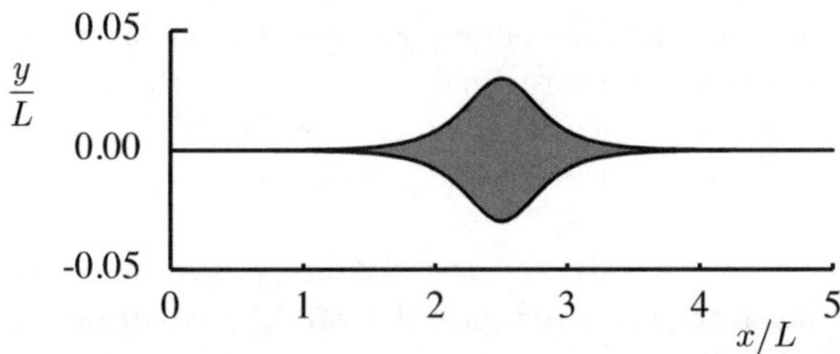

Fig. 9.1. Boundary layer on a flat plate with a hump

9.1.2 Numerical Method

The numerical solution rests upon the method proposed by Veldman [109] to a large extent. This method has been mentioned in Sect. 7.3.

The flow in the boundary layer is calculated with (8.24) and the inviscid flow is computed by a panel method [14]. As the flow includes a separated region, the boundary layer equations must be strongly coupled to the inviscid flow equations. In order to achieve this coupling, the interaction law given by the Hilbert integral (Sect. 7.3) is used as an intermediary which enables us to obtain a numerically well-conditioned problem.

A method of solution of boundary layer equations strongly coupled to the Hilbert integral has been implemented by Roget [79].

In this method u_{10} is replaced by the more standard notation u_e, and the standard boundary layer equations (8.24) whose dimensionalized form is

$$\left. \begin{aligned} \frac{\partial u}{\partial x} + \frac{\partial v}{\partial y} &= 0 \\ u\frac{\partial u}{\partial x} + v\frac{\partial u}{\partial y} &= u_e \frac{du_e}{dx} + \frac{\mu}{\varrho}\frac{\partial^2 u}{\partial y^2} \end{aligned} \right\} \tag{9.2}$$

are solved together with the Hilbert integral in the domain $[x_a, x_b]$

$$\delta u_e = \frac{1}{\pi} \int_{x_a}^{x_b} \frac{v_b}{x - \xi} \, \mathrm{d}\xi \, , \quad v_b(\xi) = \frac{\mathrm{d}}{\mathrm{d}\xi} \left[u_e(\xi)\delta_1(\xi) \right] \, . \tag{9.3}$$

In these equations, u_e is split into two parts

$$u_e(x) = \widehat{u}_e(x) + \delta u_e(x) \, , \tag{9.4}$$

where $\widehat{u}_e$ is an input which results partially from the application of the panel method [13] in which the displacement effect is taken into account.

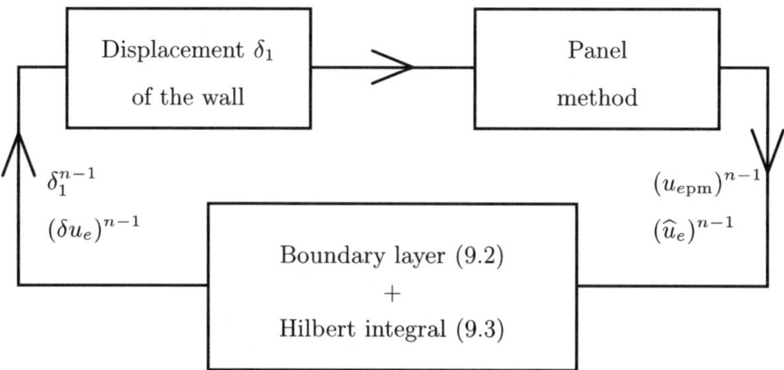

Fig. 9.2. Calculation method

An iterative method [3] solves the whole problem as depicted in Fig. 9.2. In the first iteration, the velocity $\widehat{u}_e$ is equal to the velocity calculated by the panel method applied to the real geometry, i.e. by taking into account the presence of the hump but without the boundary layer effect. In the following iterations, the velocity u_e in the boundary layer equations (9.2) is decomposed as

$$(u_e)^n = (u_{epm})^{n-1} - (\delta u_e)^{n-1} + (\delta u_e)^n \, , \quad n = 1, 2, \cdots, N \, .$$

Here, $(u_{epm})^{n-1}$ is the velocity calculated by the panel method around a body obtained by displacing the real body by the displacement thickness $(\delta_1)^{n-1}$ and δu_e is the correction of velocity given by the Hilbert integral. Therefore, at a given iteration, we have

$$(\widehat{u}_e)^{n-1} = (u_{epm})^{n-1} - (\delta u_e)^{n-1} \, .$$

When the process converges, i.e. when the difference $\left[(\delta u_e)^n - (\delta u_e)^{n-1}\right]$ between the velocity corrections δu_e between two successive iterations is very small, *the influence of the Hilbert integral disappears* so that the velocity u_e is the velocity u_{epm} calculated by the panel method with boundary layer effects. Finally, the Hilbert integral is simply used as a *calculation intermediary* to provide a strong coupling between the boundary layer equations and the inviscid flow equations so as to calculate separated flows.

9.1.3 Results

Figure 9.3 shows the effect of the hump and of the boundary layer on the wall velocity calculated in inviscid flow. We have

$$u_e = V_\infty + \Delta u_e \ .$$

The variation of velocity Δu_e labelled "without boundary layer" represents the influence of the hump alone whereas the variation of velocity Δu_e labelled "with boundary layer" represents the combined influence of the hump and of the boundary layer. For the chosen example, the boundary layer effect is of the same order as the purely geometrical effect of the wall deformation. We observe that the boundary layer reduces the velocity overshot induced near the top of the hump.

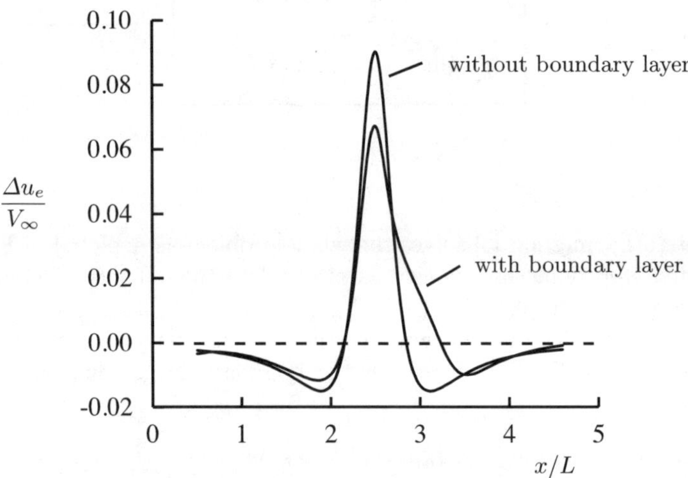

Fig. 9.3. Boundary layer on a flat plate with a hump: wall velocity of the inviscid flow

Figures 9.4 and 9.5 show the evolutions of the displacement thickness δ_1 and of the skin-friction coefficient C_f. For comparison, the evolution of these parameters on a flat plate as obtained by Blasius' solution [20] are also given; we have

$$\delta_1 = 1.721\frac{x}{\sqrt{\mathcal{R}_x}} \ , \quad C_f = \frac{0.664}{\sqrt{\mathcal{R}_x}} \ , \quad \mathcal{R}_x = \frac{\varrho u_e x}{\mu} \ .$$

The comparisons to Blasius' solution show the strong influence of the presence of the hump on the evolution of the boundary layer, but this influence is local since, upstream and downstream as well, the boundary layer characteristics are very close to their behaviour on a flat plate. The presence of a separated region characterized by negative values of the skin-friction is also noted.

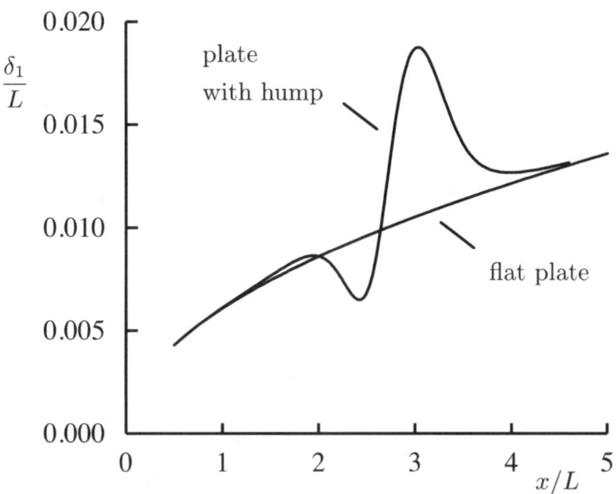

Fig. 9.4. Boundary layer on a flat plate with a hump: displacement thickness

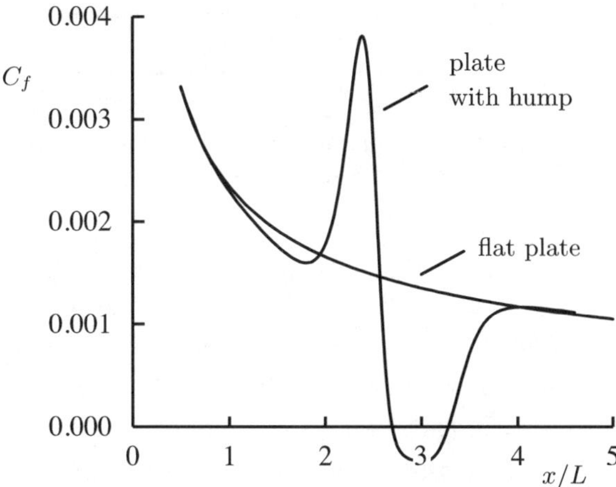

Fig. 9.5. Boundary layer on a flat plate with a hump: skin-friction coefficient

Note 9.1. A detailed discussion of the flow on a flat wall deformed locally by a blunt backward facing step is given in [86]. The results were obtained by the triple deck theory and by an IBL method. In certain cases, in the presence of separation bubbles, the solution is not unique. The results show the existence of a branch along which a short region of separation is associated with the theory of marginal separation [99]. In addition, the results show the formation of a branch with a longer separated region meeting the massive separation regime.

9.2 Application to Aerodynamic Flows

In this section, we present the application of IBL theory to aerodynamic flows. The first application considers the laminar flow around a flat plate of finite length at zero angle of attack. The second application is concerned with high Reynolds number flows around airfoils.

9.2.1 Flat Plate of Finite Length

The nature of the flow near the trailing edge of an airfoil has long been a subject of both theoretical and practical interest. The problem exhibits a singularity intriguing to the theoretician, and the question of finite Reynolds number effects on aerodynamic forces is of significant importance.

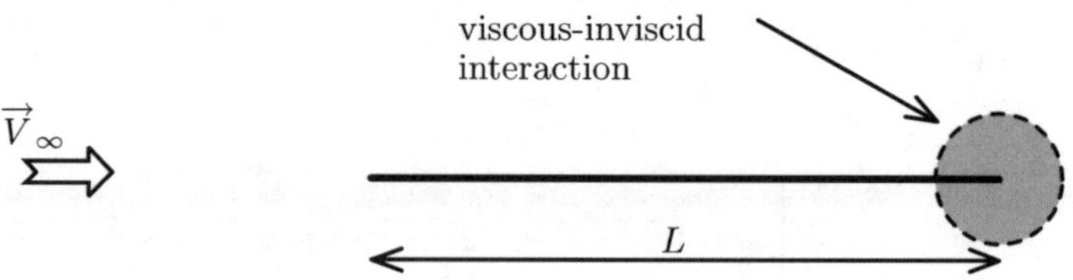

Fig. 9.6. Flow around a flate plate of finite length

In this subsection, we consider a laminar flow around a flat plate at zero angle of attack (Fig. 9.6). A difficulty arises due to the change of boundary condition near the trailing edge; along the wall, the no-slip condition $u = 0$ applies whereas along the wake centerline we have the condition of symmetry $\dfrac{\partial u}{\partial y} = 0$. The problem was first studied within the framework of boundary layer theory by Goldstein [39], who showed that the continuation of Blasius' flat plate solution beyond the trailing edge required introduction of a thin sublayer along the wake centerline whose thickness varies like $x^{1/3}$, where x is the distance from the trailing edge (see Problem 7-5). At the point of origin of this sublayer ($x = 0$), the transverse velocity v is singular, much as it is at the leading edge of the plate. In a higher order theory, this large magnitude of v would produce a similarly large pressure gradient in the external inviscid flow, which in turn would produce a significant change in the skin-friction near the trailing edge.

A number of studies have been made to construct such a higher order theory for this region, as discussed by Van Dyke [107]. However, none of these older theories properly solved the problem; more accurate solutions than those provided by simple perturbations series are required. A direct solution of the Navier-Stokes equations is possible and does not exhibit any sign of singularity [87, 88]. The understanding of the problem was disclosed by Stewartson [95] and Messiter [69] with the triple deck theory; they showed that the clue is the *interaction between the viscous flow and the inviscid external flow*. Another alternative is to solve the IBL problem [110]. In Chapter 8, the IBL theory has been fully justified with the use of SCEM and, in Chap. 10, it is is shown that IBL contains the first order triple deck theory.

The results presented in this subsection have been obtained by solving the boundary layer equations (9.2) in association with the Hilbert integral (9.3). This set of equations is sufficient to reproduce the viscous-inviscid interaction around the trailing edge. The numerical technique is described in details in [13, 14] and computer programs are available.

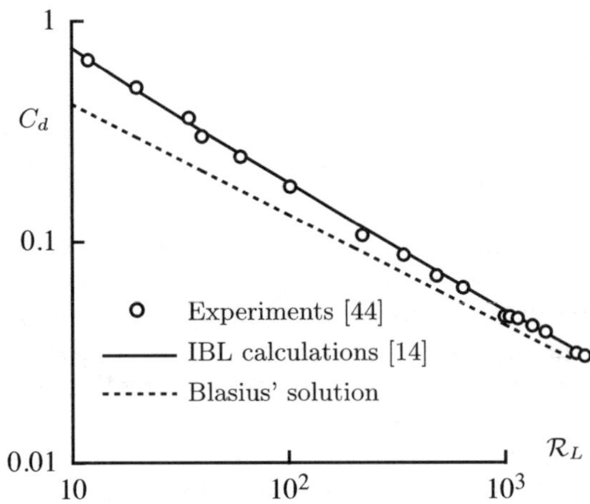

Fig. 9.7. Drag coefficient of a flate plate of finite length

Figure 9.7 shows the evolution of the drag coefficient of a flat plate as a function of the Reynods number. The length of the plate is L and the freestream velocity is V_∞.

The Reynolds number is defined by

$$\mathcal{R}_L = \frac{\varrho V_\infty L}{\mu} \, . \tag{9.5}$$

The drag coefficient is defined by

$$C_d = \frac{F}{\frac{1}{2}\varrho V_\infty^2 L} \,, \tag{9.6}$$

where F is the force applied, per unit span, to the two faces of the plate. For a flat plate at zero angle of attack, the aerodynamic forces are due to the skin-friction since the pressure forces are zero. From integration of the skin-friction along the wall, Blasius' solution gives

$$C_d = \frac{1.328}{\mathcal{R}_L^{1/2}} \,.$$

However, due to the viscous-inviscid interaction, the skin-friction is higher near the trailing edge. According to the solution of the triple deck equations, the drag coefficient is given by [94]

$$C_d = \frac{1.328}{\mathcal{R}_L^{1/2}} + \frac{2.66}{\mathcal{R}_L^{7/8}} \,. \tag{9.7}$$

The results of IBL calculations shown in Fig. 9.7 are in very good agreement with experimental results, even at relatively low Reynolds number. The agreement with the triple deck theory, given by (9.7), is also very good; the error is 2.5% at a Reynolds number $\mathcal{R}_L = 10$ and decreases for increasing values of the Reynolds number. At the scale of Fig. 9.7, the difference between the IBL calculations and triple deck results is practically negligible.

9.2.2 Airfoils at High Reynolds Numbers

The general principle of IBL calculations of the flow around an airfoil at high Reynolds number is close to the one described in Subsect. 9.1.2. However, additional complexities are present. The flow is no longer laminar all along the airfoil and laminar-turbulent transition criteria must be implemented. A turbulence model is also required to calculate the boundary layer and the wake. The viscous-inviscid interaction is taken into account with the concept of blowing velocity on the airfoil and a jump in the normal velocity component to represent the wake effects. An extensive evaluation of the interaction procedure for aerodynamic flows was conducted by Cebeci [13]; the numerical method is discussed in detail in [13, 14] and computer programs are available.

In this subsection, only a sample of results are discussed to give an idea of the efficiency of IBL methods.

Results for an NACA 0012 airfoil, with calculations including the wake effects, are shown in Figs. 9.8 and 9.9, with Fig. 9.8 corresponding to a chord Reynolds number of 3×10^6 and Fig. 9.9 to Reynolds numbers of 6×10^6 and 9×10^6. The chord Reynolds number $\mathcal{R}_c$ is defined by

$$\mathcal{R}_c = \frac{\varrho V_\infty c}{\mu} \,, \tag{9.8}$$

where V_∞ is the freestream velocity and c is the chord length of the airfoil. The lift and drag coefficients are defined by

$$C_\ell = \frac{F_\ell}{\frac{1}{2}\varrho V_\infty^2 c} \,, \tag{9.9a}$$

$$C_d = \frac{F_d}{\frac{1}{2}\varrho V_\infty^2 c} \,, \tag{9.9b}$$

where F_ℓ and F_d are the components of the aerodynamic force exerted on the airfoil per unit span; the drag force is the component of the aerodynamic force aligned with the freestream velocity and the lift force is normal to it.

Figure 9.8 shows the variation of the lift and drag coefficients of the NACA 0012 airfoil for a chord Reynolds number of 3×10^6. As can be seen from Fig. 9.8a, viscous effects considerably reduce the maximum value of the lift coefficient of the airfoil, which occurs at a stall angle of around $16°$, and the calculated results agree well with measurements [13].

Figure 9.8b shows the variation of the drag coefficient with lift coefficient. As can be seen, the measurements of drag coefficients do not extend beyond an angle of attack of $12°$ and at smaller angles agree well with the calculations. It must be noted that the calculations can be performed beyond the stall angle of attack which implies that the boundary layer is separated. For an angle of attack of $18°$, separation extends on the rear part of the airfoil over more than 50% of the chord length. The nature of the lift-drag curve at higher angles of attack is interesting with the expected increase in drag coefficient and reduction in lift coefficient for post-stall angles.

Figures 9.8 and 9.9 show the effect of Reynolds number on the lift coefficient. In accord with the measurements, the calculation method satisfactorily accounts for the effects of Reynolds number. The results show that the maximum lift coefficient, $(C_\ell)_{\mathrm{max}}$ increases with increase in Reynolds number. For a given angle of attack, this is due to a delay of separation when the Reynolds number increases, i.e. separation moves towards the trailing edge. At low angle of attack, the effects of Reynolds number on the lift curve are very small, at least in the range of Reynolds number investigated here.

For purpose of comparison and to evaluate the viscous effects, results of inviscid flow calculations are plotted in Figs. 9.8 and 9.9. The linear lift curve of slope 2π is calculated by the thin airfoil theory [71]. The thin airfoil theory is an approximate inviscid flow theory which gives a smaller slope of the lift curve than an exact inviscid theory. Figures 9.8 and 9.9 show also the results obtained with the Hess-Smith panel method, which is a numerical solution of the exact inviscid flow equations [13]. With a positive angle of attack, the boundary layer is thicker on the upper surface than on the lower surface so that the effective angle of attack is smaller, in comparison with the inviscid theory. Then, the viscous effects have a tendency to reduce the slope of the lift curve and, by coincidence, the slope of the experimental curve is very

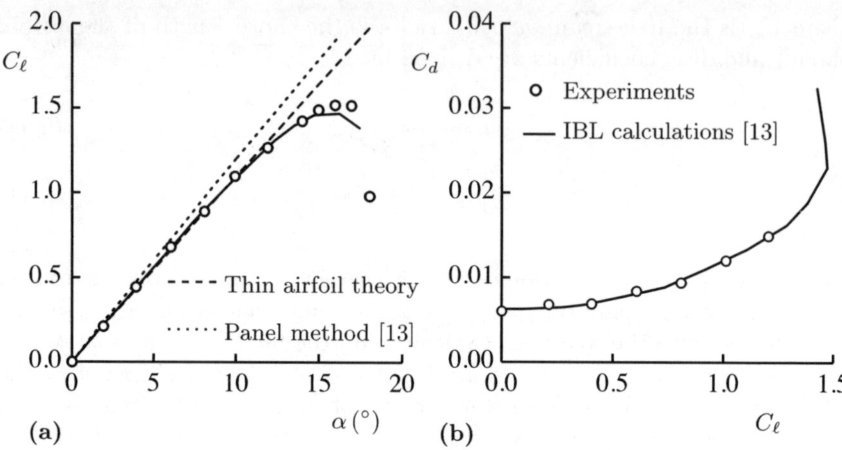

Fig. 9.8. Comparison between calculated and experimental values of (a) C_ℓ vs α, and (b) C_d vs C_ℓ. NACA 0012 airfoil at $\mathcal{R}_c = 3 \times 10^6$

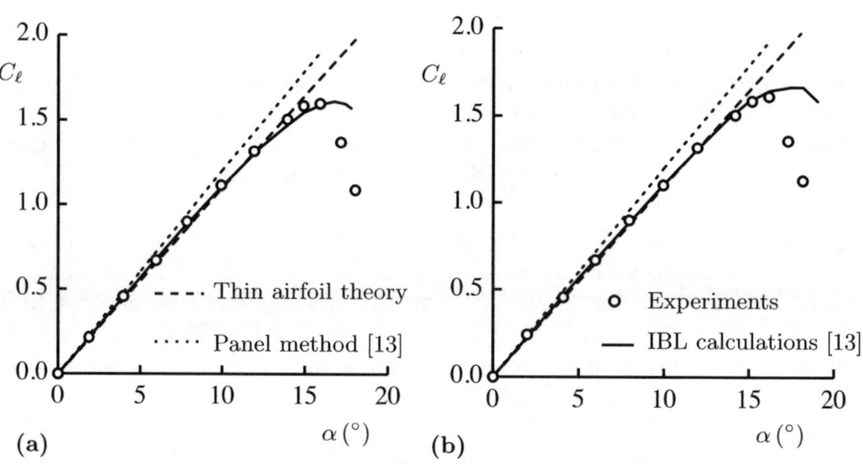

Fig. 9.9. Effect of Reynolds number on the lift coefficient. NACA 0012 airfoil at (a) $\mathcal{R}_c = 6 \times 10^6$, and (b) $\mathcal{R}_c = 9 \times 10^6$

close to 2π for the range of Reynolds number presented in Figs. 9.8, 9.9. The IBL method reproduces these viscous effects very well. For a cambered airfoil, the viscous effects are even larger. The in-depth discussion of these questions requires a more detailed analysis of airfoil data, for example, the pressure distributions on the lower and upper surfaces. Obviously, inviscid theories are not able to predict the stall which is due to boundary layer separation. The accurate prediction of stall is a very difficult problem. Inviscid theories also predict that the drag of an airfoil is zero, which is known as *d'Alembert's paradox*. The correct prediction of drag requires to include the viscous effects which are well accounted for in the IBL theory.

9.3 Influence of a Rotational External Flow

We want to calculate the boundary layer developing on a flat plate with a *rotational* oncoming flow (Fig. 9.10) and use the IBL model. Then, the Euler equations, the generalized boundary layer equations and their interaction must be solved simultaneously.

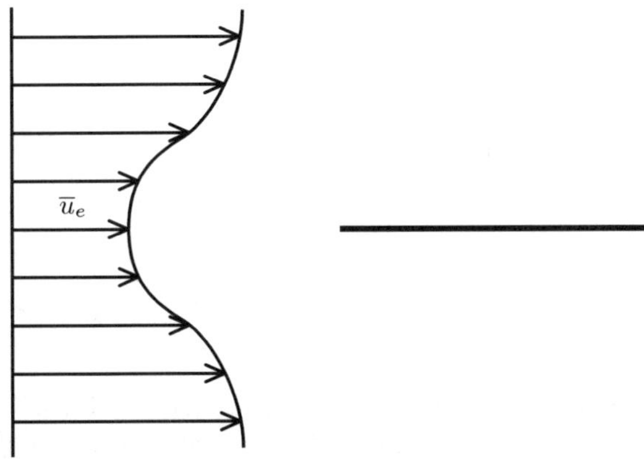

Fig. 9.10. Boundary layer on a flat plate with rotational external flow

9.3.1 Inviscid Flow

The characteristics u_1, v_1, p_1 of the outer flow are decomposed into a non perturbed component $\overline{u}_e$, $\overline{v}_e$, $\overline{p}_e$ and a perturbed component $\widetilde{u}_e$, $\widetilde{v}_e$, $\widetilde{p}_e$

$$u_1 = \overline{u}_e + \widetilde{u}_e , \tag{9.10a}$$

$$v_1 = \overline{v}_e + \widetilde{v}_e , \tag{9.10b}$$

$$p_1 = \overline{p}_e + \widetilde{p}_e . \tag{9.10c}$$

The non perturbed part is obtained without any boundary layer effect and satisfies the Euler equations. The perturbed part represents the boundary layer effect. Assuming that the perturbation is small, the perturbed part satisfies the linearized Euler equations.

We introduce the *stream-functions* $\overline{\psi}$ and $\widetilde{\psi}$

$$\overline{u}_e = \frac{\partial \overline{\psi}}{\partial y} , \ \overline{v}_e = -\frac{\partial \overline{\psi}}{\partial x} ,$$

$$\widetilde{u}_e = \frac{\partial \widetilde{\psi}}{\partial y} , \ \widetilde{v}_e = -\frac{\partial \widetilde{\psi}}{\partial x} ,$$

and the z-component of *vorticities* $\bar{\omega}_e$ and $\tilde{\omega}_e$ (vorticity is the curl of velocity)

$$\bar{\omega}_e = \frac{\partial \bar{v}_e}{\partial x} - \frac{\partial \bar{u}_e}{\partial y} \, ,$$

$$\tilde{\omega}_e = \frac{\partial \tilde{v}_e}{\partial x} - \frac{\partial \tilde{u}_e}{\partial y} \, .$$

The stagnation pressure $\bar{p}_e + \frac{1}{2}(\bar{u}_e^2 + \bar{v}_e^2)$ is constant along a streamline of the non perturbed flow and the vorticity is related to the variation of the stagnation pressure between streamlines

$$\bar{p}_e + \frac{1}{2}(\bar{u}_e^2 + \bar{v}_e^2) = \bar{f}(\bar{\psi}) \, , \tag{9.11a}$$

$$\bar{\omega}_e = -\frac{\mathrm{d}\bar{f}(\bar{\psi})}{\mathrm{d}\bar{\psi}} \, . \tag{9.11b}$$

The second equation above is deduced from (9.11a) and Euler equations. Assuming that the perturbation vanishes at usptream infinity, we have [105]

$$\tilde{p}_e + \bar{u}_e\tilde{u}_e + \bar{v}_e\tilde{v}_e = \tilde{\psi}\frac{\mathrm{d}\bar{f}(\bar{\psi})}{\mathrm{d}\bar{\psi}} \, , \tag{9.12a}$$

$$\tilde{\omega}_e = -\tilde{\psi}\frac{\mathrm{d}^2\bar{f}(\bar{\psi})}{\mathrm{d}\bar{\psi}^2} \, . \tag{9.12b}$$

Equations (9.12a, 9.12b) are linearized forms of (9.11a, 9.11b). The first equation is an integral form of linearized Euler equations obtained by integration along a streamline of the non perturbed flow. The second equation is deduced from (9.12a) and linearized Euler equations. If the non perturbed flow is irrotational, the vorticity $\tilde{\omega}_e$ of the perturbation is zero.

Vorticity is related to stream-function by

$$\tilde{\omega}_e = -\triangle\tilde{\psi} \, ,$$

so that the stream-function perturbation satisfies the equation

$$\triangle\tilde{\psi} = \frac{\mathrm{d}^2\bar{f}(\bar{\psi})}{\mathrm{d}\bar{\psi}^2}\tilde{\psi} \, .$$

In the examples discussed in Subsect. 9.3.3, the external flow is such that

$$\bar{u}_e = \bar{u}_e(y) \, , \quad \bar{v}_e = 0 \, , \tag{9.13}$$

and the equation for the stream-function perturbation becomes

$$\triangle\tilde{\psi} = \frac{1}{\bar{u}_e}\frac{\mathrm{d}^2\bar{u}_e}{\mathrm{d}y^2}\tilde{\psi} \, . \tag{9.14}$$

The pressure perturbation is given by

$$\tilde{p}_e = -\bar{u}_e\tilde{u}_e + \tilde{\psi}\frac{\mathrm{d}\bar{u}_e}{\mathrm{d}y} \, . \tag{9.15}$$

9.3.2 Method of Resolution

The inviscid and viscous equations are solved sequentially and their coupling is accounted for iteratively (Fig. 9.11) so that, when the process is converged, the following conditions are achieved

$$\lim_{y\to\infty} (u - u_1) = 0 \ , \quad \lim_{y\to\infty} (v - v_1) = 0 \ .$$

This procedure works well for flows without separation but would be inappropriate for flows with separation.

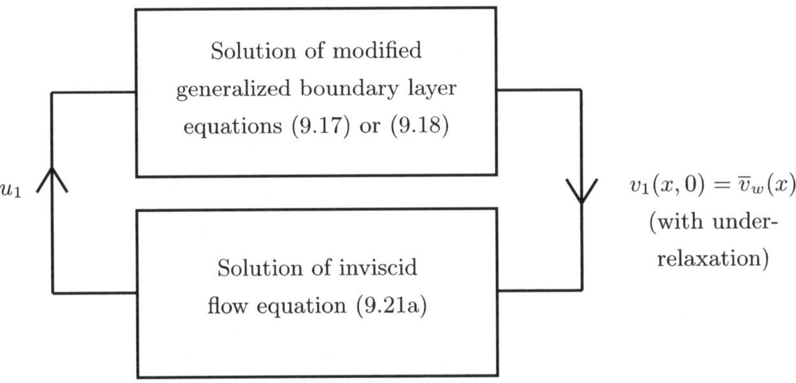

Fig. 9.11. Global iterative procedure

Modified Boundary Layer Equations

In agreement with Fig. 9.11, the boundary layer equations are solved in the direct mode: the input of boundary layer equations is a distribution of velocity u_1. Obviously, during the iterative cycle, this distribution of u_1 is not frozen since u_1 depends on the boundary layer effect on the inviscid flow. The problem to solve is to satisfy the two conditions $\lim_{y\to\infty} (u - u_1) = 0$ and $\lim_{y\to\infty} (v - v_1) = 0$ when convergence is achieved. In order to avoid numerical problems with the boundary layer equations when the solutions are not yet converged, it is convenient to replace v_1 by a modified velocity component $\overline{v}$ [29] defined by

$$\overline{v} = v + \int_y^\infty \frac{\partial(u_1 - u)}{\partial x} \, dy \ , \tag{9.16}$$

so that the velocity $\overline{v}$ satisfies the same continuity equation as v_1

$$\frac{\partial u_1}{\partial x} + \frac{\partial \overline{v}}{\partial y} = 0 \ .$$

The generalized boundary layer equations (8.17) are replaced by

$$\left.\begin{array}{l} \dfrac{\partial u}{\partial x} + \dfrac{\partial v}{\partial y} = 0 \\[2ex] u\dfrac{\partial u}{\partial x} + v\dfrac{\partial u}{\partial y} = u_1\dfrac{\partial u_1}{\partial x} + \overline{v}\dfrac{\partial u_1}{\partial y} + \dfrac{1}{\mathcal{R}}\dfrac{\partial^2(u-u_1)}{\partial y^2} \end{array}\right\} \cdot \qquad (9.17)$$

Equation (9.16) shows that condition $\lim\limits_{y\to\infty}(\overline{v}-v) = 0$ is satisfied if $\lim\limits_{y\to\infty}(u_1-u) = 0$. Thus, the modified momentum equation is identically satisfied beyond the boundary layer edge. This is a very important feature of the numerical method.

Denoting the value of v_1 at the wall by v_w, we note that $\overline{v}$ is equal to v_1 only if the value of v_w is correctly estimated (Fig. 9.12). When the iterative procedure has not yet converged, conditions $u = u_1$ and $\overline{v} = v$ are satisfied beyond the boundary layer edge but condition $\overline{v} = v_1$ is not necessarily satisfied. An iterative method is implemented to adjust v_w in order to get $\overline{v} = v_1$. This point is the heart of the viscous-inviscid interaction.

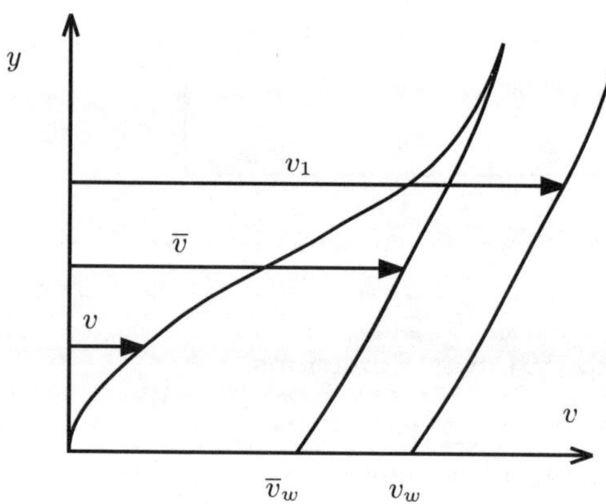

Fig. 9.12. Distributions of $\overline{v}$ and v_1 when the value of v_w is not correctly adjusted

The first order generalized boundary layer equations are simpler to treat than the second order equations because the velocity v_1 is not present in the equations. Therefore it is not needed to modify the equations

$$\left.\begin{array}{l} \dfrac{\partial u}{\partial x} + \dfrac{\partial v}{\partial y} = 0 \\[2ex] u\dfrac{\partial u}{\partial x} + v\dfrac{\partial u}{\partial y} - v\dfrac{\partial u_1}{\partial y} = u_1\dfrac{\partial u_1}{\partial x} + \dfrac{1}{\mathcal{R}}\dfrac{\partial^2(u-u_1)}{\partial y^2} \end{array}\right\} \cdot \qquad (9.18)$$

Condition $\lim\limits_{y\to\infty}(u_1-u) = 0$ is sufficient to satisfy identically the momentum equation beyond the boundary layer edge.

Solution of Boundary Layer Equations

The boundary layer equations are solved by considering that the input is a velocity field $u_1(x, y)$. After discretization according to a finite difference method, the equations are solved step by step, by a marching method from upstream to downstream. At a given station x_i, a first estimate of $v(x_i, y)$ and $\overline{v}(x_i, y)$ is assumed to be known, for example from the distributions of $v(x_{i-1}, y)$ and $\overline{v}(x_{i-1}, y)$ calculated at the previous station. A first estimate of $u(x_i, y)$ is obtained by solving the momentum equation discretized at station x_i. An updated estimate of $v(x_i, y)$ is calculated from the discretized continuity equation with the wall condition $v = 0$. Then, the value of $\overline{v}(x_i, y)$ is updated from the continuity equation

$$\overline{v}(x_i, y) = v(x_i, y) + \left[\int_y^\infty \frac{\partial(u_1 - u)}{\partial x} \, dy \right]_{x=x_i} . \tag{9.19}$$

We note that the velocity $\overline{v}$ is calculated by integrating the continuity equation from the outside boundary towards the wall in order to ensure that $\overline{v} = v$ beyond the boundary layer edge. In fact, in the numerical method, the limit at infinity is replaced by a boundary at finite distance from the wall located beyond the boundary layer edge.

If required, the calculation is repeated at station x_i with the updated distributions of u, v and $\overline{v}$ in order to resolve the non-linearity of the momentum equation. Then, the calculation proceeds to the next station x_{i+1}.

It is noted that the estimate of $\overline{v}_w(x_i) = \overline{v}(x_i, 0)$ is

$$\overline{v}_w(x_i) = \left[\frac{d}{dx} \int_0^\infty (u_1 - u) \, dy \right]_{x=x_i} . \tag{9.20}$$

Viscous-Inviscid Interaction

The resolution of boundary layer equations is a part of the iterative procedure required to account for the viscous-inviscid interaction (Fig. 9.11).

With the small perturbation hypothesis, the inviscid flow is calculated by solving Poisson's equation (9.14)

$$\frac{\partial^2 \widetilde{\psi}}{\partial x^2} + \frac{\partial^2 \widetilde{\psi}}{\partial y^2} = \frac{\widetilde{\psi}}{\overline{u}_e} \frac{d^2 \overline{u}_e}{dy^2} , \tag{9.21a}$$

with the wall condition

$$\widetilde{\psi}(x, 0) = - \int_{-\infty}^x v_1(\xi, 0) \, d\xi , \tag{9.21b}$$

and

$$v_1(\xi, 0) = \overline{v}_w(\xi) .$$

The value of $\overline{v}_w$ is obtained from (9.20).

Poisson's equation (9.21a) is solved numerically by means of a finite difference method on a rectangular mesh with a five point stencil. The discrete equations are solved iteratively, column by column, with over-relaxation.

The iterative procedure of Fig. 9.11 is continued until convergence is achieved by introducing an under-relaxation on $\overline{v}_w$.

9.3.3 Flows Studied

The chosen examples are the same as those calculated by another approach, called defect boundary layer formulation [5, 7, 8]. The flows are defined below by the non perturbed velocity $\overline{u}_e$.

Flow I. Vorticity is uniform

$$\overline{u}_e = 1 + 60y \ . \tag{9.22}$$

Flow II. The velocity distribution has a slope discontinuity

$$\begin{aligned} \overline{u}_e &= 1 + 60y \text{ if } y \leq 0.005 \ , \\ \overline{u}_e &= 1.3 \qquad \text{if } y \geq 0.005 \ . \end{aligned} \tag{9.23}$$

Flow III. Near the wall, the shear is negative and vanishes away from the wall. The slope of the velocity distribution is continuous

$$\begin{aligned} \overline{u}_e &= 125y^2 - 20y + 1 \text{ if } y \leq 0.08 \ , \\ \overline{u}_e &= 0.2 \qquad\qquad \text{if } y \geq 0.08 \ . \end{aligned} \tag{9.24}$$

Flow IV. The shear decreases continuously as the distance to the wall increases

$$\overline{u}_e = 0.85 + \sqrt{0.0225 + 18y} \ . \tag{9.25}$$

9.3.4 Results

Wall Shear Stress and Velocity Profiles

All the results have been obtained for a Reynolds number $\mathcal{R} = \mathcal{R}_{x=1} = 10^6$. The Reynolds number $\mathcal{R}_x$ is based on the reference velocity V and on the distance along the plate from the leading edge. The abscissa x is reduced by the reference length L as given in (8.2).

The lower parts of Figs. 9.13–9.16 show the velocity profiles $u(y)$ calculated at station $x = 0.9$ for the converged solution of the second order IBL model. The profiles of the inviscid flow velocity $u_1(y)$ obtained for the converged solution are also plotted in the same graph. The difference between the velocity $\overline{u}_e$ of the non perturbed inviscid flow and the velocity u_1 represents the influence of the boundary layer, i.e. the displacement effect. Figure 9.17 shows the profiles of the normal velocity components v and v_1 corresponding to u and u_1, respectively.

In general, we observe that the velocity profiles u and u_1 (Figs. 9.13–9.16) on one hand, and the profiles of v and v_1 (Fig. 9.17) on the other hand match perfectly beyond the boundary layer edge. Functions $u(x, y)$ and $v(x, y)$ which are solutions of the generalized boundary layer equations coupled to the inviscid equations form a UVA of the velocity field in the whole flow.

Figures 9.13–9.16 give the evolution of the skin-friction coefficient C_f defined by reducing the wall shear stress by $\frac{1}{2}\varrho V^2$. For the purpose of comparison, the value obtained in Blasius' solution is represented with the label "flat plate"

$$\frac{C_f}{2}\sqrt{\mathcal{R}_x} = 0.332 \ .$$

On these figures, different results are plotted

- the labels "1st order IBL" and "2nd order IBL" refer to the IBL model of first order (Sect. 8.2) or of second order (Sect. 8.3),
- the label "converged" corresponds to the converged results obtained with the iterative procedure described in Subsect. 9.3.2,
- the label "1st iteration" means that the generalized boundary layer equations of first order given by (9.18) or the second order modified equations given by (9.17) are solved with $u_1 = \overline{u}_e$,
- the label "Navier-Stokes" refers to the numerical solutions of the Navier-Stokes equations by Brazier [7].

The comparison of the skin-friction coefficient with Blasius' solution shows the strong effect of the external vorticity. When the standard boundary layer equations are used, the condition at the boundary layer edge is $u \to \overline{u}_e(0)$. Now, in all these examples, we have $\overline{u}_e(0) = 1$. Therefore, the solution to the standard equations is Blasius' solution and the results are not affected by external vorticity.

Generally, the second order IBL results are in better agreement with the Navier-Stokes results than those obtained with the first order IBL. The first order IBL model is not sufficient to account for the external vorticity effects.

For a non perturbed flow with a uniform shear

$$\overline{u}_e = 1 + \omega y \ . \tag{9.26a}$$

Van Dyke's second order theory gives [106] (see Problem 9-1)

$$\frac{C_f}{2}\sqrt{\mathcal{R}_x} = 0.332 + 3.126\omega\sqrt{\frac{x}{\mathcal{R}}} \ , \tag{9.26b}$$

with $\omega = 60$ for Flow I. Compared with the Navier-Stokes solutions, the results obtained with Van Dyke's theory overestimate the external vorticity effects. The second order IBL model shows better agreement. This model contains Van Dyke's model but differs from it by terms of order ε^2 (Chap. 10).

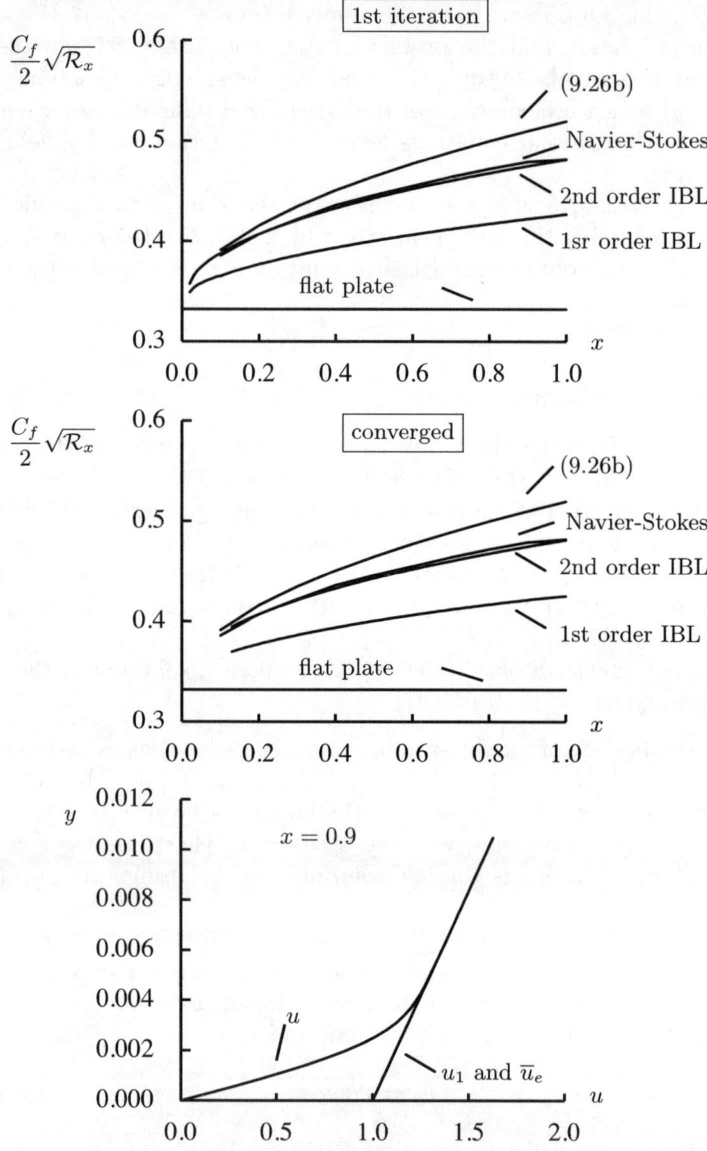

Fig. 9.13. Flow I: $\overline{u}_e = 1 + 60y$

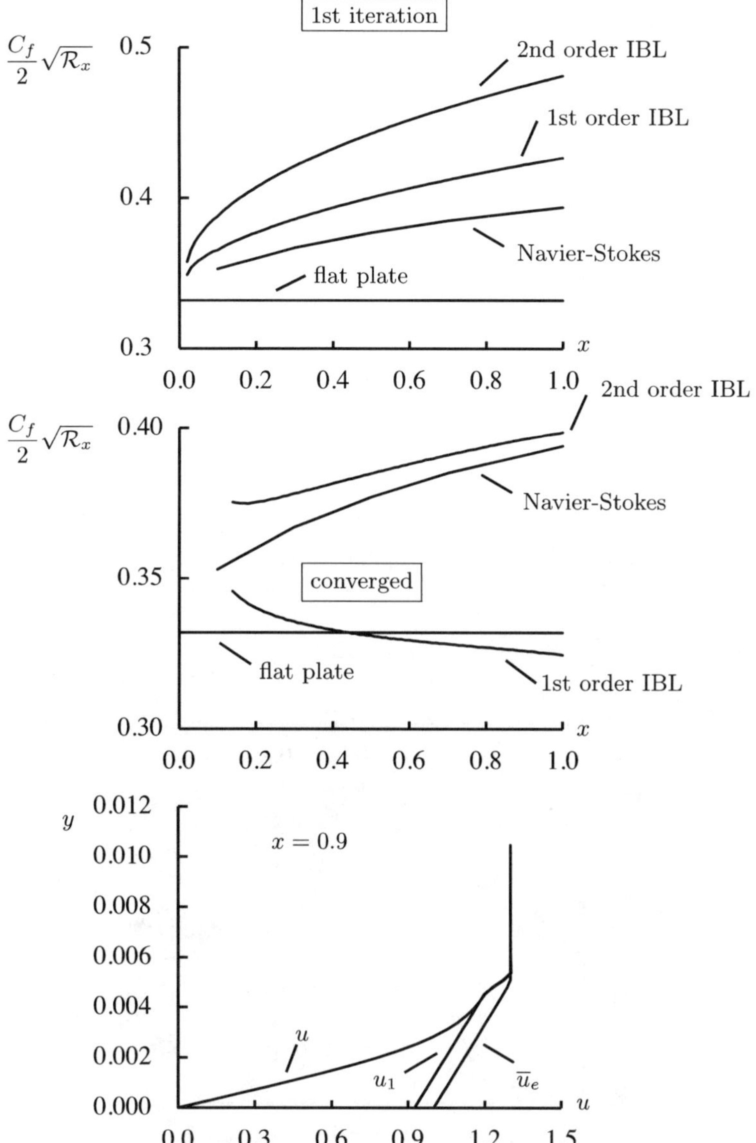

Fig. 9.14. Flow II: $\overline{u}_e = 1 + 60y$ if $y \leq 0.005$; $\overline{u}_e = 1.3$ if $y \geq 0.005$

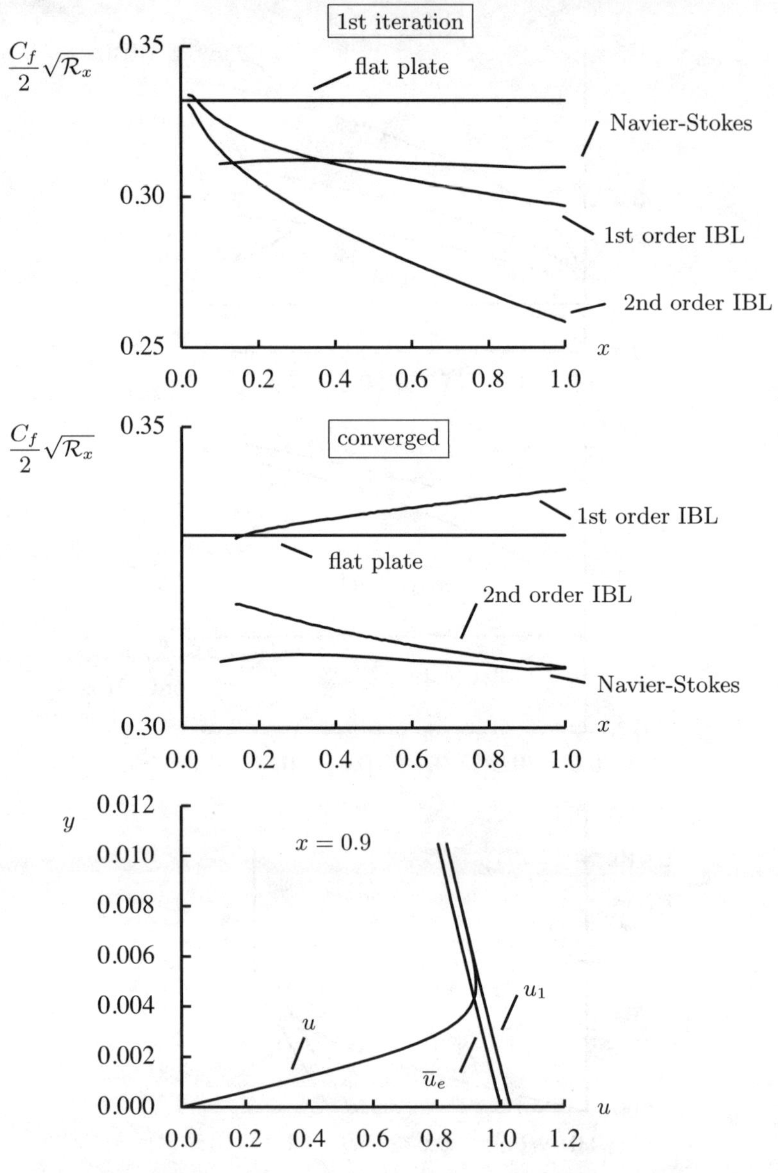

Fig. 9.15. Flow III: $\overline{u}_e = 125y^2 - 20y + 1$ if $y \leq 0.08$; $\overline{u}_e = 0.2$ if $y \geq 0.08$

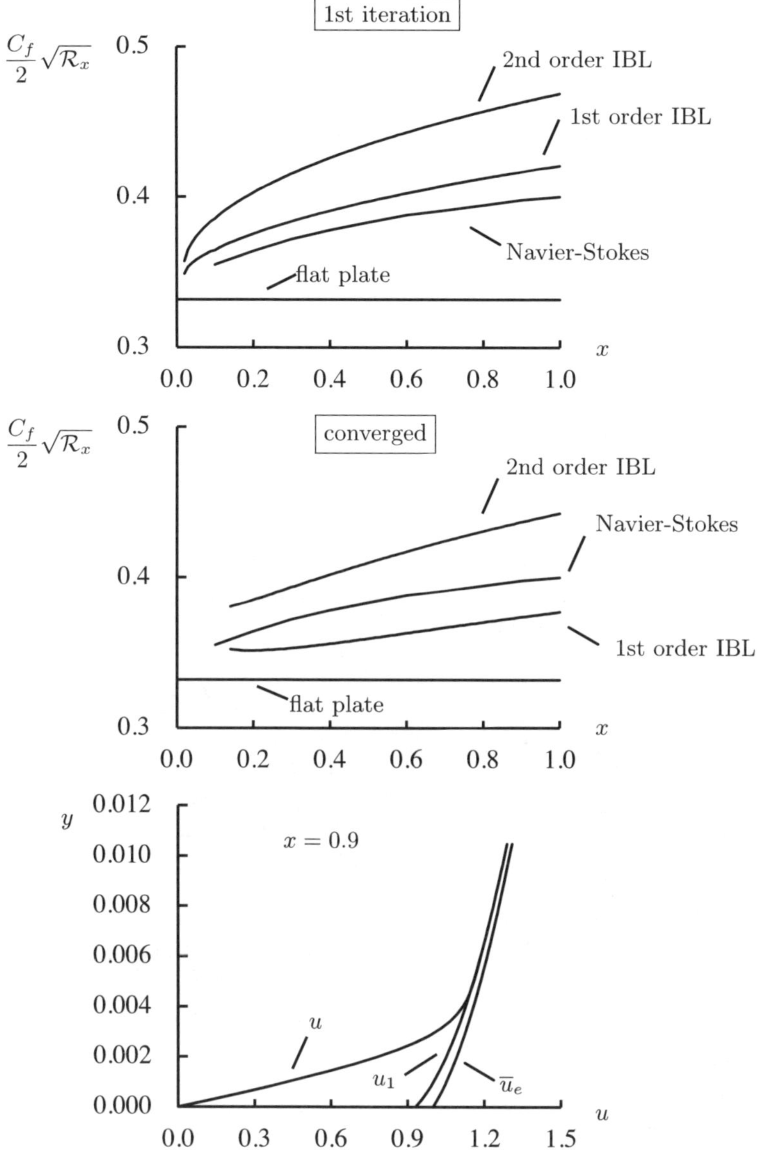

Fig. 9.16. Flow IV: $\overline{u}_e = 0.85 + \sqrt{0.0225 + 18y}$

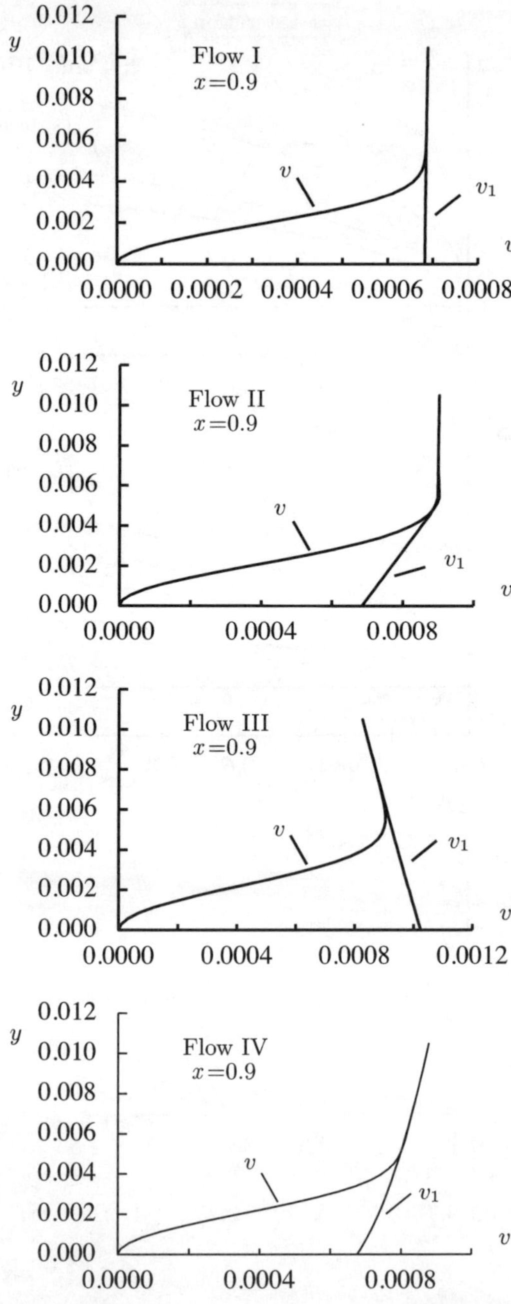

Fig. 9.17. Profiles of velocity normal to the wall

Another difference is that Van Dyke's model is *hierarchical* whereas the IBL model is *interactive*. In Van Dyke's model, the inviscid and viscous flow equations are solved alternately according to the following sequence: i) we solve the first order inviscid flow equations with the slip condition at the wall, ii) we solve the first order standard boundary layer equations, iii) we solve the second order inviscid flow equations in which the boundary layer effects are taken into account, iv) we solve the second order boundary layer equations. With the interactive model, the inviscid and viscous flow equations cannot be solved sequentially due to the boundary conditions which impose a strong coupling of equations; it is necessary to solve all the equations together. A salient advantage of the interactive model is that the calculation of separated flows is possible under the condition to use an appropriate numerical method, for example the method described in Sect. 9.1. By contrast, the application of Van Dyke's model is restricted by Goldstein's singularity and it is impossible to continue the calculations downstream of separation point.

Displacement Effects

Displacement effects are observed by comparing the results labelled "1st iteration" and "converged" or by comparing the distributions of velocities $\overline{u}_e$ and u_1 plotted at station $x = 0.9$ (Figs. 9.13–9.16).

The very weak displacement effect for Flow I is associated with the fact that the right hand side of Poisson's equation (9.21a) is zero, which is not the case for the other flows. An approximate analytical solution, given below, enables us to analyze more deeply the displacement effect.

Equation (9.21a) is present in many problems, for example in the study of the stability of parallel flows [86], in the study of singular perturbation problems of channel flows or in the study of the development of a liquid jet emerging from a two-dimensional channel [103]. An approximate analytical solution can be obtained by seeking a solution in the form

$$\widetilde{\psi} = \widetilde{\psi}(x,0)f(y) \ .$$

Equation (9.21a) becomes

$$\frac{\mathrm{d}^2\widetilde{\psi}(x,0)}{\mathrm{d}x^2}f + \widetilde{\psi}(x,0)\frac{\mathrm{d}^2 f}{\mathrm{d}y^2} = \frac{\widetilde{\psi}(x,0)}{\overline{u}_e}\frac{\mathrm{d}^2\overline{u}_e}{\mathrm{d}y^2}f \ . \tag{9.27}$$

For the application discussed here, the behaviour of $\widetilde{\psi}(x,0)$ is nearly proportional to $x^{1/2}$; this behaviour would be exact for Blasius' boundary layer. With a variation of $\widetilde{\psi}(x,0)$ in $x^{1/2}$, the first term of the left hand side of (9.27) is negligible if

$$x^2 \gg \frac{\overline{u}_e}{\left|\dfrac{\mathrm{d}^2\overline{u}_e}{\mathrm{d}y^2}\right|} \ . \tag{9.28}$$

Then, (9.27) becomes

$$\frac{\mathrm{d}^2 f}{\mathrm{d}y^2} = \frac{f}{\overline{u}_e} \frac{\mathrm{d}^2 \overline{u}_e}{\mathrm{d}y^2} \; . \tag{9.29}$$

Integrating we get

$$f \frac{\mathrm{d}\overline{u}_e}{\mathrm{d}y} - \frac{\mathrm{d}f}{\mathrm{d}y} \overline{u}_e = A \; .$$

A possible solution, if existing, giving a velocity perturbation vanishing at infinity is

$$f = -A\overline{u}_e \int_0^y \frac{1}{\overline{u}_e^2} \, \mathrm{d}y + \frac{\overline{u}_e}{\overline{u}_e(0)} \; ,$$

with

$$A = \frac{1}{\left(\dfrac{\mathrm{d}\overline{u}_e}{\mathrm{d}y}\right)_{y\to\infty} \displaystyle\int_0^\infty \frac{1}{\overline{u}_e^2} \, \mathrm{d}y + \dfrac{1}{\overline{u}_e(y\to\infty)}} \frac{1}{\overline{u}_e(0)} \left(\frac{\mathrm{d}\overline{u}_e}{\mathrm{d}y}\right)_{y\to\infty} \; ,$$

and

$$\widetilde{u}_e = \left[-A\left(\frac{\mathrm{d}\overline{u}_e}{\mathrm{d}y} \int_0^y \frac{1}{\overline{u}_e^2} \, \mathrm{d}y + \frac{1}{\overline{u}_e}\right) + \frac{1}{\overline{u}_e(0)} \frac{\mathrm{d}\overline{u}_e}{\mathrm{d}y} \right] \widetilde{\psi}(x,0) \; .$$

If $\left(\dfrac{\mathrm{d}\overline{u}_e}{\mathrm{d}y}\right)_{y\to\infty} = 0$, the solution is

$$f = \frac{\overline{u}_e}{\overline{u}_e(0)} \; ; \quad \widetilde{u}_e = \widetilde{\psi}(x,0) \frac{1}{\overline{u}_e(0)} \frac{\mathrm{d}\overline{u}_e}{\mathrm{d}y} \; . \tag{9.30}$$

This solution shows the relation between the displacement effect and the distribution of $\overline{u}_e$. In order to investigate its accuracy, the approximate solution has been calculated with $\overline{u}_e(0) = 1$ and $\widetilde{\psi}(x,0) = -1.72\sqrt{\overline{u}_e(0)\nu x}$ which would be the distribution of $\widetilde{\psi}(x,0)$ for Blasius' boundary layer. Comparisons with the numerical solution of Poisson's equation (9.21a) are given in Fig. 9.18. The agreement is excellent for Flow III; for $y \geq 0.08$, the analytical solution gives $\widetilde{u}_e = 0$, which is not strictly correct but the numerical solution yields very small values of $\widetilde{u}_e$. For Flow IV, the agreement is less good but it must be observed that condition imposed by (9.28) cannot be satisfied for too large values of y.

Equation (9.30) shows that if the plate on which the boundary layer develops is infinite, the perturbation $\widetilde{u}_e$ becomes infinite as $x \to \infty$ so that the hypothesis of small perturbations is no longer satisfied and the solution is not valid. The flow calculations have been performed for a limited domain.

The general solution of Poisson's equation (9.21a) includes eigensolutions [103] which are represented neither in the numerical solution nor in the approximate analytical solution. These eigensolutions have not been studied.

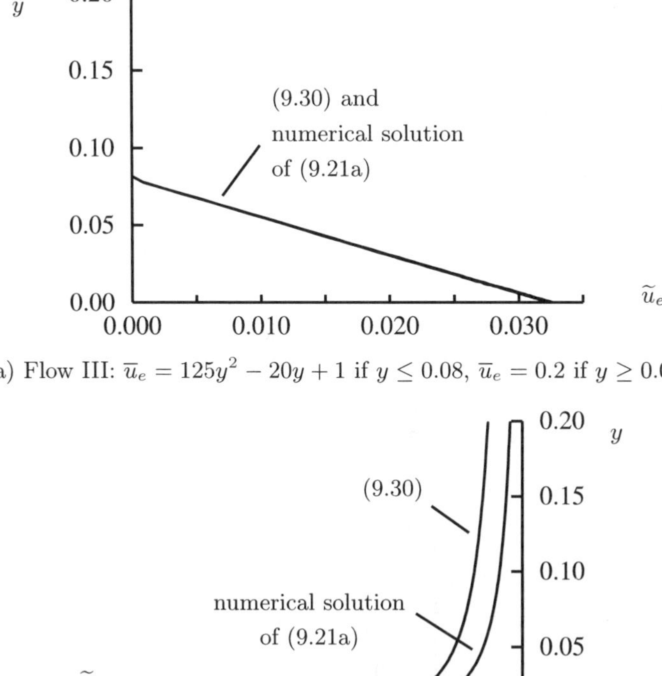

a) Flow III: $\overline{u}_e = 125y^2 - 20y + 1$ if $y \leq 0.08$, $\overline{u}_e = 0.2$ if $y \geq 0.08$

b) Flow IV: $\overline{u}_e = 0.85 + \sqrt{0.0225 + 18y}$

Fig. 9.18. Comparison of the approximate analytical solution and of the numerical solution at station $x = 0.9$

For Flow II, the distribution of $\overline{u}_e$ has a slope discontinuity. The solution of Poisson's equation (9.21a) exhibits a line of discontinuity which is characterized by i) the discontinuity of $\widetilde{u}_e$, ii) the continuity of $\widetilde{\psi}$, iii) the continuity of $\widetilde{v}_e$, iv) the continuity of $\widetilde{p}_e$. From these properties, (9.15) provides a relation between the jumps of $\widetilde{u}_e$ and $\dfrac{\mathrm{d}\overline{u}_e}{\mathrm{d}y}$ through the discontinuity line

$$[\widetilde{u}_e] = \frac{\widetilde{\psi}}{\overline{u}_e}\left[\frac{\mathrm{d}\overline{u}_e}{\mathrm{d}y}\right] , \tag{9.31}$$

where $[\widetilde{u}_e]$ is the jump of $\widetilde{u}_e$ through the discontinuity line. Numerically, for Flow II, along the discontinuity line at station $x = 0.9$, the value of $\widetilde{\psi}$ is $\widetilde{\psi} = -1.618\,10^{-3}$; with $\left[\dfrac{\mathrm{d}\overline{u}_e}{\mathrm{d}y}\right] = -60$ and $\overline{u}_e = 1.3$. Then, the theoretical value of $[\widetilde{u}_e]$ is $[\widetilde{u}_e] = 7.47\,10^{-2}$. Numerically, we obtain $[\widetilde{u}_e] = 7.41\,10^{-2}$ which is reasonably close to the theoretical value.

Limitations of the Model

Figure 9.19 shows a comparison of viscous terms $\dfrac{1}{\mathcal{R}}\dfrac{\partial^2 u}{\partial y^2}$ and $\dfrac{1}{\mathcal{R}}\dfrac{\mathrm{d}^2\overline{u}_e}{\mathrm{d}y^2}$ at station $x = 0.9$ for the different calculated flows. These two terms are present in the original momentum equation and the term $\dfrac{1}{\mathcal{R}}\dfrac{\mathrm{d}^2\overline{u}_e}{\mathrm{d}y^2}$ has been neglected in the IBL model. This hypothesis is justified for Flow I since $\dfrac{1}{\mathcal{R}}\dfrac{\mathrm{d}^2\overline{u}_e}{\mathrm{d}y^2} = 0$. For Flow IV, the jutification is less satisfactory. A limitation of the model is seen here. We assumed that $\overline{u}_e$ satisfies the Euler equations, but it is also necessary that $\overline{u}_e$ satisfies the Navier-Stokes equations with a good approx-

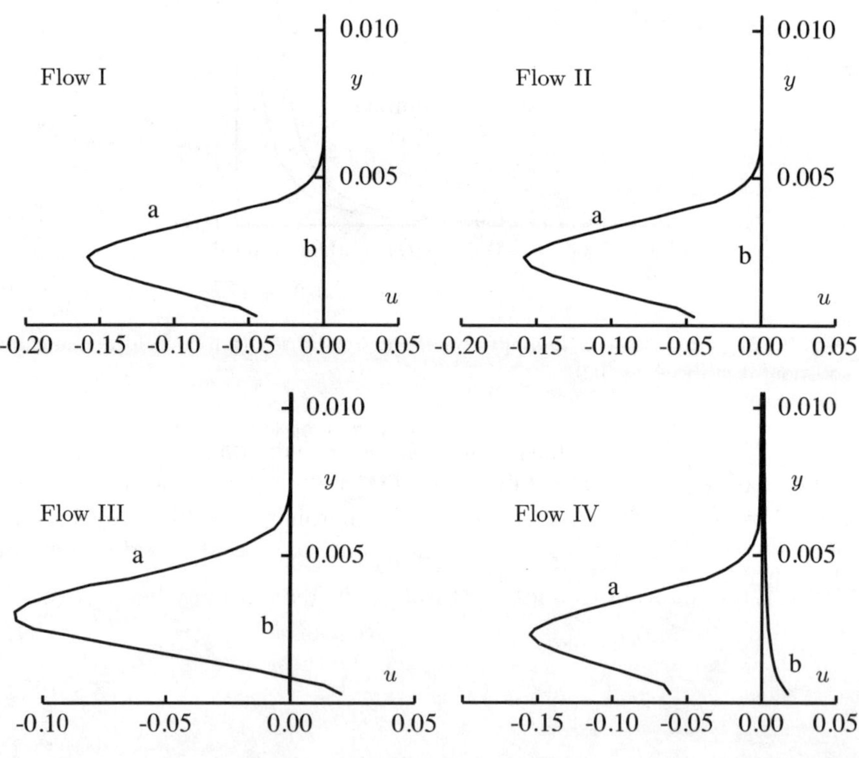

Fig. 9.19. Comparison of viscous terms at station $x = 0.9$: a) $\dfrac{1}{\mathcal{R}}\dfrac{\partial^2 u}{\partial y^2}$; b) $-\dfrac{1}{\mathcal{R}}\dfrac{\mathrm{d}^2\overline{u}_e}{\mathrm{d}y^2}$

imation. Otherwise, the viscous terms associated with this flow have a non negligible contribution which can affect the validity of the approach.

Let us remember that in a divergence-free field we have

$$\triangle\, \boldsymbol{V} = -\,\mathbf{curl}\,(\mathbf{curl}\,\boldsymbol{V})\,. \tag{9.32}$$

This equation shows that the viscous terms are zero if the curl of vorticity is zero. If the first approximation of the inviscid flow is irrotational, the Navier-Stokes equations are satisfied. However, this is no longer true if the inviscid flow is rotational. An exception is when the vorticity is uniform in the field as for Flow I. When the Reynolds number goes to infinity, this problem disapppears because the ratio of the term $\dfrac{1}{\mathcal{R}}\dfrac{\mathrm{d}^2\overline{u}_e}{\mathrm{d}y^2}$ to the boundary layer viscous term goes to zero. Therefore, for finite Reynolds numbers, the IBL model has a limitation when the viscous terms associated with the inviscid flow is not negligible in the sense discussed above.

9.4 Conclusion

For irrotational external flows, the generalized boundary layer equations simplify if their validity is restricted to the boundary layer region. Then, the standard boundary layer equations apply while the interaction with the inviscid flow equations is maintained. The notion of interaction is essential for the calculation of flows with separation. Within the framework of the standard boundary layer theory, the hierarchy between the viscous and inviscid equations leads us to Goldstein's singularity at separation which prevents us from continuing the solution downstream. With IBL, the velocity normal to the wall must satisfy a condition which breaks the hierarchy between the inviscid flow and the boundary layer. This property has a very close counterpart in the triple deck theory which ensures the absence of hicrarchy between the decks.

Then, with appropriate numerical methods, it is possible to calculate flows including those with separation. The application to aerodynamic flows, performed by different authors, have demonstrated that IBL methods are very efficient tools and competitive compared with Navier-Stokes solvers, in terms of accuracy and robustness.

When the boundary layer is fed by a rotational freestream, it is required to use second order IBL. This model accounts for the effects observed with the Navier-Stokes solution very well as far as the variations of the inviscid flow velocity do not induce too large viscous terms.

Problems

9-1. This problem results in Van Dyke's second order theory for a sheared external flow [106]. We consider a flow at high Reynolds number past a flat plate. The flow is laminar, incompressible, two-dimensional, steady. The dimensionless Navier-Stokes equations are

$$\frac{\partial \mathcal{U}}{\partial x} + \frac{\partial \mathcal{V}}{\partial y} = 0 \, ,$$

$$\mathcal{U}\frac{\partial \mathcal{U}}{\partial x} + \mathcal{V}\frac{\partial \mathcal{U}}{\partial y} = -\frac{\partial \mathcal{P}}{\partial x} + \varepsilon^2 \left(\frac{\partial^2 \mathcal{U}}{\partial x^2} + \frac{\partial^2 \mathcal{U}}{\partial y^2} \right) \, ,$$

$$\mathcal{U}\frac{\partial \mathcal{V}}{\partial x} + \mathcal{V}\frac{\partial \mathcal{V}}{\partial y} = -\frac{\partial \mathcal{P}}{\partial y} + \varepsilon^2 \left(\frac{\partial^2 \mathcal{V}}{\partial x^2} + \frac{\partial^2 \mathcal{V}}{\partial y^2} \right) \, ,$$

with

$$\varepsilon^2 = \frac{1}{Re} = \frac{\nu}{VL} \, ,$$

where the Reynolds number Re is based on the reference quantities V and L. The coordinate along the wall is x and the coordinate normal to it is y; the x- and y-velocity components are $\mathcal{U}$ and $\mathcal{V}$ respectively; the pressure is $\mathcal{P}$. The wall is defined by $y = 0$.

At upstream infinity, the flow is given by

$$u_0 = 1 + ay \, ,$$

where a is a constant.

We study the flow with MMAE.

The outer expansion is

$$\mathcal{U} = u_0 + \delta_1 u_1 + \cdots \, ,$$
$$\mathcal{V} = \delta_1 v_1 + \cdots \, ,$$
$$\mathcal{P} = \delta_1 p_1 + \cdots \, ,$$

where $\delta_1(\varepsilon) \prec 1$ is an order function.

The inner expansion is

$$\mathcal{U} = U_1 + \Delta_2 U_2 + \cdots \, ,$$
$$\mathcal{V} = \varepsilon(V_1 + \Delta_2 V_2 + \cdots) \, ,$$
$$\mathcal{P} = \Delta_2^* P_2 + \cdots \, ,$$

where $\Delta_2(\varepsilon)$ and $\Delta_2^*(\varepsilon)$ are two order functions and U_1, V_1, U_2, V_2, P_2 are functions of x and $Y = y/\varepsilon$.

1. Write the outer equations for u_1, v_1, p_1.

2. Write the boundary layer equations for U_1, V_1. Reduce the problem to a differential equation for f by putting $\eta = Y/\sqrt{2x}$ and $U_1 = f'(\eta) = \dfrac{\mathrm{d}f}{\mathrm{d}\eta}$.

Give the boundary conditions and the matching conditions.

We assume that the behaviour of f as $\eta \to \infty$ is

$$f(\eta) \underset{\eta \to \infty}{\cong} \eta - \beta_0 + \text{EST} ,$$

with $\beta_0 = 1.21678$.

3. Write the matching on $\mathcal{V}$ and deduce δ_1 and $v_1(x, 0)$.

4. Find the behaviour of the solution of outer equations as $y \to 0$ in the form

$$u_1(x, y) = a_1(x)y + \cdots ,$$
$$v_1(x, y) = b_0(x) + b_2(x)y^2 + \cdots ,$$
$$p_1(x, y) = c_0(x) + c_1(x)y + c_2(x)y^2 + \cdots ,$$

by assuming that $c_0(0) = 0$ and $a_1(\infty) = 0$. Functions a_1, b_0, b_2, c_0, c_1, c_2 will be determined precisely.

5. Write the matching on $\mathcal{P}$; deduce Δ_2^* and $P_2(x, \infty)$.

6. Write the matching on $\mathcal{U}$; deduce Δ_2 and the behaviour of U_2 as $Y \to \infty$.

7. Give the second order boundary layer equations.

10 Regular Forms of Interactive Boundary Layer

In Chap. 8, the application of the successive complementary expansion method, SCEM, to high Reynolds number flows past streamlined bodies led us to the notion of interactive boundary layer, IBL, to first and to second order. The salient feature of IBL is to ensure the interaction between the equations describing the inviscid and viscous flows.

When the external flow is irrotational, it has been shown that the IBL equations reduce, in the boundary layer region, to Prandtl's equations but their interaction with the inviscid equations is maintained thanks to the matching condition on the velocity normal to the wall.

The interactive nature of these models is essential for the calculation of flows with separation.

In this Chapter, we address the following issue: how are these methods positioned with respect to the standard approximations of Navier-Stokes equations, such that Prandtl's theory, Van Dyke's second order theory, triple deck theory which are the major references in the modelling of aerodynamics? We show that IBL contains the standard models mentioned above.

The method used to prove this result is to start with IBL formulation and seek regular expansions since the common feature to Prandtl's, Van Dyke's and triple deck theories is precisely to be expressed by means of regular expansions. Obviously, this study is performed with the hypotheses corresponding to each particular case. It is possible to follow another approach in which the gauges or the scales are a part of the solution but, for the sake of simplicity, we prefer to consider that the definitions of the asymptotic sequences and of the scales are known.

Figure 10.1 shows the main levels in the classification of models. The objective of this chapter is to prove the elements leading to the different approximations of Navier-Stokes equations.

10.1 Second Order Boundary Layer Model

We want to show that the second order IBL *contains* Van Dyke's second order model when the external flow is rotational. The procedure consists of reformulating the IBL model by means of *regular expansions*.

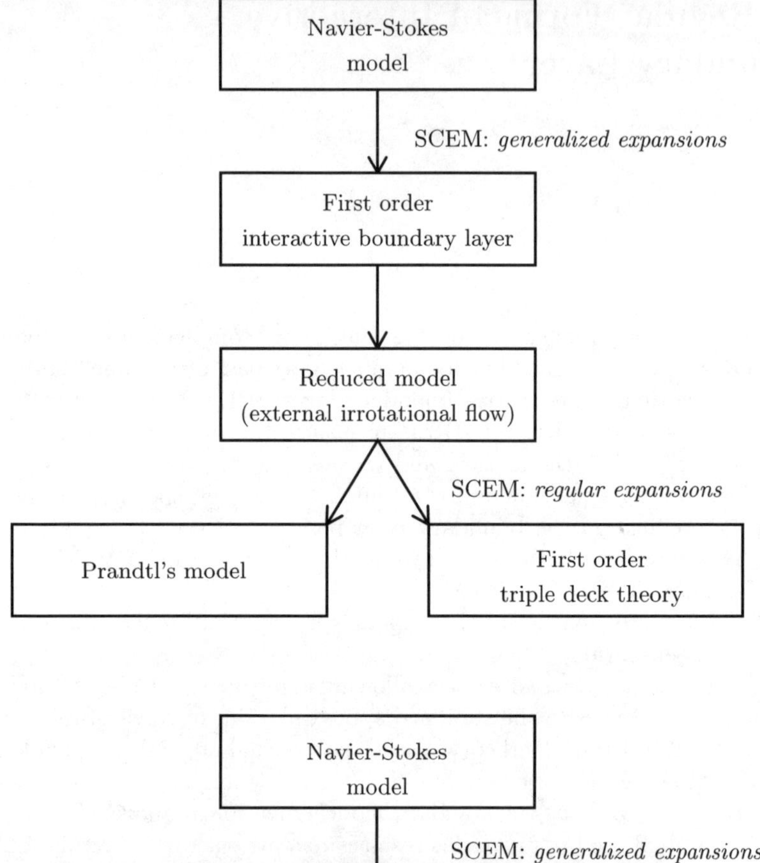

Fig. 10.1. Approximations at high Reynolds number

10.1.1 Second Order Interactive Boundary Layer Model

Let us remember that the second order IBL model yields the following UVAs (Sect. 8.3)

$$u = u_1(x, y, \varepsilon) + U_1(x, Y, \varepsilon) , \tag{10.1a}$$

$$v = v_1(x, y, \varepsilon) + \varepsilon V_1(x, Y, \varepsilon) , \tag{10.1b}$$

$$p = p_1(x, y, \varepsilon) + \varepsilon^2 P_1(x, Y, \varepsilon) . \tag{10.1c}$$

The boundary layer variable Y is

$$Y = \frac{y}{\varepsilon} \quad \text{with} \quad \varepsilon^2 = \frac{1}{\mathcal{R}} = \frac{\mu}{\varrho V L} ,$$

where V and L are reference quantities (see (8.2)).

It has been shown that the flow defined by u, v, p satisfy the generalized boundary layer equations (8.17).

$$\left.\begin{aligned}
\frac{\partial u}{\partial x} + \frac{\partial v}{\partial y} &= 0 \\
u\frac{\partial u}{\partial x} + v\frac{\partial u}{\partial y} &= u_1\frac{\partial u_1}{\partial x} + v_1\frac{\partial u_1}{\partial y} + \frac{1}{\mathcal{R}}\frac{\partial^2(u - u_1)}{\partial y^2}
\end{aligned}\right\} , \tag{10.2}$$

and that the flow defined by u_1, v_1, p_1 satisfy the Euler equations

$$\left.\begin{aligned}
\frac{\partial u_1}{\partial x} + \frac{\partial v_1}{\partial y} &= 0 \\
u_1\frac{\partial u_1}{\partial x} + v_1\frac{\partial u_1}{\partial y} &= -\frac{\partial p_1}{\partial x} \\
u_1\frac{\partial v_1}{\partial x} + v_1\frac{\partial v_1}{\partial y} &= -\frac{\partial p_1}{\partial y}
\end{aligned}\right\} . \tag{10.3}$$

The boundary conditions are

$$\left.\begin{aligned}
\text{at the wall}: \quad u &= 0 \quad , \quad v = 0 \\
y \to \infty \quad : \quad u - u_1 &\to 0 , \quad v - v_1 \to 0
\end{aligned}\right\} . \tag{10.4}$$

Conditions at infinity are also imposed for the field described by the Euler equations.

10.1.2 Van Dyke's Second Order Model

For the sake of simplicity, we assume that the form of the regular expansions is known. Strictly, this assumption is not necessary and the form of the expansions can be demonstrated [105]; this work, however, is not really useful and would make the discussion cumbersome.

For the external flow, we seek regular AEs in the form

$$u_1 = \mathrm{E}_0^1\, u_1 + \cdots ,$$
$$v_1 = \mathrm{E}_0^2\, v_1 + \cdots ,$$
$$p_1 = \mathrm{E}_0^1\, p_1 + \cdots ,$$

where E_0^i is an outer expansion operator to order $\mathrm{O}(\varepsilon^i)$.

The flow described by u_1, v_1, p_1 satisfies the Euler equations. Then, with

$$\mathrm{E}_0^1\, u_1 = \hat{u}_1(x, y) + \varepsilon \hat{u}_2(x, y) , \tag{10.5a}$$
$$\mathrm{E}_0^2\, v_1 = \hat{v}_1(x, y) + \varepsilon \hat{v}_2(x, y) + \varepsilon^2 \hat{v}_3(x, y) , \tag{10.5b}$$
$$\mathrm{E}_0^1\, p_1 = \hat{p}_1(x, y) + \varepsilon \hat{p}_2(x, y) . \tag{10.5c}$$

it is clear that the flow defined by $\hat{u}_1$, $\hat{v}_1$, $\hat{p}_1$ satisfies the Euler equations and that the flow defined by $\hat{u}_2$, $\hat{v}_2$, $\hat{p}_2$ satisfies the linearized Euler equations.

To the considered order, UVAs given by (10.1a–10.1c) write

$$u = \mathrm{E}_0^1\, u_1 + U_1 + \cdots , \tag{10.6a}$$
$$v = \mathrm{E}_0^2\, v_1 + \varepsilon V_1 + \cdots , \tag{10.6b}$$
$$p = \mathrm{E}_0^1\, p_1 + \cdots . \tag{10.6c}$$

Restricting the validity of these expansions to the boundary layer region, with $y = \varepsilon Y$, we have

$$u = \mathrm{E}_1^1\, \mathrm{E}_0^1\, u_1 + U_1 + \cdots ,$$
$$v = \mathrm{E}_1^2\, \mathrm{E}_0^2\, v_1 + \varepsilon V_1 + \cdots ,$$

where E_1^i is an expansion operator in the boundary layer to order $\mathrm{O}(\varepsilon^i)$.

Note 10.1. Assuming that $\hat{u}_1$, $\hat{v}_1$ and $\hat{u}_2$, $\hat{v}_2$ are expandable in Taylor series near the wall, we could write relations of the type

$$\mathrm{E}_1^1\, \mathrm{E}_0^1\, u_1 = \hat{u}_1(x, 0) + \varepsilon Y \left(\frac{\partial \hat{u}_1}{\partial y} \right)_{y=O} + \varepsilon \hat{u}_2(x, 0) .$$

This writing is not only not always possible (for example for a channel flow) but also is not required.

Then, we define U and V by

$$U = \mathrm{E}_1^1\, \mathrm{E}_0^1\, u_1 + U_1 ,$$
$$\varepsilon V = \mathrm{E}_1^2\, \mathrm{E}_0^2\, v_1 + \varepsilon V_1 .$$

UVAs given by (10.6a–10.6b) write

$$u = \mathrm{E}_0^1\, u_1 - \mathrm{E}_1^1\, \mathrm{E}_0^1\, u_1 + U + \cdots , \tag{10.7a}$$
$$v = \mathrm{E}_0^2\, v_1 - \mathrm{E}_1^2\, \mathrm{E}_0^2\, v_1 + \varepsilon V + \cdots . \tag{10.7b}$$

If we seek regular expansions for U and V in the form

$$\mathrm{E}_1^1\, U = \bar{u}_1(x, Y) + \varepsilon \bar{u}_2(x, Y) \, , \tag{10.8a}$$

$$\mathrm{E}_1^1\, V = \bar{v}_1(x, Y) + \varepsilon \bar{v}_2(x, Y) \, , \tag{10.8b}$$

UVAs given by (10.7a–10.7b) write

$$u = \mathrm{E}_0^1\, u_1 - \mathrm{E}_1^1\, \mathrm{E}_0^1\, u_1 + \mathrm{E}_1^1\, U + \cdots \, , \tag{10.9a}$$

$$v = \mathrm{E}_0^2\, v_1 - \mathrm{E}_1^2\, \mathrm{E}_0^2\, v_1 + \mathrm{E}_1^2\, \varepsilon V + \cdots \, , \tag{10.9b}$$

or, in the boundary layer,

$$u = \mathrm{E}_1^1\, U + \cdots \, , \tag{10.10a}$$

$$v = \varepsilon\, \mathrm{E}_1^1\, V + \cdots \, . \tag{10.10b}$$

With these expressions, we write the boundary conditions. At the wall, we have

$$Y = 0 : \quad u = 0 \, , \quad v = 0 \, , \tag{10.11}$$

hence

$$Y = 0 : \quad \mathrm{E}_1^1\, U = 0 \, , \quad \mathrm{E}_1^1\, V = 0 \, . \tag{10.12}$$

Note 10.2. It can happen that terms such that $\mathrm{E}_0^1\, u_1$ or $\mathrm{E}_0^2\, v_1$ are not bounded at the wall, but terms such that $\mathrm{E}_1^1(\mathrm{E}_0^1\, u_1 - \mathrm{E}_1^1\, \mathrm{E}_0^1\, u_1)$ or $\mathrm{E}_1^2(\mathrm{E}_0^2\, v_1 - \mathrm{E}_1^2\, \mathrm{E}_0^2\, v_1)$ are identically equal to zero.

In particular, we deduce

$$Y = 0 : \quad \bar{u}_1 = 0 \, , \quad \bar{u}_2 = 0 \, , \quad \bar{v}_1 = 0 \, , \quad \bar{v}_2 = 0 \, . \tag{10.13}$$

The conditions

$$Y \to \infty : \quad u - u_1 \to 0 \, , \quad v - v_1 \to 0 \tag{10.14}$$

can be written as

$$Y \to \infty : \quad \mathrm{E}_1^1\, U - \mathrm{E}_1^1\, \mathrm{E}_0^1\, u_1 = 0 \, , \quad \mathrm{E}_1^2\, \varepsilon V - \mathrm{E}_1^2\, \mathrm{E}_0^2\, v_1 = 0 \, , \tag{10.15}$$

whence

$$Y \to \infty : \quad \mathrm{E}_1^1\, U - \mathrm{E}_1^1(\hat{u}_1 + \varepsilon \hat{u}_2) = 0 \, , \quad \mathrm{E}_1^1\, \varepsilon V - \mathrm{E}_1^1(\hat{v}_1 + \varepsilon \hat{v}_2) = 0 \, . \tag{10.16}$$

Assuming the existence of Taylor series expansions in the neighbourhood of $y = 0$, we obtain to first order

$$\lim_{Y \to \infty} \bar{u}_1 = \hat{u}_1(x, 0) \, , \tag{10.17a}$$

$$\hat{v}_1(x, 0) = 0 \, , \tag{10.17b}$$

and, to second order

$$\lim_{Y\to\infty}\left[\overline{u}_2 - Y\left(\frac{\partial \hat{u}_1}{\partial y}\right)_{y=0}\right] = \hat{u}_2(x,0)\ , \tag{10.18a}$$

$$\lim_{Y\to\infty}\left[\overline{v}_1 - Y\left(\frac{\partial \hat{v}_1}{\partial y}\right)_{y=0}\right] = \hat{v}_2(x,0)\ . \tag{10.18b}$$

Replacing u and v by means of (10.10a, 10.10b) in the boundary layer, (10.2) become

$$\frac{\partial\, \mathrm{E}_1^1\, U}{\partial x} + \frac{\partial\, \mathrm{E}_1^1\, V}{\partial Y} = 0\ , \tag{10.19a}$$

$$\mathrm{E}_1^1\, U\frac{\partial\, \mathrm{E}_1^1\, U}{\partial x} + \mathrm{E}_1^1\, V\frac{\partial\, \mathrm{E}_1^1\, U}{\partial Y} = -\frac{\partial\, \mathrm{E}_1^1\, p_1}{\partial x} + \frac{\partial^2\, \mathrm{E}_1^1\, U}{\partial Y^2}\ . \tag{10.19b}$$

If the wall curvature is small, the term $\dfrac{\partial \hat{p}_1}{\partial y}$ is o(1) at the wall. Then, it can also be shown that the term $\dfrac{\partial^2 \hat{u}_1}{\partial x \partial y}$ is o(1) at the wall. Under these conditions, we can write

$$\mathrm{E}_1^1\, p_1 = \hat{p}_1(x,0) + \varepsilon \hat{p}_2(x,0)\ . \tag{10.20}$$

To first order, (10.19a–10.19b) yield Prandtl's boundary layer equations (see Problem 8-1)

$$\frac{\partial \overline{u}_1}{\partial x} + \frac{\partial \overline{v}_1}{\partial Y} = 0\ , \tag{10.21a}$$

$$\overline{u}_1\frac{\partial \overline{u}_1}{\partial x} + \overline{v}_1\frac{\partial \overline{u}_1}{\partial Y} = -\frac{\mathrm{d}\hat{p}_1(x,0)}{\mathrm{d}x} + \frac{\partial^2 \overline{u}_1}{\partial Y^2}\ , \tag{10.21b}$$

with

$$-\frac{\mathrm{d}\hat{p}_1(x,0)}{\mathrm{d}x} = \hat{u}_1(x,0)\frac{\mathrm{d}\hat{u}_1(x,0)}{\mathrm{d}x}\ . \tag{10.21c}$$

Moreover, conditions given by (10.13) and (10.17a) are the usual boundary conditions in Prandtl's theory.

To second order, we obtain the linearized boundary layer equations

$$\frac{\partial \overline{u}_2}{\partial x} + \frac{\partial \overline{v}_2}{\partial Y} = 0\ , \tag{10.22a}$$

$$\overline{u}_1\frac{\partial \overline{u}_2}{\partial x} + \overline{u}_2\frac{\partial \overline{u}_1}{\partial x} + \overline{v}_1\frac{\partial \overline{u}_2}{\partial Y} + \overline{v}_2\frac{\partial \overline{u}_1}{\partial Y} = -\frac{\mathrm{d}\hat{p}_2(x,0)}{\mathrm{d}x} + \frac{\partial^2 \overline{u}_2}{\partial Y^2}\ , \tag{10.22b}$$

with

$$-\frac{\mathrm{d}\hat{p}_2(x,0)}{\mathrm{d}x} = \left[\hat{u}_1\frac{\partial\hat{u}_2}{\partial x} + \hat{u}_2\frac{\partial\hat{u}_1}{\partial x} + \hat{v}_2\frac{\partial\hat{u}_1}{\partial y}\right]_{y=0}. \tag{10.22c}$$

With the boundary conditions given by (10.13) and (10.17a, 10.17b, 10.18a, 10.18b), we recover exactly Van Dyke's second order model.

It can be shown that the streamwise momentum equation is satisfied as $Y \to \infty$ if $\left(\dfrac{\partial^2\hat{u}_1}{\partial x\partial y}\right)_{y=0}$ is negligible.

Thus, *the second order IBL contains Van Dyke's second order model.* The two models are not strictly equivalent but the differences are due to terms which are neglected in the regular expansions.

We can show that the first order IBL *does not contain* Van Dyke's second order model when the external flow is *rotational.* However, for an *irrotational* external flow, the first order IBL *contains* Van Dyke's second order model; the study conducted in Sect. 8.5 lets us anticipate this result.

10.2 Triple Deck Model

10.2.1 Flow on a Flat Plate with a Small Hump

We consider a laminar, incompressible, two-dimensional flow past a flat plate at high Reynolds number. The oncoming flow is uniform and, therefore, irrotational. We assume that the perturbation produced by the boundary layer on the inviscid flow is of order ε for the velocity components and for the pressure. In addition, we assume that a small deformation of the wall induces a perturbation formally of the same order (in fact, the perturbation could be stronger). A local separated zone can be present.

The reduced IBL model developed in Sect. 8.5 is well suited to study this problem. We want to show that this model contains the first order triple deck model.

The velocities, the lengths and the pressure are nondimensionalized with reference quantities V, L and ϱV^2. Here, the reference velocity is equal to the freestream velocity and the reference length L is the distance between the plate leading edge and the location of the hump (Fig. 10.2). The Reynolds number $\mathcal{R}$ is

$$\mathcal{R} = \frac{\varrho V L}{\mu}.$$

Let us remember that the proposed UVAs are given by (8.22a–8.22c)

$$\left.\begin{aligned}
u_{\mathrm{a}} &= U + u_1 - u_{10} \\[4pt]
v_{\mathrm{a}} &= \varepsilon V + v_1 - v_{10} + y u_{1x0} \\[4pt]
\left(\frac{\partial p}{\partial y}\right)_{\mathrm{a}} &= \varepsilon\frac{\partial P}{\partial Y} + p_{1y} - p_{1y0} - y p_{1yy0}
\end{aligned}\right\}, \tag{10.23}$$

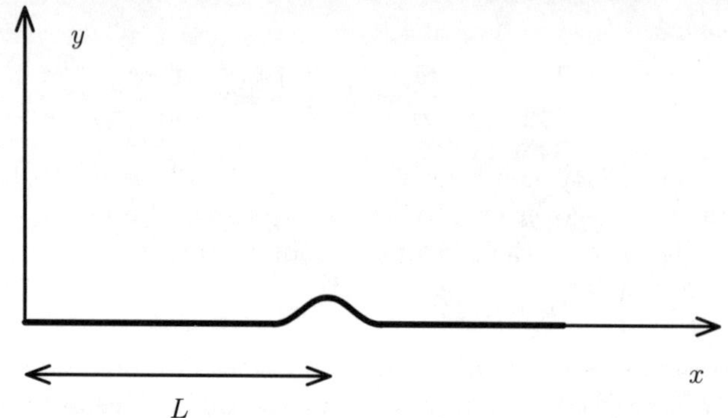

Fig. 10.2. Flow on a flat plate deformed by a hump

with

$$u_{1x} = \frac{\partial u_1}{\partial x} \;,\quad p_{1y} = \frac{\partial p_1}{\partial y} \;,\quad p_{1yy} = \frac{\partial^2 p_1}{\partial y^2} \;,$$

and the index "0" denotes a value at the wall.

Functions u_1, v_1, p_1 describe the flow outside the boundary layer whereas functions U, V, P describe the flow in the boundary layer. The boundary layer variable is Y

$$Y = \frac{y}{\varepsilon} \;,$$

with

$$\varepsilon^2 = \frac{1}{\mathcal{R}} \;.$$

In the boundary layer, the generalized boundary layer equations reduce to (8.24) and (8.25)

$$\frac{\partial U}{\partial x} + \frac{\partial V}{\partial Y} = 0 \;, \tag{10.24a}$$

$$U\frac{\partial U}{\partial x} + V\frac{\partial U}{\partial Y} = u_{10}u_{1x0} + \frac{\partial^2 U}{\partial Y^2} \;, \tag{10.24b}$$

$$U\frac{\partial V}{\partial x} + V\frac{\partial V}{\partial Y} = -\frac{\partial P}{\partial Y} + \frac{\partial^2 V}{\partial Y^2} \;. \tag{10.24c}$$

Functions u_1, v_1, p_1 satisfy the Euler equations

$$\frac{\partial u_1}{\partial x} + \frac{\partial v_1}{\partial y} = 0 \;, \tag{10.25a}$$

$$u_1\frac{\partial u_1}{\partial x} + v_1\frac{\partial u_1}{\partial y} = -\frac{\partial p_1}{\partial x} \;, \tag{10.25b}$$

$$u_1\frac{\partial v_1}{\partial x} + v_1\frac{\partial v_1}{\partial y} = -\frac{\partial p_1}{\partial y} \;. \tag{10.25c}$$

In addition, we have the wall boundary conditions (8.26a–8.26b)

$$Y = 0 : \quad U = 0 \,, \quad V = 0 \,, \tag{10.26}$$

and the conditions at the boundary layer edge (8.26c–8.26d)

$$Y \to \infty : \quad U \to u_{10} \,, \quad \lim_{Y \to \infty} (V + Y u_{1x0}) = \frac{v_{10}}{\varepsilon} \,. \tag{10.27}$$

Finally, at infinity, the flow is uniform.

10.2.2 Regular Expansions

The model discussed in Subsect. 10.2.1 is based on generalized expansions. We want to show that this model, reformulated with regular expansions, contains the triple deck model. Obviously, it is required to choose the scales and the gauges adapted to this theory. It would be possible to demonstrate all these results but, to simplify the presentation, the scales and gauges are assumed to be known (see Appendix IV and Problem 10-3). Thus, the triple deck theory consists of studying the flow in the neighbourhood of point x_0 where the bounday layer is perturbed and the length scale is $\varepsilon^{3/4}$, so that the appropriate streamwise variable is

$$X = \frac{x - x_0}{\varepsilon^{3/4}} \,. \tag{10.28}$$

With the chosen reference quantities, we have $x_0 = 1$.

Upper Deck

In the outer deck, the normal to the wall variable is

$$Y^* = \frac{y}{\varepsilon^{3/4}} \,. \tag{10.29}$$

From the triple deck theory, functions u_1, v_1, p_1 associated with the external flow are normalized as

$$u_1(x, y, \varepsilon) = 1 + \varepsilon^{1/2} U^*(X, Y^*, \varepsilon) \,, \tag{10.30a}$$
$$v_1(x, y, \varepsilon) = \varepsilon^{1/2} V^*(X, Y^*, \varepsilon) \,, \tag{10.30b}$$
$$p_1(x, y, \varepsilon) = \varepsilon^{1/2} P^*(X, Y^*, \varepsilon) \,. \tag{10.30c}$$

In the expression of u_1, the value 1 is introduced because, at infinity, we have $u_1 \to 1$. Then, (10.25a–10.25c) become

$$\frac{\partial U^*}{\partial X} + \frac{\partial V^*}{\partial Y^*} = 0 \,,$$

$$(1 + \varepsilon^{1/2} U^*)\frac{\partial U^*}{\partial X} + \varepsilon^{1/2} V^* \frac{\partial U^*}{\partial Y^*} = -\frac{\partial P^*}{\partial X} \,,$$

$$(1 + \varepsilon^{1/2} U^*)\frac{\partial V^*}{\partial X} + \varepsilon^{1/2} V^* \frac{\partial V^*}{\partial Y^*} = -\frac{\partial P^*}{\partial Y^*} \,.$$

The regular expansions of U^*, V^* et P^* are

$$U^*(X, Y^*, \varepsilon) = U_1^*(X, Y^*) + \varepsilon^{1/4} U_2^*(X, Y^*) + \cdots ,$$
$$V^*(X, Y^*, \varepsilon) = V_1^*(X, Y^*) + \varepsilon^{1/4} V_2^*(X, Y^*) + \cdots ,$$
$$P^*(X, Y^*, \varepsilon) = P_1^*(X, Y^*) + \varepsilon^{1/4} P_2^*(X, Y^*) + \cdots ,$$

and, obviously, we recover the upper deck equations to first order

$$\frac{\partial U_1^*}{\partial X} + \frac{\partial V_1^*}{\partial Y^*} = 0 , \tag{10.31a}$$

$$\frac{\partial U_1^*}{\partial X} = -\frac{\partial P_1^*}{\partial X} , \tag{10.31b}$$

$$\frac{\partial V_1^*}{\partial X} = -\frac{\partial P_1^*}{\partial Y^*} , \tag{10.31c}$$

and to second order

$$\frac{\partial U_2^*}{\partial X} + \frac{\partial V_2^*}{\partial Y^*} = 0 , \tag{10.32a}$$

$$\frac{\partial U_2^*}{\partial X} = -\frac{\partial P_2^*}{\partial X} , \tag{10.32b}$$

$$\frac{\partial V_2^*}{\partial X} = -\frac{\partial P_2^*}{\partial Y^*} . \tag{10.32c}$$

UVAs given by (10.23) become

$$\left. \begin{aligned}
u_{\mathrm{a}} &= U + \varepsilon^{1/2}(U^* - U_0^*) \\[4pt]
v_{\mathrm{a}} &= \varepsilon^{1/2}(V^* - V_0^* + Y^* U_{X0}^*) + \varepsilon V \\[4pt]
\left(\frac{\partial p}{\partial y}\right)_{\mathrm{a}} &= \varepsilon^{-1/4}\left[\frac{\partial P^*}{\partial Y^*} - \left(\frac{\partial P^*}{\partial Y^*}\right)_{Y^*=0} - Y^*\left(\frac{\partial^2 P^*}{\partial Y^{*2}}\right)_{Y^*=0}\right] \\
&\quad + \varepsilon\frac{\partial P}{\partial Y}
\end{aligned} \right\} , \tag{10.33}$$

where

$$U_0^* = U^*(X, 0, \varepsilon) , \quad V_0^* = V^*(X, 0, \varepsilon) , \quad U_{X0}^* = \left(\frac{\partial U^*}{\partial X}\right)_{Y^*=0} .$$

From (10.30a–10.30b) and conditions given by (10.27), we can write

$$\lim_{Y\to\infty} U = 1 + \varepsilon^{1/2} U_0^* , \tag{10.34a}$$

$$\lim_{Y\to\infty}\left(V + \varepsilon^{-1/4} Y U_{X0}^*\right) = \varepsilon^{-1/2} V_0^* . \tag{10.34b}$$

The conditions of uniform flow at infinity yield

$$Y^* \to \infty : \quad U_1^* = 0 , \quad U_2^* = 0 , \quad V_1^* = 0 , \quad V_2^* = 0 . \tag{10.35}$$

Main Deck and Lower Deck

These two regions correspond to the boundary layer region defined in the reduced IBL model. From the triple deck theory, the flow characteristics U, V, P are written as

$$U(x,Y,\varepsilon) = U_0(x,Y) + \varepsilon^{1/4}\widehat{U}(X,Y,\varepsilon) \,, \tag{10.36a}$$

$$V(x,Y,\varepsilon) = V_0(x,Y) + \varepsilon^{-1/2}\widehat{V}(X,Y,\varepsilon) \,, \tag{10.36b}$$

$$P(x,Y,\varepsilon) = P_0(x,Y) + \varepsilon^{-3/2}\widehat{P}(X,Y,\varepsilon) \,, \tag{10.36c}$$

where U_0 and V_0 are the velocity components of the non perturbed boundary layer, i.e. Blasius' solution.

Substituting (10.36a–10.36c) into (10.24a–10.24c) and taking into account that U_0 and V_0 satisfy Blasius' equations, we obtain

$$\frac{\partial\widehat{U}}{\partial X} + \frac{\partial\widehat{V}}{\partial Y} = 0 \,, \tag{10.37a}$$

$$U_0\frac{\partial\widehat{U}}{\partial X} + \widehat{V}\frac{\partial U_0}{\partial Y} + \varepsilon^{1/4}\left(\widehat{U}\frac{\partial\widehat{U}}{\partial X} + \widehat{V}\frac{\partial\widehat{U}}{\partial Y}\right) + \varepsilon^{3/4}\left(\widehat{U}\frac{\partial U_0}{\partial x} + V_0\frac{\partial\widehat{U}}{\partial Y}\right)$$

$$= \varepsilon^{1/4}\left(1 + \varepsilon^{1/2}U_0^*\right)U_{X0}^* + \varepsilon^{3/4}\frac{\partial^2\widehat{U}}{\partial Y^2} \,, \tag{10.37b}$$

$$\varepsilon^{1/4}U_0\frac{\partial\widehat{V}}{\partial X} = -\frac{\partial\widehat{P}}{\partial Y} + \mathrm{O}(\varepsilon^{1/2}) \,. \tag{10.37c}$$

UVAs given by (10.33) become

$$\left.\begin{array}{l}
u_{\mathrm{a}} = U_0 + \varepsilon^{1/4}\widehat{U} + \varepsilon^{1/2}(U^* - U_0^*) \\[2mm]
v_{\mathrm{a}} = \varepsilon^{1/2}(\widehat{V} + V^* - V_0^* + Y^*U_{X0}^*) + \varepsilon V_0 \\[2mm]
\left(\dfrac{\partial p}{\partial y}\right)_{\mathrm{a}} = \varepsilon^{-1/4}\left[\dfrac{\partial P^*}{\partial Y^*} - \left(\dfrac{\partial P^*}{\partial Y^*}\right)_{Y^*=0}\right] \\[3mm]
\quad -\varepsilon^{-1/4}Y^*\left(\dfrac{\partial^2 P^*}{\partial Y^{*2}}\right)_{Y^*=0} + \varepsilon^{-1/2}\dfrac{\partial\widehat{P}}{\partial Y} + \varepsilon\dfrac{\partial P_0}{\partial Y}
\end{array}\right\} \,. \tag{10.38}$$

From (10.36a–10.36b) and boundary conditions given by (10.34a–10.34b), we can write

$$\lim_{Y\to\infty}\widehat{U} = \varepsilon^{1/4}U_0^* \,, \tag{10.39a}$$

$$\lim_{Y\to\infty}\left(\widehat{V} + \varepsilon^{1/4}YU_{X0}^* + \varepsilon^{1/2}V_0\right) = V_0^* \,. \tag{10.39b}$$

The wall conditions yield

$$\widehat{U} = 0 \,, \quad \widehat{V} = 0 \,. \tag{10.40}$$

The following conclusions are drawn

- Equations (10.31a–10.31c) are identical to order 1 equations of the upper deck in the triple deck theory.
- Equations (10.37a–10.37c) contain the set of equations (IV.7a–IV.7c) which contains itself the order 1 equations of the main deck and of the lower deck.
- Boundary conditions given by (10.35), (10.39a), (10.39b), (10.40) are identical to the boundary conditions in the triple deck theory.

It is concluded that the first order triple deck model is contained in the reduced IBL model of Subsect. 10.2.1. Now, this model is contained itself in the first order IBL model (Sect. 8.5). Finally, it has been proved that *the first order IBL contains the first order triple deck model.*

On the other hand, the second order triple deck model *is not included* in the IBL model, even in the second order model. In the IBL model, the pressure term which is present in the streamwise momentum equation is obtained from the Euler equations. In the triple deck theory, the pressure $\widetilde{P}_2$ which is present in the streamwise momentum equation of the lower deck is constant in the $\widetilde{Y}$-direction and matches with the main deck pressure $P_2(X, 0)$ which is not given by the upper deck equations but by the main deck equations (Appendix IV).

The term $\dfrac{\partial \widetilde{P}_2}{\partial X}$ cannot be recovered from the second order IBL model.

10.3 Summary of Approximations of Navier-Stokes Equations

In a first step (Fig. 10.1), SCEM, applied to high Reynolds number flows, leads us to the first and second order IBL models thanks to the implementation of generalized expansions. The viscous-inviscid interaction is one of the most salient features of these models.

IBL models simplify in different circumstances. For example, when the external flow is irrotational, the same reduced model can be used and obtained by starting from the first or from the second order IBL model. It has been shown that the equations reduce to Prandtl's equations in the boundary layer. However, the interaction with the inviscid flow is maintained.

The reduced model contains Prandtl's boundary layer and the first order triple deck model.

Van Dyke's second order model is contained in the second order IBL but not in the first order IBL when the external flow is rotational.

10.4 Conclusion

The first and second order IBL models and the reduced model take into account a mutual action between the viscous and inviscid flows. This interaction results mainly from the matching of the normal to the wall velocity between the two flow regions.

Prandtl's model, Van Dyke's model and triple deck model are degeneracies of the interactive boundary layer, IBL, models obtained with regular expansions. With Prandtl's model and Van Dyke's model, the interaction is replaced by a hierarchy between the sets of equations describing the inviscid flow and the boundary layer. This modification results from the shift of orders of magnitude on the normal to the wall velocity in the invisicid flow and in the boundary layer. In the triple deck theory, the normal velocity recovers the same order in the upper deck and in the main deck. This choice of gauges is essential to treat separated flows.

Problems

10-1. We consider the equation

$$\varepsilon^3 \frac{\mathrm{d}^2 y}{\mathrm{d}x^2} + x^3 \frac{\mathrm{d}y}{\mathrm{d}x} + (x^3 - \varepsilon)y = 0 \ ,$$

with

$$y(0) = \alpha \ , \quad y(1) = \beta \ .$$

1. At first, MMAE is applied.

   Give an outer approximation of the solution.

   As the coefficient of $\dfrac{\mathrm{d}y}{\mathrm{d}x}$ is positive, we expect the existence of a boundary layer in the neighbourhood of $x = 0$. Give the thickness of this boundary layer. Determine the corresponding approximation. Show that the matching with the outer approximation is impossible.

   We deduce that an intermediate layer is required. Show that the thickness of this layer is $\varepsilon^{1/2}$.

   Give the complete solution in the form of a composite approximation.
2. Apply SCEM by imposing the exact boundary conditions.
3. By means of SCEM in its generalized form, give a two layer model such that the lower layer contains the lower and the intermediate layers of the regular form of SCEM.

10-2. We study the boundary layer on a flat plate deformed locally by a two-dimensional indentation.

All the quantities are nondimensionalized by using the freestream velocity V_∞ and the abscissa L_0 of the hump location. We have

$$x' = \frac{x^*}{L_0} \ , \quad y' = \frac{y^*}{L_0} \ , \quad u' = \frac{u^*}{V_\infty} \ , \quad v' = \frac{v^*}{V_\infty} \ , \quad p' = \frac{p^*}{\varrho V_\infty^2} \ .$$

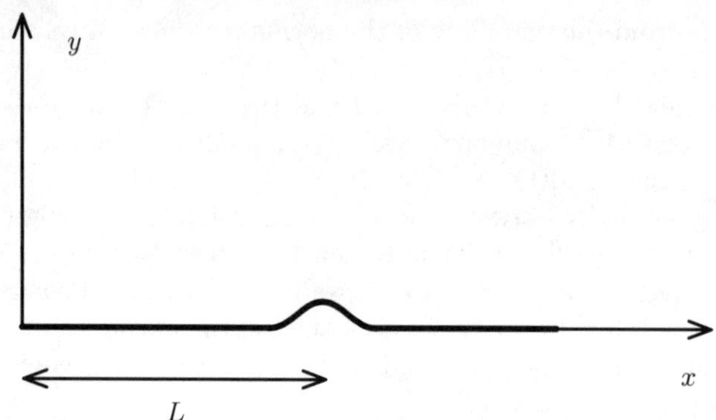

Fig. 10.3. Flow on a flat plate deformed by a hump

We define the small parameter ε by

$$\varepsilon = Re^{-1/m} \ , \quad Re = \frac{V_\infty L_0}{\nu} \ ,$$

where m is arbitrary $(m > 0)$; the value of m does not influence the results.
The equation of the hump is $y' = F(x')$.

Prandtl's transformation consists of the following change of variable

$$(x', y') \longmapsto [x = x', \, y = y' - F(x')] \ ,$$

which enables us, with the coordinates (x, y), to write the wall equation as
$y = 0$.

The following change on the velocity components is also introduced

$$u = u' \ , \quad v = v' - \frac{\mathrm{d}F}{\mathrm{d}x'} u' \ .$$

With these transformations, the Navier-Stokes equations become
• Continuity equation

$$\frac{\partial u}{\partial x} + \frac{\partial v}{\partial y} = 0 \ ,$$

• x-momentum equation

$$u\frac{\partial u}{\partial x} + v\frac{\partial u}{\partial y} = -\frac{\partial p}{\partial x} + \frac{\partial p}{\partial y}\frac{\mathrm{d}F}{\mathrm{d}x} + \varepsilon^m \frac{\partial^2 u}{\partial x^2} - 2\varepsilon^m \frac{\partial^2 u}{\partial x \partial y}\frac{\mathrm{d}F}{\mathrm{d}x}$$

$$+\varepsilon^m \frac{\partial^2 u}{\partial y^2}\left(\frac{\mathrm{d}F}{\mathrm{d}x}\right)^2 - \varepsilon^m \frac{\partial u}{\partial y}\frac{\mathrm{d}^2 F}{\mathrm{d}x^2} + \varepsilon^m \frac{\partial^2 u}{\partial y^2} \ ,$$

- y-momentum equation

$$u\frac{\partial v}{\partial x} + u^2\frac{\mathrm{d}^2 F}{\mathrm{d}x^2} + v\frac{\partial v}{\partial y} + u\frac{\mathrm{d}F}{\mathrm{d}x}\frac{\partial u}{\partial x} + v\frac{\partial u}{\partial y}\frac{\mathrm{d}F}{\mathrm{d}x} = -\frac{\partial p}{\partial y}$$

$$+2\varepsilon^m\frac{\mathrm{d}^2 F}{\mathrm{d}x^2}\frac{\partial u}{\partial x} + \varepsilon^m\frac{\partial^2 u}{\partial y^2}\left(\frac{\mathrm{d}F}{\mathrm{d}x}\right)^3 - 2\varepsilon^m\frac{\partial^2 v}{\partial x\partial y}\frac{\mathrm{d}F}{\mathrm{d}x} + \varepsilon^m u\frac{\mathrm{d}^3 F}{\mathrm{d}x^3}$$

$$-\varepsilon^m\frac{\partial v}{\partial y}\frac{\mathrm{d}^2 F}{\mathrm{d}x^2} - 2\varepsilon^m\frac{\partial^2 u}{\partial x\partial y}\left(\frac{\mathrm{d}F}{\mathrm{d}x}\right)^2 + \varepsilon^m\frac{\partial^2 u}{\partial y^2}\frac{\mathrm{d}F}{\mathrm{d}x} + \varepsilon^m\frac{\mathrm{d}F}{\mathrm{d}x}\frac{\partial^2 u}{\partial x^2}$$

$$+\varepsilon^m\frac{\partial^2 v}{\partial y^2}\left(\frac{\mathrm{d}F}{\mathrm{d}x}\right)^2 - 3\varepsilon^m\frac{\mathrm{d}^2 F}{\mathrm{d}x^2}\frac{\partial u}{\partial y}\frac{\mathrm{d}F}{\mathrm{d}x} + \varepsilon^m\frac{\partial^2 v}{\partial x^2} + \varepsilon^m\frac{\partial^2 v}{\partial y^2}\ .$$

We study the case where the indentation height is of order $\varepsilon^{2m/3}$ and the indentation length is of order $\varepsilon^{m/2}$. Thus, the hump equation has the form

$$y' = \varepsilon^{2m/3} f\left(\frac{x'}{\varepsilon^{m/2}}\right)\ .$$

Without any indentation (flat wall) the flow is said non perturbed. To first order, the solution is Blasius' boundary layer. To the order studied here, it suffices to consider the boundary layer at $x' = 1$ ($x^* = L_0$) given by the velocity profile

$$u = U_0(Y) \quad \text{with} \quad Y = \frac{y}{\varepsilon^{m/2}}\ .$$

As $Y \to 0$, we have

$$U_0 = \lambda Y\ .$$

The study of the flow perturbation in the neighbourhood of the indentation does not require, to the considered order, to take into account the variations of Blasius' boundary layer in the streamwise direction.

The structure proposed by Mauss [64, 65], Nayfeh [73], Smith [93] comprises two decks. The main deck is the continuation of Blasius' boundary layer; the appropriate variables are

$$X = \frac{x}{\varepsilon^{m/2}}\ , \quad \overline{Y} = Y = \frac{y}{\varepsilon^{m/2}}\ .$$

In the lower deck, the appropriate variables are

$$X = \frac{x}{\varepsilon^{m/2}}\ , \quad \widetilde{Y} = \frac{y}{\varepsilon^{2m/3}}\ .$$

Therefore, we observe that the lower deck has a thickness of the same order as the indentation. On the other hand, the perturbation does not reach a region thicker than the oncoming boundary layer; the inviscid region is not affected.

In the main deck, the expansions are

$$u = U_o(Y) + \varepsilon^{m/6} f\frac{\mathrm{d}U_0}{\mathrm{d}\overline{Y}} + \varepsilon^{m/3}\overline{U}_2 + \cdots\ ,$$

$$v = -\varepsilon^{m/6}\frac{\mathrm{d}f}{\mathrm{d}X}U_0 + \varepsilon^{m/3}\overline{V}_2 + \cdots ,$$

$$p = \varepsilon^{m/3}\overline{P}_2 + \cdots .$$

In the lower deck, the expansions are

$$u = \varepsilon^{m/6}\left(\lambda\widetilde{Y} + \widetilde{U}_1\right) + \cdots ,$$

$$v = \varepsilon^{m/3}\widetilde{V}_1 + \cdots ,$$

$$p = \varepsilon^{m/3}\widetilde{P}_1 + \cdots .$$

Give the equations for $\overline{U}_2$, $\overline{V}_2$, $\overline{P}_2$ and the equations for $\widetilde{U}_1$, $\widetilde{V}_1$, $\widetilde{P}_1$.
Give the boundary conditions and the matching conditions.
Show that the two sets of equations are strongly coupled.

10-3. The flow on a flat plate deformed by a small two-dimensional hump is described by a structure which depends on the hump dimensions [22, 79, 80].

All the quantities are nondimensionalized by using the freestream velocity V_∞ and the abscissa L_0 of the hump location. We have

$$x' = \frac{x^*}{L_0} , \quad y' = \frac{y^*}{L_0} , \quad u' = \frac{u^*}{V_\infty} , \quad v' = \frac{v^*}{V_\infty} , \quad p' = \frac{p^*}{\varrho V_\infty^2} .$$

We define the small parameter ε by

$$\varepsilon = Re^{-1/m} , \quad Re = \frac{V_\infty L_0}{\nu} ,$$

where m is arbitrary ($m > 0$); the value of m does not influence the results.
The hump equation is $y' = F(x')$.
The study is performed with Prandtl's transformation

$$(x', y') \longmapsto [x = x', y = y' - F(x')] ,$$

and

$$u = u' , \quad v = v' - \frac{\mathrm{d}F}{\mathrm{d}x'}u' .$$

The streamwise extent of the perturbation is ε^α so that the variable adapted to the study is

$$X = \frac{x}{\varepsilon^\alpha} ,$$

and the hump height is of order ε^β. Therefore, the hump equation has the form

$$y' = \varepsilon^\beta f(X) .$$

In the map (α, β), four significant zones are defined (Fig. 10.4). They are delimited by different straight lines whose meaning is given below.

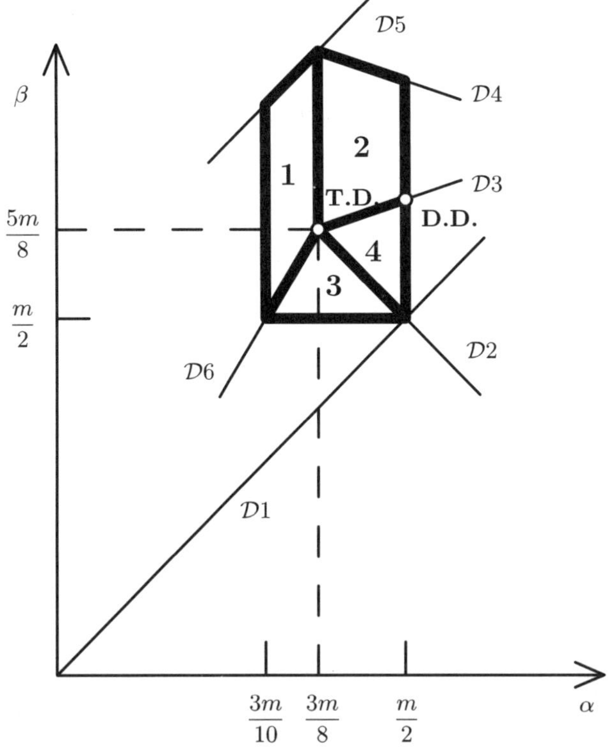

Fig. 10.4. Delimitation of different zones around the triple deck; **T.D.**: triple deck, **D.D.**: double deck

- $\mathcal{D}1$: $\beta = \alpha$. The height of the hump must be smaller than its length ($\beta > \alpha$) otherwise, as the Reynolds number tends towards infinity, the hump becomes very steep.
- $\beta = m/2$. For $\beta > m/2$, the hump height is smaller than the thickness of the oncoming boundary layer.
- $\alpha = 3m/8$. This line defines the boundary between zone 1 and zone 2 which differ by the mode of resolution. The direct mode applies in zone 1 and the inverse mode in zone 2.
- $\mathcal{D}5$: $\beta = \alpha + m/2$. If $\beta > \alpha + m/2$, the perturbations are small with respect to the second order of the standard boundary layer theory. For example, the first order of the pressure in zone 1 is $\varepsilon^{\beta - \alpha}$ whereas the second order in the boundary layer is $\varepsilon^{m/2}$.
- $\alpha = 3m/10$. This boundary is given by the study of second order terms and defines the hierarchy between U_1^* and $\overline{U}_2$. Along the straight line $\mathcal{D}6$, the point of abscissa $\alpha = 3m/10$ corresponds to $\beta = m/2$ which is the limit of the hump height.

- $\mathcal{D}6$: $\beta = 5\alpha/3$. This line defines the boundary for the linearity of lower deck equations between zones 1 and 3. In zone 1, the lower deck equations are linear whereas in zone 3, they are non linear.
- $\mathcal{D}3$: $\beta = \alpha/3 + m/2$. This line defines the boundary for the linearity of lower deck equations between zones 2 and 4. In zone 2, the lower deck equations are linear whereas in zone 4, they are non linear.
- $\alpha = m/2$. The upper deck dimension is of order ε^α. The line $\alpha = m/2$ defines the minimum of the streamwise hump extent which is supposed to be larger than the thickness of the oncoming boundary layer; this condition implies the existence of an upper deck thicker than Blasius' boundary layer which is of order $\varepsilon^{m/2}$.
- $\mathcal{D}4$: $\beta = m - \alpha/3$. If $\beta > m - \alpha/3$ the perturbations are large compared to the second order of the standard boundary layer theory. For example, the first order of the pressure in zone 2 is $\varepsilon^{\beta+\alpha/3-m/2}$ whereas the second order of the standard boundary layer is $\varepsilon^{m/2}$.
- $\mathcal{D}2$: $\beta = -\alpha + m$. This line defines the boundary of zones 3 and 4 which differ by the mode of resolution of equations: direct mode in zone 3 and inverse mode in zone 4.

In the different zones, the expansions and the equations are given below.

Zone 1

$$Y^* = \frac{y}{\varepsilon^\alpha} \, , \quad \overline{Y} = Y = \frac{y}{\varepsilon^{m/2}} \, , \quad \widetilde{Y} = \frac{y}{\varepsilon^{\alpha/3+m/2}} \, .$$

Upper deck

$$u = 1 + \varepsilon^{\beta-\alpha} U_1^* + \cdots \, , \qquad \frac{\partial U_1^*}{\partial X} + \frac{\partial V_1^*}{\partial Y^*} = 0 \, ,$$

$$v = \varepsilon^{\beta-\alpha} V_1^* + \cdots \, , \qquad \frac{\partial U_1^*}{\partial X} = -\frac{\partial P_1^*}{\partial X} \, ,$$

$$p = \varepsilon^{\beta-\alpha} P_1^* + \cdots \, , \qquad \frac{\partial V_1^*}{\partial X} + \frac{\mathrm{d}^2 f}{\mathrm{d}X^2} = -\frac{\partial P_1^*}{\partial Y^*} \, .$$

Main deck

$$u = U_0(Y) + \varepsilon^{\beta-4\alpha/3} \, \overline{U}_1 + \varepsilon^{\beta-8\alpha/3+m/2} \overline{U}_2 + \cdots \, , \qquad \frac{\partial \overline{U}_1}{\partial X} + \frac{\partial \overline{V}_1}{\partial \overline{Y}} = 0 \, ,$$

$$v = \varepsilon^{\beta-7\alpha/3+m/2} \, \overline{V}_1 + \cdots \, , \qquad U_0 \frac{\partial \overline{U}_1}{\partial X} + \overline{V}_1 \frac{\mathrm{d}U_0}{\mathrm{d}\overline{Y}} = 0 \, ,$$

$$p = \varepsilon^{\beta-\alpha} \, \overline{P}_1 \, , \qquad \frac{\partial \overline{P}_1}{\partial \overline{Y}} = 0 \, .$$

Lower deck

$$u = \varepsilon^{\alpha/3} \lambda \widetilde{Y} + \varepsilon^{\beta-4\alpha/3} \, \widetilde{U}_1 + \cdots \, , \qquad \frac{\partial \widetilde{U}_1}{\partial X} + \frac{\partial \widetilde{V}_1}{\partial \widetilde{Y}} = 0 \, ,$$

$$v = \varepsilon^{m/2-2\alpha+\beta} \, \widetilde{V}_1 + \cdots \, , \qquad \lambda \widetilde{Y} \frac{\partial \widetilde{U}_1}{\partial X} + \lambda \widetilde{V}_1 = -\frac{\partial \widetilde{P}_1}{\partial X} + \frac{\partial^2 \widetilde{U}_1}{\partial \widetilde{Y}^2} \, ,$$

$$p = \varepsilon^{\beta-\alpha} \, \widetilde{P}_1 + \cdots \, , \qquad \frac{\partial \widetilde{P}_1}{\partial \widetilde{Y}} = 0 \, .$$

Zone 2

$$Y^* = \frac{y}{\varepsilon^\alpha} \ , \quad \overline{Y} = Y = \frac{y}{\varepsilon^{m/2}} \ , \quad \widetilde{Y} = \frac{y}{\varepsilon^{\alpha/3+m/2}} \ .$$

Upper deck

$$u = 1 + \varepsilon^{\beta+\alpha/3-m/2}U_2^* + \cdots \ ,$$

$$v = -\varepsilon^{\beta-\alpha}\frac{\mathrm{d}f}{\mathrm{d}X} + \varepsilon^{\beta+\alpha/3-m/2}V_2^* + \cdots \ ,$$

$$p = \varepsilon^{\beta+\alpha/3-m/2}P_2^* + \cdots \ ,$$

$$\frac{\partial U_2^*}{\partial X} + \frac{\partial V_2^*}{\partial Y^*} = 0 \ ,$$

$$\frac{\partial U_2^*}{\partial X} = -\frac{\partial P_2^*}{\partial X} \ ,$$

$$\frac{\partial V_2^*}{\partial X} = -\frac{\partial P_2^*}{\partial Y^*} \ .$$

Main deck

$$u = U_0(Y) + \varepsilon^{\beta-m/2}f(X)\frac{\mathrm{d}U_0}{\mathrm{d}Y}$$
$$+\varepsilon^{\beta+4\alpha/3-m}\overline{U}_2 + \cdots \ ,$$

$$v = -\varepsilon^{\beta-\alpha}\frac{\mathrm{d}f}{\mathrm{d}X}U_0(Y) + \varepsilon^{\beta+\alpha/3-m/2}\overline{V}_2 + \cdots \ ,$$

$$p = \varepsilon^{\beta+\alpha/3-m/2}\overline{P}_2 + \cdots \ ,$$

$$\frac{\partial \overline{U}_2}{\partial X} + \frac{\partial \overline{V}_2}{\partial \overline{Y}} = 0 \ ,$$

$$U_0\frac{\partial \overline{U}_2}{\partial X} + \overline{V}_2\frac{\mathrm{d}U_0}{\mathrm{d}\overline{Y}} = 0 \ ,$$

$$\frac{\partial \overline{P}_2}{\partial \overline{Y}} = 0 \ .$$

Lower deck

$$u = \varepsilon^{\alpha/3}\lambda\widetilde{Y} + \varepsilon^{\beta-m/2}\widetilde{U}_1 + \cdots \ ,$$

$$v = \varepsilon^{\beta-2\alpha/3}\widetilde{V}_1 + \cdots \ ,$$

$$p = \varepsilon^{\beta+\alpha/3-m/2}\widetilde{P}_1 + \cdots \ ,$$

$$\frac{\partial \widetilde{U}_1}{\partial X} + \frac{\partial \widetilde{V}_1}{\partial \widetilde{Y}} = 0 \ ,$$

$$\lambda\widetilde{Y}\frac{\partial \widetilde{U}_1}{\partial X} + \lambda\widetilde{V}_1 = -\frac{\partial \widetilde{P}_1}{\partial X} + \frac{\partial^2 \widetilde{U}_1}{\partial \widetilde{Y}^2} \ ,$$

$$\frac{\partial \widetilde{P}_1}{\partial \widetilde{Y}} = 0 \ .$$

Zone 3

$$Y^* = \frac{y}{\varepsilon^\alpha} \ , \quad \overline{Y} = Y = \frac{y}{\varepsilon^{m/2}} \ , \quad \widetilde{Y} = \frac{y}{\varepsilon^{(3\alpha-\beta+2m)/4}} \ .$$

Upper deck

$$u = 1 + \varepsilon^{\beta-\alpha}U_1^* + \cdots \ ,$$

$$v = \varepsilon^{\beta-\alpha}V_1^* + \cdots \ ,$$

$$p = \varepsilon^{\beta-\alpha}P_1^* + \cdots \ ,$$

$$\frac{\partial U_1^*}{\partial X} + \frac{\partial V_1^*}{\partial Y^*} = 0 \ ,$$

$$\frac{\partial U_1^*}{\partial X} = -\frac{\partial P_1^*}{\partial X} \ ,$$

$$\frac{\partial V_1^*}{\partial X} + \frac{\mathrm{d}^2 f}{\mathrm{d}X^2} = -\frac{\partial P_1^*}{\partial Y^*} \ .$$

Main deck

$$u = U_0 + \varepsilon^{(\beta-\alpha)/2}\overline{U}_1 + \cdots , \qquad \frac{\partial \overline{U}_1}{\partial X} + \frac{\partial \overline{V}_1}{\partial \overline{Y}} = 0 ,$$

$$v = \varepsilon^{(\beta-3\alpha+m)/2}\overline{V}_1 + \cdots , \qquad U_0\frac{\partial \overline{U}_1}{\partial X} + \overline{V}_1\frac{\mathrm{d}U_0}{\mathrm{d}\overline{Y}} = 0 ,$$

$$p = \varepsilon^{\beta-\alpha}\overline{P}_1 + \cdots , \qquad \frac{\partial \overline{P}_1}{\partial \overline{Y}} = 0 .$$

Lower deck

$$u = \varepsilon^{(\beta-\alpha)/2}\widetilde{U}_1 + \varepsilon^{(3\alpha-\beta)/4}\lambda\widetilde{Y} + \cdots , \qquad \frac{\partial \widetilde{U}_1}{\partial X} + \frac{\partial \widetilde{V}_1}{\partial \widetilde{Y}} = 0 ,$$

$$v = \varepsilon^{(\beta-3\alpha+2m)/4}\widetilde{V}_1 + \cdots , \qquad \widetilde{U}_1\frac{\partial \widetilde{U}_1}{\partial X} + \widetilde{V}_1\frac{\partial \widetilde{U}_1}{\partial \widetilde{Y}} = -\frac{\partial \widetilde{P}_1}{\partial X} + \frac{\partial^2 \widetilde{U}_1}{\partial \widetilde{Y}^2} ,$$

$$p = \varepsilon^{\beta-\alpha}\widetilde{P}_1 + \cdots , \qquad \frac{\partial \widetilde{P}_1}{\partial \widetilde{Y}} = 0 .$$

Zone 4

$$Y^* = \frac{y}{\varepsilon^\alpha} , \quad \overline{Y} = Y = \frac{y}{\varepsilon^{m/2}} , \quad \widetilde{Y} = \frac{y}{\varepsilon^{(2\alpha-2\beta+3m)/4}} .$$

Upper deck

$$u = 1 + \varepsilon^{2\beta-m}U_2^* + \cdots , \qquad \frac{\partial U_2^*}{\partial X} + \frac{\partial V_2^*}{\partial Y^*} = 0 ,$$

$$v = -\varepsilon^{\beta-\alpha}\frac{\mathrm{d}f}{\mathrm{d}X} + \varepsilon^{2\beta-m}V_2^* + \cdots , \qquad \frac{\partial U_2^*}{\partial X} = -\frac{\partial P_2^*}{\partial X} ,$$

$$p = \varepsilon^{2\beta-m}P_2^* + \cdots , \qquad \frac{\partial V_2^*}{\partial X} = -\frac{\partial P_2^*}{\partial Y^*} .$$

Main deck

$$u = U_0 + \varepsilon^{\beta-m/2}f(X)\frac{\mathrm{d}U_0}{\mathrm{d}\overline{Y}} \qquad \frac{\partial \overline{U}_2}{\partial X} + \frac{\partial \overline{V}_2}{\partial \overline{Y}} = 0 ,$$

$$+ \varepsilon^{2\beta+\alpha-3m/2}\overline{U}_2 + \cdots ,$$

$$v = -\varepsilon^{\beta-\alpha}\frac{\mathrm{d}f}{\mathrm{d}X}U_0 + \varepsilon^{2\beta-m}\overline{V}_2 + \cdots , \qquad U_0\frac{\partial \overline{U}_2}{\partial X} + \overline{V}_2\frac{\mathrm{d}U_0}{\mathrm{d}\overline{Y}} = 0 ,$$

$$p = \varepsilon^{2\beta-m}\overline{P}_2 + \cdots , \qquad \frac{\partial \overline{P}_2}{\partial \overline{Y}} = 0 .$$

Lower deck

$$u = \varepsilon^{\beta - m/2}\widetilde{U}_1$$

$$\frac{\partial \widetilde{U}_1}{\partial X} + \frac{\partial \widetilde{V}_1}{\partial \widetilde{Y}} = 0 \, ,$$

$$+\varepsilon^{(2\alpha - 2\beta + m)/4}\lambda\widetilde{Y} + \cdots \, ,$$

$$v = \varepsilon^{(2\beta - 2\alpha + m)/4}\widetilde{V}_1 + \cdots \, ,$$

$$\widetilde{U}_1\frac{\partial \widetilde{U}_1}{\partial X} + \widetilde{V}_1\frac{\partial \widetilde{U}_1}{\partial \widetilde{Y}} = -\frac{\partial \widetilde{P}_1}{\partial X} + \frac{\partial^2 \widetilde{U}_1}{\partial \widetilde{Y}^2} \, ,$$

$$p = \varepsilon^{2\beta - m}\widetilde{P}_1 + \cdots \, .$$

$$\frac{\partial \widetilde{P}_1}{\partial \widetilde{Y}} = 0 \, .$$

We want to apply these different models to a flow on a flat plate ($f = 0$); the perturbation is due to wall mass transfer instead of a hump. Wall mass transfer is characterized by a velocity v whose order of magnitude is given by the order of the same velocity component in the lower deck. For example, in zone 1, we have

$$y = 0 : \quad v = \varepsilon^{m/2 - 2\alpha + \beta}V_w(X) \, ,$$

with $V_w \neq 0$ on a length of order ε^α.

Analyze the solution in the different zones of the map (α, β).

11 Turbulent Boundary Layer

The three-dimensional unsteady Navier-Stokes equations given in Appendix I also apply to turbulent flow if the values of dependent variables are understood as instantaneous values. A direct approach to solving the equations for turbulent flows is to solve them for specific boundary conditions and initial values that include time-dependent quantities. Mean values are needed in most practical cases, so an ensemble of solutions of time-dependent equations is required. Even for the most restricted cases, this approach, referred to as direct numerical simulation, becomes a difficult and extremely expensive computing problem because the unsteady eddy motions of turbulence appear over a wide range. The usual procedure is to average the equations rather than their solutions [77]. The averaged Navier-Stokes equations are also called Reynolds equations. This procedure is used here to study the turbulent boundary layer.

The standard asymptotic analysis decomposes the boundary layer into outer and inner regions, whose properties are discussed in this chapter. One of the most important feature is the overlap region in which the velocity profile follows a logarithmic law.

The successive complementary expansion method, SCEM, is applied by taking the scales determined in the standard analysis. As in laminar flow, an IBL model is obtained. In addition, the study of the contribution of the inner region enables us to construct simply an approximation of the velocity profile valid in the whole boundary layer, as far as the velocity profile is known in the outer region. Numerical results are presented for a flat plate boundary layer at different Reynolds numbers.

11.1 Results of the Standard Asymptotic Analysis

11.1.1 Averaged Navier-Stokes Equations

The study of incompressible turbulent flows is addressed by defining a mean flow from a *statistical average* of velocity and of pressure.

The instantaneous flow is decomposed into a mean and fluctuating flow

$$\widetilde{\mathcal{U}}_i = \mathcal{U}_i + \mathcal{U}_i' \, ,$$
$$\widetilde{\mathcal{P}} = \mathcal{P} + \mathcal{P}' \, .$$

An orthonormal axis system is used. The x-axis is along the wall and the y-axis is normal to it. All the quantities are dimensionless. The coordinates x and y are reduced by the reference length L, the velocity components by a reference velocity V, the pressure by ϱV^2, the turbulent stresses by ϱV^2. In fact, the mean flow scales are chosen to define the reference quantities V and L.

In two-dimensional, incompressible, steady flow (on the average), the *averaged Navier-Stokes equations* or *Reynolds equations* are [14, 21]

$$\frac{\partial \mathcal{U}}{\partial x} + \frac{\partial \mathcal{V}}{\partial y} = 0 \, , \tag{11.1a}$$

$$\mathcal{U}\frac{\partial \mathcal{U}}{\partial x} + \mathcal{V}\frac{\partial \mathcal{U}}{\partial y} = -\frac{\partial \mathcal{P}}{\partial x} + \frac{\partial}{\partial x}\left(\mathcal{T}_{xx} + \frac{1}{\mathcal{R}}\frac{\partial \mathcal{U}}{\partial x}\right) + \frac{\partial}{\partial y}\left(\mathcal{T}_{xy} + \frac{1}{\mathcal{R}}\frac{\partial \mathcal{U}}{\partial y}\right) \, , \tag{11.1b}$$

$$\mathcal{U}\frac{\partial \mathcal{V}}{\partial x} + \mathcal{V}\frac{\partial \mathcal{V}}{\partial y} = -\frac{\partial \mathcal{P}}{\partial y} + \frac{\partial}{\partial x}\left(\mathcal{T}_{xy} + \frac{1}{\mathcal{R}}\frac{\partial \mathcal{V}}{\partial x}\right) + \frac{\partial}{\partial y}\left(\mathcal{T}_{yy} + \frac{1}{\mathcal{R}}\frac{\partial \mathcal{V}}{\partial y}\right) \, , \tag{11.1c}$$

where $\mathcal{R}$ is the Reynolds number

$$\mathcal{R} = \frac{\varrho V L}{\mu} \, .$$

and the *turbulent stresses* $\mathcal{T}_{ij}$ are defined from the correlations between velocity fluctuations

$$\mathcal{T}_{ij} = - <\mathcal{U}'_i \mathcal{U}'_j> \, .$$

The turbulent stresses appear when the Navier-Stokes equations are averaged and they are a consequence of the non-linearity of the convection terms.

11.1.2 Scales

The results presented in this section are based, to a large extent, on a large amount of experimental data from which a consistent theoretical description has been developed to reproduce the observations and in which the notion of *turbulence scales* plays an essential role. Therefore, the issue is addressed without resting on a well-posed mathematical frame as it is the case in laminar flow.

In a standard manner, with MMAE, the flow is decomposed in two regions: the inviscid region and the boundary layer. The former is treated separately and provides us with the necessary data to calculate the boundary layer. The boundary layer is described by a two-layer structure [67, 115] consisting of: i) an outer layer characterized by the thickness δ and ii) an inner layer whose thickness is of order $\dfrac{\nu}{u_\tau}$ ($\nu = \mu/\varrho$) with u_τ denoting the *friction velocity*

$$u_\tau = \sqrt{\frac{\tau_w}{\varrho}} \, ,$$

and τ_w the wall shear stress.

The *turbulence velocity scale* – denoted by $\boldsymbol{u}$ – is identical in the outer region and in the inner region and is of the order of the friction velocity u_τ. In the outer region, the turbulence length scale, of the order of δ, is denoted by $\boldsymbol{\ell}$ whereas in the inner region, the length scale is $\nu/\boldsymbol{u}$.

In the outer region, we assume that *the time scale of the transport due to turbulence ($\boldsymbol{\ell}/\boldsymbol{u}$) is of the same order as the time scale of mean flow convection.* We can view this hypothesis as the counterpart, for turbulent flows, of the hypothesis used for a laminar boundary layer according to which the viscosity time scale is of the same order as the convection time scale (Subsect. 7.1.1). If the reference quantities V and L are chosen as velocity and length scales of the mean flow, we deduce

$$\frac{\ell}{L} = \frac{u}{V} \; . \tag{11.2}$$

The asymptotic analysis introduces the small parameters ε and $\hat{\varepsilon}$ which define, with dimensionless variables, the order of the thicknesses of the outer and inner layers

$$\varepsilon = \frac{\ell}{L} \; , \tag{11.3}$$

$$\hat{\varepsilon} = \frac{\nu}{uL} \; . \tag{11.4}$$

Using (11.2), we have

$$\varepsilon\hat{\varepsilon}\mathcal{R} = 1 \; . \tag{11.5}$$

With the skin-friction law (11.14), the following relation between the gauge ε and the Reynolds number holds

$$\varepsilon = \mathrm{O_S}\left(\frac{1}{\ln\mathcal{R}}\right) \; . \tag{11.6}$$

In particular, we deduce that, for any positive n

$$\varepsilon^n \succ \hat{\varepsilon} \succ \frac{1}{\mathcal{R}} \; .$$

The variables appropriate to the study of each region are

$$\text{Outer region}: \quad \eta = \frac{y}{\varepsilon} \; , \tag{11.7a}$$

$$\text{Inner region}: \quad \hat{y} = \frac{y}{\hat{\varepsilon}} \; . \tag{11.7b}$$

11.1.3 Structure of the Flow

The whole flow is described by a three-layer structure: the external region which is inviscid to first orders, the outer and inner regions of the boundary layer.

The results are stated here by assuming that *the wall curvature effects are negligible* [21].

External Region

In this region, the expansions are

$$\mathcal{U} = \bar{u}_0(x,y) + \varepsilon\bar{u}_1(x,y) + \cdots ,$$
$$\mathcal{V} = \bar{v}_0(x,y) + \varepsilon\bar{v}_1(x,y) + \cdots ,$$
$$\mathcal{P} = \bar{p}_0(x,y) + \varepsilon\bar{p}_1(x,y) + \cdots ,$$
$$\mathcal{T}_{ij} = 0 .$$

It follows that $\bar{u}_0$, $\bar{v}_0$, $\bar{p}_0$ satisfy the Euler equations and $\bar{u}_1$, $\bar{v}_1$, $\bar{p}_1$ satisfy the linearized Euler equations.

Matching velocity v to order ε with the outer region of the boundary layer yields

$$\bar{v}_{0w} = 0 ,$$
$$\bar{v}_{1w} = \lim_{\eta\to\infty}\left[v_0 - \eta\left(\frac{\partial\bar{v}_0}{\partial y}\right)_w\right] ,$$

where the index "w" denotes the wall.

The first condition enables us to calculate the flow defined by $\bar{u}_0$, $\bar{v}_0$, $\bar{p}_0$. Taking into account (11.9b) and the continuity equation, the second condition gives $\bar{v}_{1w} = 0$. Then, with the condition that $\bar{u}_1$, $\bar{v}_1$ and $\bar{p}_1$ vanish at infinity, everywhere in the external region we have

$$\bar{u}_1 = 0 ; \quad \bar{v}_1 = 0 ; \quad \bar{p}_1 = 0 .$$

Outer Region of the Boundary Layer

In the outer region of the boundary layer, the expansions are

$$\mathcal{U} = u_0(x,\eta) + \varepsilon u_1(x,\eta) + \cdots , \tag{11.8a}$$
$$\mathcal{V} = \varepsilon\left[v_0(x,\eta) + \varepsilon v_1(x,\eta) + \cdots\right] , \tag{11.8b}$$
$$\mathcal{P} = p_0(x,\eta) + \varepsilon p_1(x,\eta) + \cdots , \tag{11.8c}$$
$$\mathcal{T}_{ij} = \varepsilon^2 \tau_{ij,1}(x,\eta) + \cdots . \tag{11.8d}$$

The expansion of $\mathcal{V}$ is chosen in such a way that the continuity equation keeps its standard form to any order. The expansion of the turbulent stresses imply that their dominant order of magnitude is ε^2, i.e. the friction velocity is actually a *turbulence velocity scale*.

The equations for u_0, v_0 et p_0 are

$$\frac{\partial u_0}{\partial x} + \frac{\partial v_0}{\partial \eta} = 0 ,$$
$$u_0\frac{\partial u_0}{\partial x} + v_0\frac{\partial u_0}{\partial \eta} = -\frac{\partial p_0}{\partial x} ,$$
$$0 = \frac{\partial p_0}{\partial \eta} .$$

A solution which matches with the inviscid flow is

$$u_0 = u_e \,, \tag{11.9a}$$

$$v_0 = -\eta \frac{\mathrm{d}u_e}{\mathrm{d}x} \,, \tag{11.9b}$$

where u_e is the inviscid flow velocity at the wall

$$u_e = \bar{u}_{0w} \,.$$

Moreover, the pressure p_0 is constant over the thickness of the outer region and is equal to the inviscid flow pressure at the wall

$$p_0 = \bar{p}_{0w} \,.$$

Therefore, we have

$$\frac{\mathrm{d}p_0}{\mathrm{d}x} = -u_e \frac{\mathrm{d}u_e}{\mathrm{d}x} \,.$$

Neglecting the wall curvature effects (see Subsect. 10.1.2), the equations for u_1, v_1 and p_1 are

$$\frac{\partial u_1}{\partial x} + \frac{\partial v_1}{\partial \eta} = 0 \,, \tag{11.10a}$$

$$u_1 \frac{\mathrm{d}u_e}{\mathrm{d}x} + u_e \frac{\partial u_1}{\partial x} - \eta \frac{\mathrm{d}u_e}{\mathrm{d}x} \frac{\partial u_1}{\partial \eta} = -\frac{\partial p_1}{\partial x} + \frac{\partial \tau_{xy,1}}{\partial \eta} \,, \tag{11.10b}$$

$$0 = \frac{\partial p_1}{\partial \eta} \,. \tag{11.10c}$$

With the hypothesis that the wall curvature effects are negligible, it can be shown that $p_1 = 0$.

Inner Region of the Boundary Layer

It is necessary to introduce an inner region, otherwise the no-slip condition at the wall is not satisfied. In this region, the expansions are

$$\mathcal{U} = \varepsilon \hat{u}_1(x, \hat{y}) + \cdots \,, \tag{11.11a}$$

$$\mathcal{V} = \hat{\varepsilon}(\varepsilon \hat{v}_1 + \cdots) \,, \tag{11.11b}$$

$$\mathcal{P} = \hat{p}_0 + \varepsilon \hat{p}_1 + \cdots \,, \tag{11.11c}$$

$$\mathcal{T}_{ij} = \varepsilon^2 \hat{\tau}_{ij,1} + \cdots \,. \tag{11.11d}$$

The expansion chosen for $\mathcal{U}$ shows that the order of the streamwise velocity is ε. With dimensionalized variables, this means that the velocity scale is the friction velocity. This essential hypothesis, consistent with experimental results, implies the logarithmic matching between the outer and inner regions of the boundary layer.

The pressure $\hat{p}_0$ is constant along a normal to the wall and is equal to the pressure p_0 in the outer region

$$\hat{p}_0 = p_0 = \bar{p}_{0w} \ .$$

The equations for $\hat{u}_1$, $\hat{v}_1$ and $\hat{p}_1$ are

$$\frac{\partial \hat{u}_1}{\partial x} + \frac{\partial \hat{v}_1}{\partial \hat{y}} = 0 \ , \tag{11.12a}$$

$$0 = \frac{\partial}{\partial \hat{y}} \left(\hat{\tau}_{xy,1} + \frac{\partial \hat{u}_1}{\partial \hat{y}} \right) \ , \tag{11.12b}$$

$$0 = \frac{\partial \hat{p}_1}{\partial \hat{y}} \ . \tag{11.12c}$$

The matching of the pressure to order ε between the outer and inner regions of the boundary layer gives $\hat{p}_1 = 0$.

From (11.12b), the total stress – sum of the viscous stress and of the turbulent stress – is constant along a normal to the wall.

The matching between the outer region and the inner region on the velocity $\mathcal{U}$ (expansions given by (11.8a) and (11.11a)) raises a difficulty due to the absence of a term of order $\mathrm{O_S}(1)$ in the inner expansion. The solution rests upon a *logarithmic evolution* of the velocity in the overlap region (see Problems 11-1, 11-2 and 11-3)

$$u_1 = A \ln \eta + C_1 \quad \text{as} \quad \eta \to 0 \ , \tag{11.13a}$$

$$\hat{u}_1 = A \ln \hat{y} + C_2 \quad \text{as} \quad \hat{y} \to \infty \ . \tag{11.13b}$$

The law for $\hat{u}_1$ corresponds to the universal law of the wall, where A and C_2 do not depend on the conditions under which the boundary layer develops (Reynolds number, pressure gradient). Constant A corresponds to the inverse of von Kármán's constant.

Then, in the overlap region, the equality of velocity in the outer and inner regions gives (see Problem 11-4)

$$u_e + \varepsilon(A \ln \eta + C_1) = \varepsilon(A \ln \hat{y} + C_2) \ ,$$

or

$$\frac{u_e}{\varepsilon} = A \ln \frac{\varepsilon}{\hat{\varepsilon}} + C_2 - C_1 \ . \tag{11.14}$$

This equation represents the *skin-friction law*. Expressed with dimensionalized variables, this law takes the standard form

$$\frac{u_e}{u_\tau} = \frac{1}{\chi} \ln \frac{u_\tau \delta}{\nu} + B \ , \tag{11.15}$$

where $\chi \simeq 0.4$ is von Kármán's constant and B depends on the pressure gradient.

This relation and the logarithmic variation of velocity in the overlap region are the keys of the asymptotic structure of the turbulent boundary layer.

11.2 Application of SCEM

The method employed to construct a UVA consists of seeking a first approximation corresponding to the external region of the flow. Afterwards, this approximation is corrected in the outer region of the boundary layer and finally, a UVA is obtained by taking into account the contribution of the inner region of the boundary layer.

11.2.1 First Approximation

We seek a first approximation in the form

$$\mathcal{U} = u_1^*(x, y, \varepsilon) + \cdots , \tag{11.16a}$$

$$\mathcal{V} = v_1^*(x, y, \varepsilon) + \cdots , \tag{11.16b}$$

$$\mathcal{P} = p_1^*(x, y, \varepsilon) + \cdots , \tag{11.16c}$$

$$\mathcal{T}_{ij} = 0 . \tag{11.16d}$$

Putting these expansions in (11.1a–11.1c) and neglecting $\mathrm{O}(1/\mathcal{R})$ terms, it can be shown that u_1^*, v_1^*, p_1^* satisfy the Euler equations. As in laminar flow (Subsect. 8.1.1), it is required to complement the above approximation because the no-slip condition at the wall cannot be fulfilled. Moreover, the wall condition for v_1^* is not known.

11.2.2 Contribution of the Outer Region of the Boundary Layer

A correction to the previous approximation is introduced in the form of a contribution of the outer region of the boundary layer

$$\mathcal{U} = u_1^*(x, y, \varepsilon) + \varepsilon U_1(x, \eta, \varepsilon) + \cdots , \tag{11.17a}$$

$$\mathcal{V} = v_1^*(x, y, \varepsilon) + \varepsilon^2 V_1(x, \eta, \varepsilon) + \cdots , \tag{11.17b}$$

$$\mathcal{P} = p_1^*(x, y, \varepsilon) + \Delta(\varepsilon) P_1(x, \eta, \varepsilon) + \cdots , \tag{11.17c}$$

$$\mathcal{T}_{ij} = \varepsilon^2 \tau_{ij,1}(x, \eta, \varepsilon) + \cdots . \tag{11.17d}$$

The gauges for the velocity and for the Reynolds stresses are chosen according to the standard asymptotic analysis. The gauge $\Delta(\varepsilon)$ is determined by examining the y-momentum equation.

Gauge for the Pressure

Taking into account the Euler equations, the y-momentum equation can be written as

$$\varepsilon U_1 \frac{\partial v_1^*}{\partial x} + \varepsilon^2 u_1^* \frac{\partial V_1}{\partial x} + \varepsilon^3 U_1 \frac{\partial V_1}{\partial x} + \varepsilon^2 V_1 \frac{\partial v_1^*}{\partial y} + \varepsilon v_1^* \frac{\partial V_1}{\partial \eta} + \varepsilon^3 V_1 \frac{\partial V_1}{\partial \eta}$$

$$= -\frac{\Delta}{\varepsilon} \frac{\partial P_1}{\partial \eta} + \varepsilon^2 \frac{\partial \tau_{xy,1}}{\partial x} + \frac{1}{\mathcal{R}} \frac{\partial^2 v_1^*}{\partial x^2} + \frac{\varepsilon^2}{\mathcal{R}} \frac{\partial^2 V_1}{\partial x^2} + \varepsilon \frac{\partial \tau_{yy,1}}{\partial \eta} + \frac{1}{\mathcal{R}} \frac{\partial^2 v_1^*}{\partial y^2} + \frac{1}{\mathcal{R}} \frac{\partial^2 V_1}{\partial \eta^2} .$$

In the boundary layer, by using the continuity equation, the Taylor series expansion of v_1^* when $y \ll 1$ yields

$$v_1^* = v_{1\,y=0}^* + y \left(\frac{\partial v_1^*}{\partial y} \right)_{y=0} + \cdots$$

$$= v_{1\,y=0}^* - y \left(\frac{\partial u_1^*}{\partial x} \right)_{y=0} + \cdots$$

$$= v_{1\,y=0}^* - \varepsilon\eta \left(\frac{\partial u_1^*}{\partial x} \right)_{y=0} + \cdots .$$

The condition of zero velocity at the wall implies that $v_{1\,y=0}^*$ is $\mathrm{O}(\varepsilon^2)$ in order to balance the term $\varepsilon^2 V_1$ because the next term $\varepsilon\hat{\varepsilon}\widehat{V}_1$ of the expansion of v is smaller. It follows that, in the outer region of the boundary layer, v_1^* is $\mathrm{O}(\varepsilon)$. Then, the dominant term of the y-momentum equation is $\varepsilon\dfrac{\partial \tau_{yy,1}}{\partial \eta}$. It is concluded that Δ is $\mathrm{O}(\varepsilon^2)$. We set

$$\Delta = \varepsilon^2 . \tag{11.18}$$

Then, the y-momentum equation becomes

$$-\frac{\partial P_1}{\partial \eta} + \frac{\partial \tau_{yy,1}}{\partial \eta} = \mathrm{O}(\varepsilon) .$$

According to the SCEM principle, as $\eta \to \infty$, we must have $P_1 \to 0$ and $\tau_{yy,1} \to 0$. Then, neglecting terms of order $\mathrm{O}(\varepsilon)$, we have

$$-P_1 + \tau_{yy,1} = 0 . \tag{11.19}$$

Continuity Equation

Taking into account the continuity equation relating u_1^* and v_1^*, we have

$$\frac{\partial U_1}{\partial x} + \frac{\partial V_1}{\partial \eta} = 0 . \tag{11.20}$$

x-Momentum Equation

Substituting expansions given by (11.17a–11.17d) in (11.1b) and taking into account the Euler equations for u_1^*, v_1^*, p_1^*, we get

$$\varepsilon U_1 \frac{\partial u_1^*}{\partial x} + \varepsilon u_1^* \frac{\partial U_1}{\partial x} + \varepsilon^2 U_1 \frac{\partial U_1}{\partial x} + \varepsilon^2 V_1 \frac{\partial u_1^*}{\partial y} + v_1^* \frac{\partial U_1}{\partial \eta} + \varepsilon^2 V_1 \frac{\partial U_1}{\partial \eta}$$

$$= -\varepsilon^2 \frac{\partial P_1}{\partial x} + \varepsilon^2 \frac{\partial \tau_{xx,1}}{\partial x} + \frac{1}{\mathcal{R}} \frac{\partial^2 u_1^*}{\partial x^2} + \frac{\varepsilon}{\mathcal{R}} \frac{\partial^2 U_1}{\partial x^2} + \varepsilon \frac{\partial \tau_{xy,1}}{\partial \eta}$$

$$+ \frac{1}{\mathcal{R}} \frac{\partial^2 u_1^*}{\partial y^2} + \frac{1}{\varepsilon\mathcal{R}} \frac{\partial^2 U_1}{\partial \eta^2} . \tag{11.21}$$

If $O(\varepsilon^2)$ terms are neglected, (11.21) becomes

$$U_1\frac{\partial u_1^*}{\partial x} + u_1^*\frac{\partial U_1}{\partial x} + \frac{v_1^*}{\varepsilon}\frac{\partial U_1}{\partial \eta} = \frac{\partial \tau_{xy,1}}{\partial \eta} \, . \tag{11.22}$$

Note 11.1. Associated to Euler equations for u_1^* et v_1^*, this equation is a UVA over the domain formed by the outer region of the boundary layer and by the inviscid flow region. Equation (11.22) reduces to (11.10b) if the following two hypotheses are made: i) the velocities u_1^* and v_1^* are expanded in Taylor series in the neighbourhood of $y = 0$ which is justified by the fact that, in the boundary layer, $y = \varepsilon\eta$ is very small compared to unity, ii) we assume that the velocity v_1^* vanishes at the wall. With these two hypotheses, u_1^* can be replaced by its wall value u_e in (11.22) and $\frac{v_1^*}{\varepsilon}$ can be replaced by $-\eta\dfrac{\mathrm{d}u_e}{\mathrm{d}x}$ in the same equation. In conformity with the notion of regular expansion, the hypotheses and the results of MMAE are recovered.

If $O(\varepsilon^3)$ terms are neglected, taking into account (11.19), (11.21) becomes

$$U_1\frac{\partial u_1^*}{\partial x} + u_1^*\frac{\partial U_1}{\partial x} + \varepsilon U_1\frac{\partial U_1}{\partial x} + \varepsilon V_1\frac{\partial u_1^*}{\partial y} + \frac{v_1^*}{\varepsilon}\frac{\partial U_1}{\partial \eta} + \varepsilon V_1\frac{\partial U_1}{\partial \eta}$$
$$= \frac{\partial \tau_{xy,1}}{\partial \eta} + \varepsilon\left(\frac{\partial \tau_{xx,1}}{\partial x} - \frac{\partial \tau_{yy,1}}{\partial x}\right) \, . \tag{11.23}$$

For both models, described by (11.22) or (11.23), the boundary conditions as $\eta \to \infty$ are

$$\eta \to \infty : \quad U_1 \to 0 \, , \quad V_1 \to 0 \, .$$

With these conditions and by taking into account the vanishing of the turbulent stresses in the inviscid flow, we observe that (11.22) or (11.23) are perfectly satisfied as $\eta \to \infty$.

The wall boundary conditions are given later when the contribution of the inner region is studied.

Equation (11.23) can be written in a form closer to the usual boundary layer equations. We define

$$\overline{U} = u_1^* + \varepsilon U_1 \, ,$$
$$\overline{V} = v_1^* + \varepsilon^2 V_1 \, ,$$
$$\overline{T}_{ij} = \varepsilon^2 \tau_{ij,1} \, .$$

Equations (11.20) and (11.23) become

$$\frac{\partial \overline{U}}{\partial x} + \frac{\partial \overline{V}}{\partial y} = 0 \, , \tag{11.24a}$$

$$\overline{U}\frac{\partial \overline{U}}{\partial x} + \overline{V}\frac{\partial \overline{U}}{\partial y} = u_1^*\frac{\partial u_1^*}{\partial x} + v_1^*\frac{\partial u_1^*}{\partial y} + \frac{\partial \overline{T}_{xy}}{\partial y} + \frac{\partial}{\partial x}(\overline{T}_{xx} - \overline{T}_{yy}) \, . \tag{11.24b}$$

11.2.3 Contribution of the Inner Region of the Boundary Layer

We seek a *UVA* in the form

$$\mathcal{U} = u_1^*(x, y, \varepsilon) + \varepsilon U_1(x, \eta, \varepsilon) + \varepsilon \widehat{U}_1(x, \hat{y}, \varepsilon) + \cdots , \tag{11.25a}$$

$$\mathcal{V} = v_1^*(x, y, \varepsilon) + \varepsilon^2 V_1(x, \eta, \varepsilon) + \varepsilon \hat{\varepsilon} \widehat{V}_1(x, \hat{y}, \varepsilon) + \cdots , \tag{11.25b}$$

$$\mathcal{P} = p_1^*(x, y, \varepsilon) + \varepsilon^2 P_1(x, \eta, \varepsilon) + \widehat{\Delta}(\varepsilon) \widehat{P}_1(x, \hat{y}, \varepsilon) + \cdots , \tag{11.25c}$$

$$\mathcal{T}_{ij} = \varepsilon^2 \tau_{ij,1}(x, \eta, \varepsilon) + \varepsilon^2 \hat{\tau}_{ij,1}(x, \hat{y}, \varepsilon) + \cdots . \tag{11.25d}$$

The gauge $\widehat{\Delta}$ is determined by examining the y-momentum equation.

Gauge for the Pressure

Substituting expansions given by (11.25a–11.25d) in (11.1c), we obtain

$$\varepsilon U_1 \frac{\partial v_1^*}{\partial x} + \varepsilon \widehat{U}_1 \frac{\partial v_1^*}{\partial x} + \varepsilon^2 u_1^* \frac{\partial V_1}{\partial x} + \varepsilon^3 U_1 \frac{\partial V_1}{\partial x} + \varepsilon^3 \widehat{U}_1 \frac{\partial V_1}{\partial x}$$

$$+ \hat{\varepsilon} \varepsilon u_1^* \frac{\partial \widehat{V}_1}{\partial x} + \hat{\varepsilon} \varepsilon^2 U_1 \frac{\partial \widehat{V}_1}{\partial x} + \hat{\varepsilon} \varepsilon^2 \widehat{U}_1 \frac{\partial \widehat{V}_1}{\partial x} + \varepsilon^2 V_1 \frac{\partial v_1^*}{\partial y} + \hat{\varepsilon} \varepsilon \widehat{V}_1 \frac{\partial v_1^*}{\partial y}$$

$$+ \varepsilon v_1^* \frac{\partial V_1}{\partial \eta} + \varepsilon^3 V_1 \frac{\partial V_1}{\partial \eta} + \hat{\varepsilon} \varepsilon^2 \widehat{V}_1 \frac{\partial V_1}{\partial \eta} + \varepsilon v_1^* \frac{\partial \widehat{V}_1}{\partial \hat{y}} + \varepsilon^3 V_1 \frac{\partial \widehat{V}_1}{\partial \hat{y}} + \hat{\varepsilon} \varepsilon^2 \widehat{V}_1 \frac{\partial \widehat{V}_1}{\partial \hat{y}}$$

$$= -\varepsilon \frac{\partial P_1}{\partial \eta} - \frac{\widehat{\Delta}}{\hat{\varepsilon}} \frac{\partial \widehat{P}_1}{\partial \hat{y}} + \varepsilon^2 \frac{\partial \tau_{xy,1}}{\partial x} + \varepsilon^2 \frac{\partial \hat{\tau}_{xy,1}}{\partial x} + \frac{1}{\mathcal{R}} \frac{\partial^2 v_1^*}{\partial x^2} + \frac{\varepsilon^2}{\mathcal{R}} \frac{\partial^2 V_1}{\partial x^2} + \frac{\hat{\varepsilon} \varepsilon}{\mathcal{R}} \frac{\partial^2 \widehat{V}_1}{\partial x^2}$$

$$+ \varepsilon \frac{\partial \tau_{yy,1}}{\partial \eta} + \frac{\varepsilon^2}{\hat{\varepsilon}} \frac{\partial \hat{\tau}_{yy,1}}{\partial \hat{y}} + \frac{1}{\mathcal{R}} \frac{\partial^2 v_1^*}{\partial y^2} + \frac{1}{\mathcal{R}} \frac{\partial^2 V_1}{\partial \eta^2} + \frac{\varepsilon}{\hat{\varepsilon} \mathcal{R}} \frac{\partial^2 \widehat{V}_1}{\partial \hat{y}^2} .$$

In this equation, the dominant terms are of order $\varepsilon^2/\hat{\varepsilon}$. This leads us to choose

$$\widehat{\Delta} = \varepsilon^2 , \tag{11.26}$$

and the y-momentum equation becomes

$$-\frac{\partial \widehat{P}_1}{\partial \hat{y}} + \frac{\partial \hat{\tau}_{yy,1}}{\partial \hat{y}} = 0 . \tag{11.27}$$

Continuity Equation

For the velocity components $\widehat{U}_1$ and $\widehat{V}_1$, we have

$$\frac{\partial \widehat{U}_1}{\partial x} + \frac{\partial \widehat{V}_1}{\partial \hat{y}} = 0 . \tag{11.28}$$

x-Momentum Equation

Substituting expansions given by (11.25a–11.25d) in (11.1b) and taking into account the Euler equations for u_1^*, v_1^*, p_1^*, we obtain

$$
\varepsilon U_1 \frac{\partial u_1^*}{\partial x} + \varepsilon \widehat{U}_1 \frac{\partial u_1^*}{\partial x} + \varepsilon u_1^* \frac{\partial U_1}{\partial x} + \varepsilon^2 U_1 \frac{\partial U_1}{\partial x} + \varepsilon^2 \widehat{U}_1 \frac{\partial U_1}{\partial x}
$$

$$
+ \varepsilon u_1^* \frac{\partial \widehat{U}_1}{\partial x} + \varepsilon^2 U_1 \frac{\partial \widehat{U}_1}{\partial x} + \varepsilon^2 \widehat{U}_1 \frac{\partial \widehat{U}_1}{\partial x} + \varepsilon^2 V_1 \frac{\partial u_1^*}{\partial y} + \hat{\varepsilon}\varepsilon \widehat{V}_1 \frac{\partial u_1^*}{\partial y}
$$

$$
+ v_1^* \frac{\partial U_1}{\partial \eta} + \varepsilon^2 V_1 \frac{\partial U_1}{\partial \eta} + \hat{\varepsilon}\varepsilon \widehat{V}_1 \frac{\partial U_1}{\partial \eta}
$$

$$
+ \frac{\varepsilon}{\hat{\varepsilon}} v_1^* \frac{\partial \widehat{U}_1}{\partial \hat{y}} + \frac{\varepsilon^3}{\hat{\varepsilon}} V_1 \frac{\partial \widehat{U}_1}{\partial \hat{y}} + \varepsilon^2 \widehat{V}_1 \frac{\partial \widehat{U}_1}{\partial \hat{y}}
$$

$$
= -\varepsilon^2 \frac{\partial P_1}{\partial x} - \varepsilon^2 \frac{\partial \widehat{P}_1}{\partial x} + \varepsilon^2 \frac{\partial \tau_{xx,1}}{\partial x} + \varepsilon^2 \frac{\partial \hat{\tau}_{xx,1}}{\partial x}
$$

$$
+ \frac{1}{\mathcal{R}} \frac{\partial^2 u_1^*}{\partial x^2} + \frac{\varepsilon}{\mathcal{R}} \frac{\partial^2 U_1}{\partial x^2} + \frac{\varepsilon}{\mathcal{R}} \frac{\partial^2 \widehat{U}_1}{\partial x^2}
$$

$$
+ \varepsilon \frac{\partial \tau_{xy,1}}{\partial \eta} + \frac{\varepsilon^2}{\hat{\varepsilon}} \frac{\partial \hat{\tau}_{xy,1}}{\partial \hat{y}} + \frac{1}{\mathcal{R}} \frac{\partial^2 u_1^*}{\partial y^2} + \frac{1}{\varepsilon \mathcal{R}} \frac{\partial^2 U_1}{\partial \eta^2} + \frac{\varepsilon}{\hat{\varepsilon}^2 \mathcal{R}} \frac{\partial^2 \widehat{U}_1}{\partial \hat{y}^2} . \tag{11.29}
$$

In the above equation, we can eliminate the terms already taken into account in the first approximaton of the contribution of the boundary layer outer region; these terms correspond to the terms of (11.22). We obtain

$$
\varepsilon \widehat{U}_1 \frac{\partial u_1^*}{\partial x} + \varepsilon^2 U_1 \frac{\partial U_1}{\partial x} + \varepsilon^2 \widehat{U}_1 \frac{\partial U_1}{\partial x}
$$

$$
+ \varepsilon u_1^* \frac{\partial \widehat{U}_1}{\partial x} + \varepsilon^2 U_1 \frac{\partial \widehat{U}_1}{\partial x} + \varepsilon^2 \widehat{U}_1 \frac{\partial \widehat{U}_1}{\partial x} + \varepsilon^2 V_1 \frac{\partial u_1^*}{\partial y} + \hat{\varepsilon}\varepsilon \widehat{V}_1 \frac{\partial u_1^*}{\partial y}
$$

$$
+ \varepsilon^2 V_1 \frac{\partial U_1}{\partial \eta} + \hat{\varepsilon}\varepsilon \widehat{V}_1 \frac{\partial U_1}{\partial \eta}
$$

$$
+ \frac{\varepsilon}{\hat{\varepsilon}} v_1^* \frac{\partial \widehat{U}_1}{\partial \hat{y}} + \frac{\varepsilon^3}{\hat{\varepsilon}} V_1 \frac{\partial \widehat{U}_1}{\partial \hat{y}} + \varepsilon^2 \widehat{V}_1 \frac{\partial \widehat{U}_1}{\partial \hat{y}}
$$

$$
= -\varepsilon^2 \frac{\partial P_1}{\partial x} - \varepsilon^2 \frac{\partial \widehat{P}_1}{\partial x} + \varepsilon^2 \frac{\partial \tau_{xx,1}}{\partial x} + \varepsilon^2 \frac{\partial \hat{\tau}_{xx,1}}{\partial x}
$$

$$
+ \frac{1}{\mathcal{R}} \frac{\partial^2 u_1^*}{\partial x^2} + \frac{\varepsilon}{\mathcal{R}} \frac{\partial^2 U_1}{\partial x^2} + \frac{\varepsilon}{\mathcal{R}} \frac{\partial^2 \widehat{U}_1}{\partial x^2}
$$

$$
+ \frac{\varepsilon^2}{\hat{\varepsilon}} \frac{\partial \hat{\tau}_{xy,1}}{\partial \hat{y}} + \frac{1}{\mathcal{R}} \frac{\partial^2 u_1^*}{\partial y^2} + \frac{1}{\varepsilon \mathcal{R}} \frac{\partial^2 U_1}{\partial \eta^2} + \frac{\varepsilon}{\hat{\varepsilon}^2 \mathcal{R}} \frac{\partial^2 \widehat{U}_1}{\partial \hat{y}^2} . \tag{11.30}
$$

In the above equation, the dominant terms are of order $O(\varepsilon^2/\hat{\varepsilon})$.

Examine the order of term $\dfrac{\varepsilon}{\hat{\varepsilon}} v_1^* \dfrac{\partial \widehat{U}_1}{\partial \hat{y}}$. In the inner region of the boundary layer, a Taylor series expansion of v_1^* $(y \ll 1)$ yields

$$v_1^* = v_{1\,y=0}^* + y \left(\frac{\partial v_1^*}{\partial y}\right)_{y=0} + \cdots$$

$$= v_{1\,y=0}^* - y \left(\frac{\partial u_1^*}{\partial x}\right)_{y=0} + \cdots$$

$$= v_{1\,y=0}^* - \hat{\varepsilon}\hat{y} \left(\frac{\partial u_1^*}{\partial x}\right)_{y=0} + \cdots .$$

Now, as already said, $v_{1\,y=0}^*$ is $\mathrm{O}(\varepsilon^2)$. It follows that v_1^* is $\mathrm{O}(\varepsilon^2)$ in the inner region of the boundary layer and term $\dfrac{\varepsilon}{\hat{\varepsilon}} v_1^* \dfrac{\partial \widehat{U}_1}{\partial \hat{y}}$ is $\mathrm{O}(\varepsilon^3/\hat{\varepsilon})$.

From the results of the standard analysis, it is known that, as $\eta \to 0$, U_1 is a logarithmic function of η. Under these conditions, with $\hat{\varepsilon}\varepsilon\mathcal{R} = 1$, we have

$$\frac{1}{\varepsilon\mathcal{R}} \frac{\partial^2 U_1}{\partial \eta^2} \sim \frac{\varepsilon^2}{\hat{\varepsilon}} \frac{1}{\hat{y}^2} .$$

This term is $\mathrm{O}(\varepsilon^2/\hat{\varepsilon})$.

Retaining only $\mathrm{O}(\varepsilon^2/\hat{\varepsilon})$ terms, the streamwise momentum equation (11.30) reduces to

$$\frac{\varepsilon^2}{\hat{\varepsilon}} \frac{\partial \hat{\tau}_{xy,1}}{\partial \hat{y}} + \frac{1}{\varepsilon\mathcal{R}} \frac{\partial^2 U_1}{\partial \eta^2} + \frac{\varepsilon}{\hat{\varepsilon}^2\mathcal{R}} \frac{\partial^2 \widehat{U}_1}{\partial \hat{y}^2} = 0 . \tag{11.31}$$

Examine now a better approximation consistent with the second approximation of the contribution of the boundary layer outer region. Taking into account (11.23), (11.29) becomes

$$\varepsilon\widehat{U}_1 \frac{\partial u_1^*}{\partial x} + \varepsilon^2\widehat{U}_1 \frac{\partial U_1}{\partial x}$$

$$+\varepsilon u_1^* \frac{\partial \widehat{U}_1}{\partial x} + \varepsilon^2 U_1 \frac{\partial \widehat{U}_1}{\partial x} + \varepsilon^2 \widehat{U}_1 \frac{\partial \widehat{U}_1}{\partial x} + \hat{\varepsilon}\varepsilon \widehat{V}_1 \frac{\partial u_1^*}{\partial y}$$

$$+\hat{\varepsilon}\varepsilon \widehat{V}_1 \frac{\partial U_1}{\partial \eta}$$

$$+\frac{\varepsilon}{\hat{\varepsilon}} v_1^* \frac{\partial \widehat{U}_1}{\partial \hat{y}} + \frac{\varepsilon^3}{\hat{\varepsilon}} V_1 \frac{\partial \widehat{U}_1}{\partial \hat{y}} + \varepsilon^2 \widehat{V}_1 \frac{\partial \widehat{U}_1}{\partial \hat{y}}$$

$$= -\varepsilon^2 \frac{\partial \widehat{P}_1}{\partial x} + \varepsilon^2 \frac{\partial \hat{\tau}_{xx,1}}{\partial x}$$

$$+\frac{1}{\mathcal{R}} \frac{\partial^2 u_1^*}{\partial x^2} + \frac{\varepsilon}{\mathcal{R}} \frac{\partial^2 U_1}{\partial x^2} + \frac{\varepsilon}{\mathcal{R}} \frac{\partial^2 \widehat{U}_1}{\partial x^2}$$

$$+\frac{\varepsilon^2}{\hat{\varepsilon}} \frac{\partial \hat{\tau}_{xy,1}}{\partial \hat{y}} + \frac{1}{\mathcal{R}} \frac{\partial^2 u_1^*}{\partial y^2} + \frac{1}{\varepsilon\mathcal{R}} \frac{\partial^2 U_1}{\partial \eta^2} + \frac{\varepsilon}{\hat{\varepsilon}^2\mathcal{R}} \frac{\partial^2 \widehat{U}_1}{\partial \hat{y}^2} . \tag{11.32}$$

Retaining $O(\varepsilon^3/\hat{\varepsilon})$ terms, the streamwise momentum equation (11.32) becomes

$$\frac{\varepsilon}{\hat{\varepsilon}}v_1^*\frac{\partial \widehat{U}_1}{\partial \hat{y}} + \frac{\varepsilon^3}{\hat{\varepsilon}}V_1\frac{\partial \widehat{U}_1}{\partial \hat{y}} = \frac{\varepsilon^2}{\hat{\varepsilon}}\frac{\partial \hat{\tau}_{xy,1}}{\partial \hat{y}} + \frac{1}{\varepsilon\mathcal{R}}\frac{\partial^2 U_1}{\partial \eta^2} + \frac{\varepsilon}{\hat{\varepsilon}^2\mathcal{R}}\frac{\partial^2 \widehat{U}_1}{\partial \hat{y}^2} \, . \tag{11.33}$$

11.3 Interactive Boundary Layer

First, we summarize the results obtained until now. The UVA has the form (11.25a–11.25d)

$$\mathcal{U} = u_1^*(x, y, \varepsilon) + \varepsilon U_1(x, \eta, \varepsilon) + \varepsilon\widehat{U}_1(x, \hat{y}, \varepsilon) + \cdots \, , \tag{11.34a}$$

$$\mathcal{V} = v_1^*(x, y, \varepsilon) + \varepsilon^2 V_1(x, \eta, \varepsilon) + \varepsilon\hat{\varepsilon}\widehat{V}_1(x, \hat{y}, \varepsilon) + \cdots \, , \tag{11.34b}$$

$$\mathcal{P} = p_1^*(x, y, \varepsilon) + \varepsilon^2 P_1(x, \eta, \varepsilon) + \varepsilon^2\widehat{P}_1(x, \hat{y}, \varepsilon) + \cdots \, , \tag{11.34c}$$

$$\mathcal{T}_{ij} = \varepsilon^2 \tau_{ij,1}(x, \eta, \varepsilon) + \varepsilon^2 \hat{\tau}_{ij,1}(x, \hat{y}, \varepsilon) + \cdots \, . \tag{11.34d}$$

According to the order of neglected terms in the x-momentum equation, we obtain a first or second order IBL model.

11.3.1 First Order Model

This model comprises (11.20), (11.22), (11.28) and (11.31)

$$\frac{\partial U_1}{\partial x} + \frac{\partial V_1}{\partial \eta} = 0 \, , \tag{11.35a}$$

$$U_1\frac{\partial u_1^*}{\partial x} + u_1^*\frac{\partial U_1}{\partial x} + \frac{v_1^*}{\varepsilon}\frac{\partial U_1}{\partial \eta} = \frac{\partial \tau_{xy,1}}{\partial \eta} \, , \tag{11.35b}$$

$$\frac{\partial \widehat{U}_1}{\partial x} + \frac{\partial \widehat{V}_1}{\partial \hat{y}} = 0 \, , \tag{11.35c}$$

$$\frac{\varepsilon^2}{\hat{\varepsilon}}\frac{\partial \hat{\tau}_{xy,1}}{\partial \hat{y}} + \frac{1}{\varepsilon\mathcal{R}}\frac{\partial^2 U_1}{\partial \eta^2} + \frac{\varepsilon}{\hat{\varepsilon}^2\mathcal{R}}\frac{\partial^2 \widehat{U}_1}{\partial \hat{y}^2} = 0 \, . \tag{11.35d}$$

In addition, u_1^* and v_1^* satisfy the Euler equations.
The boundary conditions are

$$\eta \to \infty : \quad U_1 \to 0 \, , \quad V_1 \to 0 \, , \tag{11.36a}$$

$$\hat{y} \to \infty : \quad \widehat{U}_1 \to 0 \, , \quad \widehat{V}_1 \to 0 \, , \tag{11.36b}$$

and, at the wall

$$u_1^* + \varepsilon U_1 + \varepsilon\widehat{U}_1 = 0 \, , \tag{11.37a}$$

$$v_1^* + \varepsilon^2 V_1 + \hat{\varepsilon}\varepsilon\widehat{V}_1 = 0 \, . \tag{11.37b}$$

At infinity, we also have conditions on u_1^* and v_1^*, usually corresponding to uniform flow conditions.

11.3.2 Second Order Model

This model comprises (11.20), (11.23), (11.28) and (11.33)

$$\frac{\partial U_1}{\partial x} + \frac{\partial V_1}{\partial \eta} = 0 \ , \tag{11.38a}$$

$$U_1 \frac{\partial u_1^*}{\partial x} + u_1^* \frac{\partial U_1}{\partial x} + \varepsilon U_1 \frac{\partial U_1}{\partial x} + \varepsilon V_1 \frac{\partial u_1^*}{\partial y} + \frac{v_1^*}{\varepsilon} \frac{\partial U_1}{\partial \eta} + \varepsilon V_1 \frac{\partial U_1}{\partial \eta}$$

$$= \frac{\partial \tau_{xy,1}}{\partial \eta} + \varepsilon \left(\frac{\partial \tau_{xx,1}}{\partial x} - \frac{\partial \tau_{yy,1}}{\partial x} \right) \ , \tag{11.38b}$$

$$\frac{\partial \widehat{U}_1}{\partial x} + \frac{\partial \widehat{V}_1}{\partial \hat{y}} = 0 \ , \tag{11.38c}$$

$$\frac{\varepsilon}{\hat{\varepsilon}} v_1^* \frac{\partial \widehat{U}_1}{\partial \hat{y}} + \frac{\varepsilon^3}{\hat{\varepsilon}} V_1 \frac{\partial \widehat{U}_1}{\partial \hat{y}} = \frac{\varepsilon^2}{\hat{\varepsilon}} \frac{\partial \hat{\tau}_{xy,1}}{\partial \hat{y}} + \frac{1}{\varepsilon \mathcal{R}} \frac{\partial^2 U_1}{\partial \eta^2} + \frac{\varepsilon}{\hat{\varepsilon}^2 \mathcal{R}} \frac{\partial^2 \widehat{U}_1}{\partial \hat{y}^2} \ . \tag{11.38d}$$

This system must be associated with Euler equations for u_1^* and v_1^*. The boundary conditions are identical to those of the first order model.

11.3.3 Global Model

The models presented above can be included in a global model having the following properties: i) the global model describes the outer and the inner boundary layer layer regions ii) the global model contains the first and the second order models. This model completes and supplants (11.24a–11.24b) by taking into account the contribution of the boundary layer inner region. We set

$$u = u_1^* + \varepsilon U_1 + \varepsilon \widehat{U}_1 \ ,$$
$$v = v_1^* + \varepsilon^2 V_1 + \varepsilon \hat{\varepsilon} \widehat{V}_1 \ ,$$
$$t_{ij} = \varepsilon^2 \tau_{ij,1} + \varepsilon^2 \hat{\tau}_{ij,1} \ ,$$

with

$$t_{ij} = - < u_i' u_j' > \ .$$

The equations proposed below for u and v cannot be deduced from any model established before. This is a heuristic model which can be written as

$$\frac{\partial u}{\partial x} + \frac{\partial v}{\partial y} = 0 \ , \tag{11.39a}$$

$$u \frac{\partial u}{\partial x} + v \frac{\partial u}{\partial y} = u_1^* \frac{\partial u_1^*}{\partial x} + v_1^* \frac{\partial u_1^*}{\partial y} + \frac{\partial}{\partial y} (- < u'v' >)$$

$$+ \frac{1}{\mathcal{R}} \frac{\partial^2 (u - u_1^*)}{\partial y^2} + \frac{\partial}{\partial x} (< v'^2 > - < u'^2 >) \ . \tag{11.39b}$$

To the considered order, it can be checked that system formed by (11.38a–11.38d) is recovered after expanding (11.39a–11.39b) according to the method discussed in Sect. 11.2.

Equations (11.39a–11.39b) must be associated with the Euler equations for u_1^* and v_1^*. The boundary conditions are

$$y \to \infty \quad : \quad u - u_1^* \to 0, \quad v - v_1^* \to 0 \;, \tag{11.40a}$$

$$\text{at the wall}: \quad u = 0, \qquad v = 0 \;. \tag{11.40b}$$

Note 11.2. The global heuristic model includes the case of a laminar boundary layer analyzed in Chap. 8: it suffices to set the turbulent stresses to zero.

11.3.4 Reduced Model for an Irrotational External Flow

For *an external irrotational flow*, the global model of previous Subsect. 11.3.3 takes a simplified form if the validity of equations is restricted to the boundary layer region.

As in Sect. 8.5 for a laminar boundary layer, we can use Taylor series expansions *in the boundary layer outer region*

$$u_1^* = u_{10}^* + y \left(\frac{\partial u_1^*}{\partial y} \right)_{y=0} + \cdots$$

$$= u_{10}^* + \varepsilon \eta \left(\frac{\partial u_1^*}{\partial y} \right)_{y=0} + \cdots \;,$$

$$\frac{\partial u_1^*}{\partial x} = u_{1x0}^* + y \left(\frac{\partial^2 u_1^*}{\partial x \partial y} \right)_{y=0} + \cdots \;.$$

We assume that the inviscid flow is irrotational and that the wall curvature effects are negligible. In the boundary layer outer region, it is known that $v_1^* = \mathrm{O}(\varepsilon)$ and it follows that $\dfrac{\partial u_1^*}{\partial y} = \mathrm{O}(\varepsilon)$. We also have $\dfrac{\partial^2 v_1^*}{\partial y^2} = \mathrm{O}(\varepsilon)$. In the boundary layer outer region, we obtain

$$u_1^* = u_{10}^* + \mathrm{O}(\varepsilon^2) \;,$$

$$\frac{\partial u_1^*}{\partial x} = u_{1x0}^* + \mathrm{O}(\varepsilon^2) \;,$$

$$v_1^* = v_{10}^* - y u_{1x0}^* + \mathrm{O}(\varepsilon^3) \;,$$

$$\frac{\partial^2 v_1^*}{\partial y^2} = \mathrm{O}(\varepsilon) \;.$$

Approximations (11.17a–11.17d) give

$$\mathcal{U} = u_{10}^* + \varepsilon U_1 + \cdots \;,$$

$$\mathcal{V} = v_{10}^* - y u_{1x0}^* + \varepsilon^2 V_1 + \cdots \;,$$

$$\mathcal{T}_{ij} = \varepsilon^2 \tau_{ij,1} + \cdots \;.$$

With these hypotheses, (11.38a) and (11.38b) restricted to the outer region of the boundary layer become

$$\frac{\partial U_1}{\partial x} + \frac{\partial V_1}{\partial \eta} = 0 , \tag{11.41a}$$

$$U_1 \frac{du_{10}^*}{dx} + u_{10}^* \frac{\partial U_1}{\partial x} + \varepsilon U_1 \frac{\partial U_1}{\partial x} + \frac{v_{10}^* - y u_{1x0}^*}{\varepsilon} \frac{\partial U_1}{\partial \eta} + \varepsilon V_1 \frac{\partial U_1}{\partial \eta}$$

$$= \frac{\partial \tau_{xy,1}}{\partial \eta} + \varepsilon \left(\frac{\partial \tau_{xx,1}}{\partial x} - \frac{\partial \tau_{yy,1}}{\partial x} \right) . \tag{11.41b}$$

We set

$$\overline{U} = u_{10}^* + \varepsilon U_1 ,$$
$$\overline{V} = v_{10}^* - y u_{1x0}^* + \varepsilon^2 V_1 ,$$
$$\overline{T}_{ij} = \varepsilon^2 \tau_{ij,1} .$$

Equations (11.41a) and (11.41b) can be recast as

$$\frac{\partial \overline{U}}{\partial x} + \frac{\partial \overline{V}}{\partial y} = 0 , \tag{11.42a}$$

$$\overline{U} \frac{\partial \overline{U}}{\partial x} + \overline{V} \frac{\partial \overline{U}}{\partial y} = u_{10}^* \frac{du_{10}^*}{dx} + \frac{\partial \overline{T}_{xy}}{\partial y} + \frac{\partial}{\partial x}(\overline{T}_{xx} - \overline{T}_{yy}) . \tag{11.42b}$$

In this form, these equations are very close to the equations usually employed for the outer region of the turbulent boundary layer; the only difference is the term $\frac{\partial}{\partial x}(\overline{T}_{xx} - \overline{T}_{yy})$ which is neglected most of the time. We note that this term is not present in the first order IBL model.

In the boundary layer inner region, we know that $v_1^* = \mathrm{O}(\varepsilon^2)$. Assuming that the inviscid flow is irrotational and neglecting wall curvature effects, it follows that $\frac{\partial u_1^*}{\partial y} = \mathrm{O}(\varepsilon^2)$. In this region, we also have $\frac{\partial^2 v_1^*}{\partial y^2} = \mathrm{O}(\varepsilon^2)$. Then, the Taylor series expansions of u_1^* and v_1^* show that

$$u_1^* = u_{10}^* + \mathrm{O}(\hat{\varepsilon}\varepsilon^2) ,$$
$$v_1^* = v_{10}^* - \hat{\varepsilon}\hat{y} u_{1x0}^* + \mathrm{O}(\hat{\varepsilon}^2\varepsilon^2) .$$

Moreover, we know that $u_{10}^* = \mathrm{O}(1)$ and $v_{10}^* = \mathrm{O}(\varepsilon^2)$.

Then, UVAs given by (11.25a–11.25d), written in the boundary layer, give

$$\mathcal{U} = u_{10}^* + \varepsilon U_1 + \varepsilon \widehat{U}_1 + \cdots ,$$
$$\mathcal{V} = v_{10}^* - y u_{1x0}^* + \varepsilon^2 V_1 + \hat{\varepsilon}\varepsilon \widehat{V}_1 + \cdots ,$$
$$\mathcal{T}_{ij} = \varepsilon^2 \tau_{ij,1} + \varepsilon^2 \hat{\tau}_{ij,1} + \cdots .$$

Equations (11.38c) and (11.38d), restricted to the inner region of the boundary layer, become

$$\frac{\partial \widehat{U}_1}{\partial x} + \frac{\partial \widehat{V}_1}{\partial \hat{y}} = 0 \, , \tag{11.43a}$$

$$\frac{\varepsilon}{\hat{\varepsilon}} v_{10}^* \frac{\partial \widehat{U}_1}{\partial \hat{y}} + \frac{\varepsilon^3}{\hat{\varepsilon}} V_1 \frac{\partial \widehat{U}_1}{\partial \hat{y}} = \frac{\varepsilon^2}{\hat{\varepsilon}} \frac{\partial \hat{\tau}_{xy,1}}{\partial \hat{y}} + \frac{1}{\varepsilon \mathcal{R}} \frac{\partial^2 U_1}{\partial \eta^2} + \frac{\varepsilon}{\hat{\varepsilon}^2 \mathcal{R}} \frac{\partial^2 \widehat{U}_1}{\partial \hat{y}^2} \, . \tag{11.43b}$$

We set

$$u = u_{10}^* + \varepsilon U_1 + \varepsilon \widehat{U}_1 \, , \tag{11.44a}$$

$$v = v_{10}^* - y u_{1x0}^* + \varepsilon^2 V_1 + \varepsilon \hat{\varepsilon} \widehat{V}_1 \, , \tag{11.44b}$$

$$t_{ij} = \varepsilon^2 \tau_{ij,1} + \varepsilon^2 \hat{\tau}_{ij,1} \, , \tag{11.44c}$$

with

$$t_{ij} = - < u_i' u_j' > \, .$$

Equations (11.41a, 11.41b) and (11.43a, 11.43b) are contained in the following heuristic model, valid only in the boundary layer

$$\frac{\partial u}{\partial x} + \frac{\partial v}{\partial y} = 0 \, , \tag{11.45a}$$

$$u \frac{\partial u}{\partial x} + v \frac{\partial u}{\partial y} = u_{10}^* \frac{d u_{10}^*}{dx} + \frac{\partial}{\partial y} \left(- < u'v' > \right) + \frac{1}{\mathcal{R}} \frac{\partial^2 u}{\partial y^2}$$

$$+ \frac{\partial}{\partial x} \left(< v'^2 > - < u'^2 > \right) \, , \tag{11.45b}$$

with the boundary conditions

$$y \to \infty \quad : \quad u - u_{10}^* \to 0 \, , \quad v - v_{10}^* + y u_{1x0}^* \to 0 \, , \tag{11.46a}$$

$$\text{at the wall :} \quad u = 0 \quad , \quad v = 0 \, . \tag{11.46b}$$

Generally, the contribution of term $\dfrac{\partial}{\partial x} \left(< v'^2 > - < u'^2 > \right)$ is neglected because experimental results show that the values of $< u'^2 >$ and $< v'^2 >$ are close; in a first order model, this term is not present. With this hypothesis, the equations are

$$\frac{\partial u}{\partial x} + \frac{\partial v}{\partial y} = 0 \, , \tag{11.47a}$$

$$u \frac{\partial u}{\partial x} + v \frac{\partial u}{\partial y} = u_{10}^* \frac{d u_{10}^*}{dx} + \frac{\partial}{\partial y} \left(- < u'v' > \right) + \frac{1}{\mathcal{R}} \frac{\partial^2 u}{\partial y^2} \, . \tag{11.47b}$$

Moreover, in weak coupling, i.e. if we seek regular expansions, the second boundary condition at infinity (11.46a) gives $v_{10}^* = 0$ as in laminar flow (Subsect. 10.1.2).

The other boundary conditions are

$$y \to \infty \quad : \quad u - u_{10}^* \to 0 \,, \tag{11.48a}$$

$$\text{at the wall}: \quad u = 0, \quad v = 0 \,. \tag{11.48b}$$

The usual model of turbulent boundary layer is recovered.

11.4 Approximation of the Boundary Layer: Velocity Profile

11.4.1 Formulation of the Problem

The objective is to construct, for an external irrotational flow, an approximation of the velocity profile *in the whole boundary layer*.

For this, we use (11.31) which describes the first order contribution of the inner region of the boundary layer

$$\frac{\varepsilon^2}{\hat{\varepsilon}} \frac{\partial \hat{\tau}_{xy,1}}{\partial \hat{y}} + \frac{1}{\varepsilon \mathcal{R}} \frac{\partial^2 U_1}{\partial \eta^2} + \frac{\varepsilon}{\hat{\varepsilon}^2 \mathcal{R}} \frac{\partial^2 \widehat{U}_1}{\partial \hat{y}^2} = 0 \,. \tag{11.49}$$

The solution of this equation requires the knowledge of the function $U_1(\eta)$ and the implementation of a turbulence model to describe the evolution of $\hat{\tau}_{xy,1}$. The study relies on a mixing length scheme, particularly well adapted to a flat plate flow, and on similarity solutions for the outer region of the boundary layer [70].

For the sake of simplicity, we work with the reduced model described by (11.47a–11.47b) and boundary conditions (11.48a–11.48b).

Rather than solve (11.49), it is more convenient to use an equation which gives directly the total velocity. We return to expansion given by (11.44c) and write (11.49) in the form

$$\frac{\varepsilon^2}{\hat{\varepsilon}} \frac{\partial}{\partial \hat{y}} \left(\frac{t_{xy}}{\epsilon^2} - \tau_{xy,1} \right) + \frac{1}{\varepsilon \mathcal{R}} \frac{\partial^2 U_1}{\partial \eta^2} + \frac{\varepsilon}{\hat{\varepsilon}^2 \mathcal{R}} \frac{\partial^2 \widehat{U}_1}{\partial \hat{y}^2} = 0 \,.$$

With variable y, this equation can also be written as

$$\frac{\partial t_{xy}}{\partial y} - \frac{\partial \epsilon^2 \tau_{xy,1}}{\partial y} + \frac{1}{\mathcal{R}} \frac{\partial^2}{\partial y^2} (u_{10}^* + \varepsilon U_1 + \varepsilon \widehat{U}_1) = 0 \,,$$

or

$$\frac{\partial}{\partial y} \left[t_{xy} + \frac{1}{\mathcal{R}} \frac{\partial u}{\partial y} \right] = \epsilon^2 \frac{\partial \tau_{xy,1}}{\partial y} \,, \tag{11.50}$$

where, for an external irrotational flow, we have

$$u = u_{10}^* + \varepsilon U_1 + \varepsilon \widehat{U}_1 \,.$$

In these equations, we introduced u_{10}^* which does not depend on y.

The left hand side of (11.50) represents the total stress – sum of the turbulent and viscous stresses – *in the whole boundary layer* whereas the right hand side represents the turbulent stress in the outer region. Let us integrate this equation with respect to y from the wall $y = 0$. The dimensionalized wall shear stress being τ_w, we obtain

$$t_{xy} + \frac{1}{\mathcal{R}}\frac{\partial u}{\partial y} - \frac{\tau_w}{\varrho V^2} = \epsilon^2 \tau_{xy,1} - \frac{\tau_w}{\varrho V^2} \, ,$$

because, at $y = 0$, we have

$$\frac{\tau_w}{\varrho V^2} = \frac{1}{\mathcal{R}}\frac{\partial u}{\partial y} \quad \text{and} \quad t_{xy} = 0 \, ,$$

and, on the other hand, the outer solution is such that we have as $\eta \to 0$, i.e. at $y = 0$

$$\epsilon^2 \tau_{xy,1} = \frac{\tau_w}{\varrho V^2} \, .$$

Finally, (11.49) takes the form

$$t_{xy} + \frac{1}{\mathcal{R}}\frac{\partial u}{\partial y} = \epsilon^2 \tau_{xy,1} \, .$$

Synthetically, by dividing the two members by the dimensionless wall shear stress, the above equation becomes

$$\frac{\tau}{\tau_w} = \frac{\tau_{\text{out}}}{\tau_w} \, , \tag{11.51}$$

where the left hand side represents the dimensionless total stress in the whole boundary layer and the right hand side represents the approximation of the dimensionless turbulent stress calculated in the outer region of the boundary layer.

Following the standard asymptotic theory strictly, the right hand side is equal to 1 if the solution is sought in the inner region. Indeed, τ_{out}/τ_w is a function of η. Now, we have

$$\eta = \hat{y}\frac{\hat{\varepsilon}}{\varepsilon} \, ,$$

and, for the study of the inner region, $\hat{y}$ is kept fixed and $\hat{\varepsilon}/\varepsilon \to 0$. Therefore, the value of the right hand side of (11.51) must be taken at $\eta = 0$. This value is 1 and the inner region equation is

$$\frac{\tau}{\tau_w} = 1 \, .$$

We recover the result of the standard asymptotic theory. In the application presented here, this result is not used, and we consider that τ_{out}/τ_w is a function of η.

Here, the solution of (11.51) gives *a UVA of the velocity profile in the whole boundary layer* and not only an approximation in the inner region. This point will be discussed later, but we can note now that (11.51) enables us to satisfy the boundary conditions on the total stress. Indeed, at $y = 0$ we have $\tau/\tau_w = 1$ and at $y = \delta$ we have $\tau/\tau_w = 0$. This result is due to the behaviour of the solution in the outer region (right hand side of (11.51)).

11.4.2 Turbulence Model

Coming back to *dimensionalized variables*, including the distance to the wall y, the total stress τ in the left hand side of (11.51) is

$$\tau = -\varrho < u'v' > +\mu\frac{\partial u}{\partial y} \ . \tag{11.52}$$

A turbulence model is required to express the turbulent stress $-\varrho < u'v' >$. To this end, a mixing length scheme gives [70]

$$-\varrho < u'v' > = \varrho F^2 \ell^2 \left(\frac{\partial u}{\partial y}\right)^2 \ , \tag{11.53a}$$

$$\frac{\ell}{\delta} = 0.085 \, \mathrm{th} \, \frac{\chi}{0.085} \frac{y}{\delta} \ , \quad \chi = 0.41 \ , \tag{11.53b}$$

$$F = 1 - \exp\left[-(\tau\varrho)^{1/2}\frac{\ell}{26\chi\mu}\right] \ . \tag{11.53c}$$

In the inner region, by setting $\tau = \tau_w$ and $\ell = \chi y$, the *damping function* F takes the form proposed by Van Driest [104]

$$F = 1 - \exp\left(-\frac{y^+}{26}\right) \ ,$$

where y^+ is the wall variable

$$y^+ = \frac{yu_\tau}{\nu} \ .$$

This simple model is very convenient for our purpose particularly if we restrict the study to the flow on a flat plate.

11.4.3 Outer Region

According to experimental results, in particular for a flat plate flow, the outer region of the boundary layer is well described by *similarity solutions* [16, 18, 19, 68, 84]. We assume that the *velocity defect* is a function of $\frac{y}{\delta}$ where δ is the boundary layer thickness

$$\frac{u_e - u}{u_\tau} = F'(\eta) \quad \text{with} \quad \eta = \frac{y}{\delta} \quad \text{and} \quad u_\tau = \sqrt{\frac{\tau_w}{\varrho}} \ .$$

Usually, the quantity $(u_e - u)/u_\tau$ is called velocity defect because it represents a defect of the velocity with respect to the external velocity u_e. The similarity equation of the outer region is [70] (see Problem 11-5)

$$\frac{\tau}{\tau_w} = 1 - \frac{F}{F_1} + \left(\frac{1}{F_1} + 2\beta\right)\eta F' , \tag{11.54}$$

where

$$F = \int_0^\eta F' \, d\eta , \quad F_1 = F(1) , \quad \beta = -\frac{\delta}{u_\tau}\frac{du_e}{dx} .$$

This equation is equivalent to (11.10b).

In the outer region of the boundary layer, the stress τ consists of only the turbulent stress since the viscous stress is negligible. On the other hand, the damping function is equal to 1 because $y^+ \gg 1$. Therefore, we have

$$\frac{\tau}{\tau_w} = \left(\frac{\ell}{\delta}\right)^2 F''^2 ,$$

where F'' is the derivative of F' with respect to η.

For any admissible value of the pressure gradient parameter β, the numerical solution of the similarity equation provides us with a velocity profile $F'(\eta)$ and the turbulent stress profile, i.e. with the notations of Subsect. 11.4.1 the quantity τ_{out}/τ_w.

11.4.4 Equation to Solve

Given the Reynolds number, the velocity profile in the whole boundary layer is a solution of the equation

$$\frac{\tau}{\tau_w} = \frac{\tau_{\text{out}}}{\tau_w} , \tag{11.55}$$

where the expression of τ in the left hand side is given by

$$\tau = -\rho < u'v' > +\mu\frac{\partial u}{\partial y} ,$$

and the turbulent stress is expressed by (11.53a, 11.53b, 11.53c).

In (11.55), the right hand side is given by the solution of (11.54) which is the solution of the outer region. Moreover, the Reynolds number must be fixed. The simplest is to give the value of $\dfrac{u_\tau\delta}{\nu}$ which relates directly y^+ and η

$$y^+ = \eta\frac{u_\tau\delta}{\nu} .$$

The equation to solve is a first order ordinary differential equation for $u(y)$. With the wall variables, (11.55) can be written as

$$\frac{\partial u^+}{\partial y^+} + F_c^2 \ell^{+2}\left(\frac{\partial u^+}{\partial y^+}\right)^2 = \frac{\tau_{\text{out}}}{\tau_w} , \tag{11.56}$$

with

$$\ell^+ = \frac{\ell u_\tau}{\nu} = \frac{\ell}{\delta}\frac{u_\tau \delta}{\nu} \ , \quad u^+ = \frac{u}{u_\tau} \ , \quad y^+ = \frac{y u_\tau}{\nu} \ .$$

The wall condition is $u = 0$, i.e. $u^+ = 0$ at $y^+ = 0$. At the boundary layer edge, the condition $\tau_{\mathrm{out}}/\tau_w = 0$ imposes $\dfrac{\partial u}{\partial y} = 0$; therefore, we have $\dfrac{\partial u^+}{\partial y^+} = 0$ at $y^+ = \dfrac{u_\tau \delta}{\nu}$ $(\eta = 1)$. At the boundary layer edge, the solution yields a certain value of u^+ which gives the skin-friction coefficient since we have

$$u^+_{y=\delta} = \frac{u_e}{u_\tau} = \frac{1}{\sqrt{Cf/2}} \quad \text{with} \quad \frac{Cf}{2} = \frac{\tau_w}{\varrho u_e^2} \ . \tag{11.57}$$

11.4.5 Examples of Results

The results presented in this section have been obtained for a flat plate flow $(\beta = 0)$ for different values of the Reynolds number.

The results of Fig. 11.1 show a seemingly correct evolution of the velocity *in the whole boundary layer*. We observe that the logarithmic law is present when the Reynolds number is large enough. The extent of the logarithmic

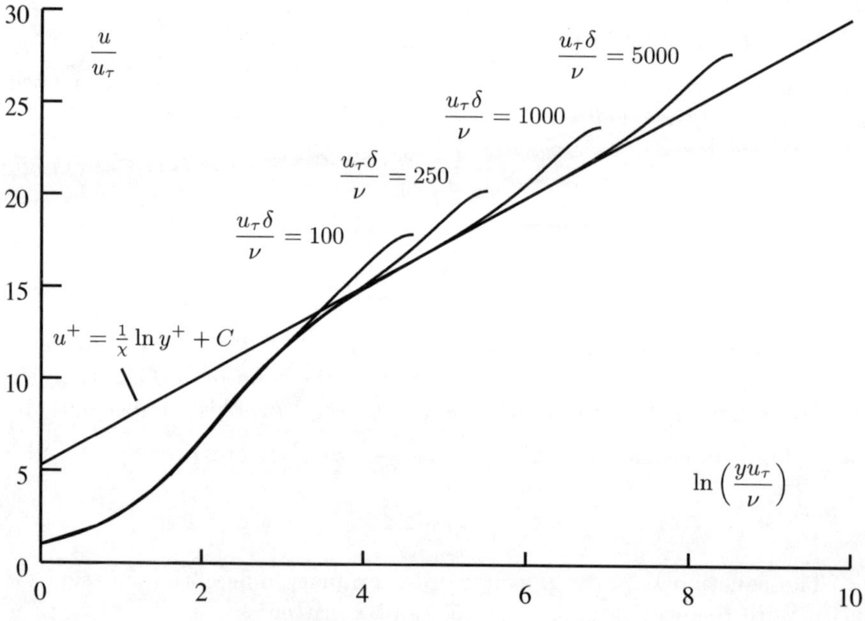

Fig. 11.1. Approximation of the velocity profiles in a flat plate turbulent boundary layer at different Reynolds numbers

region measured in wall variables increases when the Reynolds number increases. When the Reynolds number is too small, the logarithmic region disappears.

The velocity profiles in the region close to the wall are not very much sensitive to the Reynolds number. For values of $\dfrac{u_\tau \delta}{\nu} > 250$, the function $u^+(y^+)$ is practically independent of the Reynolds number for $y^+ < 50$. In this sense, the velocity law in the inner region is said *universal*. This behaviour is in agreement with the difference in the orders of magnitude between the turbulence length scales (or between the turbulent time scales) in the inner region and in the outer region of the boundary layer. The inner region has a time scale *much smaller* than the outer region. Under these conditions, the inner region acquires its own organization, independently of the parameters which govern the flow in the outer region.

The value of the skin-friction coefficient obtained from the value of u^+ at the edge of the boundary layer (see (11.57)) can be compared to the value obtained from the overlap between the law of the wall and the velocity defect law. Indeed, from the standard asymptotic analysis, in the overlap region we have simultaneously

$$\frac{u}{u_\tau} = \frac{1}{\chi} \ln \frac{y u_\tau}{\nu} + C \, , \tag{11.58a}$$

$$\frac{u_e - u}{u_\tau} = -\frac{1}{\chi} \ln \frac{y}{\delta} + D \, . \tag{11.58b}$$

The equality of velocities given by the two laws in the overlap region yields

$$\frac{u_e}{u_\tau} = \frac{1}{\chi} \ln \frac{u_\tau \delta}{\nu} + C + D \, . \tag{11.58c}$$

In the inner region, the mixing length model gives $\chi = 0.41$ and $C = 5.28$. In the outer region, for the flat plate, the solution of (11.54) gives $D = 1.76$.

Table 11.1 shows the comparison of skin-friction coefficients obtained from the two methods. Except for low values of the Reynolds number, a good

Table 11.1. Comparison of $\frac{u_\tau}{u_e}$ obtained from solution of (11.56) and from the logarithmic law (11.58c)

$u_\tau \delta / \nu$	from solution of (11.56)	law (11.58c)
5000	$3.60 \, 10^{-2}$	$3.59 \, 10^{-2}$
1000	$4.20 \, 10^{-2}$	$4.19 \, 10^{-2}$
500	$4.54 \, 10^{-2}$	$4.50 \, 10^{-2}$
250	$4.93 \, 10^{-2}$	$4.88 \, 10^{-2}$
100	$5.59 \, 10^{-2}$	$5.47 \, 10^{-2}$
50	$6.51 \, 10^{-2}$	$6.03 \, 10^{-2}$

agreement is observed which reinforces the validity of the approach used here.

It can be surprising to observe a good agreement between the two methods even when the velocity profiles of Fig. 11.1 do not exhibit a logarithmic evolution of the velocity whereas law (11.58c) rests explicity on the existence of such a logarithmic behaviour. In fact, it is not correct to seek the presence or not of a logarithmic law in Fig. 11.1. The velocity profiles obtained from SCEM must be compared to a *composite approximation* formed from outer and inner approximations obtained from MMAE. In such a representation, it is possible that the common part – which is precisely the logarithmic law – disappears even if the outer approximation and the inner approximation have a logarithmic part; this happens when the Reynolds number is not large enough.

11.5 Conclusion

In the study of the turbulent boundary layer with the standard asymptotic analysis (MMAE), a striking feature is the logarithmic overlap region between the inner and outer regions. This result is obtained without using a turbulence model but, obviously, the experimental knowledge is essential to set the starting hypotheses. Now, to calculate a boundary layer, it is necessary to implement such a model. The conclusion is that the turbulence model must be compatible with the existence of a logarithmic evolution of the velocity.

With SCEM, the issue is addressed differently because there is no overlap condition. Therefore, the result depends on the turbulence model. For the case of the flat plate flow, the numerical results show that the model used here, a simple mixing length model, leads us to the desired result. In fact, the turbulence model has been devised to achieve the right result.

Problems

11-1. A mathematical model has been proposed by Panton [76] to simulate the decomposition of the boundary layer in two regions. This is an adaptation of the model proposed by Lagerstrom to illustrate the difficulties of Stokes-Oseen's flow. Panton's model writes

$$\frac{d^2 u}{dy^2} + \frac{1}{y + \varepsilon} \frac{du}{dy} + u \frac{du}{dy} = 0 \,,$$

with the boundary conditions

$$u(0) = 0 \,, \quad \lim_{y \to \infty} u = 1 \,.$$

The solution is studied with MMAE.

1. In the outer region, we assume that the expansion has the form

$$u = f_0(y) + \delta_1(\varepsilon) f_1(y) + \cdots .$$

Give the equations for f_0 and f_1. Give the boundary conditions. It will be shown that the solution for f_0 is $f_0 = 1$. Give the solution for f_1. Do not try to determine δ_1.

2. The variable appropriate to the inner region is $\bar{y} = y/\varepsilon$. The expansion has the form

$$u = \bar{\delta}_1 \bar{f}_0(\bar{y}) + \cdots .$$

Give the equation for $\bar{f}_0$ and the boundary condition at $\bar{y} = 0$. Give the solution.

3. From the matching between the inner and outer solutions, calculate the yet undetermined constants and give δ_1. It is recalled that

$$\int_y^\infty \frac{e^{-t}}{t}\, dt \cong -\ln y - \gamma - y \quad \text{when} \quad y \to 0 ,$$

where γ is the Euler constant $\gamma = 0.57722$.

   Write the composite solution.

11-2. We consider Panton's model studied in Problem 11-1.

1. With the change of variable $y' = y + \varepsilon$, show that the problem reduces to the Stokes-Oseen's flow model proposed by Lagerstrom (Sect. 6.4).

2. The regular form of SCEM is applied here. We seek the first approximation in the form

$$u = 1 + \delta_1 F_1(y) ,$$

where δ_1 is an unknown gauge. Give the equation for F_1. Give the solution which satisfies the boundary condition as $y \to \infty$. Show that the condition at $y = 0$ cannot be satisfied. Then, we seek a UVA in the form

$$u = 1 + \delta_1 F_1(y) + \delta_1 \overline{F}_1(\bar{y}) \quad \text{with} \quad \bar{y} = \frac{y}{\varepsilon} .$$

The equation for $\overline{F}_1$ will be formulated in such a way that the solution is a function of $\bar{y}$ only.

   The constants are determined by applying the boundary conditions. The condition at $y = 0$ can be applied only after expanding the solution for F_1 as $y \to 0$ by taking into account that

$$\int_y^\infty \frac{e^{-t}}{t}\, dt \cong -\ln y - \gamma - y \quad \text{when } y \to 0 ,$$

where γ is the Euler constant $\gamma = 0.57722$.

3. SCEM is applied in its generalized form. The proposed expansion is

$$u = 1 + \delta_1 f_1 + \delta_1 \bar{f}_1 .$$

For f_1, the same equation as for F_1 will be taken. By retaining terms of order $\dfrac{\delta_1}{\varepsilon^2}$, show that the equation for $\overline{f}_1$ is

$$\frac{\mathrm{d}^2 \overline{f}_1}{\mathrm{d}\bar{y}^2} + \frac{1}{\bar{y}+1}\frac{\mathrm{d}\overline{f}_1}{\mathrm{d}\bar{y}} = \frac{1}{\bar{y}(\bar{y}+1)}\varepsilon\frac{\mathrm{d}f_1}{\mathrm{d}y} \ .$$

Do not try to solve this equation but form the equation for g

$$g = f_1 + \overline{f}_1 \ .$$

Give the solution by applying the boundary conditions at $y = 0$ and as $y \to \infty$.

11-3. According to MMAE, the inner region of the turbulent boundary layer is described by the equation

$$\frac{\tau}{\tau_w} = 1 \ .$$

In this region, with a mixing length scheme, the total stress τ is given by

$$\tau = \mu\frac{\partial u}{\partial y} + \varrho F_c^2 \ell^2 \left(\frac{\partial u}{\partial y}\right)^2 \ ,$$

with

$$F_c = 1 - e^{-y^+/26} \ , \quad \ell = \chi y \ , \quad \chi = 0.41 \ .$$

The wall variables are defined by

$$y^+ = \frac{yu_\tau}{\nu} \ , \quad u^+ = \frac{u}{u_\tau} \ , \quad u_\tau = \sqrt{\frac{\tau_w}{\varrho}} \ .$$

Write the equation of the inner region with the wall variables.
Show that the solution for $y^+ \gg 1$ has the form

$$u^+ = \frac{1}{\chi}\ln y^+ + C \ .$$

Write this equation in the form

$$\frac{\mathrm{d}u^+}{\mathrm{d}y^+} = f(y^+) \quad \text{with} \quad f(y^+) = \frac{\sqrt{1 + 4F_c^2\ell^{+2}} - 1}{2F_c^2\ell^{+2}} \ .$$

Give the boundary condition which must be prescribed.
Integrate numerically this equation between $y^+ = 0$ and $y^+ = 1000$. Plot the function $u^+(y^+)$. Estimate the value of constant C.
A simple method consists of discretizing the equation as

$$\frac{u_{n+1}^+ - u_n^+}{y_{n+1} - y_n} = f(y_{n+1/2}^+) \quad \text{with} \quad y_{n+1/2}^+ = \frac{y_{n+1} + y_n}{2} \ ,$$

where the index n refers to mesh points of a grid defined in the y-direction.

Close to the wall, the grid must be very fine. The first point near the wall must be at a distance such that $y^+ < 1$. Physically, this limit is associated with the fact that ν/u_τ represents a length scale.

Accuracy problems can arise due to the expression of function $f(y^+)$ in the neighbourhood of $y^+ = 0$. Another form can be tried

$$f = \frac{2}{\sqrt{1 + 4F_c^2 \ell^{+2}} + 1} \; .$$

It is also possible to expand f in the neighbourhood of $y^+ = 0$.

11-4. In the turbulent boundary layer, the variable appropriate to the study of the outer region is $\eta = \dfrac{y}{\delta}$. The variable appropriate to the study of the inner region is $y^+ = \dfrac{y u_\tau}{\nu}$. The small parameter of the problem is $\dfrac{u_\tau}{u_e}$.

Coles proposed to represent the velocity profile in the outer region by the formula

$$\frac{u_e - u}{u_\tau} = -\frac{1}{\chi} \ln \eta + \frac{B}{\chi} [2 - w(\eta)] \; ,$$

with

$$\eta = \frac{y}{\delta} \; , \quad u_\tau = \sqrt{\frac{\tau_w}{\varrho}} \; , \quad \chi = 0.41 \; ,$$

where δ is the boundary layer thickness and τ_w is the wall shear stress. We also have

$$w = 1 - \cos(\pi \eta) \; ,$$

and B is a constant which depends on the conditions of development of the boundary layer, for example the intensity of the pressure gradient.

On the other hand, we know that in the inner region of the turbulent boundary layer, the velocity profile follows the law of the wall

$$u^+ = f(y^+) \; , \quad u^+ = \frac{u}{u_\tau} \; , \quad y^+ = \frac{y u_\tau}{\nu} \; ,$$

and, as $y^+ \to \infty$ (in practice when $y^+ > 50$), we have

$$f(y^+) = \frac{1}{\chi} \ln y^+ + C \; , \quad C = 5.28 \; .$$

Write the matching between the outer law and the inner law. Deduce the relation between the skin-friction coefficient $C_f = \dfrac{\tau_w}{\frac{1}{2}\varrho u_e^2}$ and the Reynolds number $R_\delta = \dfrac{u_e \delta}{\nu}$. Show that $\dfrac{u_\tau}{u_e} \to 0$ as $R_\delta \to \infty$.

Give the expression of the velocity profile in the whole boundary layer by means of a composite expansion.

11-5. Two regions are distinguished in the turbulent boundary layer: the outer region and the inner region. The variable appropriate to the outer region is $\eta = \dfrac{y}{\delta}$. The variable appropriate to the inner region is $y^+ = \dfrac{y u_\tau}{\nu}$.

In the overlap region, the velocity profile takes a logarithmic form

$$\frac{u_e - u}{u_\tau} = -\frac{1}{\chi}\ln\eta + D \quad\text{as}\quad \eta \to 0 \ ,$$

$$\frac{u}{u_\tau} = \frac{1}{\chi}\ln y^+ + C \quad\text{as}\quad y^+ \to \infty \ .$$

Write the matching between the two regions and deduce the relation between $\dfrac{u_\tau}{u_e}$ and $R_\delta = \dfrac{u_e\delta}{\nu}$. Show that $\dfrac{u_\tau}{u_e} \to 0$ as $R_\delta \to \infty$.

Under certain conditions, the velocity profile in the outer region is governed by a form of self-similarity, i.e.

$$\frac{u_e - u}{u_\tau} = F'(\eta) \ ,$$

where F' is the derivative with respect to η of a function $F(\eta)$ which appears in the calculations. We take $F(0) = 0$.

It is recalled that, in the outer region, the boundary layer equations are

$$\frac{\partial u}{\partial x} + \frac{\partial v}{\partial y} = 0 \ ,$$

$$u\frac{\partial u}{\partial x} + v\frac{\partial u}{\partial y} = u_e\frac{\mathrm{d}u_e}{\mathrm{d}x} + \frac{\partial}{\partial y}\left(\frac{\tau}{\varrho}\right) \quad\text{with}\quad \tau = -\varrho < u'v' > \ .$$

From the continuity equation, express v as function of F and F'. The following notations are used

$$\gamma = \frac{u_\tau}{u_e} \ , \quad \gamma' = \frac{\mathrm{d}\gamma}{\mathrm{d}x} \ , \quad \delta' = \frac{\mathrm{d}\delta}{\mathrm{d}x} \ , \quad u'_e = \frac{\mathrm{d}u_e}{\mathrm{d}x} \ .$$

Write the momentum equation with the hypotheses given above. The following notation is used

$$\beta = -\frac{\delta \, u'_e}{\gamma \, u_e} \ .$$

If F' is a function of η only, it is necessary to have

$$\beta = \text{cst} \ , \quad \gamma = \text{cst} \ , \quad \frac{u_e}{u'_e}\frac{\gamma'}{\gamma} = \text{cst} \ , \quad \frac{u_e}{u'_e}\frac{\delta'}{\delta} = \text{cst}$$

as $R_\delta \to \infty$. Show that $\gamma \to 0$ and $\beta\dfrac{u_e}{u'_e}\dfrac{\gamma'}{\gamma} \to 0$. Simplify the momentum equation.

Integrate the so obtained equation with respect to η from $\eta = 0$. Express the quantity $\beta\left(1 + \dfrac{u_e}{u'_e}\dfrac{\delta'}{\delta}\right)$ as function of $F_1 = F(1)$ and β.

11-6. It is proposed to study the spectral density of turbulent kinetic energy in a turbulent flow. By definition, the turbulent kinetic energy is

$$k = \frac{<u_i' u_i'>}{2} \; .$$

Its spectrum $E(\xi)$ is such that

$$k = \int_0^\infty E(\xi) \, \mathrm{d}\xi \; ,$$

where ξ is the wave number.

The turbulent field can be viewed as a set of structures of different sizes. Two important ranges are distinguished: a range of low wavenumbers (large scale structures) and a range of large wavenumbers (small scale structures).

Most of the kinetic energy is contained in the range of large structures. These structures are characterized by a Reynolds number very large compared to unity. Viscosity has a negligible effect.

Viscosity is efficient in the range of small scale structures whose characteristic Reynolds number is of order 1. The role of viscosity is to dissipate kinetic energy into heat. The quantity of kinetic energy dissipated by time unit is denoted by ε, the so-called dissipation. To follow the common use, the notation ε for dissipation is employed here but this is not a small parameter.

The length scale associated with large scale structures is ℓ. The length scale associated with small scale structures is η, the so-called Kolmogorov scale. This scale is formed from ε and ν (kinematic viscosity coefficient). Determine η.

Give the form of the spectrum in the range of large scale structures knowing that the length scale is ℓ and that the scale of the turbulent kinetic energy is $\boldsymbol{u}^2$. It will be shown simply that

$$E = \boldsymbol{u}^2 \boldsymbol{\ell} F(\xi \boldsymbol{\ell}) \; .$$

Give the form of the spectrum in the range of small scale structures knowing that the governing parameters are ν and ε.

Write the matching between these two ranges. Assuming that in the overlap region the spectrum follows a power law, give the form of the spectrum as function of ε et ξ. It is noted that, in the overlap region, the influence of viscosity must disappear since this range belongs to the small scale structures and to the large scale structures as well.

Express ε as function of $\boldsymbol{u}$ and $\boldsymbol{\ell}$. Conclude.

12 Channel Flow

In this chapter, we consider a steady, two-dimensional, incompressible, laminar channel flow of a viscous fluid. At high Reynolds numbers, small wall perturbations, such as indentations, can generate adverse pressure gradients leading to flow separation. In a channel, there is no external flow region and the asymptotic models for the flow perturbations are mainly based on an inviscid rotational core flow region together with boundary layers near the walls; a comprehensive discussion of this structure can be found in Sobey [94]. The asymptotic analysis of these flows has been performed essentially by Smith [89, 90, 91, 92] and a systematic approach has been proposed later by Saintlos and Mauss [85]. More recently, the modelling of channel flow has been examined by Lagrée et al. [49] and by Lagrée and Lorthois [50].

Here, the flow is analyzed by using the successive complementary expansion method, SCEM, in which we seek a uniformly valid approximation, UVA, based on generalized asymptotic expansions [28].

12.1 Formulation of the problem

The flow is governed by the Navier-Stokes equations which are, in dimensionless form (see Sect. 8.1),

$$\frac{\partial \mathcal{U}}{\partial x} + \frac{\partial \mathcal{V}}{\partial y} = 0 , \tag{12.1a}$$

$$\mathcal{U}\frac{\partial \mathcal{U}}{\partial x} + \mathcal{V}\frac{\partial \mathcal{U}}{\partial y} = -\frac{\partial \mathcal{P}}{\partial x} + \frac{1}{\mathcal{R}}\left(\frac{\partial^2 \mathcal{U}}{\partial x^2} + \frac{\partial^2 \mathcal{U}}{\partial y^2}\right) , \tag{12.1b}$$

$$\mathcal{U}\frac{\partial \mathcal{V}}{\partial x} + \mathcal{V}\frac{\partial \mathcal{V}}{\partial y} = -\frac{\partial \mathcal{P}}{\partial y} + \frac{1}{\mathcal{R}}\left(\frac{\partial^2 \mathcal{V}}{\partial x^2} + \frac{\partial^2 \mathcal{V}}{\partial y^2}\right) . \tag{12.1c}$$

The above equations are written in an orthonormal axis-system. Coordinates x and y are reduced by the width H^* of the basic channel so that the lower wall of the channel is $y = -\frac{1}{2}$ and the upper wall is $y = \frac{1}{2}$ (Fig. 12.1).

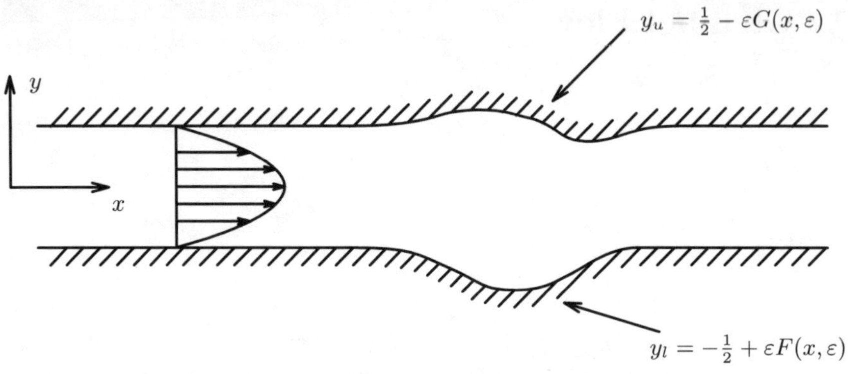

Fig. 12.1. Flow in a two-dimensional channel with deformed walls. In this figure, all quantities are dimensionless

The velocity components are reduced by a reference velocity V^* which is defined below by (12.4b). The Reynolds number is defined by

$$\mathcal{R} = \frac{\varrho^* V^* H^*}{\mu^*} \, , \tag{12.2}$$

where ϱ^*, V^*, H^* and μ^* are dimensionalized quantities.

The basic flow is the *plane Poiseuille flow* for which we have

$$\mathcal{U} = u_0 = \frac{1}{4} - y^2 \, , \tag{12.3a}$$

$$\mathcal{V} = v_0 = 0 \, , \tag{12.3b}$$

$$\mathcal{P} = p_0 = -\frac{2x}{\mathcal{R}} + p_c \, , \tag{12.3c}$$

where p_c is an arbitrary constant.

The reference velocity V^* is expressed as a function of the basic pressure gradient or, equivalently, of the dimensionalized mass flow $\mathcal{Q}^*$ per unit width of the channel. We have

$$\mathcal{Q}^* = \varrho^* V^* H^* \int_{-1/2}^{1/2} u_0 \, \mathrm{d}y \, , \tag{12.4a}$$

$$V^* = 6 \frac{\mathcal{Q}^*}{\varrho^* H^*} \, . \tag{12.4b}$$

We also have

$$\frac{\mathrm{d}p_0}{\mathrm{d}x} = -\frac{2\mu^*}{\varrho^* V^* H^*} \, . \tag{12.4c}$$

The basic flow is perturbed, for example, by wall indentations such that

$$y_l = -\frac{1}{2} + \varepsilon F(x, \varepsilon) \, , \tag{12.5a}$$

$$y_u = \frac{1}{2} - \varepsilon G(x, \varepsilon) \, , \tag{12.5b}$$

where y_l and y_u denote the lower and upper walls, respectively, and ε is a small parameter which characterizes the height of the wall indentations. We seek a solution in the form

$$\mathcal{U} = u_0(y) + \varepsilon^r u(x, y, \varepsilon) \, , \tag{12.6a}$$

$$\mathcal{V} = \varepsilon^r v(x, y, \varepsilon) \, , \tag{12.6b}$$

$$\mathcal{P} - p_c = -\frac{2x}{\mathcal{R}} + \varepsilon^s p(x, y, \varepsilon) \, , \tag{12.6c}$$

where r and s are constants yet undetermined. The Navier-Stokes equations become

$$\frac{\partial u}{\partial x} + \frac{\partial v}{\partial y} = 0 \, , \tag{12.7a}$$

$$\mathrm{L}_\varepsilon \, u = \varepsilon^r \left(u \frac{\partial u}{\partial x} + v \frac{\partial u}{\partial y} \right) + u_0 \frac{\partial u}{\partial x} + v \frac{\mathrm{d}u_0}{\mathrm{d}y}$$
$$+ \varepsilon^{s-r} \frac{\partial p}{\partial x} - \frac{1}{\mathcal{R}} \left(\frac{\partial^2 u}{\partial x^2} + \frac{\partial^2 u}{\partial y^2} \right) = 0 \, , \tag{12.7b}$$

$$\mathrm{L}_\varepsilon \, v = \varepsilon^r \left(u \frac{\partial v}{\partial x} + v \frac{\partial v}{\partial y} \right) + u_0 \frac{\partial v}{\partial x}$$
$$+ \varepsilon^{s-r} \frac{\partial p}{\partial y} - \frac{1}{\mathcal{R}} \left(\frac{\partial^2 v}{\partial x^2} + \frac{\partial^2 v}{\partial y^2} \right) = 0 \, , \tag{12.7c}$$

where the operators $\mathrm{L}_\varepsilon \, u$ and $\mathrm{L}_\varepsilon \, v$ denote the streamwise and transverse momentum equations, respectively. We have necessarily $s \geq r$. This shows that s is positive since r is positive because $\varepsilon^r u$ is a small perturbation compared to u_0. If regular AEs are used, a more thorough analysis is required to go further but, with SCEM, generalized expansions allow us to take $s = r$ without loss of generality since the order of magnitude of the pressure p is not yet determined.

At high Reynolds number, the momentum equations reduce to first order partial differential equations and we are faced to a *singular perturbation problem*. Therefore, in the core of the flow, we seek approximations which, in terms of generalized expansions, are written

$$u = u_1(x, y, \varepsilon) + \cdots \, , \tag{12.8a}$$

$$v = v_1(x, y, \varepsilon) + \cdots \, , \tag{12.8b}$$

$$p = p_1(x, y, \varepsilon) + \cdots \, . \tag{12.8c}$$

Neglecting formally terms of order $\mathrm{O}\left(\varepsilon^r, \dfrac{1}{\mathcal{R}} \right)$, the equations for *the flow perturbations* (12.7a–12.7c) reduce to inviscid flow equations

$$\frac{\partial u_1}{\partial x} + \frac{\partial v_1}{\partial y} = 0 \, , \tag{12.9a}$$

$$u_0 \frac{\partial u_1}{\partial x} + v_1 \frac{du_0}{dy} = -\frac{\partial p_1}{\partial x} , \tag{12.9b}$$

$$u_0 \frac{\partial v_1}{\partial x} = -\frac{\partial p_1}{\partial y} . \tag{12.9c}$$

It is useful to note the behaviour of the solution of (12.9a–12.9c) in the vicinity of the walls. As $y \to -1/2$, we have

$$u_1 = -2p_{10} \ln\left(\frac{1}{2} + y\right) + c_{10} + \cdots , \tag{12.10a}$$

$$v_1 = -p_{10x} + 2p_{10x}\left(\frac{1}{2} + y\right)\ln\left(\frac{1}{2} + y\right)$$
$$- \left(\frac{1}{2} + y\right)\left(2p_{10x} + c_{10x}\right) + \cdots , \tag{12.10b}$$

$$p_1 = p_{10} + \frac{1}{2}\left(\frac{1}{2} + y\right)^2 p_{10xx} + \cdots , \tag{12.10c}$$

and, as $y \to 1/2$, we have

$$u_1 = -2p_{11} \ln\left(\frac{1}{2} - y\right) + c_{11} + \cdots , \tag{12.11a}$$

$$v_1 = p_{11x} - 2p_{11x}\left(\frac{1}{2} - y\right)\ln\left(\frac{1}{2} - y\right)$$
$$+ \left(\frac{1}{2} - y\right)\left(2p_{11x} + c_{11x}\right) + \cdots , \tag{12.11b}$$

$$p_1 = p_{11} + \frac{1}{2}\left(\frac{1}{2} - y\right)^2 p_{11xx} + \cdots . \tag{12.11c}$$

In the above equations, p_{10}, p_{11}, c_{10}, c_{11} are functions of x and ε. The letter x in index denotes a derivative with respect to the streamwise variable x; for example p_{11x} means

$$p_{11x} = \frac{\partial p_{11}}{\partial x}(x, \varepsilon) .$$

12.2 Uniformly Valid Approximation

In order to satisfy the no-slip condition at the walls, two boundary layer variables are defined by

$$Y = \frac{\frac{1}{2} + y}{\varepsilon} , \tag{12.12a}$$

$$\widehat{Y} = \frac{\frac{1}{2} - y}{\varepsilon} . \tag{12.12b}$$

In terms of boundary layer variables, the boundary layer thicknesses are of order 1. Then, in the two boundary layers, we have $u_0 = O(\varepsilon)$. This leads to the choice of $r = 1$. In this way, u_0 and εu_1 are of the same order near the walls and the velocity $u_0 + \varepsilon u$ in (12.6a) can be negative. Then, the choice $r = 1$ means that we want to obtain an approximation which is able to describe separation. According to SCEM, a UVA is obtained by complementing the core approximation

$$u = U_1(x, Y, \varepsilon) + \widehat{U}_1(x, \widehat{Y}, \varepsilon) + u_1(x, y, \varepsilon) \, , \tag{12.13a}$$

$$v = \varepsilon V_1(x, Y, \varepsilon) - \varepsilon \widehat{V}_1(x, \widehat{Y}, \varepsilon) + v_1(x, y, \varepsilon) \, , \tag{12.13b}$$

$$p = \Delta(\varepsilon) P_1(x, Y, \varepsilon) + \Delta(\varepsilon) \widehat{P}_1(x, \widehat{Y}, \varepsilon) + p_1(x, y, \varepsilon) \, , \tag{12.13c}$$

where the gauge function $\Delta(\varepsilon)$ is yet undetermined. Here, the quantities (u, v, p) do not represent the exact solution but only an approximate solution. Then, if all the boundary conditions are satisfied, $\mathrm{L}_\varepsilon\, u$ and $\mathrm{L}_\varepsilon\, v$ in (12.7b, 12.7c) are not zero but small in a certain sense.

The form of approximation for v in (12.13b) is imposed by the continuity equation which must be non trivial

$$\frac{\partial U_1}{\partial x} + \frac{\partial V_1}{\partial Y} = 0 \, , \tag{12.14a}$$

$$\frac{\partial \widehat{U}_1}{\partial x} + \frac{\partial \widehat{V}_1}{\partial \widehat{Y}} = 0 \, . \tag{12.14b}$$

With this formulation, it is clear that, if (u_1, v_1) represent an approximation in the core of the flow, we have

$$Y \to \infty : \quad U_1 \to 0 \, , \quad V_1 \to 0 \, , \tag{12.15a}$$

$$\widehat{Y} \to \infty : \quad \widehat{U}_1 \to 0 \, , \quad \widehat{V}_1 \to 0 \, . \tag{12.15b}$$

Boundary conditions are required along the lower and upper walls of the channel, i.e. along the lines $Y = F(x, \varepsilon)$ and $\widehat{Y} = G(x, \varepsilon)$. Along these two walls, we have

$$Y = F(x, \varepsilon) : \quad u_0 + \varepsilon u = 0 \, , \quad v = 0 \, , \tag{12.16a}$$

$$\widehat{Y} = G(x, \varepsilon) : \quad u_0 + \varepsilon u = 0 \, , \quad v = 0 \, . \tag{12.16b}$$

With the approximation given by (12.13a, 12.13b), we have

$$Y = F(x, \varepsilon) : \quad u_0 + \varepsilon U_1 + \varepsilon u_1 = 0 \, , \quad \varepsilon V_1 + v_1 = 0 \, , \tag{12.17a}$$

$$\widehat{Y} = G(x, \varepsilon) : \quad u_0 + \varepsilon \widehat{U}_1 + \varepsilon u_1 = 0 \, , \quad -\varepsilon \widehat{V}_1 + v_1 = 0 \, . \tag{12.17b}$$

It is useful to note that, in contrast with external boundary layers (Chap. 8), according to (12.10a, 12.10b, 12.11a, 12.11b), the terms u_1 and v_1

or their y-derivatives are singular in the vicinity of the walls. This shows the great advantage of SCEM since the UVAs for u and v are perfectly regular.

It is possible that the accuracy of the approximation given by the solution of (12.9a–12.9c) is not sufficient in the core of the flow. To improve the accuracy, we neglect formally O $\left(\dfrac{1}{\mathcal{R}}\right)$ terms in (12.7a–12.7c), and we obtain

$$\frac{\partial u_1}{\partial x} + \frac{\partial v_1}{\partial y} = 0 , \tag{12.18a}$$

$$\varepsilon \left(u_1 \frac{\partial u_1}{\partial x} + v_1 \frac{\partial u_1}{\partial y} \right) + u_0 \frac{\partial u_1}{\partial x} + v_1 \frac{du_0}{dy} = -\frac{\partial p_1}{\partial x} , \tag{12.18b}$$

$$\varepsilon \left(u_1 \frac{\partial v_1}{\partial x} + v_1 \frac{\partial v_1}{\partial y} \right) + u_0 \frac{\partial v_1}{\partial x} = -\frac{\partial p_1}{\partial y} . \tag{12.18c}$$

Finally, we have two models to describe the core flow: a linear model given by (12.9a–12.9c) and a non linear model given by (12.18a–12.18c). The second model is not simple to solve. In addition, taking into account that an analysis with regular AEs is possible, the general solution of (12.18a, 12.18b) is probably singular in the vicinity of the walls even if the direct analysis based on these equations is not as simple as on (12.9a–12.9c).

To close this section, let us remember that we are seeking a UVA for the Navier-Stokes equations

$$\frac{\partial u}{\partial x} + \frac{\partial v}{\partial y} = 0 , \tag{12.19a}$$

$$\varepsilon \left(u \frac{\partial u}{\partial x} + v \frac{\partial u}{\partial y} \right) + u_0 \frac{\partial u}{\partial x} + v \frac{du_0}{dy} = -\frac{\partial p}{\partial x} + \frac{1}{\mathcal{R}} \left(\frac{\partial^2 u}{\partial x^2} + \frac{\partial^2 u}{\partial y^2} \right) , \tag{12.19b}$$

$$\varepsilon \left(u \frac{\partial v}{\partial x} + v \frac{\partial v}{\partial y} \right) + u_0 \frac{\partial v}{\partial x} = -\frac{\partial p}{\partial y} + \frac{1}{\mathcal{R}} \left(\frac{\partial^2 v}{\partial x^2} + \frac{\partial^2 v}{\partial y^2} \right) . \tag{12.19c}$$

In Sect. 12.4, we return to the question of a uniformly valid model. In the next Sect. 12.3, we seek to construct a boundary layer model coupled to the core flow equations given either by (12.9a–12.9c) or by (12.18a–12.18c). This type of model, just as in Chap. 8, is an interactive boundary layer model, IBL.

12.3 IBL Model for the Lower Wall

In order to obtain a UVA in the lower boundary layer and in the core flow, we set

$$u = U_1(x, Y, \varepsilon) + u_1(x, y, \varepsilon) , \tag{12.20a}$$

$$v = \varepsilon V_1(x, Y, \varepsilon) + v_1(x, y, \varepsilon) , \tag{12.20b}$$

$$p = \Delta(\varepsilon) P_1(x, Y, \varepsilon) + p_1(x, y, \varepsilon) , \tag{12.20c}$$

where, for the sake of simplicity of notation, the same notation (u, v, p) as for the preceding UVA given by (12.13a–12.13c) is used. In order to have the same order for the inertia terms and the viscous terms in the boundary layer, we take

$$\mathcal{R} = \frac{1}{\varepsilon^3} \, , \tag{12.21}$$

From (12.19a–12.19c), we obtain

$$\frac{\partial U_1}{\partial x} + \frac{\partial V_1}{\partial Y} = 0 \, , \tag{12.22a}$$

$$\varepsilon \left[(U_1 + u_1) \left(\frac{\partial U_1}{\partial x} + \frac{\partial u_1}{\partial x} \right) + (\varepsilon V_1 + v_1) \left(\varepsilon^{-1} \frac{\partial U_1}{\partial Y} + \frac{\partial u_1}{\partial y} \right) \right]$$

$$+ u_0 \left(\frac{\partial U_1}{\partial x} + \frac{\partial u_1}{\partial x} \right) + \frac{\mathrm{d}u_0}{\mathrm{d}y} (\varepsilon V_1 + v_1)$$

$$= -\frac{\partial p_1}{\partial x} - \Delta \frac{\partial P_1}{\partial x} + \varepsilon \left(\frac{\partial^2 U_1}{\partial Y^2} + \varepsilon^2 \frac{\partial^2 u_1}{\partial y^2} \right) + \mathrm{O}(\varepsilon^3) \, , \tag{12.22b}$$

$$\varepsilon \left[(U_1 + u_1) \left(\varepsilon \frac{\partial V_1}{\partial x} + \frac{\partial v_1}{\partial x} \right) + (\varepsilon V_1 + v_1) \left(\frac{\partial V_1}{\partial Y} + \frac{\partial v_1}{\partial y} \right) \right]$$

$$+ u_0 \left(\varepsilon \frac{\partial V_1}{\partial x} + \frac{\partial v_1}{\partial x} \right)$$

$$= -\frac{\partial p_1}{\partial y} - \Delta \varepsilon^{-1} \frac{\partial P_1}{\partial Y} + \varepsilon^2 \left(\frac{\partial^2 V_1}{\partial Y^2} + \varepsilon \frac{\partial^2 v_1}{\partial y^2} \right) + \mathrm{O}(\varepsilon^3) \, . \tag{12.22c}$$

In the above equations, it is necessary to keep terms which are apparently negligible in order to ensure that the behaviour at the wall is bounded. Now, it is essential to examine the pressure terms. From the condition on the transverse velocity given by (12.17a) and from (12.9c) or (12.18c), in the boundary layer, we have $v_1 = \mathrm{O}(\varepsilon)$ and $\dfrac{\partial p_1}{\partial y} = \mathrm{O}(\varepsilon^2)$. Then, from (12.22c), we must take $\Delta = \varepsilon^3$, otherwise the transverse momentum equation cannot be satisfied. Neglecting $\mathrm{O}(\varepsilon^3)$ terms in (12.22b), and coming back to approximations (u, v, p) expressed by (12.20a–12.20c), we obtain

$$\frac{\partial u}{\partial x} + \frac{\partial v}{\partial y} = 0 \, , \tag{12.23a}$$

$$\varepsilon \left(u \frac{\partial u}{\partial x} + v \frac{\partial u}{\partial y} \right) + u_0 \frac{\partial u}{\partial x} + v \frac{\mathrm{d}u_0}{\mathrm{d}y} = -\frac{\partial p_1}{\partial x} + \frac{1}{\mathcal{R}} \frac{\partial^2 u}{\partial y^2} \, . \tag{12.23b}$$

Similar equations can be obtained for the upper boundary layer. These equations must be solved in association with the core flow equations. Therefore, it is clear that (12.23a, 12.23b) associated with the core flow equations give an *approximation valid in the whole channel*.

If necessary, the transverse momentum equation can be used to give the transverse pressure gradient $\dfrac{\partial p}{\partial y}$. Neglecting $\mathrm{O}(\varepsilon^3)$ terms in (12.22c), we have

$$\varepsilon\left(u\frac{\partial v}{\partial x} + v\frac{\partial v}{\partial y}\right) + u_0\frac{\partial v}{\partial x} = -\frac{\partial p}{\partial y} + \frac{1}{\mathcal{R}}\frac{\partial^2 v}{\partial y^2}\,. \tag{12.24}$$

Note 12.1. To within the viscous terms which are probably very small in the core, (12.23a, 12.23b) reduce exactly to (12.18a, 12.18b) in the core. By contrast, in the core, (12.23a, 12.23b) reduce to (12.9a, 12.9b) to within $\mathrm{O}(\varepsilon)$ terms. In this sense, the model based on (12.23a, 12.23b) and (12.18a–12.18c) is more consistent than the model based on (12.23a, 12.23b) and (12.9a–12.9c).

12.4 Global IBL Model

The generalized asymptotic expansions for the velocity components are given by

$$\mathcal{U} = u_0(y) + \varepsilon u(x, y, \varepsilon) + \cdots\,, \tag{12.25a}$$
$$\mathcal{V} = \varepsilon v(x, y, \varepsilon) + \cdots\,. \tag{12.25b}$$

Let us remember that it is necessary to solve (12.23a, 12.23b) in association with the core flow equations (12.9a–12.9c) or (12.18a–12.18c) according to the desired accuracy.

Equations (12.23a, 12.23b) can be recast in the same form as Prandtl's equations if we set

$$\tilde{u} = u_0 + \varepsilon u\,, \tag{12.26a}$$
$$\tilde{v} = \varepsilon v\,, \tag{12.26b}$$
$$\tilde{p}_1 = -\frac{2x}{\mathcal{R}} + \varepsilon p_1 + p_c\,, \tag{12.26c}$$

where p_c is an arbitrary constant. Equations (12.23a, 12.23b) become

$$\frac{\partial \tilde{u}}{\partial x} + \frac{\partial \tilde{v}}{\partial y} = 0\,, \tag{12.27a}$$
$$\tilde{u}\frac{\partial \tilde{u}}{\partial x} + \tilde{v}\frac{\partial \tilde{u}}{\partial y} = -\frac{\partial \tilde{p}_1}{\partial x} + \frac{1}{\mathcal{R}}\frac{\partial^2 \tilde{u}}{\partial y^2}\,. \tag{12.27b}$$

The above equations have the same form as Prandtl's equations, but the pressure is not constant in the y-direction.

Equations (12.27a), (12.27b) are associated with boundary conditions. At the walls, the no-slip conditions are $\tilde{u} = 0$ and $\tilde{v} = 0$. It is clear that the pressure gradient must be adjusted to ensure mass flow conservation in the channel. In addition, the solution for the core flow equations requires additional conditions. These questions are discussed in Sect. 12.5.

Note 12.2. Let us note that the remainders $L_\varepsilon\, u$ and $L_\varepsilon\, v$, defined by (12.7b) and (12.7c), are

$$L_\varepsilon\, u = \left(\frac{\partial p}{\partial x} - \frac{\partial p_1}{\partial x}\right) - \frac{1}{\mathcal{R}}\frac{\partial^2 \tilde{u}}{\partial x^2}\,, \tag{12.28a}$$

$$L_\varepsilon\, v = -\frac{1}{\mathcal{R}}\frac{\partial^2 \tilde{v}}{\partial x^2}\,. \tag{12.28b}$$

The term $\left(\dfrac{\partial p}{\partial x} - \dfrac{\partial p_1}{\partial x}\right)$ is a boundary layer term which is small in the core of the flow.

Note 12.3. For axisymmetric channel flows, Lagrée et al. [49] consider a reduced form of the Navier-Stokes equations which is equivalent to (12.27a), (12.27b) but with no variation of pressure in the transverse direction: $\dfrac{\partial \tilde{p}_1}{\partial y} = 0$. A comprehensive study of this model is carried out for different channel configurations and it is shown that, for axisymmetric flows, the model has a large domain of validity.

12.5 Numerical Solution

In this section, we present a brief description of the numerical solution of the global IBL model. We consider two cases according to the equations used to calculate the core flow and, therefore, the pressure. The first case is based on (12.18a–12.18c) and, in principle, is able to simulate the wider range of channel flow configurations. The second case is more restricted since the core flow equations are those obtained in Smith's theory [94] for the case with upstream influence which is similar, to a certain extent, to the triple deck theory for an external flow (see Problems 12-1 and 12-2).

12.5.1 General Method

Let us consider the model comprising the generalized boundary layer equations (12.23a) and (12.23b), and the core flow equations (12.18a–12.18c). The full problem consists of solving the following equations

$$\frac{\partial \tilde{u}}{\partial x} + \frac{\partial \tilde{v}}{\partial y} = 0\,, \tag{12.29a}$$

$$\tilde{u}\frac{\partial \tilde{u}}{\partial x} + \tilde{v}\frac{\partial \tilde{u}}{\partial y} = -\frac{\partial \tilde{p}_1}{\partial x} + \frac{1}{\mathcal{R}}\frac{\partial^2 \tilde{u}}{\partial y^2}\,, \tag{12.29b}$$

and

$$\frac{\partial \tilde{u}_1}{\partial x} + \frac{\partial \tilde{v}_1}{\partial y} = 0\,, \tag{12.30a}$$

$$\tilde{u}_1 \frac{\partial \tilde{u}_1}{\partial x} + \tilde{v}_1 \frac{\partial \tilde{u}_1}{\partial y} = -\frac{\partial \tilde{p}_1}{\partial x} + \frac{1}{\mathcal{R}} \frac{\mathrm{d}^2 u_0}{\mathrm{d}y^2} \,, \tag{12.30b}$$

$$\tilde{u}_1 \frac{\partial \tilde{v}_1}{\partial x} + \tilde{v}_1 \frac{\partial \tilde{v}_1}{\partial y} = -\frac{\partial \tilde{p}_1}{\partial y} \,. \tag{12.30c}$$

In the above equations, we used the change of variables

$$\tilde{u} = u_0 + \varepsilon u \,, \tag{12.31a}$$

$$\tilde{v} = \varepsilon v \,, \tag{12.31b}$$

$$\tilde{p}_1 = -\frac{2x}{\mathcal{R}} + \varepsilon p_1 + p_c \,, \tag{12.31c}$$

and

$$\tilde{u}_1 = u_0 + \varepsilon u_1 \,, \tag{12.32a}$$

$$\tilde{v}_1 = \varepsilon v_1 \,. \tag{12.32b}$$

Note 12.4. In (12.30b), the viscous term $\dfrac{1}{\mathcal{R}} \dfrac{\mathrm{d}^2 u_0}{\mathrm{d}y^2}$ cancels with the term $\dfrac{2}{\mathcal{R}}$ included in the pressure term $-\dfrac{\partial \tilde{p}_1}{\partial x}$.

The two sets of equations, (12.29a,12.29b) and (12.30a–12.30c), are coupled through the pressure and must be solved simultaneously. They are subject to the wall boundary conditions

$$y = y_l : \quad \tilde{u} = 0 \,, \quad \tilde{v} = 0 \,, \tag{12.33a}$$
$$y = y_u : \quad \tilde{u} = 0 \,, \quad \tilde{v} = 0 \,. \tag{12.33b}$$

It remains to define the boundary conditions for the core flow. At infinity, in the streamwise direction, the flow perturbation vanishes so that we must have $\tilde{u}_1 - u_0 \to 0$ and $\tilde{v}_1 \to 0$. Along the walls, boundary conditions must also be prescribed. From the continuity equations (12.29a) and (12.30a), we obtain

$$\frac{\partial(\tilde{u} - \tilde{u}_1)}{\partial x} + \frac{\partial(\tilde{v} - \tilde{v}_1)}{\partial y} = 0 \,. \tag{12.34}$$

Let $y = y_c$ be a line located in the core flow. This line is arbitrary but it is important that its location is outside the wall boundary layers. Equation (12.34) is integrated with respect to y between the lower wall $y = y_l$ and the core line $y = y_c$ to obtain

$$\int_{y_l}^{y_c} \frac{\partial(\tilde{u} - \tilde{u}_1)}{\partial x} \, \mathrm{d}y + [\tilde{v} - \tilde{v}_1]_{y_l}^{y_c} = 0 \,. \tag{12.35}$$

At the wall, $y = y_l$, we have $\tilde{v} = 0$. Along the line $y = y_c$, we impose

$$\tilde{v}(y_c) = \tilde{v}_1(y_c) \,. \tag{12.36}$$

This assumption is justified by the fact that the wall boundary layers are thin and that the flow perturbation in the core is inviscid; therefore the solution for $\tilde{u}$ and $\tilde{v}$, which is assumed to be a *uniformly valid approximation*, must agree with the solution for $\tilde{u}_1$ and $\tilde{v}_1$.

From (12.35) and (12.36), we obtain

$$\tilde{v}_1(y_l) = -\int_{y_l}^{y_c} \frac{\partial(\tilde{u} - \tilde{u}_1)}{\partial x}\, dy \ . \tag{12.37a}$$

In the same way, along the upper wall $y = y_u$, we obtain

$$\tilde{v}_1(y_u) = -\int_{y_u}^{y_c} \frac{\partial(\tilde{u} - \tilde{u}_1)}{\partial x}\, dy \ . \tag{12.37b}$$

The conditions given by (12.37a) and (12.37b) are the boundary conditions imposed on $\tilde{v}_1$ along the lower and upper walls.

12.5.2 Simplified Method for the Pressure

The use of (12.30a–12.30c) to calculate the pressure is rather involved and, to produce the results discussed in Sect. 12.6, it is more convenient to use a simpler approach suggested by Smith's theory in the case of longer wall deformations (see Problem 12-2). The core flow equations are a reduced form of (12.30a–12.30c)

$$\frac{\partial(\tilde{u}_1 - u_0)}{\partial x} + \frac{\partial\tilde{v}_1}{\partial y} = 0 \ , \tag{12.38a}$$

$$u_0 \frac{\partial(\tilde{u}_1 - u_0)}{\partial x} + \tilde{v}_1 \frac{du_0}{dy} = 0 \ , \tag{12.38b}$$

$$u_0 \frac{\partial\tilde{v}_1}{\partial x} = -\frac{\partial}{\partial y}\left(\tilde{p}_1 + \frac{2x}{\mathcal{R}}\right) \ . \tag{12.38c}$$

The solution is given by

$$\tilde{u}_1 - u_0 = \widetilde{A}(x)\frac{du_0}{dy} \ , \tag{12.39a}$$

$$\tilde{v}_1 = -\frac{d\widetilde{A}}{dx}u_0 \ , \tag{12.39b}$$

$$\tilde{p}_1 + \frac{2x}{\mathcal{R}} = \widetilde{B}(x) + \frac{d^2\widetilde{A}}{dx^2}\int_0^y u_0^2(\eta)\, d\eta \ , \tag{12.39c}$$

where η is an integration variable and the arbitrary constant in the pressure is absorbed in the function $\widetilde{B}(x)$.

Note 12.5. The solution of (12.29a–12.29b) associated with (12.39a–12.39c) depends on the Reynolds number $\mathcal{R}$ which is a parameter of the solution. Then, the functions $\widetilde{A}$ and $\widetilde{B}$ depend parametrically on $\mathcal{R}$. Strictly, we should write $\widetilde{A} = \widetilde{A}(x;\mathcal{R})$ and $\widetilde{B} = \widetilde{B}(x;\mathcal{R})$. Here, the numerical solution is determined for a given Reynolds number and the dependence of $\widetilde{A}$ and $\widetilde{B}$ on $\mathcal{R}$ is omitted.

With

$$u_0 = \frac{1}{4} - y^2 \; ,$$

the pressure is given by

$$\tilde{p}_1 + \frac{2x}{\mathcal{R}} = \widetilde{B}(x) + \frac{\mathrm{d}^2 \widetilde{A}}{\mathrm{d}x^2}\left(\frac{y}{16} - \frac{y^3}{6} + \frac{y^5}{5}\right) \; . \tag{12.40}$$

In this formulation, the question is to determine the function $\widetilde{B}(x)$ and the so-called *displacement function* $\widetilde{A}(x)$. To this end, two conditions are used. The first one is to ensure mass flow conservation in the channel and the second one is given by (12.36).

More precisely, the numerical solution is obtained as described below. A step by step marching procedure from upstream to downstream is used. Several sweeps of the calculation domain are required in order to take into account the upstream influence. At a given station, as a first approximation, it is assumed that the function $\widetilde{A}(x)$ is known. The solution of the generalized boundary layer equations (12.29a)–(12.29b) is determined by iterating on the value of function $\widetilde{B}$ at the considered station in order to ensure global mass flow conservation in the channel. More precisely, the derivative $\dfrac{\mathrm{d}\widetilde{B}}{\mathrm{d}x}$, which is present in the momentum equation (12.29b), is determined. Upon convergence, an updated value of $\widetilde{A}$ is calculated. The method is very similar to the one described in Subsect. 12.5.1. The integration of the continuity equation, (12.29a) gives

$$\int_{y_l}^{y_c} \frac{\partial \tilde{u}}{\partial x}\,\mathrm{d}y + [\tilde{v}]_{y_l}^{y_c} = 0 \; , \tag{12.41}$$

where y_c is an arbitrary line in the core flow. Using (12.36) and taking into account the wall condition $\tilde{v}(y_l) = 0$, we obtain

$$\int_{y_l}^{y_c} \frac{\partial \tilde{u}}{\partial x}\,\mathrm{d}y + \tilde{v}_1(y_c) = 0 \; . \tag{12.42}$$

Taking into account (12.39b), we obtain the updated value of the displacement function, or more precisely of its derivative

$$\frac{\mathrm{d}\widetilde{A}}{\mathrm{d}x} = \frac{1}{u_0(y_c)} \int_{y_l}^{y_c} \frac{\partial \tilde{u}}{\partial x}\,\mathrm{d}y \; . \tag{12.43}$$

When the updated value of the displacement function is determined, the calculations proceed to the next station. The updated value of the displacement function is used at the next sweep. In fact, in order to ensure the convergence of the successive sweeps, an underrelaxation on the displacement function is necessary.

Another point to discuss is the discretization of the equations, in particular the discretization of the pressure term, which is important to reproduce the upstream influence [81]. In the numerical method described in this subsection, we calculate the derivative $\dfrac{\mathrm{d}\widetilde{A}}{\mathrm{d}x}$ by means of (12.43) and we discretize the quantity $\dfrac{\mathrm{d}^3\widetilde{A}}{\mathrm{d}x^3}$ to express $\dfrac{\partial \tilde{p}_1}{\partial x}$ from (12.40). To evaluate the second derivative of $\dfrac{\mathrm{d}\widetilde{A}}{\mathrm{d}x}$, we use a five point stencil, with one point upstream and three points downstream of the calculated station.

Note 12.6. If we assume that the pressure is constant in a cross-section, $\dfrac{\partial \tilde{p}_1}{\partial y} = 0$, the method described in this subsection is valid by assuming that the function $\widetilde{A}$ is zero. Only one sweep is required to solve the equations. A similar numerical method was used by Lagrée et al. [49]. This approach is useful to calculate flows in channels with symmetrically deformed walls. We note that separated flows can be calculated but no upstream influence exists. This means that, upstream of any change of wall geometry, the basic flow is not perturbed.

12.6 Application of the Global IBL model

In this section, we present results obtained with the global IBL model (Sect. 12.4). More precisely, we solve the system of generalized boundary layer equations (12.29a) and (12.29b) associated with (12.39c) and (12.43) which give the evolution of the pressure and the distribution of the displacement function. The numerical method is described in Subsect. 12.5.2.

In Subsect. 12.6.1, results are presented with a discussion on the numerical procedure. Comparisons between IBL results and elements of Smith's theory are described in Subsect. 12.6.2, and comparisons between IBL results and Navier-Stokes solutions are discussed in Subsect. 12.6.3.

12.6.1 Discussion of the Numerical Procedure

In the numerical procedure described in Subsect. 12.5.2, the main point is (12.43) which allows the determination of the displacement function. The question addressed here is to know if the results are sensitive to the location $y = y_c$ where (12.43) is applied. This equation results from (12.36)

$$\tilde{v}(y_c) = \tilde{v}_1(y_c) \,. \tag{12.44}$$

The prescription of condition given by (12.44) is derived from the assumption that the UVA obtained by solving (12.29a) and (12.29b) associated with (12.39c) and (12.43) agrees with the core flow solution given by (12.39a–12.39b). It must be noted, however, that i) (12.44) is not a matching condition

On the transverse velocity component, and ii) the solution for $\tilde{v}_1$ is symmetrical about the axis $y = 0$ while there is no reason for $\tilde{v}$ to be symmetrical. This point is discussed later in this subsection, see (12.47).

To check the relevance of the numerical procedure, different calculations were performed by varying the value of y_c in (12.43) for a channel whose upper wall is flat ($y_u = 1/2$) and lower wall is given by

$$y_l = -\frac{1}{2} + h \exp\left[-\left(\frac{x}{L}\right)^2\right] , \tag{12.45}$$

where h and L are nondimensional scaling factors and x is the streamwise coordinate reduced by the width H^* of the non deformed channel, with $h = -0.1$ and $L^2 = 0.5$. The values of Reynolds number $\mathcal{R}$ defined by (12.2) are $\mathcal{R} = 10^2$ (Fig. 12.2) and $\mathcal{R} = 10^3$ (Fig. 12.3).

The skin-friction coefficient shown in Figs. 12.2 and 12.3 is defined by

$$\frac{C_f}{2} = \frac{\tau_w}{\varrho^* V^{*2}} , \tag{12.46}$$

where τ_w is the dimensionalized wall shear stress and ϱ^*, V^* are the reference quantities used to define the Reynolds number from (12.2). With the same notation as in (12.29a) and (12.29b), on the lower wall we have

$$\tau_w = \frac{\mu^* V^*}{H^*} \mu \left(\frac{\partial \tilde{u}}{\partial y}\right)_{y=y_l} ,$$

and, on the upper wall,

$$\tau_w = -\frac{\mu^* V^*}{H^*} \mu \left(\frac{\partial \tilde{u}}{\partial y}\right)_{y=y_u} .$$

In the above equations, μ is the viscosity coefficient reduced by μ^*, so that $\mu = 1$.

The results for $y_c = 0$, Fig. 12.2a, and for $y_c = -0.2$, Fig. 12.2b, with a Reynolds number $\mathcal{R} = 10^2$, show that the evolution of the derivative of the displacement function $\dfrac{\mathrm{d}\tilde{A}}{\mathrm{d}x}$ is sensitive to the value of y_c, but the evolution of the skin-friction is less sensitive. It is noted that the influence of the value of y_c is weaker at higher Reynolds number (Figs. 12.3a and 12.3b).

Taking into account the fact that $\tilde{v}_1$ is symmetrical about the axis $y = 0$ but not $\tilde{v}$, we tried to apply (12.36) for the symmetrical part of $\tilde{v}$, that is

$$\frac{\tilde{v}(y_c) + \tilde{v}(-y_c)}{2} = \tilde{v}_1(y_c) , \tag{12.47}$$

and (12.36) becomes

$$\frac{\mathrm{d}\tilde{A}}{\mathrm{d}x} = \frac{1}{2u_0(y_c)} \left[\int_{y_l}^{y_c} \frac{\partial \tilde{u}}{\partial x}\, \mathrm{d}y + \int_{y_l}^{-y_c} \frac{\partial \tilde{u}}{\partial x}\, \mathrm{d}y\right] . \tag{12.48}$$

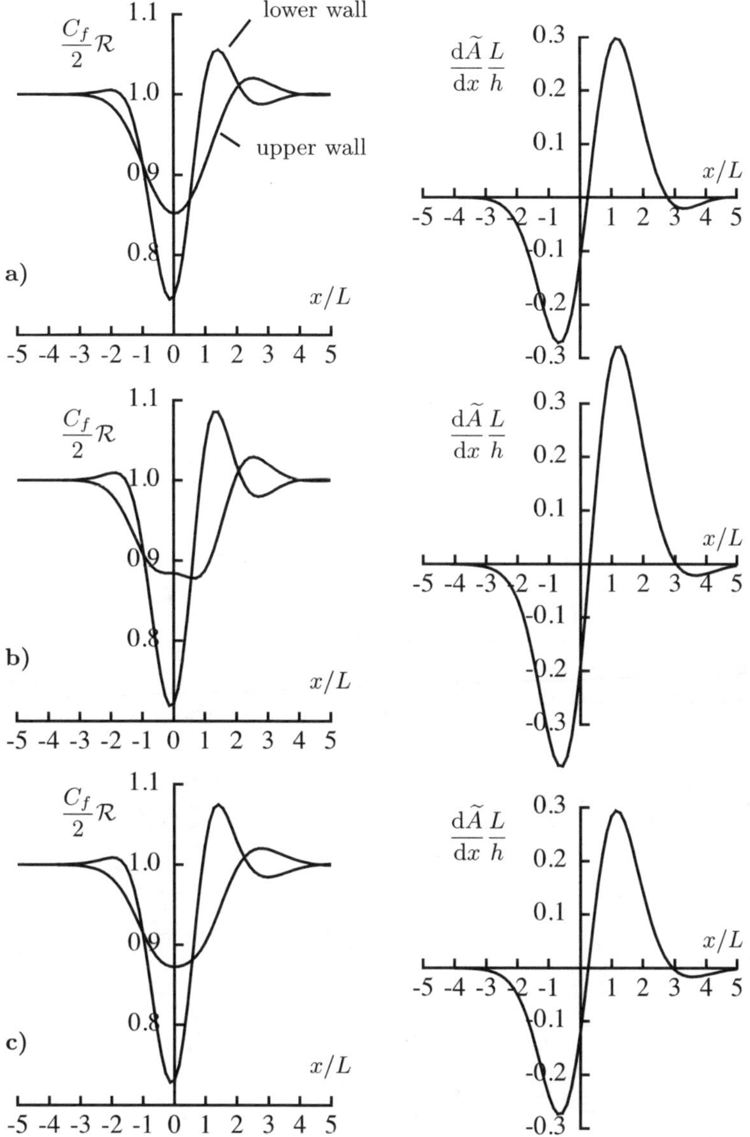

Fig. 12.2. Influence of the location of the line $y = y_c$, $\mathcal{R} = 10^2$, **a)** (12.43) with $y_c = 0$, **b)** (12.43) with $y_c = -0.2$, **c)** (12.48) with $\tilde{y}_c = -0.2$

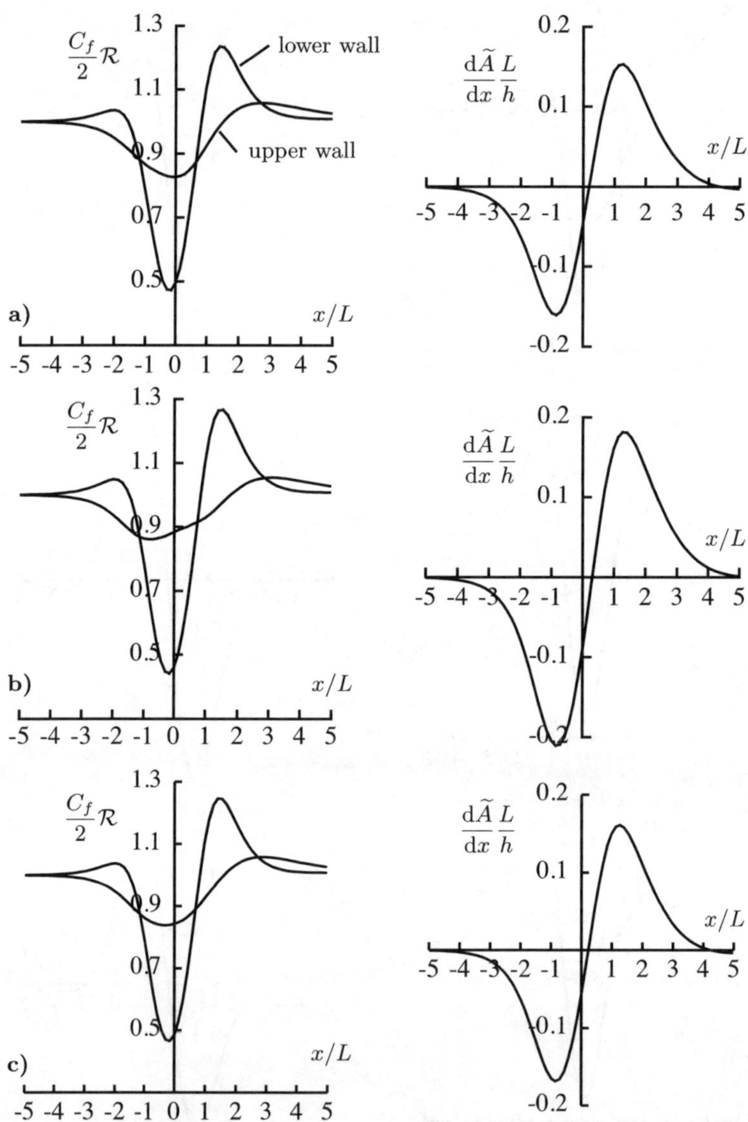

Fig. 12.3. Influence of the location of the line $y = y_c$, $\mathcal{R} = 10^3$, **a)** (12.43) with $y_c = 0$, **b)** (12.43) with $y_c = -0.2$, **c)** (12.48) with $y_c = -0.2$

Figures **12.2c** and **12.3c** show that the results obtained with $y_c = -0.2$ in (12.48) are very close to those obtained with $y_c = 0$ (Figs. **12.2a** and **12.3a**).

It is concluded that, with (12.48), the results are practically independent from the location of the line $y = y_c$ which is very satisfactory but difficult to justify (12.48), except by invoking symmetry arguments and also the linearity of the solution for small perturbations.

12.6.2 Comparisons with Smith's theory

Smith's theory

In Smith's theory, we consider the flow in a two-dimensional channel whose walls are slightly deformed (Fig.12.4). Far upstream of the deformation, we assume that the flow is fully developed, i.e. we have a Poiseuille flow. The analysis uses the method of matched asymptotic expansions which implies *regular expansions* [89, 90, 91].

Different flow regimes are obtained according to the scales of the wall indentation. The equations of the lower and upper walls are

$$y_l = -\frac{1}{2} + hf\left(\frac{x}{L}\right)\,, \tag{12.49a}$$

$$y_u = \frac{1}{2} - hg\left(\frac{x}{L}\right)\,. \tag{12.49b}$$

In these equations, y_l, y_u and x are nondimensionalized with the width H^* of the non deformed channel. The scale h of the nondimensional height of the wall perturbation is a small parameter defined in the same way as in Sect. 12.1

$$\varepsilon = h = \frac{h^*}{H^*}\,. \tag{12.50}$$

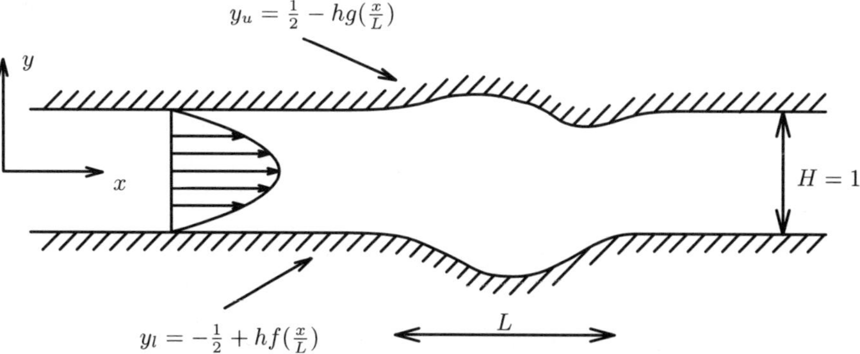

Fig. 12.4. Asymptotic analysis of a two-dimensional channel with deformed walls. In this figure, all quantities are dimensionless

The inverse of the longitudinal length scale $L = L^*/H^*$ of the wall indentation is also a small parameter. It is assumed that the two small parameters are related by

$$\frac{1}{L} = \varepsilon^{\alpha} , \tag{12.51}$$

where α is a positive number. Smith characterizes the indentation by its slope δ given by

$$\delta = \frac{h}{L} = \varepsilon^{1+\alpha} . \tag{12.52}$$

From the analysis of Problems 12-1 and 12-2, we have

$$\varepsilon = \mathcal{R}^{-1/(3+\alpha)} , \tag{12.53}$$

where the Reynolds number $\mathcal{R}$ is defined by (12.2). The length scales of the wall indentation are such that

$$L = \mathcal{R}^{\alpha/(3+\alpha)} , \tag{12.54a}$$
$$h = \mathcal{R}^{-1/(3+\alpha)} , \tag{12.54b}$$
$$\frac{h}{L} = \mathcal{R}^{-(1+\alpha)/(3+\alpha)} . \tag{12.54c}$$

Two flow regimes are of interest here for asymmetric channels. One regime is such that

$$\alpha > \frac{1}{2} , \tag{12.55}$$

and

$$\mathcal{R}^{-1} \prec \delta \prec \mathcal{R}^{-3/7} . \tag{12.56}$$

It is shown that *the pressure is constant in a cross-section*. Flow separation is possible but there is no upstream influence. This means that the effect of the wall indentation is relatively weak.

The other regime corresponds to the value $\alpha = 1/2$. In a way, this case corresponds to the triple deck theory for external flows. We have

$$L = \mathcal{R}^{1/7} , \tag{12.57a}$$
$$h = \mathcal{R}^{-2/7} , \tag{12.57b}$$
$$\frac{h}{L} = \mathcal{R}^{-3/7} . \tag{12.57c}$$

For $\alpha = 1/2$, the slope of the wall indentation is larger than for $\alpha > 1/2$. The pressure is no longer constant in a cross-section and its variation is described by (12.39c). As for $\alpha > 1/2$, flow separation is possible. A fundamental difference with the previous case is that *upstream influence* is possible. According to a linearized theory, the influence of any wall perturbation extends a distance upstream of order $\mathcal{R}^{1/7}$ times the channel width.

Comparison between IBL model and Smith's theory

In order to compare IBL results with Smith's theory, four sets of calculations were performed with the IBL model. The upper wall of the channel was assumed to be flat and the lower wall was deformed by a trough.

The equations for the walls are (Fig 12.4)

$$f = \exp\left[-\left(\frac{x}{L}\right)^2\right] , \tag{12.58a}$$

$$g = 0 , \tag{12.58b}$$

which is in agreement with the form required by the theory and given by (12.49a–12.49b).

The four sets of calculations are defined by the geometry of the trough:

Case I

$$L = L_0\left[\frac{\mathcal{R}}{\mathcal{R}_0}\right]^{1/7} , \quad h = h_0\left[\frac{\mathcal{R}}{\mathcal{R}_0}\right]^{-2/7} , \tag{12.59a}$$

$$L_0 = 2.236 \quad , \quad h_0 = -0.5 . \tag{12.59b}$$

Case II

$$L = L_0\left[\frac{\mathcal{R}}{\mathcal{R}_0}\right]^{1/4} , \quad h = h_0\left[\frac{\mathcal{R}}{\mathcal{R}_0}\right]^{-1/4} , \tag{12.60a}$$

$$L_0 = 2.236 \quad , \quad h_0 = -0.5 . \tag{12.60b}$$

Case III

$$L = L_0\left[\frac{\mathcal{R}}{\mathcal{R}_0}\right]^{1/7} , \quad h = h_0\left[\frac{\mathcal{R}}{\mathcal{R}_0}\right]^{-2/7} , \tag{12.61a}$$

$$L_0 = 0.707 \quad , \quad h_0 = -0.57 . \tag{12.61b}$$

Case IV

$$L = L_0\left[\frac{\mathcal{R}}{\mathcal{R}_0}\right]^{1/4} , \quad h = h_0\left[\frac{\mathcal{R}}{\mathcal{R}_0}\right]^{-1/4} , \tag{12.62a}$$

$$L_0 = 0.707 \quad , \quad h_0 = -0.57 . \tag{12.62b}$$

For each set of calculations, three values of the Reynolds number are considered: $\mathcal{R} = 10^2$, $\mathcal{R} = 10^3$, $\mathcal{R} = 10^4$. The Reynolds number $\mathcal{R}_0$ is $\mathcal{R}_0 = 10^2$.

It must be noted that the variations of the length scales L and h are in agreement with (12.54a) and (12.54b) by taking $\alpha = 1/2$ or $\alpha = 1$.

For Cases I and II, the dimensions were chosen to have a weak perturbation and the pressure calculated by the IBL model is nearly constant in a cross-section of the channel. The family of troughs corresponding to $\alpha = 1$ (Case II) was taken as a typical representative in the range $\alpha > 1/2$. For Cases III and IV, the geometry of troughs was chosen to produce a flow with incipient separation.

Figures 12.5–12.8 show the evolution of the reduced skin-friction along the walls of the channel. The skin-friction coefficient is defined by (12.46).

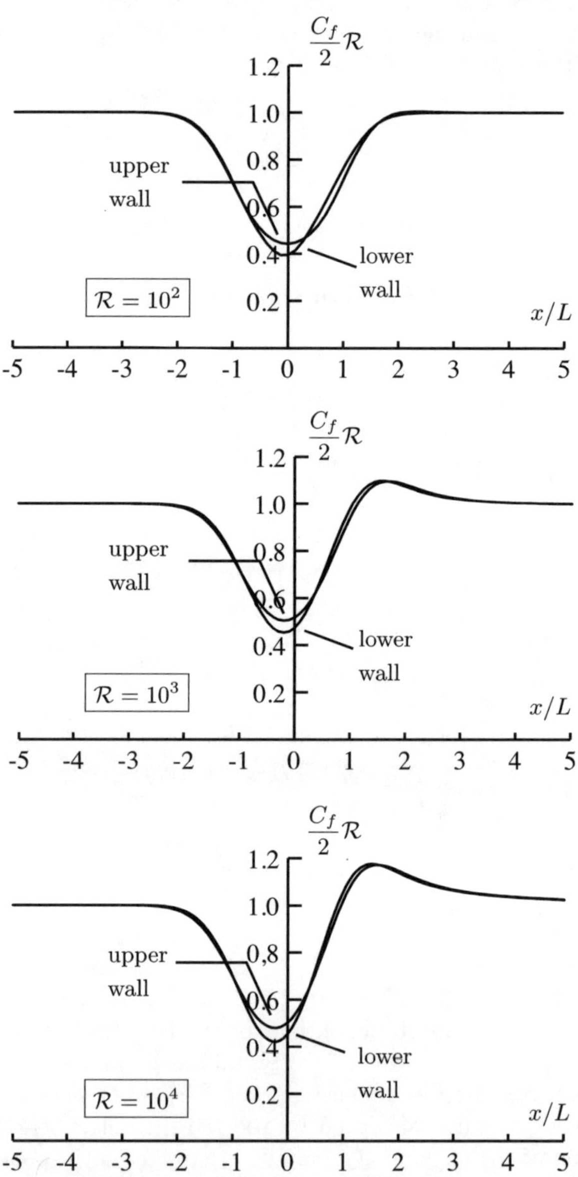

Fig. 12.5. Application of IBL. Evolution of skin-friction in Case I

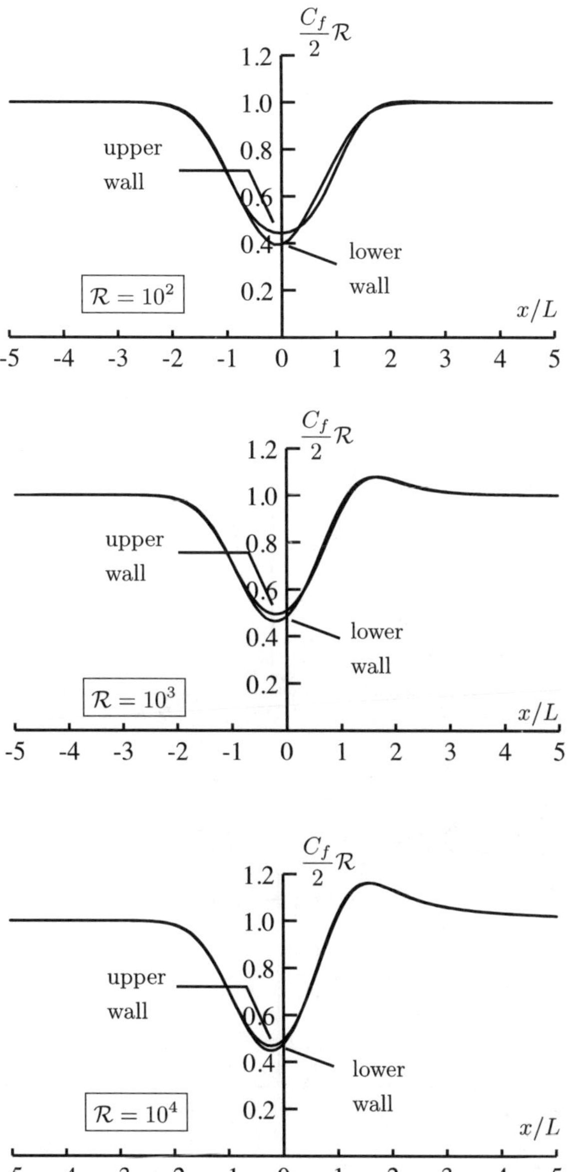

Fig. 12.6. Application of IBL. Evolution of skin-friction in Case II

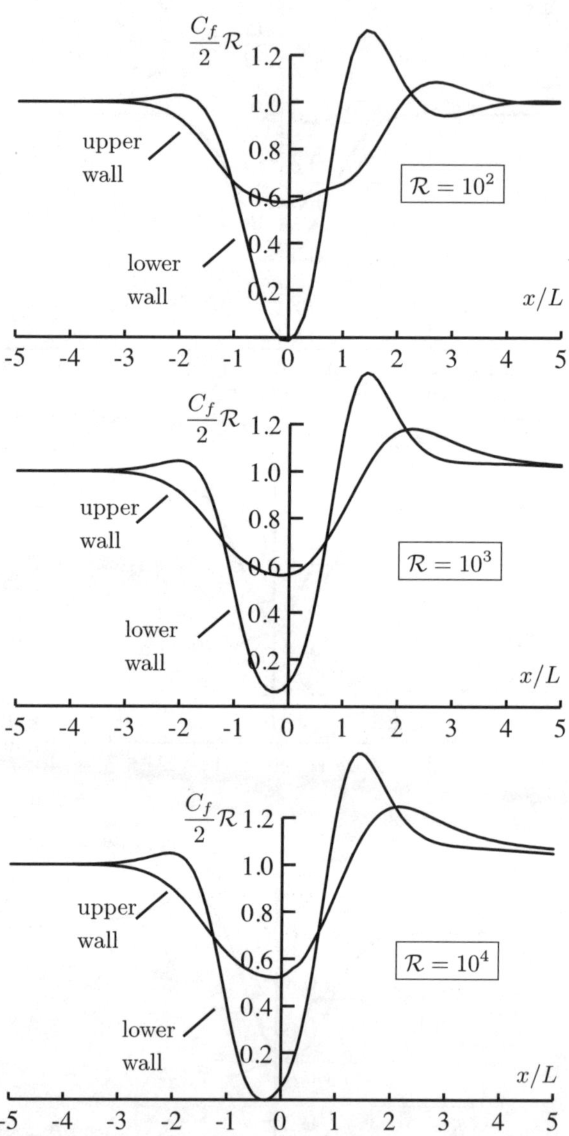

Fig. 12.7. Application of IBL. Evolution of skin-friction in Case III

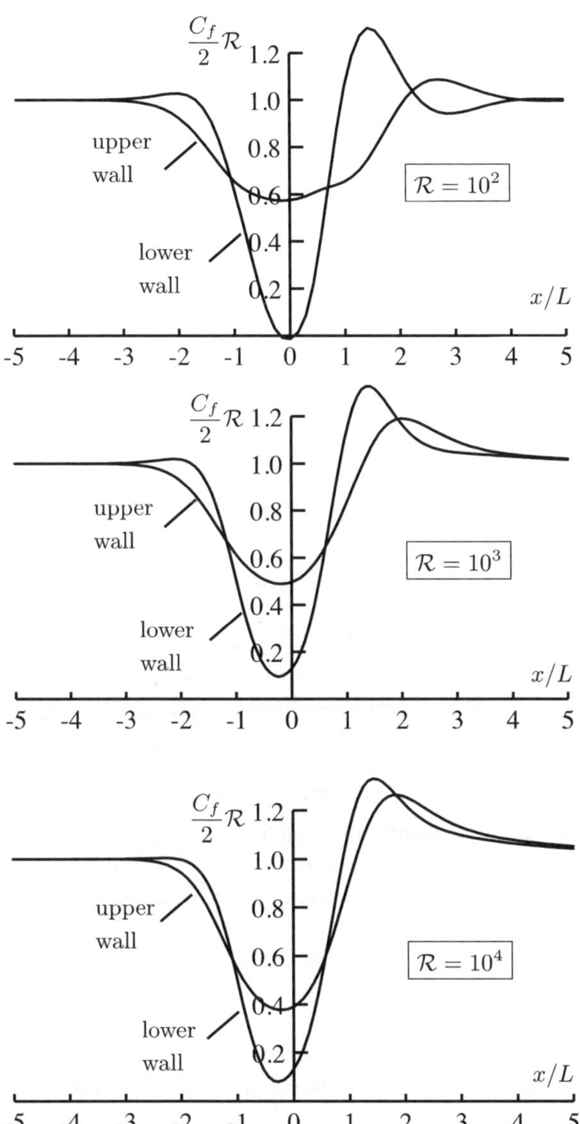

Fig. 12.8. Application of IBL. Evolution of skin-friction in Case IV

For a given set of calculations, *if IBL results are in agreement with Smith's theory*, the evolutions of $\dfrac{C_f}{2}\mathcal{R}$ as function of x/L along the lower wall or the upper wall should be independent of the Reynolds number, exactly as in the theory.

In Cases I and II, the results plotted in Figs. 12.5 and 12.6 show that the theory is reasonably well satisfied for $\alpha = 1/2$ and $\alpha = 1$. These two cases correspond to families of troughs which produce a weak perturbation.

For Case III (Fig. 12.7), the agreement with Smith's theory is also fairly good, specially for the two higher Reynolds numbers $\mathcal{R} = 10^3$ and $\mathcal{R} = 10^4$. On the contrary, for Case IV (Fig. 12.8), the evolution of the skin-friction along the upper wall depends on the Reynolds number and Smith's theory is not satisfied. The wall indentation produces a significant variation of pressure in a cross-section and the value $\alpha = 1$ in Smith's theory is not appropriate.

The conclusion of this subsection is that IBL results are in good agreement with Smith's theory, at least for $\alpha \geq 1/2$ which covers a wide range of variation of the dimensions of the wall deformation. It is interesting to note that relatively low Reynolds numbers were tested and it seems that the IBL model produces reasonably correct results.

12.6.3 Comparison with Navier-Stokes Solutions

In order to assess the validity of the proposed global IBL model, comparisons with Navier-Stokes solutions are presented in this subsection. The IBL model is based on the system of generalized boundary layer equations (12.29a) and (12.29b) associated with (12.39c) and (12.43) which give the evolution of the pressure and the distribution of the displacement function. The Navier-Stokes solutions were obtained by Dechaume who developed a highly accurate solver [27]. A spectral method based on Legendre polynomials has been implemented and the solution involves a domain decomposition of Dirichlet-Neuman type. A technique of velocity-pressure decoupling is used. For the time integration, the time derivatives are expressed by an implicit Euler scheme, the nonlinear terms and the pressure boundary conditions are extrapolated. The resulting linear systems are solved by successive diagonalisations.

For these comparisons, the flow is calculated in a channel whose upper wall is flat and the lower wall is deformed in the domain $-L/2 \leq x \leq L/2$ according to

$$y_l = -\frac{1}{2} + \frac{h}{2}\left[1 + \cos\left(\frac{2\pi x}{L}\right)\right] \tag{12.63}$$

Outside the domain $-L/2 \leq x \leq L/2$, the lower wall is flat. In (12.63), y_l, x and L are nondimensionalized with the width H^* of the non deformed channel.

Three cases have been selected:

Case I

$$h = -0.3 \,, \quad L = 4 \,, \quad \mathcal{R} = 1000 \,, \tag{12.64}$$

Case II

$$h = -0.16 \,, \quad L = 1 \,, \quad \mathcal{R} = 1000 \,, \tag{12.65}$$

Case III

$$h = 0.36 \,, \quad L = 4 \,, \quad \mathcal{R} = 1000 \,. \tag{12.66}$$

In the first two cases, the lower wall is deformed by a trough, and in the third case, the lower wall is deformed by a bump. The characteristics of the wall indentations were chosen to produce a flow which is close to separation somewhere on the lower wall. For all cases, the flow was calculated in a domain much larger than the wall deformation so that a non perturbed flow is recovered at the inlet and outlet sections.

Using the definition of the skin-friction coefficient in (12.46), the evolution of $\dfrac{C_f}{2}\mathcal{R}$ is plotted as function of x in Figs 12.9–12.11. The overall agreement between the global IBL results and the Navier-Stokes solutions is very satisfactory. Considering that the Navier-Stokes results are reference solutions, the shape of the curves and the level of the skin-friction are well predicted by the global IBL model. In the case of the shorter trough ($L = 1$, $h = -0.16$), Case II, the global IBL model leads to a very irregular evolution of the skin-friction on the upper wall whereas the Navier-Stokes solution is smooth. This case is on the limit of validity of the global IBL model. The main reason is probably that the law (12.39c) for the pressure variation in a cross-section is too simple. The pressure variations induced by the lower wall deformation are not well transmitted to the upper wall when the indentation is too severe. In such a case, it is expected that a more elaborated model for the pressure variations is required, for example the model given by (12.30a–12.30c). Let us note, however, that the asymptotic theory is established for large Reynolds numbers and for wall indentations whose height is small and length is large compared to the channel width. Even if these conditions are not satisfied a priori in the test cases, the agreement of IBL results with Navier-Stokes solutions is strikingly good. When the flow perturbations induced by the wall deformation are weak, the cross-section pressure variations are very small and it is sufficient to assume that $\dfrac{\partial p}{\partial y} = 0$. Then, the evolutions of the skin-friction on the upper and lower walls are identical as illustrated in Figs. 12.5 and 12.6. For more severe wall deformations, the hypothesis of a constant pressure in a cross-section does not hold and (12.39c) can be used to calculate the pressure variations. In this case, for a non symmetric wall deformation, the skin-friction evolutions along the upper and lower walls are not the same. However, if the wall deformation is too severe, the difference between the skin-friction evolutions along the upper and lower walls becomes

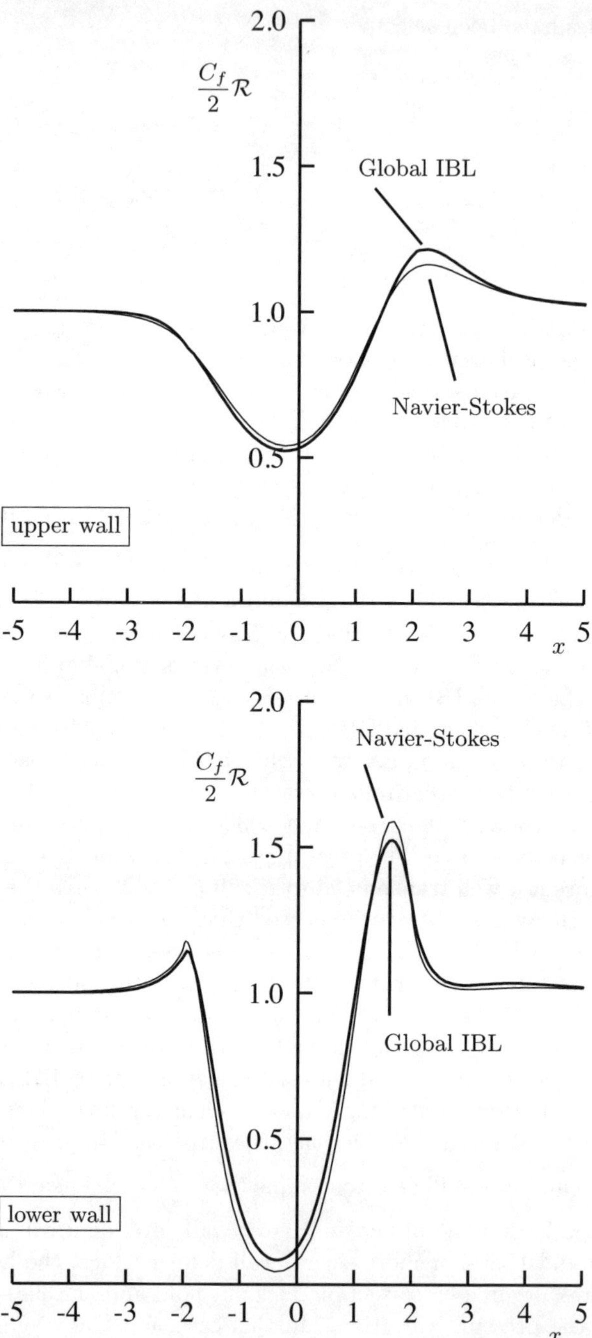

Fig. 12.9. Comparisons between global IBL results and Navier-Stokes solutions. Case I ($\mathcal{R} = 1000$, $h = -0.3$, $L = 4$)

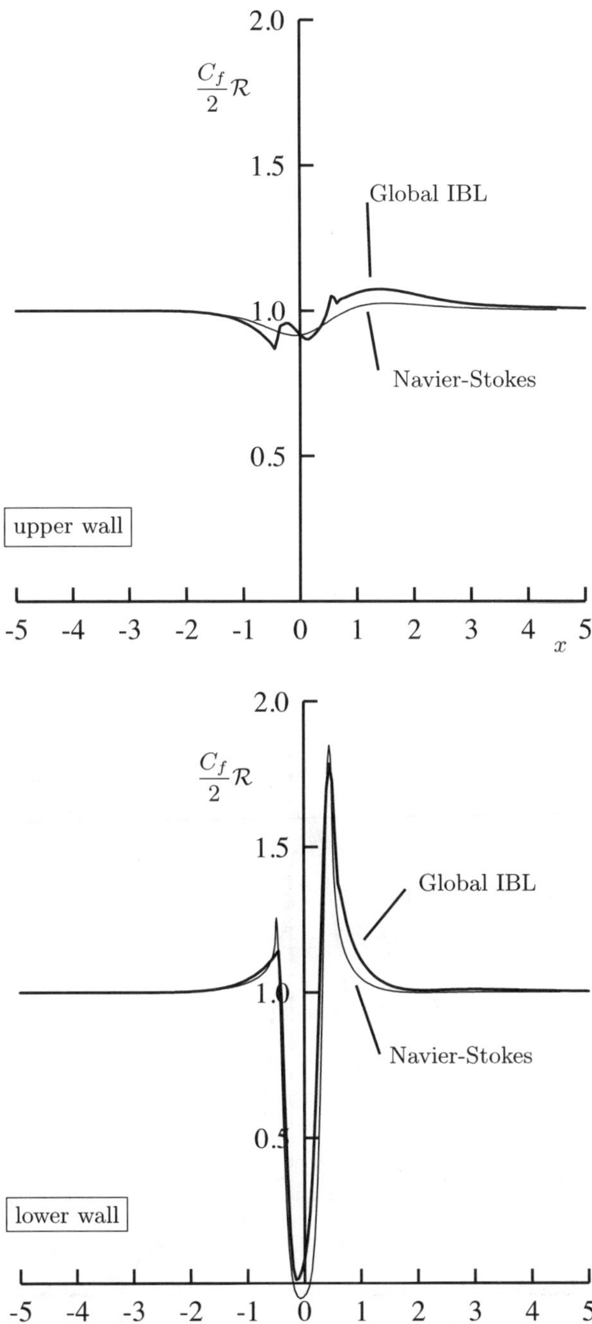

Fig. 12.10. Comparisons between global IBL results and Navier-Stokes solutions. Case II ($\mathcal{R} = 1000$, $h = -0.16$, $L = 1$)

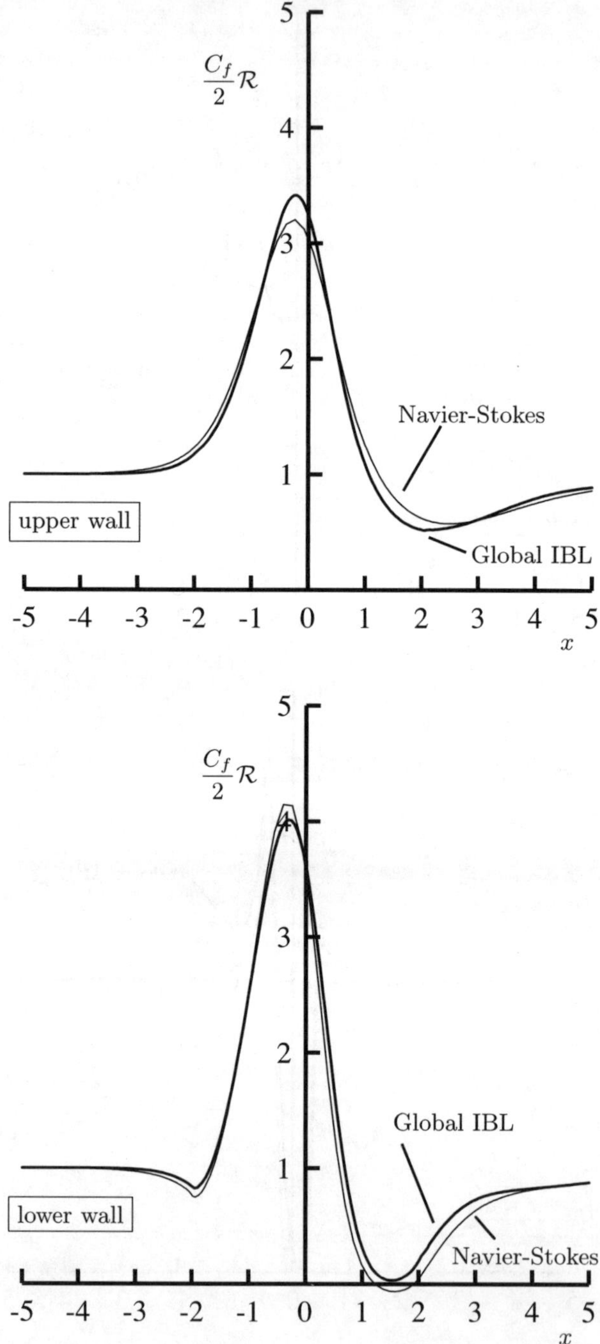

Fig. 12.11. Comparisons between global IBL results and Navier-Stokes solutions. Case III ($\mathcal{R} = 1000$, $h = 0.36$, $L = 4$)

very large as shown in Fig 12.10 for Case II and (12.39c) is not sufficient to reproduce the pressure variations correctly.

12.7 Conclusion

Different approximations of Navier-Stokes equations for the study of high Reynolds number flows in a two-dimensional channel with deformed walls are obtained by applying SCEM.

The flow perturbations are described by an inviscid flow model in the core which is strongly coupled to generalized boundary layer equations valid in the whole channel. Finally, we obtain a global *interactive boundary layer* model. As in the study of external flows, SCEM proved to be a very fruitful tool for analyzing the flow structure.

A simplified model for the pressure variations has been implemented numerically. Essentially, this model for the pressure is based on Smith's theory which is the equivalent of the triple deck theory for external flows. IBL results obtained with this simplified pressure equation are in very good agreement with Smith's theory, at least as far as boundary layer characteristics are concerned, and also with the Navier-Stokes solutions. Even with relatively severe wall indentations and not very large Reynolds numbers, the global IBL model produces satisfactory results. It is expected that even better results can be obtained with a more refined model for the core flow.

It should be noted that SCEM offers interesting perspectives with the construction of a UVA. In fact, $(\tilde{u}_1, \tilde{v}_1)$ is not necessarily an approximation in the flow core whereas $(\tilde{u}, \tilde{v})$ gives a UVA. Perhaps, this is why the symmetrization of $\tilde{v}$ with (12.47) gives a better result. This is an unusual implementation of SCEM with respect to more standard methods.

Problems

12-1. *Smith's theory.* We consider the laminar flow in a two-dimensional channel whose walls are slightly deformed. The equations of the lower and upper walls are

$$y_l = -\frac{1}{2} + \varepsilon^\beta F(X)\,,$$

$$y_u = \frac{1}{2} - \varepsilon^\beta G(X)\,,$$

where

$$X = x\varepsilon^\alpha\,.$$

The exponents α and $\beta > 0$ characterize the longitudinal and transverse length scales of the wall indentations. It is assumed that ε is a small parameter.

We are looking for approximations of the channel flow at high Reynolds number for which separation is possible. Regular expansions are used.

1. The flow satisfies (12.1a–12.1c). We seek a solution in the form

$$\mathcal{U} = u_0(y) + \varepsilon^r u(X, y, \varepsilon) \,,$$
$$\mathcal{V} = \varepsilon^{r+\alpha} v(X, y, \varepsilon) \,,$$
$$\mathcal{P} - p_c = -\frac{2x}{\mathcal{R}} + \varepsilon^s p(X, y, \varepsilon) \,,$$

where p_c is an arbitrary constant. We assume that the small parameter ε is related to the Reynolds number $\mathcal{R}$ by

$$\mathcal{R} = \varepsilon^{-m} \,,$$

where m is a positive number. In fact, the value of m has no influence on the final results. It can be convenient to choose $m = 1$.

Write (12.1a–12.1c) by using the variables X and y and show that the Navier-Stokes equations reduce to first order partial differential equations when $m > |\alpha|$.

2. To study the boundary layer in the vicinity of the lower wall, we use the variable

$$Y = \frac{y + \frac{1}{2}}{\varepsilon^\beta} \,.$$

We want to satisfy the following conditions
a. matching with $u_0(y)$,
b. conservation of mass,
c. same order of magnitude of pressure and viscous terms in the longitudinal
    momentum equation
Show that we must set

$$U = \frac{\mathcal{U}}{\varepsilon^\beta} \,,$$
$$V = \frac{\mathcal{V}}{\varepsilon^{2\beta+\alpha}} \,,$$
$$P = \frac{\mathcal{P} - p_c}{\varepsilon^{-\alpha-\beta+m}} \,.$$

Write (12.1a–12.1c) with the variables X, Y.

The scale of the slope of the wall indentation is characterized by $\alpha + \beta$. Show that, with $\alpha + \beta > 0$, the Navier-Stokes equations reduce to boundary layer type equations if

$$3\beta = m - \alpha \,.$$

3. Show that, in the lower wall boundary layer I, the inner regular asymptotic expansions are

$$\mathcal{U} = \varepsilon^{(m-\alpha)/3} U_1(X, Y) + \cdots \,,$$

$$V = \varepsilon^{(2m+\alpha)/3} V_1(X, Y) + \cdots ,$$
$$P - p_c = \varepsilon^{(2m-2\alpha)/3} P_1(X, Y) + \cdots .$$

Write the corresponding equations.

4. Show that in the middle deck M (core flow), the pressure can be written

$$P - p_c = \varepsilon^{(2m-2\alpha)/3} p_1(X, y) ,$$

with the matching condition

$$\lim_{y \to -1/2} p_1(X, y) = P_1(X) .$$

5. Regular expansions are used in the middle deck

$$\mathcal{U} = u_0(y) + \varepsilon^r u_1(X, y) + \cdots ,$$
$$\mathcal{V} = \varepsilon^{r+\alpha} v_1(X, y) + \cdots .$$

Show that

$$r \le \frac{2}{3}(m - \alpha) .$$

We assume that

$$r < \frac{2}{3}(m - \alpha) .$$

Show that

$$u_1 = A(X) \frac{du_0}{dy} ,$$
$$v_1 = -\frac{dA}{dX} u_0 ,$$

where $A(X)$ is an undetermined function.

6. With the condition $r < 2(m - \alpha)/3$, show that a pertinent theory implies that the couple (r, α) is in the hatched triangle shown in Fig. 12.12 (it is useful to use the MVDP for longitudinal velocity).

7. For a given value of α, the most significant indentation corresponds to the largest perturbation in the longitudinal velocity. Give the corresponding value of r for $0 < \alpha < m$.

12-2. *Global problem.* This problem is the continuation of Problem 12-1 in which we want to study the upper wall boundary layer and the solution for the flow in the whole channel.

1. Assuming $0 < \alpha \le m/7$, show that the pressure in the core is given by

$$p_1(X, y) = P_1(X) + \frac{A''(X)}{60} \left(12y^5 - 10y^3 + \frac{15}{4} y + 1 \right) ,$$

and that the pressure in the upper wall boundary layer is given by

$$\overline{P}_1(X) = P_1(X) + \frac{A''(X)}{30} .$$

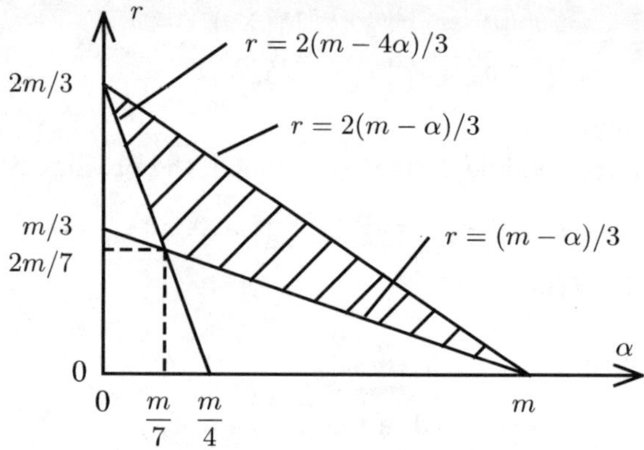

Fig. 12.12. Domain covered by the study in the (r, α)-plane

2. Prandtl's transformation for the lower wall is given by

$$Z = Y - F(X) \,,$$
$$W = V_1 - F'(X)U_1 \,.$$

Write the lower wall boundary layer problem.

3. To study the upper wall boundary layer, we introduce the variable

$$\overline{Y} = \frac{\frac{1}{2} - y}{\varepsilon^{(m-\alpha)/3}}$$

We seek a regular expansion in the form

$$\mathcal{U} = \varepsilon^{(m-\alpha)/3}\overline{U}_1(X, \overline{Y}) + \cdots \,,$$
$$\mathcal{V} = -\varepsilon^{(2m+\alpha)/3}\overline{V}_1(X, \overline{Y}) + \cdots \,.$$

Prandtl's transformation for the upper wall is given by

$$\overline{Z} = \overline{Y} - G(X) \,,$$
$$\overline{W} = \overline{V}_1 - G'(X)\overline{U}_1 \,.$$

Show that the upper wall boundary layer problem is similar to the lower wall problem.

4. We suppose now that $m/7 < \alpha < m$. Show that

$$A(X) = \frac{1}{2}\left[G(X) - F(X)\right] \,.$$

5. Interpret the classification of cases studied above in terms of Reynolds number $\mathcal{R}$, characteristic length $L = \varepsilon^{-\alpha}$ and slope of indentation $\delta = \varepsilon^{\alpha+\beta}$.

12-3. *Study of the symmetric case.* This problem is a continuation of Problems 12-1 and 12-2. The equations of the wall indentations are written

$$y_l = -\frac{1}{2} + \varepsilon F(X)\,,$$

$$y_u = \frac{1}{2} - \varepsilon G(X)\,.$$

This means that the small parameter is taken as being the thickness of the boundary layer. It means also that the value of β considered in Problem 12-1 is now $\beta = 1$.

1. Show that Fig. 12.12 becomes Fig. 12.13.

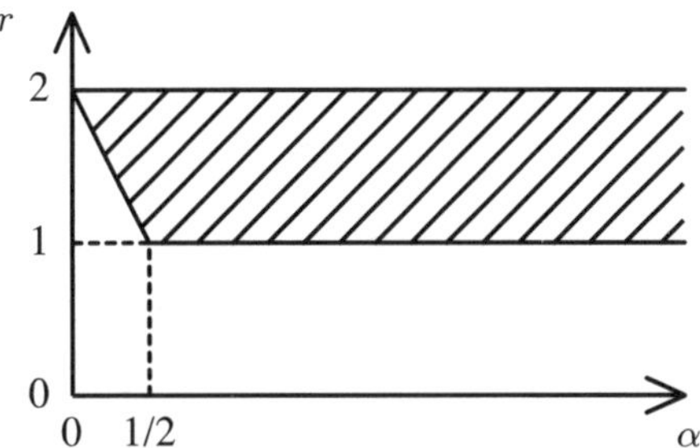

Fig. 12.13. Domain covered by the study in the $(r,\ \alpha)$-plane $(\beta = 1)$

Express the Reynolds number $\mathcal{R}$ and the slope δ of the wall indentations as functions of ε and α.

2. Write the core flow equations with r and α. Show that the gauge of the longitudinal velocity for the second order problem is ε^2.

3. Calculate the analytical solution of the second order problem for $\alpha > 1/2$.

4. Study the matching with the lower wall boundary layer.

5. Write the solution for a symmetric channel.

13 Conclusion

This book contains two distinct parts.

Chapters 2–6 describe a relatively new approach of singular perturbation problems including one or more *boundary layers*. In a certain way, a tool is proposed to teachers for presenting as simply and rigourously as possible the necessary material required to study these difficult problems. Simultaneously, for teachers, students or scientists, we present an indepth analysis of the methods devoted to the study of boundary layers. The most popular method is the *method of matched asymptotic expansions*, MMAE. In Chap. 5, which is the heart of this analysis, we show that the idea of *uniformly valid approximation*, UVA, enables us not only to understand how all the standard methods are based on this idea, but also how new and more efficient methods can be implemented. These developments are associated with the nature of what we call a *generalized asymptotic expansion*; this name is chosen in order to make a clear distinction with the more commonly used concept of *regular asymptotic expansion* or with the even more restricted concept of Poincaré expansion. This extension enables us to produce UVAs with a much better accuracy than with a regular expansion. Then, as this is seen with ordinary differential equations in Chap. 6, the calculation results show a better accuracy of approximations for not extremely small values of the small parameter. We can also produce UVAs which cannot be obtained with MMAE, showing definitely the wider applicability of the method we call the *successive complementary expansion method*, SCEM.

This is well illustrated in the second part of this book which is devoted to the application of SCEM to boundary layer problems in fluid mechanics.

A brief discussion of the *triple deck* theory in Chap. 7 for an incompressible flow enables us to evaluate the limitations of this theory when the *interactive boundary layer* formulation, IBL, is presented. The successive degeneracies of Navier-Stokes equations are analyzed and should allow the reader to understand the reasons leading to the different models. The applications to flows including the effects of external vorticity or to turbulent boundary layers show the advantage of methods based on UVAs. The use of generalized asymptotic expansions with SCEM is particularly fruitful and produce valuable results even if the Reynolds number in not extremely large.

Thus, SCEM reveals to be not only an attractive practical tool, but also a pertinent theoretical means of analysis since SCEM supplies a rational basis for justifying and constructing IBL methods whose efficiency has been widely shown by different authors, specially in aerodynamics. Thanks to SCEM, the asymptotic analysis of boundary layer problems is refined.

Other applications in fluid mechanics can be considered. A few of them are studied in this book, others are simply mentioned. The authors think that interesting progress can be obtained in various fields such as flow stability, laminar-turbulent transition or flow control. Obviously, if fluid mechanics is at the heart of the meeting of the two authors, the first part of this book is independent of this natural and historical application. This part concerns many fields in physics when the mathematical models are constructed around small parameters leading to singular perturbations of boundary layer type. These models require a detailed analysis of the structure of the solutions which, if not carried out, can be detrimental to the numerical simulation.

We do hope that this book, besides the benefit for teachers and students, will encourage scientists to develop these methods in traditional domains of fluid mechanics and also to find applications in other practical fields. In all the concerned disciplines of physics, to find an original and practical asymptotic technique becomes a valuable achievement.

I Navier-Stokes Equations

We consider an incompressible flow of Newtonian fluid in which the gravity forces are neglected. We assume that the velocities are small with respect to the sound celerity so that the Mach number is very small compared to unity. We also assume that the temperature variations are very small compared to the characteristic temperature of fluid. With these hypotheses, the state law is

$$\varrho = \mathrm{Cst} \ . \tag{I.1}$$

The density is uniform in space and does not vary in time. Thus, for a flow of perfect gas, the state law is indeed (I.1) and not the standard law of perfect gases.

With the hypothesis of Newtonian fluid, the viscous stress tensor within the flow is expressed by means of a linear function of the rate of stress tensor. Taking into account the hypothesis of incompressibility, we have

$$\bar{\bar{\tau}} = 2\mu\bar{\bar{S}} \ , \tag{I.2}$$

where $\bar{\bar{\tau}}$ is the viscous stress tensor, $\bar{\bar{S}}$ is the rate of strain tensor and μ is the dynamic viscosity coefficient. The kinematic viscosity coefficient ν can also be used

$$\nu = \frac{\mu}{\varrho} \ . \tag{I.3}$$

We assume that the viscosity coefficients μ and ν are uniform in space.

In an orthonormal axis system, the viscous stress tensor components are

$$\tau_{ij} = \mu \left(\frac{\partial u_i}{\partial x_j} + \frac{\partial u_j}{\partial x_i} \right) \ , \tag{I.4}$$

where u_i represents the x_i-velocity component.

The equations of fluid mechanics comprise the continuity equation or mass conservation equation and the momentum equation which expresses Newton's second law [51, 86].

In tensor notation, the equations of fluid mechanics (Navier-Stokes equations) are

$$\mathbf{div}\,\boldsymbol{u} = 0 \ , \tag{I.5a}$$

$$\varrho \frac{\mathrm{d}\boldsymbol{u}}{\mathrm{d}t} = \mathbf{div}(\bar{\bar{\tau}} - p\bar{\bar{I}}) \ . \tag{I.5b}$$

where $\dfrac{\mathrm{d}}{\mathrm{d}t}$ represents the substantial derivative, p is the pressure and $\bar{\bar{I}}$ is the identity tensor.

In an orthonormal axis system, these equations become

$$\frac{\partial u_i}{\partial x_i} = 0 \, , \tag{I.6a}$$

$$\varrho \frac{\partial u_i}{\partial t} + \varrho u_j \frac{\partial u_i}{\partial x_j} = -\frac{\partial p}{\partial x_i} + \frac{\partial \tau_{ij}}{\partial x_j} \, . \tag{I.6b}$$

If the flow is two-dimensional and steady, the equations are

$$\frac{\partial u}{\partial x} + \frac{\partial v}{\partial y} = 0 \, , \tag{I.7a}$$

$$\varrho u \frac{\partial u}{\partial x} + \varrho v \frac{\partial u}{\partial y} = -\frac{\partial p}{\partial x} + \mu \frac{\partial^2 u}{\partial x^2} + \mu \frac{\partial^2 u}{\partial y^2} \, , \tag{I.7b}$$

$$\varrho u \frac{\partial v}{\partial x} + \varrho v \frac{\partial v}{\partial y} = -\frac{\partial p}{\partial y} + \mu \frac{\partial^2 v}{\partial x^2} + \mu \frac{\partial^2 v}{\partial y^2} \, , \tag{I.7c}$$

where u and v are the x- and y-velocity components respectively.

Choosing a reference velocity V_r and a reference length L_r, we set

$$X = \frac{x}{L_r} \, , \quad Y = \frac{y}{L_r} \, , \quad U = \frac{u}{V_r} \, , \quad V = \frac{v}{V_r} \, , \quad P = \frac{p}{\varrho V_r^2} \, ,$$

and we define the Reynolds number $\mathcal{R}$ by

$$\mathcal{R} = \frac{\varrho V_r L_r}{\mu} \, .$$

In dimensionless form, the Navier-Stokes equations are

$$\frac{\partial U}{\partial X} + \frac{\partial V}{\partial Y} = 0 \, , \tag{I.8a}$$

$$U \frac{\partial U}{\partial X} + V \frac{\partial U}{\partial Y} = -\frac{\partial p}{\partial X} + \frac{1}{\mathcal{R}} \frac{\partial^2 U}{\partial X^2} + \frac{1}{\mathcal{R}} \frac{\partial^2 U}{\partial Y^2} \, , \tag{I.8b}$$

$$U \frac{\partial V}{\partial X} + V \frac{\partial V}{\partial Y} = -\frac{\partial p}{\partial Y} + \frac{1}{\mathcal{R}} \frac{\partial^2 V}{\partial X^2} + \frac{1}{\mathcal{R}} \frac{\partial^2 V}{\partial Y^2} \, . \tag{I.8c}$$

II Elements of Two-Dimensional Linearized Aerodynamics

The problems of linearized aerodynamics, or thin airfoil theory, are very close to those encountered in the solution of the upper deck of the triple deck theory. Thus, it is useful to review a few results.

We consider an inviscid, two-dimensional, incompressible, irrotational flow.

An airfoil produces a *small disturbance* in a uniform flow whose velocity is V_∞ [61, 71].

The cartesian, orthonormal axis system is chosen in such a way that the x-axis is parallel to the freestream velocity. The velocity components are

$$U = ||V_\infty|| + u \,,$$
$$V = v \,,$$

where u et v are the disturbances produced by the airfoil.

The velocity potential φ, defined by

$$u = \frac{\partial \varphi}{\partial x} \,, \tag{II.1a}$$

$$v = \frac{\partial \varphi}{\partial y} \,, \tag{II.1b}$$

satisfies Laplace's equation

$$\frac{\partial^2 \varphi}{\partial x^2} + \frac{\partial^2 \varphi}{\partial y^2} = 0 \,.$$

The pressure coefficient is given by

$$C_p = \frac{P - P_\infty}{\frac{1}{2}\varrho V_\infty^2} = -2\frac{u}{V_\infty} \,.$$

The linearity of the problem enables us to decompose the airfoil in a thick symmetrical airfoil at zero angle of attack and in a zero-thickness airfoil. The two corresponding problems are discussed in the next sections.

The airfoil is defined in the domain $-l/2 \le x \le l/2$. The perturbation potential is

$$\varphi = \varphi_{\mathrm{nl}} + \varphi_{\mathrm{l}} \,,$$

where index "nl" refers to the symmetrical thick airfoil (non lifting case) and index "l" refers to the zero-thickness airfoil (lifting case).

The boundary conditions are linearized so that they are defined on the segment (S) $[-l/2 \leq x \leq l/2, y = 0]$. Segment (S) is the support of singularity elements which enable us to satisfy the boundary conditions and which give the solution.

II.1 Thickness Problem (Non Lifting Case)

The distribution of singularity elements must be such that the velocity component $v = \dfrac{\partial \varphi}{\partial y}$ is discontinuous through segment (S) since

$$\frac{\partial \varphi_{\mathrm{nl}}}{\partial y}(x, 0^{\pm}) = \pm V_{\infty} \delta_{\mathrm{nl}}(x) \qquad \text{for} \qquad -\frac{l}{2} \leq x \leq \frac{l}{2} \,,$$

where $y = 0^{+}$ denotes the airfoil upper surface and $y = 0^{-}$ the lower surface, δ_{nl} denotes the slope of the airfoil upper surface.

This problem is modelled by means of *sources* distributed on segment (S). The sources strength per unit length is $\sigma'(x) = 2V_{\infty} \delta_{\mathrm{nl}}(x)$.

At any point $(x_a, 0^{\pm})$ of the airfoil, the velocity field is given by

$$\frac{u_{\mathrm{nl}}^{\pm}(x_a)}{V_{\infty}} = \frac{1}{\pi} \oint_{-l/2}^{l/2} \frac{\delta_{\mathrm{nl}}(x_0)}{x_a - x_0} \, \mathrm{d}x_0 \,, \tag{II.2a}$$

$$\frac{v_{\mathrm{nl}}^{\pm}(x_a)}{V_{\infty}} = \pm \delta_{\mathrm{nl}}(x_a) \,. \tag{II.2b}$$

At any point outside the airfoil surface the velocity field is given by

$$\frac{u_{\mathrm{nl}}(x, y)}{V_{\infty}} = \frac{1}{\pi} \int_{-l/2}^{l/2} \frac{\delta_{\mathrm{nl}}(x_0)(x - x_0)}{(x - x_0)^2 + y^2} \, \mathrm{d}x_0 \,, \tag{II.3a}$$

$$\frac{v_{\mathrm{nl}}(x, y)}{V_{\infty}} = \frac{y}{\pi} \int_{-l/2}^{l/2} \frac{\delta_{\mathrm{nl}}(x_0)}{(x - x_0)^2 + y^2} \, \mathrm{d}x_0 \,. \tag{II.3b}$$

Obviously, the velocity component v is zero on the axis $y = 0$ outside segment (S).

The distribution of pressure coefficient on a thick symmetrical airfoil is symmetrical

$$C_{p_{\mathrm{nl}}}^{+}(x_a) = C_{p_{\mathrm{nl}}}^{-}(x_a) = -\frac{2}{\pi} \oint_{-l/2}^{l/2} \frac{\delta_{\mathrm{nl}}(x_0)}{x_a - x_0} \, \mathrm{d}x_0 \,.$$

II.2 Zero-Thickness Problem (Lifting Case)

The distribution of singularity elements must be such that the velocity component $v = \dfrac{\partial \varphi}{\partial y}$ is continuous through segment (S) since

$$\frac{\partial \varphi_1}{\partial y}(x, 0^\pm) = V_\infty \delta_1(x) \qquad \text{for} \qquad -\frac{l}{2} \le x \le \frac{l}{2} \,.$$

This problem is modelled by means of *vortices* distributed on segment (S). Their strength $\gamma'(x)$ is such that

$$V_\infty \delta_1(x) = \frac{1}{2\pi} \fint_{-l/2}^{l/2} \frac{\gamma'(x_0)}{x - x_0} \, \mathrm{d}x_0 \,.$$

At any point $(x_a, 0^\pm)$ of the airfoil, the velocity field is given by

$$\frac{u_1^\pm(x_a)}{V_\infty} = \mp \frac{\gamma'(x_a)}{2V_\infty} \,, \tag{II.4a}$$

$$\frac{v_1^\pm(x_a)}{V_\infty} = \frac{1}{2\pi} \fint_{-l/2}^{l/2} \frac{\gamma'(x_0)/V_\infty}{x_a - x_0} \, \mathrm{d}x_0 \,. \tag{II.4b}$$

At any point outside the airfoil, the velocity field is given by

$$\frac{u_1(x, y)}{V_\infty} = -\frac{y}{2\pi} \int_{-l/2}^{l/2} \frac{\gamma'(x_0)/V_\infty}{(x - x_0)^2 + y^2} \, \mathrm{d}x_0 \,, \tag{II.5a}$$

$$\frac{v_1(x, y)}{V_\infty} = \frac{1}{2\pi} \int_{-l/2}^{l/2} \frac{(x - x_0)\gamma'(x_0)/V_\infty}{(x - x_0)^2 + y^2} \, \mathrm{d}x_0 \,. \tag{II.5b}$$

We note that the velocity component u vanishes on the axis $y = 0$ outside segment (S).

The distribution of pressure coefficient on a zero-thickness airfoil is skew-symmetrical

$$C_{p_1}^+(x_a) = -C_{p_1}^-(x_a) = \frac{\gamma'(x_a)}{V_\infty} \,.$$

The pressure coefficient on the airfoil is related to the airfoil slope by

$$\delta_1(x_a) = \frac{1}{2\pi} \fint_{-l/2}^{l/2} \frac{C_{p_1}^+(x_0)}{x_a - x_0} \, \mathrm{d}x_0 \,.$$

Note II.1. The non lifting case can be solved easily if the airfoil shape is known, i.e. if the velocity distribution $v_{\mathrm{nl}}(x_a)$ is known, since the source distribution $\sigma'(x_a)$ is deduced directly. This problem is called direct thickness problem (non lifting).

Along the line $y = 0$, outside the airfoil, the value of v is zero. In the whole field, the pressure is such that

$$C_p = -2\frac{u}{V_\infty} \,.$$

Now, we have

$$u = \frac{\partial \varphi}{\partial x} \, ,$$

whence

$$\int_{-\infty}^{+\infty} u(\xi, y) \, \mathrm{d}\xi = [\varphi(x, y)]_{x \to -\infty}^{x \to +\infty} = 0 \, ,$$

since $\varphi(x, y)$ vanishes as $x \to \pm\infty$. Therefore, we have

$$\int_{-\infty}^{+\infty} C_p(\xi, y) \, \mathrm{d}\xi = 0 \, .$$

In particular, we have

$$\int_{-\infty}^{+\infty} C_p(\xi, 0) \, \mathrm{d}\xi = 0 \, .$$

Note II.2. The zero-thickness problem can be solved easily if the distribution of pressure $C_p(x_a)$ on the airfoil is known since the vortex distribution $\gamma'(x_a)$ is deduced directly. This problem is called inverse lifting problem. Along the line $y = 0$, the pressure coefficient is zero outside the airfoil. As the pressure distribution on the airfoil is arbitrary, in general we have

$$\int_{-l/2}^{l/2} C_p(\xi, 0) \, d\xi = \int_{-\infty}^{+\infty} C_p(\xi, 0) \, d\xi \neq 0 \, .$$

Note II.3. We could think of obtaining the solution of the inverse thickness problem by inverting (II.2a) (see Appendix III). In this way, the source distribution could be calculated as function of a given velocity distribution u (or a given pressure distribution) along the line $y = 0$. However, it must be noted that the inverse formula involves the velocity distribution u all along the line $y = 0$ and not only on segment (S) since the value of u does not vanish outside segment (S). Moreover, the distribution of u along the line $y = 0$ cannot be arbitrary since its integral with respect to x must be zero. In practice, it is necessary to use (II.2a) to calculate the source distribution from a given velocity (or pressure) distribution on segment (S) [61]. A solution can be sought by expanding the complex velocity in Laurent series whose coefficients are unknown, the form of the expansion being guided by results obtained from the exact theory.

To solve the direct lifting problem, we can think of inverting (II.4b) (see Appendix III) to calculate the vortex distribution from a given shape of the airfoil. However, the inverse formula involves the distribution of v all along the line $y = 0$ and not only on segment (S) since the value of v does not vanish outside segment (S) although the vortex distribution is zero outside segment (S). We also note that, if the vortex distribution vanishes outside segment (S), the velocity component v along the line $y = 0$ behaves, in general, like $1/x$ as $x \to \pm\infty$, as far as $\int_{-l/2}^{l/2} \gamma'(x) \, \mathrm{d}x \neq 0$. In practice, it is necessary to use (II.4b) to calculate the vortex distribution from a given shape of the airfoil [61]. As for the inverse non lifting problem, a solution can be sought by expanding the complex velocity in Laurent series whose coefficients are unknown.

III Solutions of the Upper Deck of the Triple Deck Theory

III.1 Two-Dimensional Flow

We consider the incompressible flow defined in the upper deck of the triple deck theory. The disturbances of velocity components u and v, the pressure disturbance p, the coordinates x and y are dimensionless. The equations are

$$\frac{\partial u}{\partial x} + \frac{\partial v}{\partial y} = 0 \, , \tag{III.1a}$$

$$\frac{\partial u}{\partial x} = -\frac{\partial p}{\partial x} \, , \tag{III.1b}$$

$$\frac{\partial v}{\partial x} = -\frac{\partial p}{\partial y} \, . \tag{III.1c}$$

These equations are identical to the dimensionless equations of linearized aerodynamics (Appendix II). Here, the equations are to solve in the half-plane $y \geq 0$. Along the line $y = 0$, either a distribution $v(x,0)$, or a distribution $u(x,0)$ can be prescribed.

At infinity, we assume that the disturbances vanish

$$u \to 0, \quad v \to 0, \quad p \to 0 \quad \text{as} \quad x \to \pm\infty \quad \text{or} \quad y \to \infty \, .$$

The flow is irrotational since we are concerned with the perturbation of a uniform inviscid flow. The x-momentum equation shows that $p + u = F(y)$. Now, as $x \to -\infty$, we have $p = 0$ and $u = 0$. It is deduced that $F(y) = 0$ and that $p + u = 0$. Then, we obtain

$$\frac{\partial u}{\partial y} = -\frac{\partial p}{\partial y} \, .$$

Then, the y-momentum equation gives

$$\frac{\partial v}{\partial x} = \frac{\partial u}{\partial y} \, ,$$

which shows directly that the perturbation of the flow is irrotational.

Below, the results are obtained by a Fourier transform method. The results of Appendix II could also be applied.

The Fourier transform $\widehat{F}(\alpha, y)$ of a function $f(x, y)$ and the inversion integral are defined by the following formulas

$$\widehat{F}(\alpha, y) = \int_{-\infty}^{+\infty} f(x, y) \, e^{-2\mathrm{i}\pi x \alpha} \, \mathrm{d}x \ ,$$

$$f(x, y) = \int_{-\infty}^{+\infty} \widehat{F}(\alpha, y) \, e^{2\mathrm{i}\pi x \alpha} \, \mathrm{d}\alpha \ .$$

The Fourier transform of $\dfrac{\partial f}{\partial x}$ is $2\mathrm{i}\pi\alpha\widehat{F}$

$$\widehat{\frac{\partial f}{\partial x}} = 2\mathrm{i}\pi\alpha\widehat{F} \ .$$

Obviously, we assume that the so-defined functions exist: the Fourier transform of f, its inverse and the Fourier transform of the derivative $\dfrac{\partial f}{\partial x}$. In particular, we must have $f \to 0$ as $|x| \to \infty$. It will be required to assume that the Fourier transforms of f and of $\dfrac{\partial f}{\partial x}$ ($f = u$ ou $f = v$) exist. These conditions are fulfilled if (sufficient condition) f is continuous and if the integrals of $|f|$ and of $\left| \dfrac{\partial f}{\partial x} \right|$ exist.

From the equations written in physical space, we obtain the following equations in Fourier space

$$2\mathrm{i}\pi\alpha\widehat{u} + \frac{\partial\widehat{v}}{\partial y} = 0 \ ,$$

$$\widehat{u} = -\widehat{p} \ ,$$

$$2\mathrm{i}\pi\alpha\widehat{v} = -\frac{\partial\widehat{p}}{\partial y} \ .$$

The equation for $\widehat{v}$ is deduced

$$-4\pi^2\alpha^2\widehat{v} + \frac{\partial^2\widehat{v}}{\partial y^2} = 0 \ .$$

The solution is

$$\widehat{v} = K_1 \, e^{2\pi\alpha y} + K_2 \, e^{-2\pi\alpha y} \ .$$

Let $\widehat{v}_0$ be the Fourier transform of the velocity component v at $y = 0$

$$\widehat{v}_0(\alpha) = \widehat{v}(\alpha, 0) \ .$$

In order that the velocity component v vanish as $y \to \infty$, the solution writes

$$\alpha \leq 0 \quad : \quad \widehat{v} = \widehat{v}_0 \, e^{2\pi\alpha y} \ ,$$

$$\alpha \geq 0 \quad : \quad \widehat{v} = \widehat{v}_0 \, e^{-2\pi\alpha y} \ ,$$

or

$$\widehat{v} = \widehat{v}_0 \, e^{-2\pi|\alpha|y} \ ,$$

and we obtain

$$\widehat{u} = -\mathrm{i}\,\mathrm{sgn}(\alpha)\widehat{v}_0\,\mathrm{e}^{-2\pi|\alpha|y} \ .$$

The solution can also be expressed as function of the Fourier transform $\widehat{u}_0$ of the velocity component u at $y = 0$

$$\widehat{u} = \widehat{u}_0\,\mathrm{e}^{-2\pi|\alpha|y} \ ,$$
$$\widehat{v} = \mathrm{i}\,\mathrm{sgn}(\alpha)\widehat{u}_0\,\mathrm{e}^{-2\pi|\alpha|y} \ .$$

To return to physical space, it is required to know the following formulas

$$-\mathrm{i}\pi\,\mathrm{sgn}(\alpha)\,\mathrm{e}^{-2\pi|\alpha|y} = \int_{-\infty}^{+\infty} \frac{x}{x^2 + y^2}\,\mathrm{e}^{-2\mathrm{i}\pi\alpha x}\,\mathrm{d}x \ ,$$

$$\pi\,\mathrm{e}^{-2\pi|\alpha|y} = \int_{-\infty}^{+\infty} \frac{y}{x^2 + y^2}\,\mathrm{e}^{-2\mathrm{i}\pi\alpha x}\,\mathrm{d}x \ .$$

For $y \neq 0$, we obtain

$$u(x,y) = \frac{1}{\pi} \int_{-\infty}^{+\infty} \frac{v(\xi,0)(x-\xi)}{(x-\xi)^2 + y^2}\,\mathrm{d}\xi \ , \tag{III.2a}$$

$$v(x,y) = \frac{1}{\pi} \int_{-\infty}^{+\infty} \frac{v(\xi,0)y}{(x-\xi)^2 + y^2}\,\mathrm{d}\xi \ , \tag{III.2b}$$

$$u(x,y) = \frac{1}{\pi} \int_{-\infty}^{+\infty} \frac{u(\xi,0)y}{(x-\xi)^2 + y^2}\,\mathrm{d}\xi \ , \tag{III.2c}$$

$$v(x,y) = -\frac{1}{\pi} \int_{-\infty}^{+\infty} \frac{u(\xi,0)(x-\xi)}{(x-\xi)^2 + y^2}\,\mathrm{d}\xi \ . \tag{III.2d}$$

In the above formulas, the velocity component u can be replaced by $-p$. Along the line $y = 0$, the following results are obtained

$$u(x,0) = -p(x,0) = \frac{1}{\pi}\!\!\fint_{-\infty}^{+\infty} \frac{v(\xi,0)}{x-\xi}\,\mathrm{d}\xi \ , \tag{III.3a}$$

$$v(x,0) = -\frac{1}{\pi}\!\!\fint_{-\infty}^{+\infty} \frac{u(\xi,0)}{x-\xi}\,\mathrm{d}\xi = \frac{1}{\pi}\!\!\fint_{-\infty}^{+\infty} \frac{p(\xi,0)}{x-\xi}\,\mathrm{d}\xi \ . \tag{III.3b}$$

All these results show that if the distribution of $u(\xi,0)$ (or $v(\xi,0)$) is known, the fields of u and v can be calculated. The data cannot be arbitrary since, at least, all the integrals must be defined.

Note III.1. Given a velocity distribution $v(\xi,0) \neq 0$ in a bounded interval, the problem is equivalent to the thick symmetrical airfoil problem (non lifting case) presented in Appendix II. Then, we have

$$\int_{-\infty}^{+\infty} u\,\mathrm{d}x = 0 \ , \quad \int_{-\infty}^{+\infty} p\,\mathrm{d}x = 0 \ .$$

This result is also obtained by integrating directly (III.2a).

Given a velocity distribution $u(\xi, 0) \neq 0$ (or $p(\xi, 0) \neq 0$) in a bounded interval, the problem is equivalent to the zero-thickness airfoil problem (lifting case) presented in Appendix II. *In general*, we then have

$$\int_{-\infty}^{+\infty} u \, \mathrm{d}x \neq 0 \ , \quad \int_{-\infty}^{+\infty} p \, \mathrm{d}x \neq 0 \ .$$

Note III.2. The solution to the problem could be sought in the complex plane by introducing the complex velocity $g = u - iv$. Indeed, it has been shown that u and $-v$ satisfy the Cauchy-Riemann conditions. In addition, in the half-plane $y \geq 0$, function g cannot have singularity. The complex velocity g is a holomorphic function of $z = x + iy$ for $y \geq 0$. Applying Kramers-Kronig's relations [4],

$$\Re g(z) = -\frac{1}{\pi} \oint_{-\infty}^{+\infty} \frac{\Im g(z)}{x - \xi} \, \mathrm{d}\xi \ , \quad \Im g(z) = \frac{1}{\pi} \oint_{-\infty}^{+\infty} \frac{\Re g(z)}{x - \xi} \, \mathrm{d}\xi \ ,$$

we recover exactly (III.3a–III.3b) relating u and v at $y = 0$. This result is valid if we assume that g tends towards zero at infinity in the half-plane $y \geq 0$.

III.2 Three-Dimensional Flow

For the velocity disturbances u, v and w, and the pressure disturbance p, the upper deck equations write

$$\frac{\partial u}{\partial x} + \frac{\partial v}{\partial y} + \frac{\partial w}{\partial z} = 0 \ , \tag{III.4a}$$

$$\frac{\partial u}{\partial x} = -\frac{\partial p}{\partial x} \ , \tag{III.4b}$$

$$\frac{\partial v}{\partial x} = -\frac{\partial p}{\partial y} \ , \tag{III.4c}$$

$$\frac{\partial w}{\partial x} = -\frac{\partial p}{\partial z} \ . \tag{III.4d}$$

These equations are to be solved in the half-space $y \geq 0$. Along the surface $y = 0$, a velocity distribution $v(x, z, 0)$ is prescribed. Regarding the boundary conditions, two cases are studied below

1. The velocity disturbances u, v, w and the pressure disturbance p vanish at infinity ($x \to \pm\infty$ or $y \to \infty$).
2. The conditions are identical to the previous case except for the velocity disturbances v and w: at downstream infinity, these velocity disturbances do not vanish but we assume only that the derivatives $\dfrac{\partial v}{\partial x}$ and $\dfrac{\partial w}{\partial x}$ vanish.

III.2.1 Zero Perturbations at Infinity

We assume that the Fourier transforms with respect to x and z of the perturbations of velocity and pressure exist. The Fourier transform $\widehat{f}(\alpha, \gamma, y)$ of a function $f(x, z, y)$ and the inversion integral are

$$\widehat{f}(\alpha, \gamma, y) = \int_{-\infty}^{+\infty} \int_{-\infty}^{+\infty} f(x, z, y)\, e^{-2i\pi(\alpha x + \gamma z)}\ \mathrm{d}x\, \mathrm{d}z \ ,$$

$$f(x, z, y) = \int_{-\infty}^{+\infty} \int_{-\infty}^{+\infty} \widehat{f}(\alpha, \gamma, y)\, e^{2i\pi(\alpha x + \gamma z)}\ \mathrm{d}\alpha\, \mathrm{d}\gamma \ .$$

The Fourier transforms of derivatives with respect to x and z, if they exist, are given by

$$\widehat{\frac{\partial f}{\partial x}} = 2i\pi\alpha\widehat{f} \ , \ \widehat{\frac{\partial f}{\partial z}} = 2i\pi\gamma\widehat{f} \ .$$

We take the Fourier transform of (III.4a), (III.4b), (III.4c) and (III.4d)

$$2i\pi\alpha\widehat{u} + \frac{\partial\widehat{v}}{\partial y} + 2i\pi\gamma\widehat{w} = 0 \ ,$$

$$2i\pi\alpha\widehat{u} = -2i\pi\alpha\widehat{p} \ ,$$

$$2i\pi\alpha\widehat{v} = -\frac{\partial\widehat{p}}{\partial y} \ ,$$

$$2i\pi\alpha\widehat{w} = -2i\pi\gamma\widehat{p} \ .$$

The equation for $\widehat{v}$ results

$$\frac{\partial^2\widehat{v}}{\partial y^2} - 4\pi^2(\alpha^2 + \gamma^2)\widehat{v} = 0 \ .$$

With the conditions of vanishing perturbations at infinity, we obtain

$$\widehat{v} = \widehat{v}_0\, e^{-2\pi R y} \quad \text{with} \quad R = \sqrt{\alpha^2 + \gamma^2} \ ,$$

where $\widehat{v}_0$ represents the Fourier transform of v at $y = 0$.

Then, we obtain

$$\widehat{u} = -i\frac{\alpha}{R}\widehat{v}_0\, e^{-2\pi R y} \ ,$$

$$\widehat{w} = -i\frac{\gamma}{R}\widehat{v}_0\, e^{-2\pi R y} \ ,$$

$$\widehat{p} = i\frac{\alpha}{R}\widehat{v}_0\, e^{-2\pi R y} \ .$$

To express this solution in physical space, the following transforms are used

$$-\frac{i\alpha}{\sqrt{\alpha^2 + \gamma^2}}\, e^{-2\pi R y} = \frac{1}{2\pi} \int_{-\infty}^{\infty} \int_{-\infty}^{\infty} \frac{x}{(x^2 + z^2 + y^2)^{3/2}}\, e^{-2i\pi(\alpha x + \gamma z)}\ \mathrm{d}x\, \mathrm{d}z \ ,$$

$$\mathrm{e}^{-2\pi Ry} = \frac{1}{2\pi} \int_{-\infty}^{\infty} \int_{-\infty}^{\infty} \frac{y}{(x^2 + z^2 + y^2)^{3/2}} \, \mathrm{e}^{-2\mathrm{i}\pi(\alpha x + \gamma z)} \, \mathrm{d}x \, \mathrm{d}z \,,$$

$$-\frac{\mathrm{i}\gamma}{\sqrt{\alpha^2 + \gamma^2}} \, \mathrm{e}^{-2\pi Ry} = \frac{1}{2\pi} \int_{-\infty}^{\infty} \int_{-\infty}^{\infty} \frac{z}{(x^2 + z^2 + y^2)^{3/2}} \, \mathrm{e}^{-2\mathrm{i}\pi(\alpha x + \gamma z)} \, \mathrm{d}x \, \mathrm{d}z \,,$$

and we obtain, for $y \neq 0$

$$u = \frac{1}{2\pi} \int_{-\infty}^{\infty} \int_{-\infty}^{\infty} \frac{[v(\xi, \eta, 0)]\,(x - \xi)}{\left((x - \xi)^2 + (z - \eta)^2 + y^2\right)^{3/2}} \, \mathrm{d}\xi \, \mathrm{d}\eta \,, \qquad \text{(III.5a)}$$

$$v = \frac{1}{2\pi} \int_{-\infty}^{\infty} \int_{-\infty}^{\infty} \frac{[v(\xi, \eta, 0)]\,y}{\left((x - \xi)^2 + (z - \eta)^2 + y^2\right)^{3/2}} \, \mathrm{d}\xi \, \mathrm{d}\eta \,, \qquad \text{(III.5b)}$$

$$w = \frac{1}{2\pi} \int_{-\infty}^{\infty} \int_{-\infty}^{\infty} \frac{[v(\xi, \eta, 0)]\,(z - \eta)}{\left((x - \xi)^2 + (z - \eta)^2 + y^2\right)^{3/2}} \, \mathrm{d}\xi \, \mathrm{d}\eta \,, \qquad \text{(III.5c)}$$

$$p = -\frac{1}{2\pi} \int_{-\infty}^{\infty} \int_{-\infty}^{\infty} \frac{[v(\xi, \eta, 0)]\,(x - \xi)}{\left((x - \xi)^2 + (z - \eta)^2 + y^2\right)^{3/2}} \, \mathrm{d}\xi \, \mathrm{d}\eta \,. \qquad \text{(III.5d)}$$

Along the line $y = 0$, we have

$$p(x, z, 0) = -\frac{1}{2\pi} \fint_{-\infty}^{\infty} \fint_{-\infty}^{\infty} \frac{[v(\xi, \eta, 0)]\,(x - \xi)}{\left((x - \xi)^2 + (z - \eta)^2\right)^{3/2}} \, \mathrm{d}\xi \, \mathrm{d}\eta \,. \qquad \text{(III.6)}$$

Note III.3. The problem studied in this section is equivalent to the problem of small disturbances induced in a uniform flow by a three-dimensional finite wing symmetrical with respect to $y = 0$ for which the lift is zero. The velocity potential φ is such that

$$u = \frac{\partial \varphi}{\partial x} \,,$$

and we have

$$\int_{-\infty}^{+\infty} u \, \mathrm{d}x = [\varphi]_{-\infty}^{+\infty} = 0 \,,$$

since the potential vanishes at infinity (upstream and downstream).

We also have

$$\int_{-\infty}^{+\infty} p \, \mathrm{d}x = 0 \,.$$

These results can also be obtained directly by integrating (III.5a) and (III.5d) with respect to x.

III.2.2 Non Zero Cross-Flow Perturbations at Downstream Infinity

At infinity ($x \to \pm\infty$ or $y \to \infty$), we assume that the perturbations of velocity u and pressure p vanish. At downstream infinity, we assume that only $\dfrac{\partial v}{\partial x}$

and $\dfrac{\partial w}{\partial x}$ vanish whereas at upstream infinity and also as $y \to \infty$ we assume that the velocity perturbations v and w vanish.

We assume that the following Fourier transforms exist

$$\widehat{u}(\alpha, \gamma, y) = \int_{-\infty}^{+\infty} \int_{-\infty}^{+\infty} u(x, z, y) \, \mathrm{e}^{-2\mathrm{i}\pi(\alpha x + \gamma z)} \, \mathrm{d}x \, \mathrm{d}z \ ,$$

$$\widehat{p}(\alpha, \gamma, y) = \int_{-\infty}^{+\infty} \int_{-\infty}^{+\infty} p(x, z, y) \, \mathrm{e}^{-2\mathrm{i}\pi(\alpha x + \gamma z)} \, \mathrm{d}x \, \mathrm{d}z \ ,$$

$$\widehat{dv}(\alpha, \gamma, y) = \int_{-\infty}^{+\infty} \int_{-\infty}^{+\infty} \frac{\partial v}{\partial x} \, \mathrm{e}^{-2\mathrm{i}\pi(\alpha x + \gamma z)} \, \mathrm{d}x \, \mathrm{d}z \ ,$$

$$\widehat{dw}(\alpha, \gamma, y) = \int_{-\infty}^{+\infty} \int_{-\infty}^{+\infty} \frac{\partial w}{\partial x} \, \mathrm{e}^{-2\mathrm{i}\pi(\alpha x + \gamma z)} \, \mathrm{d}x \, \mathrm{d}z \ .$$

The continuity equation is differentiated with respect to x

$$\frac{\partial^2 u}{\partial x^2} + \frac{\partial}{\partial y}\left(\frac{\partial v}{\partial x}\right) + \frac{\partial}{\partial z}\left(\frac{\partial w}{\partial x}\right) = 0 \ .$$

By taking the Fourier transform of this equation and of (III.4b), (III.4c) and (III.4d), we obtain

$$-4\pi^2 \alpha^2 \widehat{u} + \frac{\partial \widehat{dv}}{\partial y} + 2\mathrm{i}\pi\gamma \widehat{dw} = 0 \ ,$$

$$2\mathrm{i}\pi\alpha\widehat{u} = -2\mathrm{i}\pi\alpha\widehat{p} \ ,$$

$$\widehat{dv} = -\frac{\partial \widehat{p}}{\partial y} \ ,$$

$$\widehat{dw} = -2\mathrm{i}\pi\gamma\widehat{p} \ ,$$

and

$$\frac{\partial^2 \widehat{dv}}{\partial y^2} - 4\pi^2(\alpha^2 + \gamma^2)\widehat{dv} = 0 \ .$$

We set

$$R = \sqrt{\alpha^2 + \gamma^2} \ .$$

With condition $\dfrac{\partial v}{\partial x} \to 0$ as $y \to \infty$, the solution is

$$\widehat{dv} = \widehat{dv_0} \, \mathrm{e}^{-2\pi R y} \ ,$$

where $\widehat{dv_0}$ represents the Fourier transform of the derivative of v with respect to x at $y = 0$. Now, we have

$$\widehat{dv} = -\frac{\partial \widehat{p}}{\partial y} = \widehat{dv_0} \, \mathrm{e}^{-2\pi R y} \ .$$

With the condition of vanishing pressure as $y \to \infty$ and integrating with respect to y, we obtain

$$\widehat{p} = \frac{1}{2\pi R}\widehat{dv_0}\, e^{-2\pi Ry} \ .$$

We also have

$$\widehat{u} = -\frac{1}{2\pi R}\widehat{dv_0}\, e^{-2\pi Ry} \ ,$$

$$\widehat{dw} = -\mathrm{i}\frac{\gamma}{R}\widehat{dv_0}\, e^{-2\pi Ry} \ .$$

Using the following transforms

$$\frac{1}{\sqrt{\alpha^2 + \gamma^2}}\, e^{-2\pi Ry} = \int_{-\infty}^{\infty}\int_{-\infty}^{\infty} \frac{1}{(x^2 + z^2 + y^2)^{1/2}}\, e^{-2\mathrm{i}\pi(\alpha x + \gamma z)}\, \mathrm{d}x\, \mathrm{d}z \ ,$$

$$e^{-2\pi Ry} = \frac{1}{2\pi}\int_{-\infty}^{\infty}\int_{-\infty}^{\infty} \frac{y}{(x^2 + z^2 + y^2)^{3/2}}\, e^{-2\mathrm{i}\pi(\alpha x + \gamma z)}\, \mathrm{d}x\, \mathrm{d}z \ ,$$

$$-\frac{\mathrm{i}\gamma}{\sqrt{\alpha^2 + \gamma^2}}\, e^{-2\pi Ry} = \frac{1}{2\pi}\int_{-\infty}^{\infty}\int_{-\infty}^{\infty} \frac{z}{(x^2 + z^2 + y^2)^{3/2}}\, e^{-2\mathrm{i}\pi(\alpha x + \gamma z)}\, \mathrm{d}x\, \mathrm{d}z \ ,$$

we obtain the solution in physical space for $y \neq 0$

$$u = -\frac{1}{2\pi}\int_{-\infty}^{\infty}\int_{-\infty}^{\infty} \frac{\dfrac{\partial v}{\partial \xi}(\xi, \eta, 0)}{\sqrt{(x - \xi)^2 + (z - \eta)^2 + y^2}}\, \mathrm{d}\xi\, \mathrm{d}\eta \ , \qquad \text{(III.7a)}$$

$$\frac{\partial v}{\partial x} = \frac{1}{2\pi}\int_{-\infty}^{\infty}\int_{-\infty}^{\infty} \frac{\left[\dfrac{\partial v}{\partial \xi}(\xi, \eta, 0)\right] y}{((x - \xi)^2 + (z - \eta)^2 + y^2)^{3/2}}\, \mathrm{d}\xi\, \mathrm{d}\eta \ , \qquad \text{(III.7b)}$$

$$\frac{\partial w}{\partial x} = \frac{1}{2\pi}\int_{-\infty}^{\infty}\int_{-\infty}^{\infty} \frac{\left[\dfrac{\partial v}{\partial \xi}(\xi, \eta, 0)\right](z - \eta)}{((x - \xi)^2 + (z - \eta)^2 + y^2)^{3/2}}\, \mathrm{d}\xi\, \mathrm{d}\eta \ , \qquad \text{(III.7c)}$$

$$p = \frac{1}{2\pi}\int_{-\infty}^{\infty}\int_{-\infty}^{\infty} \frac{\dfrac{\partial v}{\partial \xi}(\xi, \eta, 0)}{\sqrt{(x - \xi)^2 + (z - \eta)^2 + y^2}}\, \mathrm{d}\xi\, \mathrm{d}\eta \ . \qquad \text{(III.7d)}$$

Along the line $y = 0$, we have

$$p(x, z, 0) = \frac{1}{2\pi}\fint_{-\infty}^{\infty}\fint_{-\infty}^{\infty} \frac{\dfrac{\partial v}{\partial \xi}(\xi, \eta, 0)}{\sqrt{(x - \xi)^2 + (z - \eta)^2}}\, \mathrm{d}\xi\, \mathrm{d}\eta \ . \qquad \text{(III.8)}$$

Note that the velocity perturbations v and w do not necessarily vanish at downstream infinity.

We also note that this solution reduces to the solution developed in Subsect. III.2.1 when the velocity perturbations v and w vanish at downstream infinity.

Note III.4. The problem studied in this section is equivalent to the problem of small disturbances produced in a uniform flow by a three-dimensional finite wing of zero thickness for which the lift is non zero. The velocity potential φ is such that

$$u = \frac{\partial \varphi}{\partial x} \ .$$

At upstream infinity, the potential φ vanishes but at downstream infinity, due to the velocity components v and w induced by the vortex sheet leaving the wing, the potential φ does not vanish. Then, we have

$$\int_{-\infty}^{+\infty} u \ \mathrm{d}x \neq 0 \ ,$$

and

$$\int_{-\infty}^{+\infty} p \ \mathrm{d}x \neq 0 \ .$$

IV Second Order Triple Deck Theory

IV.1 Main Results

We consider a steady, two-dimensional, incompressible, laminar flow on
a semi-infinite flat plate.

At distance L from the plate leading edge, the boundary layer is perturbed, for example, by a small hump at the wall. The hump can produce
boundary layer separation.

Under certain conditions specified later, the triple deck theory defines
a model able to avoid the singular behaviour of the solution of boundary
layer equations but simpler than the Navier-Stokes equations. It must be
noted that the model describes the *perturbations* of the basic flow.

The velocity components, the lengths and the pressure are nondimensionalized by means of reference quantities V, L and ϱV^2. The reference velocity
is the freestream velocity and the reference length is the length of boundary layer development from the plate leading edge up to the location of the
disturbance (Fig. IV.1).

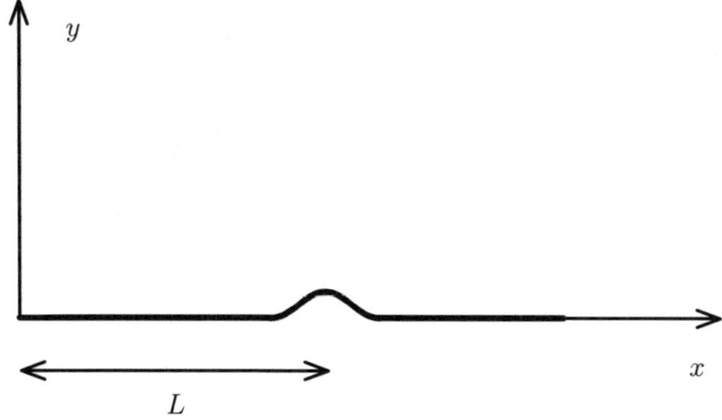

Fig. IV.1. Flow on a flat plate deformed by a hump

The flow is described by the Navier-Stokes model. In a cartesian, orthonormal axis system tied to the flat plate. The dimensionless equations

are

$$\frac{\partial u}{\partial x} + \frac{\partial v}{\partial y} = 0 \, , \tag{IV.1a}$$

$$u\frac{\partial u}{\partial x} + v\frac{\partial u}{\partial y} = -\frac{\partial p}{\partial x} + \varepsilon^2\frac{\partial^2 u}{\partial x^2} + \varepsilon^2\frac{\partial^2 u}{\partial y^2} \, , \tag{IV.1b}$$

$$u\frac{\partial v}{\partial x} + v\frac{\partial v}{\partial y} = -\frac{\partial p}{\partial y} + \varepsilon^2\frac{\partial^2 v}{\partial x^2} + \varepsilon^2\frac{\partial^2 v}{\partial y^2} \, , \tag{IV.1c}$$

where u and v are the x- and y-velocity components respectively; the y-axis is normal to the wall; p is the pressure.

The Reynolds number is defined by

$$\mathcal{R} = \frac{\varrho V L}{\mu} \, ,$$

and the small parameter ε is such that

$$\varepsilon^2 = \frac{1}{\mathcal{R}} \, .$$

Around the hump, the perturbed flow is structured in *three decks* as shown in Fig. IV.2: *a lower deck, a main deck and an upper deck.*

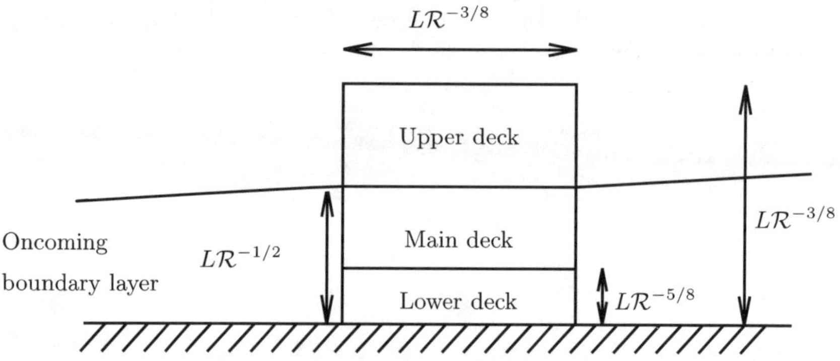

Fig. IV.2. Triple deck structure

The streamwise and transverse length scales of the perturbed region are of order $LR^{-3/8}$. Within the perturbed region, three decks are identified. The lower deck thickness is $LR^{-5/8}$; the viscous effects are significant in it. The main deck is the continuation of the oncoming boundary layer. The main deck thickness is $LR^{-1/2}$ and the visccous effects, for the perturbations, are negligible. In the upper deck, the viscous effects are negligible; its thickness is $LR^{-3/8}$.

The triple deck theory describes, for example, the flow around a hump whose thickness is of order $L\mathcal{R}^{-5/8}$ and whose length is of order $L\mathcal{R}^{-3/8}$. It is essential to understand that the theory is valid for a hump *whose dimensions vary with the Reynolds number* and tend towards zero as the Reynolds number tends towards infinity.

In each deck, the appropriate variables are

$$\text{Upper deck: } X = \varepsilon^{-3/4}(x - x_0) \, , \; Y^* = \varepsilon^{-3/4}y \, , \qquad \text{(IV.2a)}$$

$$\text{Main deck: } \; X = \varepsilon^{-3/4}(x - x_0) \, , \;\; Y = \varepsilon^{-1}y \, , \qquad \text{(IV.2b)}$$

$$\text{Lower deck: } X = \varepsilon^{-3/4}(x - x_0) \, , \;\; \widetilde{Y} = \varepsilon^{-5/4}y \, , \qquad \text{(IV.2c)}$$

where $x_0 = 1$ is the location of the disturbance.

The non perturbed flow is described by the solution of Blasius' equation

$$2f''' + ff'' = 0 \quad \text{with} \quad U_0(x, Y) = f'(\eta) \quad \text{and} \quad \eta = Y x^{-1/2} \, ,$$

and it is known that

$$f \cong \eta - \beta_0 + \text{EST} \quad \text{as} \quad Y \to \infty \, ,$$

or

$$V_0 \cong \frac{1}{2}\beta_0 x^{-1/2} + \text{EST} \quad \text{as} \quad Y \to \infty \, .$$

It is also known that

$$U_0 \cong \lambda Y + \mathrm{O}(Y^4) \quad \text{as} \quad Y \to 0 \quad \text{with} \quad \lambda = \lambda_0 x^{-1/2} \, .$$

The expansions appropriate to each deck are
- Upper deck

$$u = 1 + \varepsilon^{1/2}U_1^*(X, Y^*) + \varepsilon^{3/4}U_2^*(X, Y^*) + \cdots \, ,$$

$$v = \varepsilon^{1/2}V_1^*(X, Y^*) + \varepsilon^{3/4}V_2^*(X, Y^*) + \cdots \, ,$$

$$p = \varepsilon^{1/2}P_1^*(X, Y^*) + \varepsilon^{3/4}P_2^*(X, Y^*) + \cdots \, .$$

- Main deck

$$u = U_0(Y) + \varepsilon^{1/4}U_1(X, Y) + \varepsilon^{1/2}U_2(X, Y) + \cdots \, ,$$

$$v = \varepsilon^{1/2}V_1(X, Y) + \varepsilon^{3/4}V_2(X, Y) + \cdots \, ,$$

$$p = \varepsilon^{1/2}P_1(X, Y) + \varepsilon^{3/4}P_2(X, Y) + \cdots \, .$$

- Lower deck

$$u = \varepsilon^{1/4}\lambda_0\widetilde{Y} + \varepsilon^{1/4}\widetilde{U}_1(X, \widetilde{Y}) + \varepsilon^{1/2}\widetilde{U}_2(X, \widetilde{Y}) + \cdots \, ,$$

$$v = \varepsilon^{3/4}\widetilde{V}_1(X,\widetilde{Y}) + \varepsilon\widetilde{V}_2(X,\widetilde{Y}) + \cdots ,$$

$$p = \varepsilon^{1/2}\widetilde{P}_1(X,\widetilde{Y}) + \varepsilon^{3/4}\widetilde{P}_2(X,\widetilde{Y}) + \cdots .$$

The equations are

• Upper deck

- Order 1

$$\frac{\partial U_1^*}{\partial X} + \frac{\partial V_1^*}{\partial Y^*} = 0 ,$$

$$\frac{\partial U_1^*}{\partial X} = -\frac{\partial P_1^*}{\partial X} ,$$

$$\frac{\partial V_1^*}{\partial X} = -\frac{\partial P_1^*}{\partial Y^*} .$$

- Order 2

$$\frac{\partial U_2^*}{\partial X} + \frac{\partial V_2^*}{\partial Y^*} = 0 ,$$

$$\frac{\partial U_2^*}{\partial X} = -\frac{\partial P_2^*}{\partial X} ,$$

$$\frac{\partial V_2^*}{\partial X} = -\frac{\partial P_2^*}{\partial Y^*} .$$

• Main deck

- Order 1

$$\frac{\partial U_1}{\partial X} + \frac{\partial V_1}{\partial Y} = 0 ,$$

$$U_0\frac{\partial U_1}{\partial X} + V_1\frac{\partial U_0}{\partial Y} = 0 ,$$

$$\frac{\partial P_1}{\partial Y} = 0 .$$

- Ordre 2

$$\frac{\partial U_2}{\partial X} + \frac{\partial V_2}{\partial Y} = 0 ,$$

$$U_0\frac{\partial U_2}{\partial X} + V_2\frac{\partial U_0}{\partial Y} = -\frac{\partial P_1}{\partial X} - U_1\frac{\partial U_1}{\partial X} - V_1\frac{\partial U_1}{\partial Y} ,$$

$$U_0\frac{\partial V_1}{\partial X} = -\frac{\partial P_2}{\partial Y} .$$

• Lower deck

- Order 1

$$\frac{\partial \widetilde{U}_1}{\partial X} + \frac{\partial \widetilde{V}_1}{\partial \widetilde{Y}} = 0 ,$$

$$\left(\lambda_0\widetilde{Y}+\widetilde{U}_1\right)\frac{\partial\widetilde{U}_1}{\partial X}+\widetilde{V}_1\left(\lambda_0+\frac{\partial\widetilde{U}_1}{\partial\widetilde{Y}}\right)=-\frac{\partial\widetilde{P}_1}{\partial X}+\frac{\partial^2\widetilde{U}_1}{\partial\widetilde{Y}^2}\,,$$

$$\frac{\partial\widetilde{P}_1}{\partial\widetilde{Y}}=0\,.$$

- Order 2

$$\frac{\partial\widetilde{U}_2}{\partial X}+\frac{\partial\widetilde{V}_2}{\partial\widetilde{Y}}=0\,,$$

$$\left(\lambda_0\widetilde{Y}+\widetilde{U}_1\right)\frac{\partial\widetilde{U}_2}{\partial X}+\widetilde{V}_1\frac{\partial\widetilde{U}_2}{\partial\widetilde{Y}}+\widetilde{U}_2\frac{\partial\widetilde{U}_1}{\partial X}+\widetilde{V}_2\left(\lambda_0+\frac{\partial\widetilde{U}_1}{\partial\widetilde{Y}}\right)=-\frac{\partial\widetilde{P}_2}{\partial X}+\frac{\partial^2\widetilde{U}_2}{\partial\widetilde{Y}^2}\,,$$

$$\frac{\partial\widetilde{P}_2}{\partial\widetilde{Y}}=0\,.$$

In the main deck, the order 1 solution is

$$U_1=A_1(X)U_0'(Y)\quad\text{with}\quad U_0'(Y)=\frac{\mathrm{d}U_0}{\mathrm{d}Y}\,,$$

$$V_1=-A_1'(X)U_0(Y)\quad\text{with}\quad A_1'(X)=\frac{\mathrm{d}A_1}{\mathrm{d}X}\,,$$

where function $A_1(X)$ is an unknown of the problem.

To order 2, the solution in the main deck is

$$U_2=-P_1(1+\varPhi')+A_2U_0'+\frac{A_1^2}{2}\varPsi'\,,$$

$$V_2=\frac{\mathrm{d}P_1}{\mathrm{d}X}(Y+\varPhi)-A_2'U_0-A_1A_1'\varPsi\,,$$

$$P_2=H(X)+A_1''x^{1/2}(2f''+ff'-2\lambda_0)\,,$$

where $A_2(X)$ is the second order displacement function which is an unknown of the problem. Function $H(X)$ is obtained from the matching with the lower deck. Functions $\varPhi$ and $\varPsi$ are obtained as solutions of equations

$$U_0\varPhi'-U_0'\varPhi=1-U_0+YU_0'\,,$$

$$U_0\varPsi'-U_0'\varPsi=-(U_0')^2+U_0U_0''\,,$$

with

$$\varPhi'=\frac{\mathrm{d}\varPhi}{\mathrm{d}Y}\,,\,\varPsi'=\frac{\mathrm{d}\varPsi}{\mathrm{d}Y}\,.$$

Functions $\varPhi$ and $\varPsi$ are

$$\varPhi(\zeta)=U_0\int_\infty^\zeta\frac{1-U_0+YU_0'}{U_0^2}\,\mathrm{d}Y\,,$$

$$\varPsi(\zeta)=U_0\int_\infty^\zeta\frac{U_0U_0''-(U_0')^2}{U_0^2}\,\mathrm{d}Y\,.$$

From equations for Φ and Ψ, we obtain

$$\Phi \cong -\frac{1}{\lambda_0} + o(Y) \quad \text{as} \quad Y \to 0 \, ,$$

$$\Psi \cong \lambda_0 + o(Y) \quad \text{as} \quad Y \to 0 \, .$$

We also have

$$U_2 \cong -P_1 + \text{EST} \quad \text{as} \quad Y \to \infty \, ,$$

$$V_2 \cong \left(Y \frac{dP_1}{dX} - A_2' \right) + \text{EST} \quad \text{as} \quad Y \to \infty \, ,$$

$$P_2 \cong H(X) + A_1'' \left(Y - (\beta_0 + 2\lambda_0) \right) + \text{EST} \quad \text{as} \quad Y \to \infty \, ,$$

$$U_2 \cong \lambda_0 A_2 - P_1 + \cdots \quad \text{as} \quad Y \to 0 \, ,$$

$$V_2 \cong -\frac{1}{\lambda_0} \frac{dP_1}{dX} - \lambda_0 A_1 A_1' + \cdots \quad \text{as} \quad Y \to 0 \, ,$$

$$P_2(X,0) = H(X) \, .$$

The different matching between the decks give the following results

$$\lim_{\widetilde{Y} \to \infty} \widetilde{U}_1 = \lambda_0 A_1 \, ,$$

$$\lim_{\widetilde{Y} \to \infty} \widetilde{U}_2 = \lambda_0 A_2 - P_1 \, ,$$

$$P_1^*(X,0) = P_1(X) \, ,$$

$$P_1(X) = \widetilde{P}_1(X) \, ,$$

$$P_2(X,0) = \widetilde{P}_2(X) \, ,$$

$$P_2^*(X,0) = \widetilde{P}_2(X) - A_1''(X)(\beta_0 + 2\lambda_0) \, ,$$

$$V_1^*(X,0) = -A_1'(X) \, ,$$

$$V_2^*(X,0) = -A_2'(X) \, .$$

We obtain

$$\frac{dP_1}{dX} = -\frac{\partial U_1^*(X,0)}{\partial X} \, .$$

The solution in the upper deck gives in particular

$$V_1^*(X,Y^*) = \frac{1}{\pi} \int_{-\infty}^{\infty} \frac{(X - \xi)\widetilde{P}_1(\xi)}{(X - \xi)^2 + (Y^*)^2} \, d\xi \, ,$$

hence the interaction law

$$A_1'(X) = -\frac{1}{\pi} \fint_{-\infty}^{\infty} \frac{\widetilde{P}_1(\xi)}{X - \xi} \, d\xi \, .$$

IV.2 Global Model for the Main Deck and the Lower Deck

With index "0" denoting the basic flow, we seek the perturbed flow in the form

$$u(x, y, \varepsilon) = U_0(x, Y) + \varepsilon^{1/4}\overline{U}(X, Y, \varepsilon)\,,$$
$$v(x, y, \varepsilon) = \varepsilon V_0(x, Y) + \varepsilon^{1/2}\overline{V}(X, Y, \varepsilon)\,,$$
$$p(x, y, \varepsilon) = \varepsilon^2 P_0(x, Y) + \varepsilon^{1/2}\overline{P}(X, Y, \varepsilon)\,.$$

Substituting these expansions in the Navier-Stokes equations and taking into account Blasius' boundary layer equations, we obtain

$$\frac{\partial \overline{U}}{\partial X} + \frac{\partial \overline{V}}{\partial Y} = 0\,, \tag{IV.3a}$$

$$U_0 \frac{\partial \overline{U}}{\partial X} + \overline{V}\frac{\partial U_0}{\partial Y} + \varepsilon^{1/4}\left(\overline{U}\frac{\partial \overline{U}}{\partial X} + \overline{V}\frac{\partial \overline{U}}{\partial Y}\right) + \varepsilon^{3/4}\left(\overline{U}\frac{\partial U_0}{\partial X} + V_0\frac{\partial \overline{U}}{\partial Y}\right)$$
$$= -\varepsilon^{1/4}\frac{\partial \overline{P}}{\partial X} + \varepsilon^{3/4}\frac{\partial^2 \overline{U}}{\partial Y^2} + \mathrm{O}(\varepsilon^{5/4})\,, \tag{IV.3b}$$

$$U_0 \frac{\partial \overline{V}}{\partial X} + \varepsilon^{1/4}\left(\overline{U}\frac{\partial \overline{V}}{\partial X} + \overline{V}\frac{\partial \overline{V}}{\partial Y}\right) + \varepsilon^{3/4}\left(\overline{V}\frac{\partial V_0}{\partial Y} + V_0\frac{\partial \overline{V}}{\partial Y}\right)$$
$$= -\varepsilon^{-1/4}\frac{\partial \overline{P}}{\partial Y} + \varepsilon^{3/4}\frac{\partial^2 \overline{V}}{\partial Y^2} + \mathrm{O}(\varepsilon^{5/4})\,. \tag{IV.3c}$$

For the main deck, the regular expansion is

$$\overline{U}(X, Y, \varepsilon) = U_1(X, Y) + \varepsilon^{1/4}U_2(X, Y) + \cdots\,, \tag{IV.4a}$$
$$\overline{V}(X, Y, \varepsilon) = V_1(X, Y) + \varepsilon^{1/4}V_2(X, Y) + \cdots\,, \tag{IV.4b}$$
$$\overline{P}(X, Y, \varepsilon) = P_1(X, Y) + \varepsilon^{1/4}P_2(X, Y) + \cdots\,. \tag{IV.4c}$$

For the lower deck, the regular expansion is

$$\overline{U}(X, Y, \varepsilon) = \widetilde{U}_1(X, \widetilde{Y}) + \varepsilon^{1/4}\widetilde{U}_2(X, \widetilde{Y}) + \cdots\,, \tag{IV.5a}$$
$$\overline{V}(X, Y, \varepsilon) = \varepsilon^{1/4}\widetilde{V}_1(X, \widetilde{Y}) + \varepsilon^{1/2}\widetilde{V}_2(X, \widetilde{Y}) + \cdots\,, \tag{IV.5b}$$
$$\overline{P}(X, Y, \varepsilon) = \widetilde{P}_1(X, \widetilde{Y}) + \varepsilon^{1/4}\widetilde{P}_2(X, \widetilde{Y}) + \cdots\,. \tag{IV.5c}$$

Substituting these expansions in (IV.3a–IV.3c), the results of the triple deck theory in the main deck and in the lower deck are exactly recovered to second order. The matching conditions between the main deck and the lower deck are also identical.

In fact, to recover the results of the second order triple deck theory in the main deck and in the lower deck, a more restricted system than (IV.3a–IV.3c) can be used

$$\frac{\partial \overline{U}}{\partial X} + \frac{\partial \overline{V}}{\partial Y} = 0\,, \tag{IV.6a}$$

$$U_0 \frac{\partial \overline{U}}{\partial X} + \overline{V} \frac{\mathrm{d}U_0}{\mathrm{d}Y} + \varepsilon^{1/4} \left(\overline{U} \frac{\partial \overline{U}}{\partial X} + \overline{V} \frac{\partial \overline{U}}{\partial Y} \right) = -\varepsilon^{1/4} \frac{\partial \overline{P}}{\partial X} + \varepsilon^{3/4} \frac{\partial^2 \overline{U}}{\partial Y^2} \, , \quad \text{(IV.6b)}$$

$$U_0 \frac{\partial \overline{V}}{\partial X} = -\varepsilon^{-1/4} \frac{\partial \overline{P}}{\partial Y} \, . \quad \text{(IV.6c)}$$

Substituting expansions given by (IV.4a–IV.4c) and (IV.5a–IV.5c) in these equations, the results of the triple deck theory in the main deck and in the lower deck are exactly recovered to second order. The matching conditions between the main deck and the lower deck are also identical.

It must be noted that we have

$$\frac{\partial P_1}{\partial Y} = 0 \, ,$$

$$\frac{\partial P_1}{\partial X} = -\frac{\partial U_1^*(X,0)}{\partial X} \, ,$$

$$\frac{\partial \widetilde{P}_1}{\partial \widetilde{Y}} = 0 \, ,$$

$$\frac{\partial \widetilde{P}_2}{\partial \widetilde{Y}} = 0 \, .$$

In order to recover the results of the main deck and of the lower deck to first order, the following system is sufficient

$$\frac{\partial \overline{U}}{\partial X} + \frac{\partial \overline{V}}{\partial Y} = 0 \, , \quad \text{(IV.7a)}$$

$$U_0 \frac{\partial \overline{U}}{\partial X} + \overline{V} \frac{\mathrm{d}U_0}{\mathrm{d}Y} + \varepsilon^{1/4} \left(\overline{U} \frac{\partial \overline{U}}{\partial X} + \overline{V} \frac{\partial \overline{U}}{\partial Y} \right) = \varepsilon^{1/4} \frac{\partial U_1^*(X,0)}{\partial X}$$

$$+ \varepsilon^{3/4} \frac{\partial^2 \overline{U}}{\partial Y^2} \, , \quad \text{(IV.7b)}$$

$$\frac{\partial \overline{P}}{\partial Y} = 0 \, . \quad \text{(IV.7c)}$$

V Behaviour of an Asymptotic Expansion

V.1 Formulation of the Problem

We consider a singular perturbation problem in which two significant regions have been identified. The studied function is defined in a domain such that $x \geq 0$. We assume that the singularity is located in the neighbourhood of point $x = 0$. In the outer region, the appropriate variable is x and in the inner region, the appropriate variable is X

$$X = \frac{x}{\nu(\varepsilon)} \quad \text{with} \quad \nu \prec 1 .$$

The outer and inner expansions are

$$\Phi_0 = \mathrm{E}_0\, \Phi = \sum_{i=1}^{m} \delta_i\,(\varepsilon)\, \varphi_i\,(x) \ , \tag{V.1}$$

$$\Phi_1 = \mathrm{E}_1\, \Phi = \sum_{i=1}^{m} \delta_i\,(\varepsilon)\, \psi_i\,(X) \ , \tag{V.2}$$

where, by definition, E_0 and E_1 are expansion operators to order δ_m.

As $x \to 0$, the behaviour of functions $\varphi_i\,(x)$ is

$$\varphi_i\,(x) = \sum_{j=1}^{m_i} a_{ij}\,\Delta_{ij}\,(x) + \mathrm{o}\,[\Delta_{im_i}\,(x)] \ , \tag{V.3}$$

where a_{ij} is a series of constants and Δ_{ij} is a sequence of gauge functions such that

$$\Delta_{ij}\,(x) = x^p \left(\ln \frac{1}{x} \right)^q , \tag{V.4}$$

where p and q are real numbers.

In the next sections, it is shown that

$$\mathrm{E}_0\, \mathrm{E}_1\, \Phi_0 = \mathrm{E}_1\, \mathrm{E}_0\, \Phi_0 \ ,$$

$$\mathrm{E}_0\, \mathrm{E}_1\, \Phi_1 = \mathrm{E}_1\, \mathrm{E}_0\, \Phi_1 \ .$$

In fact, only the proof of the first equality is given. A similar proof applies to the second equality.

V.2 Study of the Gauge Functions

In a first step, it is shown that

$$\mathrm{E}_0^* \, \mathrm{E}_1^* \, \Delta_{ij}(x) = \mathrm{E}_1^* \, \mathrm{E}_0^* \, \Delta_{ij}(x) \,, \tag{V.5}$$

where E_0^* and E_1^* are outer and inner expansion operators to order δ^* such that $\delta^*(\varepsilon) \preceq 1$. To simplify the notations, we set

$$\Delta(x) = \Delta_{ij}(x) \,.$$

Obviously, we have

$$\mathrm{E}_0^* \, \Delta(x) = \Delta(x) \,. \tag{V.6}$$

In order to determine $\mathrm{E}_1^* \, \Delta(x)$, we first form $\Delta(\nu X)$

$$
\begin{aligned}
\Delta(\nu X) &= (\nu X)^p \left(\ln \frac{1}{\nu X} \right)^q \\
&= (\nu X)^p \left(\ln \frac{1}{\nu} \right)^q \left(1 - \frac{\ln X}{\ln \frac{1}{\nu}} \right)^q \,.
\end{aligned}
$$

This function is studied as $\varepsilon \to 0$. Assuming that $\dfrac{\ln X}{\ln \frac{1}{\nu}} < 1$, which is always possible if ν is small enough for a fixed value of X, a Taylor series expansion of the last term of the right hand side is taken

$$\Delta(\nu X) = (\nu X)^p \left(\ln \frac{1}{\nu} \right)^q \left[1 - q \frac{\ln X}{\ln \frac{1}{\nu}} + \cdots + \alpha_n \left(\frac{\ln X}{\ln \frac{1}{\nu}} \right)^n + \cdots \right] \,, \tag{V.7}$$

where α_n is the non explicitly expressed coefficient of the corresponding term and n is a positive integer.

According to the value of p, two cases are considered.

First case: $p < 0$. For any value of q and n, we have

$$\nu^p \left(\ln \frac{1}{\nu} \right)^{q-n} \to \infty \quad \text{as} \quad \nu \to 0 \,.$$

In order to calculate $\mathrm{E}_1^* \, \Delta$, it is necessary to keep all the terms present in expansion (V.7). Therefore, we have

$$\mathrm{E}_1^* \, \Delta(\nu X) = \Delta(\nu X) \,,$$

or, with the variable x

$$\mathrm{E}_1^* \, \Delta(\nu X) = \Delta(x) \,.$$

Obviously, we obtain

$$\mathrm{E}_0^* \, \mathrm{E}_1^* \, \Delta(x) = \Delta(x) \,.$$

Finally, we have

$$E_0^* \, E_1^* \, \Delta(x) = E_1^* \, E_0^* \, \Delta(x) \ .$$

Second case: $p \geq 0$**.** According to the order of magnitude of $\nu^p \left(\ln \dfrac{1}{\nu} \right)^q$ with respect to δ^*, two possibilities must be considered.

First possibility: $\nu^p \left(\ln \dfrac{1}{\nu} \right)^q \prec \delta^*$**.** From expression (V.7), the asymptotically largest term in the expansion of $\Delta(\nu X)$ is asymptotically smaller than δ^*. We obtain

$$E_1^* \, \Delta(x) = 0 \ ,$$

and therefore

$$E_0^* \, E_1^* \, \Delta(x) = 0 \ .$$

We also have

$$E_0^* \, E_1^* \, \Delta(x) = E_1^* \, E_0^* \, \Delta(x) \ .$$

Second possibility: $\nu^p \left(\ln \dfrac{1}{\nu} \right)^q \succeq \delta^*$**.**

In the series of expression (V.7), we keep the terms whose exponent is such that

$$\nu^p \left(\ln \dfrac{1}{\nu} \right)^{q-n} \succeq \delta^* \ .$$

We assume that this property is satisfied for $n \leq N$. We obtain

$$E_1^* \, \Delta(\nu X) = (\nu X)^p \left(\ln \frac{1}{\nu} \right)^q \left[1 - q \frac{\ln X}{\ln \frac{1}{\nu}} + \cdots + \alpha_N \left(\frac{\ln X}{\ln \frac{1}{\nu}} \right)^N \right] \ .$$

In order to calculate $E_0^* \, E_1^* \, \Delta$, the variable x is used

$$E_1^* \, \Delta(\nu X) = (x)^p \left(\ln \frac{1}{\nu} \right)^q \left[1 - q \frac{\ln x + \ln \frac{1}{\nu}}{\ln \frac{1}{\nu}} + \cdots + \alpha_N \left(\frac{\ln x + \ln \frac{1}{\nu}}{\ln \frac{1}{\nu}} \right)^N \right]$$

$$= (x)^p \left(\ln \frac{1}{\nu} \right)^q \left[1 - q \left(1 + \frac{\ln x}{\ln \frac{1}{\nu}} \right) + \cdots + \alpha_N \left(1 + \frac{\ln x}{\ln \frac{1}{\nu}} \right)^N \right] \ .$$

In this expression, the asymptotically smallest term is of order $\left(\ln \dfrac{1}{\nu} \right)^{q-N}$.
Taking into account that

$$\nu^p \left(\ln \frac{1}{\nu} \right)^{q-N} \succeq \delta^* \ ,$$

we obtain

$$\left(\ln\frac{1}{\nu}\right)^{q-N} \succeq \delta^* \, ,$$

since $p \geq 0$. Therefore, we have

$$\mathrm{E}_0^* \, \mathrm{E}_1^* \, \Delta(x) = (x)^p \left(\ln\frac{1}{\nu}\right)^q \left[1 - q\left(1 + \frac{\ln x}{\ln\frac{1}{\nu}}\right) + \cdots + \alpha_N \left(1 + \frac{\ln x}{\ln\frac{1}{\nu}}\right)^N \right] \, .$$

Finally, we obtain

$$\mathrm{E}_0^* \, \mathrm{E}_1^* \, \Delta(x) = \mathrm{E}_1^* \, \mathrm{E}_0^* \, \Delta(x) \, . \tag{V.8}$$

Thus, equality (V.5) is proved.

V.3 Study of the Outer Expansion

Now, we wish to show that

$$\mathrm{E}_1 \, \mathrm{E}_0 \, \Phi_0 = \mathrm{E}_0 \, \mathrm{E}_1 \, \Phi_0 \, . \tag{V.9}$$

Obviously, we have

$$\mathrm{E}_0 \, \Phi_0 = \Phi_0 \, .$$

We want to calculate $\mathrm{E}_1 \, \Phi_0$ and therefore $\mathrm{E}_1 \, \delta_i(\varepsilon)\varphi_i(x)$ in particular. We define an inner operator $\overline{\mathrm{E}}_1$ to order $\bar\delta_i$ such that

$$\delta_i\bar\delta_i = \delta_m \, .$$

As $\delta_m \preceq \delta_i$, we have

$$\bar\delta_i \preceq 1 \, .$$

We obtain

$$\mathrm{E}_1 \, \delta_i(\varepsilon)\varphi_i(x) = \delta_i(\varepsilon)\overline{\mathrm{E}}_1\varphi_i(x) \, .$$

From the behaviour (V.3) of φ_i in the neighbourhood of $x = 0$, we obtain

$$\overline{\mathrm{E}}_1\varphi_i(x) = \sum_{j=1}^{\overline{m}_i} a_{ij}\overline{\mathrm{F}}_1\Delta_{ij}(x) \, , \tag{V.10}$$

where $\overline{m}_i$ is such that we are certain to keep in $\Delta_{ij}(\nu(\varepsilon)X)$, for any j, all the terms asymptotically larger than or of the same order as $\bar\delta_i$.

To calculate $\mathrm{E}_0 \, \mathrm{E}_1 \, \Phi_0$, it is required to know $\mathrm{E}_0 \, \mathrm{E}_1 \, \delta_i(\varepsilon)\varphi_i(x)$. Now, we have

$$\mathrm{E}_0 \, \mathrm{E}_1 \, \delta_i(\varepsilon)\varphi_i(x) = \mathrm{E}_0 \, \delta_i(\varepsilon)\overline{\mathrm{E}}_1\varphi_i(x) \, .$$

We define the outer operator $\overline{E}_0$ to order $\bar{\delta}_i$, whence

$$E_0\,\delta_i(\varepsilon)\overline{E}_1\varphi_i(x) = \delta_i(\varepsilon)\overline{E}_0\overline{E}_1\varphi_i(x)$$
$$= \delta_i(\varepsilon)\sum_{j=1}^{\overline{m}_i} a_{ij}\overline{E}_0\overline{E}_1\Delta_{ij}(x)\ ,$$

or

$$E_0\,E_1\,\delta_i(\varepsilon)\varphi_i(x) = \delta_i(\varepsilon)\sum_{j=1}^{\overline{m}_i} a_{ij}\overline{E}_0\overline{E}_1\Delta_{ij}(x)\ . \tag{V.11}$$

On the other hand, let us calculate $E_1\,E_0\,\delta_i(\varepsilon)\varphi_i(x)$. Taking into account that

$$E_0\,\delta_i(\varepsilon)\varphi_i(x) = \delta_i(\varepsilon)\varphi_i(x)\ ,$$

we obtain

$$E_1\,E_0\,\delta_i(\varepsilon)\varphi_i(x) = E_1\,\delta_i(\varepsilon)\varphi_i(x) = \delta_i(\varepsilon)\overline{E}_1\varphi_i(x)\ ,$$

which gives with (V.10)

$$E_1\,E_0\,\delta_i(\varepsilon)\varphi_i(x) = \delta_i(\varepsilon)\sum_{j=1}^{\overline{m}_i} a_{ij}\overline{E}_1\Delta_{ij}(x)\ . \tag{V.12}$$

As $\bar{\delta}_i \preceq 1$, from result (V.8), we can write

$$\overline{E}_0\overline{E}_1\Delta_{ij}(x) = \overline{E}_1\overline{E}_0\Delta_{ij}(x) = \overline{E}_1\Delta_{ij}(x)\ ,$$

or finally, comparing (V.11) and (V.12)

$$E_0\,E_1\,\delta_i(\varepsilon)\varphi_i(x) = E_1\,E_0\,\delta_i(\varepsilon)\varphi_i(x)\ ,$$

and therefore

$$E_0\,E_1\,\Phi_0 = E_1\,E_0\,\Phi_0\ . \tag{V.13}$$

A similar reasoning for the inner expansion as for the outer expansion leads us to the result

$$E_0\,E_1\,\Phi_1 = E_1\,E_0\,\Phi_1\ . \tag{V.14}$$

Solutions of Problems

Chapter 2

2-1.

1. The exact solutions are

$$x = \frac{-\varepsilon \pm \sqrt{\varepsilon^2 + 4}}{2} \ .$$

We obtain

$$x^{(1)} = 1 - \frac{\varepsilon}{2} + \frac{\varepsilon^2}{8} + \cdots \ ,$$

$$x^{(2)} = -1 - \frac{\varepsilon}{2} - \frac{\varepsilon^2}{8} + \cdots \ .$$

2. We examine the iterative process

$$x_n = \pm\sqrt{1 - \varepsilon x_{n-1}} \ .$$

Starting from $x_0 = 1$, we have

$$x_1 = \sqrt{1 - \varepsilon} \ .$$

Using a Taylor series expansion, we obtain

$$x_1 = 1 - \frac{\varepsilon}{2} \ .$$

The next approximation is

$$x_2 = \sqrt{1 - \varepsilon\left(1 - \frac{\varepsilon}{2}\right)} \ .$$

Using a Taylor series expansion, we have

$$x_2 = 1 + \frac{1}{2}\left(-\varepsilon + \frac{\varepsilon^2}{2}\right) - \frac{1}{8}\left(-\varepsilon + \frac{\varepsilon^2}{2}\right)^2 + \cdots \ ,$$

i.e., to order ε^2

$$x_2 = 1 - \frac{\varepsilon}{2} + \frac{\varepsilon^2}{8} \ .$$

Similarly, starting from $x_0 = -1$, we obtain

$$x_0 = -1 \,, \quad x_1 = -1 - \frac{\varepsilon}{2} \,, \quad x_2 = -1 - \frac{\varepsilon}{2} - \frac{\varepsilon^2}{8} \,.$$

3. We set

$$x = x_0 + \varepsilon x_1 + \varepsilon^2 x_2 + \cdots \,.$$

This expression is substituted in the initial equation, by assuming that this procedure is licit

$$(x_0 + \varepsilon x_1 + \varepsilon^2 x_2 + \cdots)^2 + \varepsilon(x_0 + \varepsilon x_1 + \varepsilon^2 x_2 + \cdots) - 1 = 0 \,.$$

Expanding the squared term and equating the coefficients of like powers of ε, we obtain

$$x_0^2 - 1 = 0 \,,$$
$$2x_0 x_1 + x_0 = 0 \,,$$
$$x_1^2 + 2x_0 x_2 + x_1 = 0 \,.$$

Starting from $x_0 = 1$, we have

$$x_0 = 1 \,, \quad x_1 = -\frac{1}{2} \,, \quad x_2 = \frac{1}{8} \,.$$

Starting from $x_0 = -1$, we have

$$x_0 = -1 \,, \quad x_1 = -\frac{1}{2} \,, \quad x_2 = -\frac{1}{8} \,.$$

4. We set

$$x = x_0 + \delta_1(\varepsilon)x_1 + \delta_2(\varepsilon)x_2 + \cdots \,.$$

This expression is substituted in the initial equation

$$(x_0 + \delta_1(\varepsilon)x_1 + \delta_2(\varepsilon)x_2 + \cdots)^2 + \varepsilon\,(x_0 + \delta_1(\varepsilon)x_1 + \delta_2(\varepsilon)x_2 + \cdots) - 1 = 0 \,.$$

To first order, we have $x_0^2 - 1 = 0$. The next order is ε or δ_1 according to the relative order of ε with respect to δ_1. In order to have a significant result, we must take $\delta_1 = \varepsilon$ or, at least δ_1 must be of the same order as ε, i.e. δ_1 must behave like ε as $\varepsilon \to 0$; for the sake of simplicity, we take $\delta_1 = \varepsilon$.

The next order is δ_2 or ε^2. As previously, a significant result is obtained only by taking $\delta_2 = \varepsilon^2$.

Therefore, we have a constructive method to define the expansion of the roots of the equation.

2-2.

1. The exact solution is

$$x = \frac{-1 \pm \sqrt{1 + 4\varepsilon}}{2\varepsilon} \,.$$

The expansion writes

$$x^{(1)} = 1 - \varepsilon + 2\varepsilon^2 + \cdots ,$$
$$x^{(2)} = -\frac{1}{\varepsilon} - 1 + \varepsilon + \cdots .$$

2. With $x_0 = 1$, the first process gives successively

$$x_1 = 1 - \varepsilon , \quad x_2 = 1 - \varepsilon + 2\varepsilon^2 + \cdots .$$

With $x_0 = -\dfrac{1}{\varepsilon}$, the second process gives

$$x_1 = -\frac{1}{\varepsilon} - 1 , \quad x_2 = -\frac{1}{\varepsilon} - 1 + \varepsilon + \cdots .$$

3. We set

$$x^{(1)} = x_0^{(1)} + \varepsilon x_1^{(1)} + \varepsilon^2 x_2^{(1)} + \cdots .$$

This expression is replaced in the initial equation

$$\varepsilon \left(x_0^{(1)} + \varepsilon x_1^{(1)} + \varepsilon^2 x_2^{(1)} + \cdots \right)^2 + x_0^{(1)} + \varepsilon x_1^{(1)} + \varepsilon^2 x_2^{(1)} + \cdots - 1 = 0 .$$

Equating coefficients of like powers of ε, we obtain

$$x_0^{(1)} - 1 = 0 ,$$
$$(x_0^{(1)})^2 + x_1^{(1)} = 0 ,$$
$$2 x_0^{(1)} x_1^{(1)} + x_2^{(1)} = 0 ,$$

whence

$$x_0^{(1)} = 1 , \quad x_1^{(1)} = -1 , \quad x_2^{(1)} = 2 .$$

For the other root, we set

$$x^{(2)} = \frac{x_{-1}^{(2)}}{\varepsilon} + x_0^{(2)} + \varepsilon x_1^{(2)} + \cdots .$$

This expression is substituted in the initial equation

$$\varepsilon \left(\frac{x_{-1}^{(2)}}{\varepsilon} + x_0^{(2)} + \varepsilon x_1^{(2)} + \cdots \right)^2 + \frac{x_{-1}^{(2)}}{\varepsilon} + x_0^{(2)} + \varepsilon x_1^{(2)} + \cdots - 1 = 0 .$$

Equating coefficients of like powers of ε, we obtain

$$(x_{-1}^{(2)})^2 + x_{-1}^{(2)} = 0 ,$$
$$2 x_0^{(2)} x_{-1}^{(2)} + x_0^{(2)} - 1 = 0 ,$$
$$(x_0^{(2)})^2 + 2 x_{-1}^{(2)} x_1^{(2)} + x_1^{(2)} = 0 ,$$

whence

$$x_{-1}^{(2)} = -1 , \quad x_0^{(2)} = -1 , \quad x_1^{(2)} = 1 .$$

2-3.

1. The (complex) exact solution is

$$f = \alpha\, e^{i\lambda x} + \beta\, e^{-i\lambda x} \ .$$

We impose the boundary conditions at $x = \varepsilon$ and $x = \pi$. We find

$$\alpha = -\beta\, e^{-2i\lambda\varepsilon} \quad \text{and} \quad \lambda = \frac{n\pi}{\pi - \varepsilon}\ ,$$

where n is an integer, $n \geq 1$ since we assumed that $\lambda > 0$.

Taking the real solution, the expansion with respect to ε yields

$$f = A\left[\sin nx + \varepsilon\frac{n}{\pi}x\cos nx - \varepsilon n\cos nx\right] + \cdots\ ,$$

and

$$\lambda = n\left[1 + \frac{\varepsilon}{\pi}\right] + \cdots\ .$$

2. Substituting the proposed expansions for f and λ in the initial equation, we obtain

$$\frac{\mathrm{d}^2\varphi_0}{\mathrm{d}x^2} + \lambda_0^2\varphi_0 = 0\ ,$$

and

$$\frac{\mathrm{d}^2\varphi_1}{\mathrm{d}x^2} + \lambda_0^2\varphi_1 = -2\lambda_0\lambda_1\varphi_0\ .$$

The boundary condition at $x = \pi$ yields

$$\varphi_0(\pi) = 0\ , \quad \varphi_1(\pi) = 0\ .$$

The boundary condition at $x = \varepsilon$ yields

$$\varphi_0(0) + \varepsilon\frac{\mathrm{d}\varphi_0}{\mathrm{d}x}(0) + \varepsilon\varphi_1(0) + \cdots = 0\ ,$$

or

$$\varphi_0(0) = 0\ , \quad \varphi_1(0) = -\frac{\mathrm{d}\varphi_0}{\mathrm{d}x}(0)\ .$$

We obtain the solution for φ_0

$$\varphi_0 = A\sin nx\ , \quad \lambda_0 = n\ ,$$

where n is an integer $n \geq 1$; A is the arbitrary amplitude of the solution for $\varepsilon = 0$.

Therefore, the problem to solve for φ_1 is

$$\frac{\mathrm{d}^2\varphi_1}{\mathrm{d}x^2} + n^2\varphi_1 = -2n\lambda_1 A\sin nx\ , \quad \varphi_1(0) = -nA\ , \quad \varphi_1(\pi) = 0\ .$$

We find

$$\varphi_1 = K\sin nx - nA\cos nx + \frac{n}{\pi}A\,x\cos nx\ ,$$

and

$$\lambda_1 = \frac{n}{\pi} \, .$$

Apparently, nothing enables us to determine constant K except if we assume that the amplitude of the term in $\sin nx$ must be independent of ε; then $K = 0$.

To order ε, the solution is

$$\varphi = A \sin nx + \varepsilon \left[-nA \cos nx + \frac{n}{\pi} A \, x \cos nx \right] + \cdots ,$$

and

$$\lambda = n + \varepsilon \frac{n}{\pi} + \cdots .$$

We recover the expansion of the exact solution.

2-4.

1. Substituting the proposed expansion in equation $\triangle \psi = 0$, we obtain equations for ψ_0 and ψ_1

$$\triangle \psi_0 = 0 \, , \quad \triangle \psi_1 = 0 \, .$$

Away from the body, the flow is uniform; this flow is characterized by $\psi = U_\infty y$. Therefore, we have

$$r \to \infty : \quad \psi_0 = U_\infty r \sin \theta \quad \text{and} \quad \psi_1 = 0 \, .$$

Along the body, we have $\psi = 0$, which gives

$$\psi_0 \left[a(1 - \varepsilon sin^2\theta), \theta \right] + \varepsilon \psi_1 \left[a(1 - \varepsilon sin^2\theta), \theta \right] + \cdots = 0 \, .$$

The functions must be expanded in the neighbourhood of $r = a$. We obtain

$$\psi_0(a, \theta) + \varepsilon \left[\psi_1(a, \theta) - a \sin^2 \theta \left(\frac{\partial \psi_0}{\partial r} \right)_{r=a} \right] + \cdots = 0 \, ,$$

whence

$$\psi_0(a, \theta) = 0 \quad \text{and} \quad \psi_1(a, \theta) = a \sin^2 \theta \left(\frac{\partial \psi_0}{\partial r} \right)_{r=a} \, .$$

It is deduced that the solution for ψ_0 is the flow around a circular cylinder

$$\psi_0 = U_\infty \left(r - \frac{a^2}{r} \right) \sin \theta \, .$$

Then, we obtain

$$\psi_1(a, \theta) = 2U_\infty a \sin^3 \theta = \frac{1}{2} U_\infty a (3 \sin \theta - \sin 3\theta) \, .$$

2. The general solution of equation $\triangle \psi_1 = 0$ with condition $\psi_1 = 0$ as $r \to \infty$ is $\sum b_n r^{-n} \sin n\theta$ where n is an integer, $n > 0$. To satisfy the condition at $r = a$, we take $n = 1$ and $n = 3$ with $b_1 = \frac{3}{2} U_\infty a^2$ and $b_3 = -\frac{1}{2} U_\infty a^4$, whence

$$\psi = U_\infty \left(r - \frac{a^2}{r} \right) \sin \theta + \varepsilon \frac{U_\infty a}{2} \left(3 \frac{a}{r} \sin \theta - \frac{a^3}{r^3} \sin 3\theta \right) + \cdots \, .$$

We have

$$\psi_1(a, \theta) = \frac{1}{2} U_\infty a (3 \sin\theta - \sin 3\theta) \,.$$

3. The velocity modulus at any point of the field is

$$V = \sqrt{u^2 + v^2} = \sqrt{\left(\frac{\partial\psi}{\partial r}\right)^2 + \frac{1}{r^2}\left(\frac{\partial\psi}{\partial\theta}\right)^2} \,.$$

After calculation, by taking care of expanding the functions in the neighbourhood of $r = a$, at the body wall, we obtain to order ε

$$V = U_\infty (2 \sin\theta + \varepsilon \sin 3\theta) \,.$$

2-5.

1. The dimensionless quantities are

$$x = \frac{x^*}{a}\,, \quad y = \frac{y^*}{a}\,, \quad r = \frac{r^*}{a}\,, \quad u = \frac{u^*}{U_\infty}\,, \quad v = \frac{v^*}{U_\infty}\,, \quad \psi = \frac{\psi^*}{U_\infty a}\,, \quad \omega = \frac{\omega^* a}{U_\infty} \,.$$

The problem becomes

$$\frac{\partial^2\psi}{\partial x^2} + \frac{\partial^2\psi}{\partial y^2} = -\omega \,,$$

with the boundary conditions $\psi = 0$ at $r = 1$ and $\psi \to y + \frac{1}{3}\varepsilon y^3$ as $r \to \infty$.

At upstream infinity, we have $\omega = -\dfrac{\partial U}{\partial y} = -\dfrac{\partial^2\psi}{\partial y^2} = -2\varepsilon y$, and

$$y = \psi - \frac{1}{3}\varepsilon y^3 \,.$$

The first approximation, obtained with $\varepsilon = 0$, is $y = \psi$. Iterating, we have

$$y = \psi - \frac{1}{3}\varepsilon\psi^3 + \mathrm{O}(\varepsilon^2\psi^5) \,,$$

and

$$\omega = -2\varepsilon\psi + \frac{2}{3}\varepsilon^2\psi^3 + \mathrm{O}(\varepsilon^3\psi^5) \,.$$

2. Substituting the expansion

$$\psi = \psi_0 + \varepsilon\psi_1 + \cdots$$

in equation for ψ, we obtain

$$\triangle\psi_0 = 0 \quad \text{and} \quad \triangle\psi_1 = 2\psi_0 \,.$$

The condition $\psi = 0$ at $r = 1$ yields $\psi_0 = 0$ at $r = 1$ and $\psi_1 = 0$ at $r = 1$. The condition $\psi \to y + \varepsilon y^3/3$ as $r \to \infty$ yields

$$\psi_0 \to y \quad \text{as} \quad r \to \infty \,,$$

and
$$\psi_1 \to \frac{1}{3}y^3 \quad \text{as} \quad r \to \infty \,,$$

that is
$$\psi_0 \to r \sin\theta \quad \text{as} \quad r \to \infty \,,$$

and
$$\psi_1 \to \frac{1}{3}r^3 \sin^3\theta \quad \text{as} \quad r \to \infty \,.$$

The solution for ψ_0 represents the flow around a circular cylinder plunged in a uniform flow, i.e.
$$\psi_0 = \left(r - \frac{1}{r} \right) \sin\theta \,,$$

whence the equation for ψ_1
$$\Delta \psi_1 = 2 \left(r - \frac{1}{r} \right) \sin\theta \,.$$

A particular solution of equation
$$\Delta \psi_1 = 2r \sin\theta$$

is $\frac{1}{3}r^3 \sin^3\theta$. A particular solution of equation
$$\Delta \psi_1 = -\frac{2}{r}\sin\theta$$

is $-r\ln r \sin\theta$. The condition $\psi_1 \to \frac{1}{3}r^3 \sin^3\theta$ as $r \to \infty$ is satisfied by the first particular solution. We add the solution $\sum b_n r^{-n} \sin\theta$ of the equation without right hand side which respects the symmetry properties and which gives $\psi_1 \to 0$ at infinity; in this equation, we take $n = 1$ and $n = 3$ in order to satisfy the slip condition at the wall ($\psi = 0$). We obtain
$$\psi_1 = \frac{1}{3}r^3 \sin^3\theta - r\ln r \sin\theta - \frac{1}{4}\frac{1}{r}\sin\theta + \frac{1}{12}\frac{1}{r^3}\sin 3\theta \,,$$

whence
$$\psi = \left(r - \frac{1}{r} \right)\sin\theta + \varepsilon\left[\frac{1}{3}r^3 \sin^3\theta - r\ln r \sin\theta - \frac{1}{4}\frac{1}{r}\sin\theta + \frac{1}{12}\frac{1}{r^3}\sin 3\theta \right] + \cdots \,.$$

As $r \to \infty$, we observe that ψ_1 introduces a parasitic term in $r\ln r$ which does not tend towards zero. This term is small compared to the term in $r^3 \sin^3\theta$ but it increases faster than the term $r\sin\theta$ coming from ψ_0. In fact, this term is at the origin of a singular problem. If the expansion is continued as if it were a regular problem, it is not possible to find a solution to the next order which behaves correctly at infinity.

Chapter 3

3-1.

1. In the domain $0 \leq x < 1$, $y_0 = e^x$; in the domain $1 < x \leq 2$, $y_0 = 0$.

2. $x_0 = 1$; $\delta = \sqrt{\varepsilon}$; $Y_0 = B \int_0^X e^{-t^2} \, dt + C$.

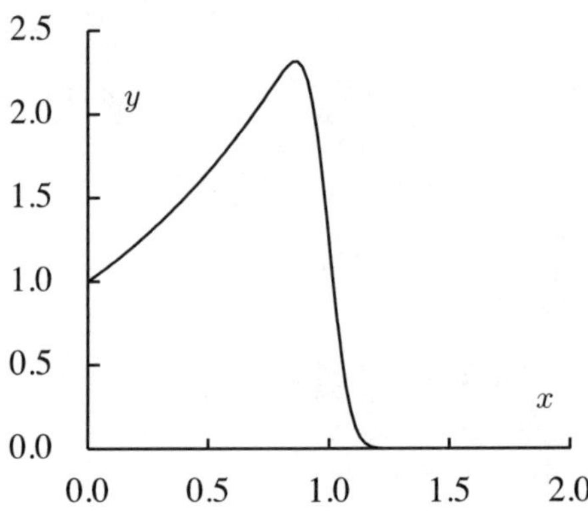

Fig. S.1. Solution for $\varepsilon = 0.01$

3. $e = -B \int_0^\infty e^{-s^2} \, ds + C \; ; \; 0 = B \int_0^\infty e^{-s^2} \, ds + C \; ; \; Y_0 = e \left(\dfrac{1}{2} - \dfrac{1}{\sqrt{\pi}} \int_0^X e^{-t^2} \, dt \right).$

See the plot of the solution in Fig. S.1.

4. In domain $0 \leq x \leq 1$, $y_{\mathrm{app}} = e^x - e \left(\dfrac{1}{2} + \dfrac{1}{\sqrt{\pi}} \int_0^X e^{-t^2} \, dt \right).$

In domain $1 \leq x \leq 2$, $y_{\mathrm{app}} = e \left(\dfrac{1}{2} - \dfrac{1}{\sqrt{\pi}} \int_0^X e^{-t^2} \, dt \right).$

3-2.

1. Outside of any boundary layer, the solution is

$$y_0(x) = \frac{C}{1 + \alpha x} \, ,$$

solution of equation

$$(1 + \alpha x) \frac{dy_0}{dx} + \alpha y_0 = 0 \, .$$

2. For $\alpha > -1$, we have $1 + \alpha x > 0$. A boundary layer exists in the neighbourhood of $x = 0$ whose thickness is ε because $(1 + \alpha x)|_{x=0} > 0$. If we set $X = x/\varepsilon$, the equation for Y_0 is

$$\frac{d^2 Y_0}{dX^2} + \frac{dY_0}{dX} = 0 \, ,$$

whose solution satisfying condition $Y_0(0) = 1$ is

$$Y_0(X) = 1 + A - A e^{-X} .$$

Moreover, $y_0(1) = 1$ implies $C = 1 + \alpha$. The matching condition gives $A = \alpha$. Finally, we have

$$y_0(x) = \frac{1 + \alpha}{1 + \alpha x} ,$$

$$Y_0(X) = 1 + \alpha - \alpha e^{-X} ,$$

$$y_{\text{app}} = \frac{1 + \alpha}{1 + \alpha x} - \alpha e^{-X} .$$

3. There are two boundary layers, one at $x = 0$ with $X = x/\varepsilon$ and the other one at $x = 1$ with $X^* = (1 - x)/\varepsilon$. Indeed, for $x < -1/\alpha$, we have $1 + \alpha x > 0$ whereas for $x > -1/\alpha$, we have $1 + \alpha x < 0$. As $y_0(x) = 0$ for $x = -1/\alpha$, $y_0(x) = 0$; we also have $Y_0(X) = 1 + A - A e^{-X}$ whereas $Y_0^*(X^*) = 1 + B - B e^{(1+\alpha)X^*}$ is solution of

$$\frac{\mathrm{d}^2 Y_0^*}{\mathrm{d}X^{*2}} - (1 + \alpha)\frac{\mathrm{d}Y_0^*}{\mathrm{d}X^*} = 0 .$$

The asymptotic matching gives $A = -1$ and $B = -1$, whence

$$y_{\text{app}} = e^{-X} + e^{(1+\alpha)X^*} .$$

3-3. Taking into account the boundary conditions, the exact solution is

$$y = e^{X^2}\left\{ 1 + (e^{-1/(2\varepsilon)} - 1)\frac{\displaystyle\int_0^X e^{-t^2}\, \mathrm{d}t}{\displaystyle\int_0^{1/\sqrt{2\varepsilon}} e^{-t^2}\, \mathrm{d}t} \right\} .$$

The expansion as $\varepsilon \to 0$, x being kept fixed, in domain $0 < x < 1$ yields

$$y = \frac{2}{\sqrt{\pi}}\frac{\varepsilon^{1/2}}{1 - x} + \cdots .$$

This expansion exhibits a singularity at $x = 0$ and another singularity at $x = 1$ since the boundary conditions are not satisfied at these points.

To establish the change of variable in the neighbourhood of $x = 0$ from the initial equation, we set

$$X = \frac{x}{\delta(\varepsilon)} .$$

We obtain

$$\frac{\varepsilon}{\delta^2}\frac{\mathrm{d}^2 y}{\mathrm{d}x^2} + \frac{1-\delta X}{\delta}\frac{\mathrm{d}y}{\mathrm{d}X} - y = 0 \ .$$

To restore the boundary layer, the second derivative term must be kept. Then, we compare this term to the other two. The solution $\delta = \varepsilon^{1/2}$ is not valid because there would be only one dominant term which is the first derivative term. We must take $\delta = \varepsilon$.

To establish the change of variable in the neighbourhood of $x = 1$ from the initial equation, we set

$$X = \frac{1-x}{\delta(\varepsilon)} \ .$$

We obtain

$$\frac{\varepsilon}{\delta^2}\frac{\mathrm{d}^2 y}{\mathrm{d}X^2} - X\frac{\mathrm{d}y}{\mathrm{d}X} - y = 0 \ .$$

The choice $\delta = \varepsilon^{1/2}$ is appropriate. We note that the initial equation does not simplify for the study of the boundary layer in the neighbourhood of $x = 1$.

Chapter 4

4-1.

$$1 \succ -\frac{1}{\ln \varepsilon} \succ \varepsilon^\nu \succ -\varepsilon \ln \varepsilon \succ \varepsilon \ .$$

4-2. i) $\varphi = o(1)$; ii) $\varphi = O_S(\varepsilon)$; iii) $\varphi = o(1)$.

4-3.

1. $e^{\varepsilon x} = 1 + O(\varepsilon)$: this approximation is uniformly valid in domain $0 \le x \le 1$, even for $x = 0$ since then $e^{\varepsilon x} = 1$.

2. $\dfrac{1}{x+\varepsilon} = O(1)$: this approximation is not uniformly valid in domain $0 \le x \le 1$ since for $x = 0$, the function is $1/\varepsilon$ which is not $O(1)$.

3. $e^{-x/\varepsilon} = o(\varepsilon^n)$ for any $n > 0$: this approximation is not uniformly valid in domain $0 \le x \le 1$ since for $x = 0$ we have $e^{-x/\varepsilon} = 1$ which is not $o(\varepsilon^n)$ for any $n > 0$.

4-4. The straightforward expansion of φ writes

$$\varphi = 1 - \varepsilon\frac{2x-1}{1-x} + \varepsilon^2\left(\frac{2x-1}{1-x}\right)^2 + \cdots \ .$$

This is not an asymptotic expansion in the domain $0 < A_1\varepsilon \le 1 - x \le A_2\varepsilon$ where A_1 and A_2 are constants independent of ε.

We can write

$$\varphi = \frac{1}{1 + \frac{\varepsilon}{1-x} - 2\varepsilon} = \frac{1}{1 + \frac{\varepsilon}{1-x}}\frac{1}{1 - \frac{2\varepsilon}{1 + \frac{\varepsilon}{1-x}}} \ ,$$

whence the expansion

$$\varphi = \frac{1}{1 + \frac{\varepsilon}{1-x}} \left[1 + \frac{2\varepsilon}{1 + \frac{\varepsilon}{1-x}} + \left(\frac{2\varepsilon}{1 + \frac{\varepsilon}{1-x}} \right)^2 + \cdots \right] .$$

This is an asymptotic expansion in the whole domain $0 \leq x \leq 1$; this is a generalized expansion.

4-5. A first integration gives

$$E_1 = \frac{e^{-x}}{x} - \int_x^\infty \frac{e^{-t}}{t^2} \, dt .$$

Repeating integrations by parts, we obtain finally

$$E_1(x) = \frac{e^{-x}}{x} \left[1 - \frac{1}{x} + \frac{2}{x^2} + \cdots + (-1)^n \frac{n!}{x^n} \right] + (-1)^{n+1}(n+1)! \int_x^\infty \frac{e^{-t}}{t^{n+2}} \, dt .$$

We have

$$\int_x^\infty \frac{e^{-t}}{t^{n+2}} \, dt < e^{-x} \int_x^\infty \frac{1}{t^{n+2}} \, dt ,$$

or

$$\int_x^\infty \frac{e^{-t}}{t^{n+2}} \, dt < \frac{e^{-x}}{(n+1)x^{n+1}} .$$

Therefore, we have

$$E_1(x) = \frac{e^{-x}}{x} \left[1 - \frac{1}{x} + \frac{2}{x^2} + \cdots + (-1)^{n-1} \frac{(n-1)!}{x^{n-1}} + O(\frac{1}{x^n}) \right] .$$

Thus, an asymptotic expansion for large x has been formed.

We set

$$R_n(x) = (-1)^n n! \int_x^\infty \frac{e^{-t}}{t^{n+1}} \, dt .$$

The expression of $R_n(x)$ tells us that

$$|R_n(x)| \to \infty \text{ as } n \to \infty \text{ with } x \text{ being kept fixed,}$$

and, from the previous calculations, with n being kept fixed, we have

$$|R_n(x)| \to 0 \text{ as } x \to \infty .$$

The series is divergent since the ratio, in absolute value, of two successive terms is n/x so that the convergence radius is

$$\frac{1}{x} = \lim_{n \to \infty} \frac{1}{n} = 0 .$$

Table S.1 and Fig. S.2 give the so-obtained approximation for $x = 3$ according to the number of terms of the expansion.

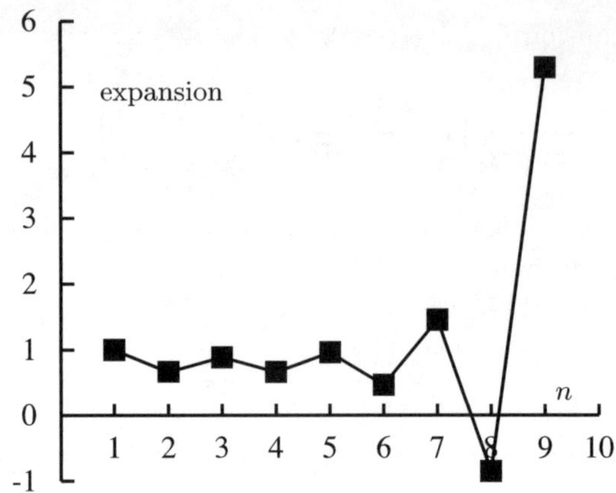

Fig. S.2. Approximation of the function $x\,\mathrm{e}^x\,E_1(x)$ for $x = 3$

Table S.1. Approximation of function $x\,\mathrm{e}^x\,E_1(x)$ for $x = 3$

n	1	2	3	4	5
Expansion	1.	0.66667	0.88889	0.66667	0.96296

n	6	7	8	9	10
Expansion	0.46914	1.4568	−0.84774	5.2977	−13.138

4-6. The change of variable $X = x/\delta$ gives the equation

$$\varepsilon\delta^2 X^2 + \delta X - 1 = 0\,.$$

1.$\delta \prec 1$: impossible because the equation would be $-1 = 0$.
2.$\delta = 1$. The equation becomes $X - 1 = 0$. The regular root is found.
3.$1 \prec \delta \prec \varepsilon^{-1}$. The equation reduces to $X = 0$. This solution is not valid.
4.$\delta \succ \varepsilon^{-1}$. The equation reduces to $X^2 = 0$. This solution is not valid.
5.$\delta = \varepsilon^{-1}$. The equation becomes $X^2 + X = 0$. The root $X = -1$ is significant.
We choose the expansion in the form

$$x = -\frac{1}{\varepsilon} + x_1 + x_2\varepsilon + \cdots\,.$$

Substituting this expansion in the initial equation gives

$$\varepsilon\left(-\frac{1}{\varepsilon} + x_1 + x_2\varepsilon + \cdots\right)^2 - \frac{1}{\varepsilon} + x_1 + x_2\varepsilon + \cdots - 1 = 0\,.$$

To order 1, the equation writes

$$-x_1 - 1 = 0 \ .$$

We deduce $x_1 = -1$. To order ε, the equation becomes

$$1 - x_2 = 0 \ .$$

We deduce $x_2 = 1$. This result can be checked with the exact solution.

4-7. We have

$$f\left[x(\varepsilon)\right] = \exp\left(\varepsilon^{-2} + 2 + \varepsilon^2\right) \ ,$$

whence

$$f\left[x(\varepsilon)\right] = \exp\left(\varepsilon^{-2}\right) e^2 \left(1 + \varepsilon^2 + \frac{1}{2}\varepsilon^4 + \cdots\right) \ .$$

If we keep only $x = 1/\varepsilon$, we obtain

$$f = \exp\left(\varepsilon^{-2}\right) \ ,$$

which is not the dominant term of the expansion of $f\left[x(\varepsilon)\right]$. A great care must be taken when calculating expansions embedded one in each other.

Chapter 5

5-1.

1. If

$$y(x, \varepsilon) = y_1(x) + \cdots \ ,$$

we have

$$y_1(x) = \frac{A}{x} e^{-x} \ ,$$

and the solution is singular at origin.

The change of variable $X = x/\varepsilon$ gives the equation

$$(1 + X)\frac{dY}{dX} + (1 + \varepsilon)Y + \varepsilon XY = 0 \ ,$$

with

$$Y(X, \varepsilon) \equiv y(x, \varepsilon) \ .$$

If

$$Y(X, \varepsilon) = Y_1(X) + \cdots \ ,$$

with the boundary condition $Y = 1$ at $X = 0$, we have

$$Y_1(X) = \frac{1}{1 + X} \ .$$

Now, to order 1, we have

$$E_0\,E_1\,Y_1 = 0\,,$$

whereas to order ε, we have

$$E_0\,E_1\,Y_1 = \frac{\varepsilon}{x}\,.$$

This lets us suppose that the outer expansion is such that

$$y(x,\varepsilon) = \varepsilon y_2(x) + \cdots\,.$$

If we set

$$Y(X,\varepsilon) = Y_1(X) + \varepsilon Y_2(X) + \cdots\,,$$

we find

$$Y_2(X) = -\frac{X}{1+X}\,,$$

where it has been assumed that $Y_2(0) = 0$.

Finally, to order ε, we have

$$E_0\,y = \varepsilon\frac{B}{x}\,\mathrm{e}^{-x}\,,$$

$$E_1\,y = \frac{1}{1+X} - \frac{\varepsilon X}{1+X}\,.$$

The application of the MVDP enables us to find

$$B = 1\,,$$

and, to this order, the UVA is

$$y_{\mathrm{a}} = \frac{\varepsilon}{x}(\mathrm{e}^{-x}-1) + \frac{1+\varepsilon}{1+X}\,.$$

2. SCEM leads us to seek a first approximation in the form

$$y_{\mathrm{a}1} = \frac{A}{x}\,\mathrm{e}^{-x} + Y_1(X,\varepsilon)\,,$$

whence the equation

$$L_\varepsilon\,y_{\mathrm{a}1} = (1+X)\frac{\mathrm{d}Y_1}{\mathrm{d}X} + (1+\varepsilon)Y_1 + \varepsilon XY_1 - \frac{\varepsilon A}{x^2}\,\mathrm{e}^{-x}\,.$$

The last term being formally of order ε^{-1} in the boundary layer, it is appropriate to set $A = \varepsilon A_0$ in order to solve the equation

$$(1+X)\frac{\mathrm{d}Y_1}{\mathrm{d}X} + Y_1 = \frac{A_0}{X^2}\,,$$

and we have

$$\mathrm{L}_\varepsilon\, y_{\mathrm{a}1} = \varepsilon(1 + X)Y_1 - \frac{A_0}{X^2}\left(\mathrm{e}^{-\varepsilon X} - 1\right) .$$

The solution is given by

$$y_{\mathrm{a}1} = \frac{\varepsilon A_0}{x}\left[\mathrm{e}^{-x} - \frac{\varepsilon}{x + \varepsilon}\right] + \frac{C_0}{1 + X} .$$

To satisfy the condition at origin, we set

$$C_0 - \varepsilon A_0 = 1 ,$$

which clearly leads us to take $A = 0$ and to solve the complete equation. Obviously, this is only a pedagogical example leading us to the exact solution

$$y = \frac{\varepsilon}{x + \varepsilon}\, \mathrm{e}^{-x} .$$

It is checked that

$$\lim_{\substack{\varepsilon \to 0 \\ x \ \text{fixed}}} \frac{y}{\varepsilon} = \frac{1}{x}\, \mathrm{e}^{-x} ,$$

whereas

$$\lim_{\substack{\varepsilon \to 0 \\ X \ \text{fixed}}} y = \frac{1}{1 + X} .$$

Note 5.1. If we use the change of variable $\bar{x} = \varepsilon x$, the initial equation becomes

$$\varepsilon(\bar{x} + \varepsilon^2)\frac{\mathrm{d}y}{\mathrm{d}\bar{x}} + \varepsilon(1 + \varepsilon)y + \bar{x}y = 0 .$$

With the boundary condition $y(0) = 1$, the solution is

$$y = \frac{\varepsilon^2}{\bar{x} + \varepsilon^2}\, \mathrm{e}^{-\bar{x}/\varepsilon} .$$

With MMAE, three layers appear. In the outer layer, the appropriate variable is $\bar{x}$; in fact, in this zone, the solution is simply $\bar{y} = 0$ to any order ε^n due to the term $\mathrm{e}^{-\bar{x}/\varepsilon}$ which appears in the exact solution. In the neighbourhood of the origin, two boundary layers form. For one of them (the middle layer), the appropriate variable is $\tilde{x} = \bar{x}/\varepsilon$ and for the other one (the inner layer), the appropriate variable is $\hat{x} = \bar{x}/\varepsilon^2$. In the middle layer, the initial equation becomes

$$(\tilde{x} + \varepsilon)\frac{\mathrm{d}\tilde{y}}{\mathrm{d}\tilde{x}} + (1 + \varepsilon)\tilde{y} + \tilde{x}\tilde{y} = 0 ,$$

with $\tilde{y}(0) = 1$. The problem reduces to the problem studied previously and it is seen that SCEM enables us to reduce the study of the two layers to the study of only one layer.

Even if the example is artificial, the advantge of SCEM is clearly demonstrated.

5-2. With the variable η, the outer expansion writes

$$\Phi = \frac{2-\alpha}{\ln\frac{1}{\varepsilon}} - \frac{1+\ln\eta}{\left(\ln\frac{1}{\varepsilon}\right)^2} \ .$$

With the variable η, the inner expansion writes

$$\Phi = \frac{1}{\ln\eta + \alpha\ln\frac{1}{\varepsilon} + 1} = \frac{1}{\alpha\ln\frac{1}{\varepsilon}\left(1 + \frac{1+\ln\eta}{\alpha\ln\frac{1}{\varepsilon}}\right)} = \frac{1}{\alpha\ln\frac{1}{\varepsilon}}\left[1 - \frac{1+\ln\eta}{\alpha\ln\frac{1}{\varepsilon}} + \cdots\right] \ .$$

Comparing the two expressions shows that with the chosen scales ε^α, it is not possible to satisfy the rule of intermediate matching.

With

$$\eta = x\ln\frac{1}{\varepsilon} = X\varepsilon\ln\frac{1}{\varepsilon} \ ,$$

the outer expansion becomes

$$\Phi = \frac{1}{\ln\frac{1}{\varepsilon}} - \frac{1+\ln\left(\eta/\ln\frac{1}{\varepsilon}\right)}{\left(\ln\frac{1}{\varepsilon}\right)^2} = \frac{1}{\ln\frac{1}{\varepsilon}} - \frac{1+\ln\eta}{\left(\ln\frac{1}{\varepsilon}\right)^2} + \cdots \ ,$$

and the inner expansion becomes

$$\Phi = \frac{1}{\ln\left[\frac{\eta}{\varepsilon\ln\frac{1}{\varepsilon}}\right] + 1} = \frac{1}{\ln\eta + \ln\frac{1}{\varepsilon} - \ln\ln\frac{1}{\varepsilon} + 1} = \frac{1}{\ln\frac{1}{\varepsilon}}\left[1 - \frac{1+\ln\eta}{\ln\frac{1}{\varepsilon}}\right] + \cdots \ .$$

With the scale $1/\left[\ln(1/\varepsilon)\right]$, the rule of intermediate matching is satisfied.

5-3.

1. With the variable η, the outer expansion writes

$$\Phi = 1 + \frac{B_1}{\ln\frac{1}{\varepsilon}}\left[-\ln\eta - (\alpha-1)\ln\frac{1}{\varepsilon} - \gamma\right]$$
$$+ \frac{B_2}{\left(\ln\frac{1}{\varepsilon}\right)^2}\left[-\ln\eta - (\alpha-1)\ln\frac{1}{\varepsilon} - \gamma\right] + \frac{B_1^2}{\left(\ln\frac{1}{\varepsilon}\right)^2}\left[-(\alpha-1)\ln\frac{1}{\varepsilon}\right] + \cdots \ ,$$

or

$$\Phi = 1 - (\alpha-1)B_1$$
$$+ \frac{1}{\ln\frac{1}{\varepsilon}}\left[-B_1\ln\eta - \gamma B_1 - (\alpha-1)B_2 - (\alpha-1)B_1^2\right] + \cdots \ .$$

With the variable η, the inner expansion writes

$$\Phi = \alpha A_1 + \frac{1}{\ln\frac{1}{\varepsilon}}\left[A_1\ln\eta + \alpha A_2\right] + \cdots \ .$$

Comparing the two expressions yields

$$\alpha A_1 = 1 - (\alpha - 1)B_1 \ ,$$
$$A_1 \ln \eta + \alpha A_2 = -B_1 \ln \eta - \gamma B_1 - (\alpha - 1)B_2 - (\alpha - 1)B_1^2 \ .$$

In order to satisfy the two equalities for any α such that $0 < \alpha < 1$, we have

$$A_1 = 1 \ , \quad B_1 = -1 \ , \quad A_2 = \gamma \ , \quad B_2 = -\gamma - 1 \ .$$

2.

$$\mathrm{E}_0^{(1)} \, \mathrm{E}_1^{(1)} \, \varPhi = A_1 \ ,$$
$$\mathrm{E}_1^{(1)} \, \mathrm{E}_0^{(1)} \, \varPhi = 1 \ .$$

The rule is satisfied since $A_1 = 1$.

$$\mathrm{E}_0^{(2)} \, \mathrm{E}_1^{(1)} \, \varPhi = A_1 + A_1 \frac{\ln x}{\ln \frac{1}{\varepsilon}}$$
$$= A_1 \frac{\ln X}{\ln \frac{1}{\varepsilon}} \ ,$$
$$\mathrm{E}_1^{(1)} \, \mathrm{E}_0^{(2)} \, \varPhi = 1 + B_1 - B_1 \frac{\ln X + \gamma}{\ln \frac{1}{\varepsilon}} \ .$$

The rule is not satisfied.

5-4. To order 1, we have

$$\mathrm{E}_0 \, y = \mathrm{e}^{1-x} \ ,$$
$$\mathrm{E}_1 \, \mathrm{E}_0 \, y = \mathrm{e} \ ,$$
$$\mathrm{E}_1 \, y = A_0(1 - \mathrm{e}^{-X}) \ ,$$
$$\mathrm{E}_0 \, \mathrm{E}_1 \, y = A_0 \ .$$

We have $A_0 = \mathrm{e}$.

To order ε, we have

$$\mathrm{E}_0 \, y = \mathrm{e}^{1-x} \left[1 + \varepsilon(1 - x)\right] \ ,$$
$$\mathrm{E}_0 \, y = \mathrm{e} \, \mathrm{e}^{-\varepsilon X} \left[1 + \varepsilon(1 - \varepsilon X)\right] \ ,$$
$$\mathrm{E}_1 \, \mathrm{E}_0 \, y = \mathrm{e}(1 - \varepsilon X + \varepsilon) \ ,$$
$$\mathrm{E}_1 \, \mathrm{E}_0 \, y = \mathrm{e}(1 - x + \varepsilon) \ ,$$

and

$$\mathrm{E}_1 \, y = \mathrm{e} \left(1 - \mathrm{e}^{-X}\right) + \varepsilon \left[(A_1 - \mathrm{e} \, X) - (A_1 + \mathrm{e} \, X)\mathrm{e}^{-X}\right] \ ,$$
$$\mathrm{E}_1 \, y = \mathrm{e} \left(1 - \mathrm{e}^{-x/\varepsilon}\right) + \varepsilon \left[(A_1 - \mathrm{e} \, \frac{x}{\varepsilon}) - (A_1 + \mathrm{e} \, \frac{x}{\varepsilon})\mathrm{e}^{-x/\varepsilon}\right] \ ,$$
$$\mathrm{E}_0 \, \mathrm{E}_1 \, y = \mathrm{e} - \mathrm{e} \, x + \varepsilon A_1 \ .$$

We obtain $A_1 = e$.

Therefore, the outer and inner expansions write

$$y = e^{1-x}\left[1 + \varepsilon(1-x)\right] + O(\varepsilon^2) \,,$$
$$y = e\left(1 - e^{-X}\right) + \varepsilon\, e\left[(1-X) - (1+X)\,e^{-X}\right] + O(\varepsilon^2) \,.$$

The UVA is obtained by forming a composite approximation

$$y_{\mathrm{a}} = \mathrm{E}_0\, y + \mathrm{E}_1\, y - \mathrm{E}_0\, \mathrm{E}_1\, y \,,$$
$$y_{\mathrm{a}} = e^{1-x} - e^{1-X} + \varepsilon\left[(1-x)\,e^{1-x} - (1+X)\,e^{1-X}\right] \,.$$

5-5.

1. If τ denotes time, we have

$$m\frac{\mathrm{d}^2 r}{\mathrm{d}\tau^2} = F_T + F_L \,,$$

where m is the mass of the spaceship and we have

$$F_T = -G\frac{mM_T}{r^2} \,, \quad F_L = G\frac{mM_L}{(d-r)^2} \,,$$

or

$$\frac{\mathrm{d}^2 r}{\mathrm{d}\tau^2} = -GM_T\frac{1}{r^2} + GM_L\frac{1}{(d-r)^2} \,.$$

With $t = \tau/T$, we obtain

$$\frac{\mathrm{d}^2 x}{\mathrm{d}t^2} = -\frac{1-\varepsilon}{x^2} + \frac{\varepsilon}{(1-x)^2} \,.$$

The radius of the Earth is taken as zero which introduces an apparent singularity at $x = 0$.

2.

$$\frac{\mathrm{d}t}{\mathrm{d}x} = \frac{\sqrt{\frac{x}{2}}}{\sqrt{1 + \varepsilon\frac{2x-1}{1-x}}} \,,$$

$$t_0(x) = \frac{\sqrt{2}}{3}x^{3/2} \,,$$

$$t_1(x) = \frac{\sqrt{2}}{3}x^{3/2} + \sqrt{\frac{x}{2}} - \frac{1}{2\sqrt{2}}\ln\frac{1+\sqrt{x}}{1-\sqrt{x}} \,.$$

We have $t_0(0) = 0$, $t_1(0) = 0$.

3.

$$T_0 = A \,,$$
$$\frac{\mathrm{d}T_1}{\mathrm{d}X} = -\frac{1}{\sqrt{2}}\sqrt{\frac{X}{1+X}} \,,$$
$$T_1 = -\frac{1}{\sqrt{2}}\sqrt{X(1+X)} + \frac{1}{\sqrt{2}}\ln\left[\sqrt{X} + \sqrt{1+X}\right] + B \,.$$

Using operators to order ε, we have

$$\mathrm{E}_1\,\mathrm{E}_0\,t = \frac{\sqrt{2}}{3} + \varepsilon\left(-\frac{X}{\sqrt{2}} + \frac{5}{3\sqrt{2}} - \frac{\ln 2}{\sqrt{2}} + \frac{1}{2\sqrt{2}}\ln\varepsilon + \frac{1}{2\sqrt{2}}\ln X\right),$$

$$\mathrm{E}_0\,\mathrm{E}_1\,t = A + \varepsilon\left(-\frac{1}{\sqrt{2}}\frac{1-x}{\varepsilon} - \frac{1}{2\sqrt{2}} + \frac{\ln 2}{\sqrt{2}} - \frac{1}{2\sqrt{2}}\ln\varepsilon\right.$$
$$\left. + \frac{1}{2\sqrt{2}}\ln(1-x) + B\right),$$

$$A = \frac{\sqrt{2}}{3},$$

$$B = \frac{13}{6\sqrt{2}} - \frac{2\ln 2}{\sqrt{2}} + \frac{1}{2\sqrt{2}}\ln\varepsilon,$$

$$t_{\mathrm{app}} = \frac{\sqrt{2}}{3}x^{3/2} + \frac{\varepsilon}{\sqrt{2}}\left\{X - \sqrt{X(1+X)} + \ln\left[\sqrt{X} + \sqrt{1+X}\right]\right.$$
$$\left. + \frac{2}{3}x^{3/2} + \sqrt{x} + \frac{1}{2} - \ln 2 - \ln(1+\sqrt{x}) + \frac{1}{2}\ln\varepsilon\right\}.$$

4. $f_0(x) = \dfrac{\sqrt{2}}{3}x^{3/2}.$

The remainder $\mathrm{L}_\varepsilon(t_{\mathrm{a}1})$ is $\mathrm{O}(\varepsilon)$ in domain $0 < A_1 \le x \le A_2 < 1$ but $\mathrm{L}_\varepsilon(t_{\mathrm{a}1})$ is $\mathrm{O}(1)$ in domain $0 < B_1 \le X \le B_2$.

$$f_1 = \frac{\sqrt{2}}{3}x^{3/2} + \sqrt{\frac{x}{2}} - \frac{1}{2\sqrt{2}}\ln\frac{1+\sqrt{x}}{1-\sqrt{x}},$$

$$F_1 = \frac{X}{\sqrt{2}} - \frac{1}{\sqrt{2}}\sqrt{X(1+X)} + \frac{1}{\sqrt{2}}\ln\left[\sqrt{X} + \sqrt{1+X}\right]$$
$$- \frac{1}{2\sqrt{2}}\ln X + \frac{1}{2\sqrt{2}} - \frac{\ln 2}{\sqrt{2}}.$$

In its regular form, SCEM yields an approximation identical to the composite form deduced from MMAE.

The remainder $\mathrm{L}_\varepsilon(t_{\mathrm{a}2})$ is $\mathrm{O}(\varepsilon^2)$ in domain $0 < A_1 \le x \le A_2 < 1$ but $\mathrm{L}_\varepsilon(t_{\mathrm{a}2})$ is $\mathrm{O}(\varepsilon)$ in domain $0 < B_1 \le X \le B_2$.

5.

$$\frac{\mathrm{d}t}{\mathrm{d}x} = \frac{\sqrt{\frac{x}{2}}}{\sqrt{1 - 2\varepsilon + \frac{\varepsilon}{1-x}}},$$

$$\frac{\mathrm{d}y_0}{\mathrm{d}x} = \sqrt{\frac{x}{2}}\frac{(1-x)^{1/2}}{(1-x+\varepsilon)^{1/2}},$$

$$\frac{\mathrm{d}y_1}{\mathrm{d}x} = \sqrt{\frac{x}{2}}\frac{(1-x)^{3/2}}{(1-x+\varepsilon)^{3/2}}.$$

6. Numerical results are given in Table S.2.

Table S.2. Numerical results

ε	numerical solution	MMAE	generalized SCEM
0.01	0.4602	0.4606	0.4602
0.1	0.4249	0.4452	0.4209
0.5	0.3927	0.6248	0.3481

5-6.

1. Substituting the outer expansion in the initial equation, we obtain

$$\frac{\mathrm{d}^2 w_0}{\mathrm{d}x^2} = -p(x) \, ,$$

whence the solution

$$w_0 = B_0 + A_0 x - \int_0^x \left(\int_0^\xi p(\lambda)\, \mathrm{d}\lambda \right) \mathrm{d}\xi \, .$$

In fact, the integral represents a double integral. Integrating first with respect to ξ, we have

$$\int_0^x \left(\int_0^\xi p(\lambda)\, \mathrm{d}\lambda \right) \mathrm{d}\xi = \int_0^x p(\lambda) \left(\int_\lambda^x \mathrm{d}\xi \right) \mathrm{d}\lambda = \int_0^x p(\lambda)(x - \lambda)\, \mathrm{d}\lambda \, .$$

In the neighbourhood of $x = 0$, using a Taylor series expansion, we obtain

$$w_0 = w_0(0) + x w_0'(0) + \frac{x^2}{2} w''(0) + \frac{x^3}{3!} w'''(0) + \mathrm{O}(x^4) \, .$$

We have

$$w_0(0) = B_0 \, , \qquad \frac{\mathrm{d}w_0}{\mathrm{d}x}(0) = A_0 \, ,$$

$$\frac{\mathrm{d}^2 w_0}{\mathrm{d}x^2}(0) = -p(0) \, , \quad \frac{\mathrm{d}^3 w_0}{\mathrm{d}x^3}(0) = -p'(0) \, ,$$

whence

$$w_0(x) = B_0 + A_0 x - p(0)\frac{x^2}{2} - p'(0)\frac{x^3}{3!} + \mathrm{O}(x^4) \quad \text{as} \quad x \to 0 \, .$$

In the same way, as $x \to 1$, we obtain

$$w_0(x) = B_0 + A_0 - \int_0^1 p(\lambda)(1 - \lambda)\, \mathrm{d}\lambda + \left[A_0 - \int_0^1 p(\lambda)\, \mathrm{d}\lambda \right] (x - 1)$$

$$-p(1)\frac{(x - 1)^2}{2} - p'(1)\frac{(x - 1)^3}{3!} + \mathrm{O}\left[(x - 1)^4\right] \, .$$

2. Substituting the inner expansion in the initial equation, we obtain

$$\frac{\varepsilon\mu_0}{\delta^4}\frac{\mathrm{d}^4 W_0}{\mathrm{d}X^4} + \frac{\varepsilon\mu_1}{\delta^4}\frac{\mathrm{d}^4 W_1}{\mathrm{d}X^4} - \frac{\mu_0}{\delta^2}\frac{\mathrm{d}^2 W_0}{\mathrm{d}X^2} - \frac{\mu_1}{\delta^2}\frac{\mathrm{d}^2 W_1}{\mathrm{d}X^2} = p(0) + X\delta p'(0) + \cdots .$$

In order to keep the order 4 derivative, we must take

$$\delta = \varepsilon^{1/2} .$$

With $\mu_0 = \varepsilon^{1/2}$, the equation for W_0 is

$$\frac{\mathrm{d}^4 W_0}{\mathrm{d}X^4} - \frac{\mathrm{d}^2 W_0}{\mathrm{d}X^2} = 0 .$$

The general solution of the equation is obtained from

$$\frac{\mathrm{d}^2 W_0}{\mathrm{d}X^2} - W_0 = -C_0 X + D_0 .$$

We have

$$W_0 = C_0 \left(X + \frac{\mathrm{e}^{-X}}{2} - \frac{\mathrm{e}^X}{2} \right) + D_0 \left(-1 + \frac{\mathrm{e}^{-X}}{2} + \frac{\mathrm{e}^X}{2} \right) .$$

The presence of a term in e^X makes impossible the matching with the outer expansion. We conclude that $C_0 = D_0$ and the solution for W_0 becomes

$$W_0 = C_0(X + \mathrm{e}^{-X} -1) .$$

3. With the expansion operators to order $\varepsilon^{1/2}$, we have

$$\mathrm{E}_1\,\mathrm{E}_0\,w = B_0 + A_0\varepsilon^{1/2}X + \varepsilon^{1/2}w_1(0) ,$$
$$\mathrm{E}_0\,\mathrm{E}_1\,w = C_0\varepsilon^{1/2}X - C_0\varepsilon^{1/2} ,$$

whence

$$B_0 = 0 \quad ; \quad A_0 = C_0 \quad ; \quad w_1(0) = -C_0 .$$

Therefore, we have

$$w_0 = C_0 x - \int_0^x p(\lambda)(x - \lambda)\,\mathrm{d}\lambda .$$

With $\nu_1 = \varepsilon^{1/2}$, the initial equation becomes

$$\varepsilon\left[\frac{\mathrm{d}^4 w_0}{\mathrm{d}x^4} + \varepsilon^{1/2}\frac{\mathrm{d}^4 w_1}{\mathrm{d}x^4}\right] - \left[\frac{\mathrm{d}^2 w_0}{\mathrm{d}x^2} + \varepsilon^{1/2}\frac{\mathrm{d}^2 w_1}{\mathrm{d}x^2}\right] = p(x) .$$

The equation for w_1 is

$$\frac{\mathrm{d}^2 w_1}{\mathrm{d}x^2} = 0 \, .$$

The solution is

$$w_1 = B_1 + A_1 x \, .$$

With $w_1(0) = -C_0$ we have $B_1 = -C_0$. The solution becomes

$$w_1 = -C_0 + A_1 x \, .$$

4. The boundary layer in the neighbourhood of $x = 1$ is similar to the boundary layer developing in the neighbourhood of $x = 0$. Repeating the same operations, we find that $\delta^+ = \varepsilon^{1/2}$ and

$$X^+ = \frac{x - 1}{\varepsilon^{1/2}} \, ,$$

with $X^+ \leq 0$.

We must take $\mu_0^+ = \varepsilon^{1/2}$ and the equation for W_0^+ is

$$\frac{\mathrm{d}^4 W_0^+}{\mathrm{d}X^{+4}} - \frac{\mathrm{d}^2 W_0^+}{\mathrm{d}X^{+2}} = 0 \, .$$

The solution has the form

$$W_0^+ = C_0^+ \left(X^+ + \frac{\mathrm{e}^{-X^+}}{2} - \frac{\mathrm{e}^{X^+}}{2} \right) + D_0^+ \left(-1 + \frac{\mathrm{e}^{-X^+}}{2} + \frac{\mathrm{e}^{X^+}}{2} \right) \, .$$

The matching with the outer expansion is impossible with a term in e^{-X^+} since $X^+ \leq 0$. Therefore, we must have $C_0^+ = -D_0^+$, whence

$$W_0^+ = C_0^+ \left(X^+ + 1 - \mathrm{e}^{X^+} \right) \, .$$

Using operators E_0 and E_1^+ to order $\varepsilon^{1/2}$, we have

$$\mathrm{E}_1^+ \, \mathrm{E}_0 \, w = C_0 - \int_0^1 p(\lambda)(1 - \lambda) \, \mathrm{d}\lambda + \left[C_0 - \int_0^1 p(\lambda) \, \mathrm{d}\lambda \right] \varepsilon^{1/2} X^+$$

$$+ \varepsilon^{1/2}(A_1 - C_0) \, ,$$

$$\mathrm{E}_0 \, \mathrm{E}_1^+ \, w = \varepsilon^{1/2} C_0^+ + \varepsilon^{1/2} C_0^+ X^+ \, .$$

We obtain

$$C_0 = -M^{(1)} \, , \quad C_0^+ = -M^{(0)} \, , \quad A_1 = -M^{(0)} - M^{(1)} \, .$$

In summary, we have the following approximations
- in the neighbourhood of $x = 0$

$$w = -\varepsilon^{1/2} M^{(1)} \left(X - 1 + e^{-X} \right) + o(\varepsilon^{1/2}) \,,$$

- outer expansion

$$w = -M^{(1)} x - \int_0^x (x - \lambda) p(\lambda) \, d\lambda + \varepsilon^{1/2} \left[M^{(1)} - (M^{(0)} + M^{(1)}) x \right] + o(\varepsilon^{1/2}) \,,$$

- in the neighbourhood of $x = 1$

$$w = -\varepsilon^{1/2} M^{(0)} \left(X^+ + 1 - e^{X^+} \right) + o(\varepsilon^{1/2}) \,.$$

Chapter 6

6-1.

1. Substituting the outer expansion in the initial equation, we obtain the equations for y_1 and y_2

$$\frac{dy_1}{dx} + y_1 = 0 \,,$$

$$\frac{dy_2}{dx} + y_2 = -\frac{d^2 y_1}{dx^2} \,.$$

With the boundary condition $y(1) = b$, we obtain the boundary conditions for y_1 and y_2

$$y_1(1) = b \,,$$

$$y_2(1) = 0 \,.$$

Thus, we have

$$y_1 = b\, e^{1-x} \,,$$

$$y_2 = b(1 - x)\, e^{1-x} \,.$$

With the change of variable $X = x/\varepsilon$, and setting $Y(X, \varepsilon) \equiv y(x, \varepsilon)$ the initial equation becomes

$$\frac{d^2 Y}{dX^2} + \frac{dY}{dX} + \varepsilon Y = 0 \,.$$

Substituting the inner expansion in this equation, we obtain the equations for Y_1 and Y_2

$$\frac{d^2 Y_1}{dX^2} + \frac{dY_1}{dX} = 0 \,,$$

$$\frac{d^2 Y_2}{dX^2} + \frac{dY_2}{dX} = -Y_1 \,.$$

Since the inner expansion describes the inner region, we can apply only the boundary condition $y(0) = a$. Then, the boundary conditions for Y_1 and Y_2 are

$$Y_1(0) = a \, ,$$
$$Y_2(0) = 0 \, ,$$

whence the solutions

$$Y_1 = a + A(1 - \mathrm{e}^{-X}) \, ,$$
$$Y_2 = B(1 - \mathrm{e}^{-X}) - (a + A)X - AX\,\mathrm{e}^{-X} \, .$$

Constants A and B are obtained by applying the MVDP. To order 1, we have

$$\mathrm{E}_1\,\mathrm{E}_0\,y = b\,\mathrm{e} \, ,$$
$$\mathrm{E}_0\,\mathrm{E}_1\,y = a + A \, ,$$

whence

$$A + a = b\,\mathrm{e} \, .$$

To order ε, we have

$$\mathrm{E}_1\,\mathrm{E}_0\,y = b\,\mathrm{e} + \varepsilon(-b\,\mathrm{e}\,X + b\,\mathrm{e}) \, ,$$
$$\mathrm{E}_0\,\mathrm{E}_1\,y = b\,\mathrm{e} + \varepsilon(B - b\,\mathrm{e}\,X) \, .$$

We obtain

$$B = b\,\mathrm{e} \, .$$

The composite approximations are obtained from

$$y_\mathrm{a} = \mathrm{E}_0\,y + \mathrm{E}_1\,y - \mathrm{E}_0\,\mathrm{E}_1\,y \, ,$$

that is

$$y_\mathrm{a1} = b\,\mathrm{e}^{1-x} + (a - b\,\mathrm{e})\,\mathrm{e}^{-X} \, ,$$
$$y_\mathrm{a2} = b\,\mathrm{e}^{1-x} + (a - b\,\mathrm{e})\,\mathrm{e}^{-X}$$
$$+ \varepsilon\left[b(1 - x)\,\mathrm{e}^{1-x} - b\,\mathrm{e}\,\mathrm{e}^{-X} + (a - b\,\mathrm{e})X\,\mathrm{e}^{-X} \right] \, .$$

2. We have

$$\mathrm{L}_\varepsilon\,y_\mathrm{a1} = (a - b\,\mathrm{e})\,\mathrm{e}^{-x/\varepsilon} + \varepsilon b\,\mathrm{e}^{1-x} \, ,$$
$$\mathrm{L}_\varepsilon\,y_\mathrm{a2} = (a - b\,\mathrm{e})x\,\mathrm{e}^{-x/\varepsilon} - \varepsilon b\,\mathrm{e}\,\mathrm{e}^{-x/\varepsilon} + (3 - x)\varepsilon^2 b\,\mathrm{e}^{1-x} \, .$$

In domain $0 \le x \le 1$, we have

$$\mathrm{L}_\varepsilon\,y_\mathrm{a1} = \mathrm{O}(1) \, ,$$
$$\mathrm{L}_\varepsilon\,y_\mathrm{a2} = \mathrm{O}(\varepsilon) \, ,$$

whereas in domain $0 < A_0 \leq x \leq 1$, we have

$$L_\varepsilon\, y_{a1} = O(\varepsilon)\,,$$
$$L_\varepsilon\, y_{a2} = O(\varepsilon^2)\,.$$

Complement. In bounded domains, it can be shown that if, in the domain of definition D of function y, we have

$$L_\varepsilon\, y_a = O(\varepsilon)\,,$$

then

$$y - y_a = O(\varepsilon)\,.$$

In other words, y_a is an approximation of y to order $O(\varepsilon)$.

Now, in domain D, we have

$$L_\varepsilon\, y_{a2} = O(\varepsilon)\,.$$

Therefore, we are certain that there exists a constant K such that

$$|y - y_{a2}| < K\varepsilon\,.$$

On the other hand, we can write

$$y - y_{a1} = y - y_{a2} + y_{a2} - y_{a1}\,.$$

As we have

$$y_{a2} - y_{a1} = O(\varepsilon)\,,$$

it is clear that there exists a constant K_1 such that

$$|y - y_{a1}| < K_1\varepsilon\,.$$

Therefore, we have demonstrated that y_{a1} is an approximation to order $O(\varepsilon)$.

In fact, with more sophisticated estimates, it can be shown directly that if, in domain $0 < A_0 \leq x \leq 1$, we have

$$L_\varepsilon\, y_a = O(\varepsilon)\,,$$

then

$$y - y_a = O(\varepsilon)\,.$$

This shows that y_{a1} is an approximation to order $O(\varepsilon)$ although $L_\varepsilon\, y_{a1} = O(1)$ in the whole domain D. In fact, here, the exact solution is available and we know that y_{a1} and y_{a2} are indeed approximations to the considered orders.

This is a much more complex chapter of the asymptotic analysis for singular perturbation problems, namely the justification of the matching principle.

6-2. The coefficient of $\dfrac{dy}{dx}$ being positive, the boundary layer is in the neighbourhood of $x = 0$.

The reduced equation is

$$\frac{dy_0}{dx} + y_0 = 0\,.$$

With $y_0(1) = 1$, the solution is

$$y_0 = e^{1-x} .$$

We seek a UVA in the form

$$y_{a1} = y_0 + Y_0(X, \varepsilon) , \quad X = \frac{x}{\delta} .$$

We are led to take $\delta = \varepsilon$. The initial equation becomes

$$\varepsilon \frac{d^2 y_0}{dx^2} + \frac{1}{\varepsilon} \frac{d^2 Y_0}{dX^2} + \frac{1}{\varepsilon} \frac{dY_0}{dX} + Y_0 = 0 .$$

Term $\varepsilon \dfrac{d^2 y_0}{dx^2}$ is of order ε in domain $0 < A_1 \leq X \leq A_2$ so that the equation for Y_0 is

$$\frac{d^2 Y_0}{dX^2} + \frac{dY_0}{dX} = 0 .$$

The solution is

$$Y_0 = \alpha + \beta\, e^{-X} .$$

The boundary conditions yield $\alpha = 0$ and $\beta = 0$.

Therefore we have

$$y_{a1} = e^{1-x} .$$

The next approximation is

$$y = y_0 + \nu y_1(x, \varepsilon) .$$

The initial equation becomes

$$\varepsilon \frac{d^2 y_0}{dx^2} + \varepsilon \nu \frac{d^2 y_1}{dx^2} + \nu \frac{dy_1}{dx} + \nu y_1 = 0 .$$

We take $\nu = \varepsilon$ and the eqution for y_1 becomes

$$\frac{dy_1}{dx} + y_1 = -\frac{d^2 y_0}{dx^2} .$$

With $y_1(1) = 0$, the solution is

$$y_1 = (1 - x)\, e^{(1-x)} .$$

We seek a UVA in the form

$$y = y_0(x) + \varepsilon y_1(x) + \varepsilon Y_1(X, \varepsilon) .$$

The initial equation becomes

$$\varepsilon^2 \frac{d^2 y_1}{dx^2} + \frac{d^2 Y_1}{dX^2} + \frac{dY_1}{dX} + \varepsilon Y_1 = 0 .$$

Term $\varepsilon^2 \dfrac{\mathrm{d}^2 y_1}{\mathrm{d}x^2}$ is of order ε^2 in domain $0 < A_1 \leq X \leq A_2$. The equation for Y_1 is

$$\frac{\mathrm{d}^2 Y_1}{\mathrm{d}X^2} + \frac{\mathrm{d}Y_1}{\mathrm{d}X} = 0 \,.$$

The solution is

$$Y_1 = A + B\,\mathrm{e}^{-X} \,.$$

The boundary conditions are such that

$$x = 0 : \quad y_1 + Y_1 = 0 \,; \quad x = 1 : \quad y_1 + Y_1 = 0 \,.$$

We obtain

$$A = \frac{\mathrm{e}^{1-1/\varepsilon}}{1 - \mathrm{e}^{-1/\varepsilon}} \,, \quad B = -\frac{\mathrm{e}}{1 - \mathrm{e}^{-1/\varepsilon}} \,,$$

whence the solution

$$y = \mathrm{e}^{1-x} + \varepsilon \left[(1-x)\,\mathrm{e}^{1-x} + \frac{\mathrm{e}^{1-1/\varepsilon} - \mathrm{e}^{1-X}}{1 - \mathrm{e}^{-1/\varepsilon}} \right] \,.$$

With the regular form of SCEM, the solution is

$$y = \mathrm{e}^{1-x} + \varepsilon \left[(1-x)\,\mathrm{e}^{1-x} - \mathrm{e}^{1-X} \right] \,.$$

6-3. To determine the variables appropriate to the boundary layers, we set $\xi = X/\delta_1(\varepsilon)$ and $\zeta = (1-x)/\delta_2(\varepsilon)$. With these changes of variables, we compare the order of magnitude of different terms in the equation. It is easily shown that the changes of variables enabling us to describe the boundary layers are

$$X = \frac{x}{\varepsilon} \,; \quad \zeta = \frac{1-x}{\varepsilon^{1/2}} \,.$$

The reduced equation is

$$(1-x)\frac{\mathrm{d}y_0}{\mathrm{d}x} - y_0 = 0 \,.$$

The solution is

$$y_0 = \frac{b}{1-x} \,.$$

The solution is complemented in the form

$$y = y_0(x) + Z_0(\zeta) \,.$$

The initial equation becomes

$$\varepsilon \frac{\mathrm{d}^2 y_0}{\mathrm{d}x^2} + \frac{\mathrm{d}^2 Z_0}{\mathrm{d}\zeta^2} - \zeta \frac{\mathrm{d}Z_0}{\mathrm{d}\zeta} - Z_0 = 0 \,.$$

Now, we have

$$\varepsilon \frac{\mathrm{d}^2 y_0}{\mathrm{d}x^2} = \frac{1}{\varepsilon^{1/2}} \frac{2b}{\zeta^3} .$$

We conclude that $b = 0$, otherwise this term of order $\varepsilon^{-1/2}$ remains alone in the equation. Therefore, we obtain

$$\frac{\mathrm{d}^2 Z_0}{\mathrm{d}\zeta^2} - \zeta \frac{\mathrm{d}Z_0}{\mathrm{d}\zeta} - Z_0 = 0 .$$

It is checked that the solution has the form

$$Z_0 = \mathrm{e}^{\zeta^2/2} \left[A + B \int_0^{\zeta/\sqrt{2}} \mathrm{e}^{-t^2} \, \mathrm{d}t \right] .$$

At $x = 1$, the condition $y = 1$ implies $A = 1$.

We seek a UVA in the form

$$y_\mathrm{a} = Z_0(\zeta) + Y_0(\xi) .$$

Substituting in the initial equation, it is shown that the equation for Y_0 is

$$\frac{\mathrm{d}^2 Y_0}{\mathrm{d}\xi^2} + \frac{\mathrm{d}Y_0}{\mathrm{d}\xi} = 0 ,$$

whence the solution

$$Y_0 = \alpha + \beta \, \mathrm{e}^{-\xi} .$$

This function must satisfy the boundary condition at $x = 0$, i.e. $\xi = 0$. Therefore, we have

$$\alpha + \beta = 1 .$$

Since the condition $y(0) = 1$ is satisfied by Y_0, the contribution of Z_0 vanishes at $x = 0$. This condition must be applied at $\zeta = 1/\varepsilon^{1/2}$. In its regular form, SCEM demands that this limit is imposed as $\zeta \to \infty$. Therefore, we obtain

$$A + B \int_0^\infty \mathrm{e}^{-t^2} \, \mathrm{d}t = 0 ,$$

or, with $A = 1$,

$$B = -\frac{2}{\sqrt{\pi}} .$$

In the same way, since the condition $y(1) = 1$ is satisfied by Z_0, the contribution of Y_0 vanishes at $x = 1$. This condition must be applied at $\xi = 1/\varepsilon$, i.e. with the regular form of SCEM as $\xi \to \infty$, whence

$$\alpha = 0 ,$$

and, with $\alpha + \beta = 1$, we have $\beta = 1$.

Finally, the solution is

$$y_{\mathrm{a}} = \mathrm{e}^{-\xi} + \mathrm{e}^{\zeta^2/2}\left[1 - \frac{2}{\sqrt{\pi}}\int_0^{\zeta/\sqrt{2}} \mathrm{e}^{-t^2}\,\mathrm{d}t\right] .$$

6-4. The equation for p_0 is obtained by setting $\varepsilon = 0$

$$\frac{\mathrm{d}(p_0 h)}{\mathrm{d}x} = 0 .$$

By taking into account the condition at $x = 0$, the solution is

$$p_0 = \frac{h_0}{h} .$$

To determine the boundary layer variable, we set $X = (1-x)/\delta(\varepsilon)$. With this change of variable, it is easily shown that the boundary layer is restored by taking $\delta = \varepsilon$. The equation for $P_0(X)$ is obtained by expanding $h(x)$ and $p_0(x)$ in the neighbourhood of $x = 1$ in order to have $P_0 = P_0(X)$. Thus, we have

$$h(x) = h(1 - \varepsilon X) = h(1) - \varepsilon X\left(\frac{\mathrm{d}h}{\mathrm{d}x}\right)_{x=1} + \cdots = 1 - \varepsilon X\left(\frac{\mathrm{d}h}{\mathrm{d}x}\right)_{x=1} + \cdots ,$$

and

$$p_0 = \frac{h_0}{h} = h_0\left[1 + \varepsilon X\left(\frac{\mathrm{d}h}{\mathrm{d}x}\right)_{x=1} + \cdots\right] .$$

Moreover, we have

$$\frac{\mathrm{d}P_0}{\mathrm{d}x} = -\frac{1}{\varepsilon}\frac{\mathrm{d}P_0}{\mathrm{d}X} .$$

In the initial equation, we substitute $p_0(x) + P_0(X)$ for p. After examining the order of magnitude of all the terms in domain $0 < A_1 \leq X \leq A_2$ where A_1 and A_2 are constants independent of ε, we obtain

$$\frac{\mathrm{d}}{\mathrm{d}X}\left[(h_0 + P_0)\frac{\mathrm{d}P_0}{\mathrm{d}X} + P_0\right] = 0 ,$$

whence

$$-X = (h_0 + C_1)\ln|P_0 - C_1| + P_0 + C_2 .$$

The boundary conditions are applied. Taking into account p_0, at $x = 1$ or $X = 0$ we must have

$$P_0(0) = 1 - h_0 .$$

The condition at $x = 0$ becomes a condition at $X = 1/\varepsilon$. With the regular form of SCEM, the condition must be prescribed as $X \to \infty$; we have

$$X \to \infty : \quad P_0 = 0 ,$$

which gives $U_1 = 0$ and we obtain

$$-X = h_0 \ln \frac{|P_0|}{|1 - h_0|} + P_0 - 1 + h_0 \ .$$

6-5.

1. The velocity components are

$$u = 1 + \varepsilon \frac{\partial \varphi_1}{\partial x} + \frac{\partial \Phi_1}{\partial S_1} + \frac{\partial \Psi_1}{\partial S_2} + \varepsilon \left[\frac{\partial \Phi_2}{\partial S_1} + \frac{\partial \Psi_2}{\partial S_2} \right] \ ,$$

$$v = \varepsilon \frac{\partial \varphi_1}{\partial y} + \frac{\partial \Phi_1}{\partial Y} + \frac{\partial \Psi_1}{\partial Y} + \varepsilon \left[\frac{\partial \Phi_2}{\partial Y} + \frac{\partial \Psi_2}{\partial Y} \right] \ .$$

2. The slip condition at the wall writes

$$v = \pm \varepsilon T'(x) u \ ,$$

and must be expressed at $y = \pm \varepsilon T(x)$. Thus, using the appropriate Taylor series expansions as $\varepsilon \to 0$, the slip condition along the upper surface becomes

$$\frac{\partial \varphi_1}{\partial y}(x, 0+) = T' - f + g \ ,$$

$$\frac{1}{\sqrt{2S_1}} \left[1 + \frac{\partial \Phi_1}{\partial S_1}(S_1, \sqrt{2S_1}) \right] = \frac{\partial \Phi_1}{\partial Y}(S_1, \sqrt{2S_1}) \ ,$$

$$\frac{1}{\sqrt{2S_1}} \left[\frac{\partial \varphi_1}{\partial x}(-1_+, 0_+) + \frac{\partial \Phi_2}{\partial S_1}(S_1, \sqrt{2S_1}) \right] = \frac{\partial \Phi_2}{\partial Y}(S_1, \sqrt{2S_1}) \ ,$$

$$-\frac{1}{\sqrt{-2S_2}} \left[1 + \frac{\partial \Psi_1}{\partial S_2}(S_2, \sqrt{-2S_2}) \right] = \frac{\partial \Psi_1}{\partial Y}(S_2, \sqrt{-2S_2}) \ ,$$

$$-\frac{1}{\sqrt{-2S_2}} \left[\frac{\partial \varphi_1}{\partial x}(1_-, 0_+) + \frac{\partial \Psi_2}{\partial S_2}(S_2, \sqrt{-2S_2}) \right] = \frac{\partial \Psi_2}{\partial Y}(S_2, \sqrt{-2S_2}) \ ,$$

with

$$S_1 > 0 \ , \quad S_2 < 0 \ .$$

The notation $\dfrac{\partial \varphi_1}{\partial x}(-1_+, 0_+)$ means that the derivative $\dfrac{\partial \varphi_1}{\partial x}$ is evaluated on the upper surface ($y = 0_+$) as $x \to -1$ with $x > -1$. Similarly, the notation $\dfrac{\partial \varphi_1}{\partial x}(1_-, 0_+)$ means that the derivative $\dfrac{\partial \varphi_1}{\partial x}$ is evaluated on the upper surface ($y = 0_+$) as $x \to 1$ with $x < 1$.

3. Each potential φ_1, Φ_1, Φ_2, Ψ_1, Ψ_2 satisfies the potential equation

$$\triangle \varphi_1 = 0 \ , \quad \triangle \Phi_1 = 0 \ , \quad \triangle \Phi_2 = 0 \ , \quad \triangle \Psi_1 = 0 \ , \quad \triangle \Psi_2 = 0 \ .$$

The solution for the potential φ_1 is given by the thin airfoil theory, but this potential does not correspond to the flow around an ellipse because the terms $-f$ and g have been added. These terms enable us to eliminate the singularities introduced by the thin airfoil theory at the leading edge and at the trailing edge.

The last four relations describing the slip condition give the flow around a parabola whose equation is $Y = \sqrt{2S_1}$ or $Y = \sqrt{-2S_2}$.

From relation

$$\frac{1}{\sqrt{2S_1}}\left[1 + \frac{\partial \Phi_1}{\partial S_1}(S_1, \sqrt{2S_1})\right] = \frac{\partial \Phi_1}{\partial Y}(S_1, \sqrt{2S_1}),$$

the potential $\Phi_1 + S_1$ corresponds to the flow around a parabola of equation $Y = \sqrt{2S_1}$. Then, the velocity components u and v on the parabola $Y = \sqrt{2S_1}$ corresponding to potential Φ_1 are

$$u = -\frac{\varepsilon^2}{\varepsilon^2 + 2(1+x)}, \qquad v = \varepsilon\frac{\sqrt{2(1+x)}}{\varepsilon^2 + 2(1+x)}.$$

The velocity components u and v on the parabola $Y = \sqrt{-2S_2}$ corresponding to potential Ψ_1 are

$$u = -\frac{\varepsilon^2}{\varepsilon^2 + 2(1-x)}, \qquad v = -\varepsilon\frac{\sqrt{2(1-x)}}{\varepsilon^2 + 2(1-x)}.$$

In equation

$$\frac{1}{\sqrt{2S_1}}\left[\frac{\partial \varphi_1}{\partial x}(-1_+, 0_+) + \frac{\partial \Phi_2}{\partial S_1}(S_1, \sqrt{2S_1})\right] = \frac{\partial \Phi_2}{\partial Y}(S_1, \sqrt{2S_1}),$$

taking into account the solution for φ_1, we have $\dfrac{\partial \varphi_1}{\partial x}(-1_+, 0_+) = 1$.

Similarly, in equation

$$-\frac{1}{\sqrt{-2S_2}}\left[\frac{\partial \varphi_1}{\partial x}(1_-, 0_+) + \frac{\partial \Psi_2}{\partial S_2}(S_2, \sqrt{-2S_2})\right] = \frac{\partial \Psi_2}{\partial Y}(S_2, \sqrt{-2S_2}),$$

we have $\dfrac{\partial \varphi_1}{\partial x}(1_-, 0_+) = 1$.

Therefore, the solution for Φ_2 is identical to the solution for Φ_1. In the same way, the solution for Ψ_2 is identical to the solution for Ψ_1.

The velocity components u and v on the parabola $Y = \sqrt{2S_1}$ corresponding to potential Φ_2 are

$$u = -\frac{\varepsilon^2}{\varepsilon^2 + 2(1+x)}, \qquad v = \varepsilon\frac{\sqrt{2(1+x)}}{\varepsilon^2 + 2(1+x)}.$$

The velocity components u and v on the parabola $Y = \sqrt{-2S_2}$ corresponding to potential Ψ_2 are

$$u = -\frac{\varepsilon^2}{\varepsilon^2 + 2(1-x)}, \qquad v = -\varepsilon\frac{\sqrt{2(1-x)}}{\varepsilon^2 + 2(1-x)}.$$

Finally, the velocity components on the ellipse and the resulting velocity are

$$u = (1 + \varepsilon)\left[1 - \frac{\varepsilon^2}{\varepsilon^2 + 2(1 + x)} - \frac{\varepsilon^2}{\varepsilon^2 + 2(1 - x)}\right] ,$$

$$v = \varepsilon\left[-\frac{x}{\sqrt{1 - x^2}} - \frac{1}{\sqrt{2(x + 1)}} + \frac{1}{\sqrt{2(1 - x)}}\right]$$

$$+ \varepsilon(1 + \varepsilon)\left[\frac{\sqrt{2(1 + x)}}{\varepsilon^2 + 2(1 + x)} - \frac{\sqrt{2(1 - x)}}{\varepsilon^2 + 2(1 - x)}\right] ,$$

$$q = \sqrt{u^2 + v^2} .$$

Figure S.3 compares the exact solution with the regular SCEM approximation. We note that the SCEM approximation does not give exactly zero velocity at the stagnation point because the expansions are regular and the boundary conditions are not applied exactly.

6-6. First, we have $\varphi_1 = \bar{\varphi}_1$ and, as $\varepsilon \to 0$, we have

$$\bar{\psi}_1 = \psi_1(X) - 1 + \mathrm{EST} ,$$

where $\varphi_1(x)$ and $\psi_1(X)$ are the functions appearing in MMAE expansions.

On the other hand, in variable X, as $\varepsilon \to 0$, we have

$$\varepsilon\frac{\mathrm{d}^2\bar{\varphi}_1}{\mathrm{d}x^2} \cong -\frac{1}{4}X^{-5/4} + \frac{2}{3}\varepsilon^{3/5}X^{-1/2} + \frac{10}{9}\varepsilon^{6/5}X^{1/4} + \cdots .$$

Expansion of $\bar{\psi}_2$. A regular expansion of $\bar{\psi}_2$ to order $\varepsilon^{2/5}$ is given by

$$\bar{\psi}_2 = \bar{F}_1(X) + \varepsilon^{2/5}(\bar{f}_1(x) + \bar{F}_2(X)) + \mathrm{o}(\varepsilon^{2/5}) , \tag{S.1}$$

where $\bar{F}_1$, $\bar{f}_1$ and $\bar{F}_2$ satisfy the following equations

$$\frac{\mathrm{d}^2\bar{F}_1}{\mathrm{d}X^2} + X^{1/4}\frac{\mathrm{d}\bar{F}_1}{\mathrm{d}X} = \psi_1 - 1 + \frac{1}{4}X^{-5/4} ,$$

$$x^{1/4}\frac{\mathrm{d}\bar{f}_1}{\mathrm{d}x} = -\frac{\mathrm{d}^2\varphi_1}{\mathrm{d}x^2} - \frac{1}{4}x^{-5/4} ,$$

$$\frac{\mathrm{d}^2\bar{F}_2}{\mathrm{d}X^2} + X^{1/4}\frac{\mathrm{d}\bar{F}_2}{\mathrm{d}X} = 0 .$$

The boundary conditions for $\bar{F}_1$, $\bar{f}_1$ and $\bar{F}_2$ are obtained by determining the outer and inner expansions of $\bar{\psi}_2$. From (S.1), we have

$$\mathrm{E}_0\,\bar{\psi}_2 = \mathrm{E}_0\,\bar{F}_1 + \varepsilon^{2/5}(\bar{f}_1 + \mathrm{E}_0\,\bar{F}_2) .$$

Now, it can be shown that

$$\bar{F}_1 \cong C_1 - \frac{1}{2}X^{-1/2} + \cdots \quad \text{as} \quad X \to \infty ,$$

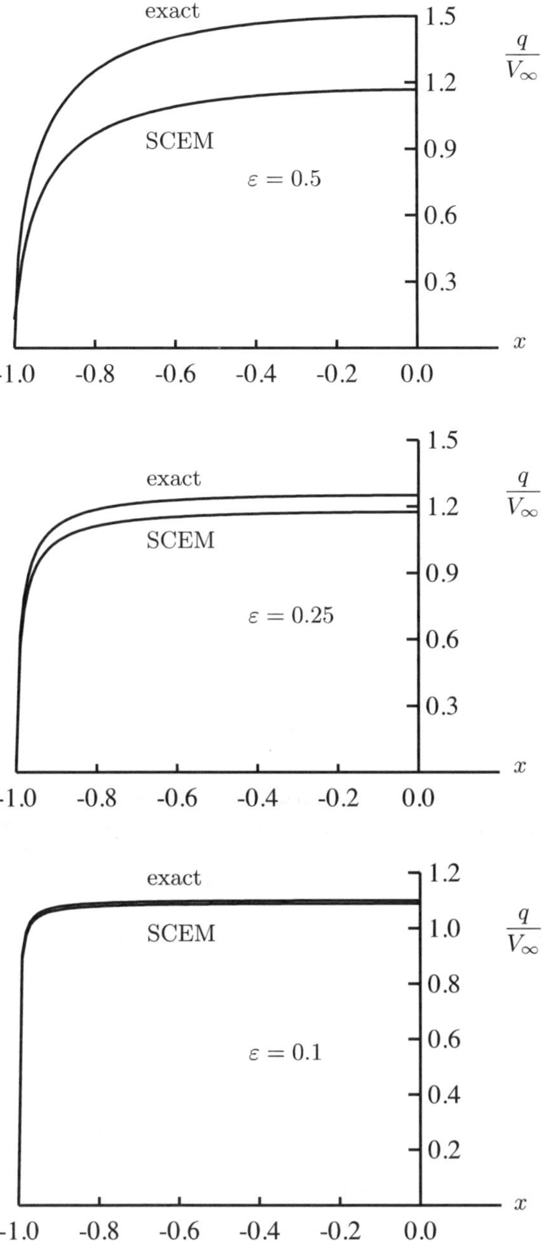

Fig. S.3. Velocity modulus on the ellipse

whence to order $\varepsilon^{2/5}$

$$\mathrm{E}_0\,\bar{F}_1 = C_1 - \frac{1}{2}\varepsilon^{2/5}x^{-1/2}\;.$$

On the other hand, to order 1, we have

$$\mathrm{E}_0\,\bar{F}_2 = C_2\;.$$

We obtain

$$\mathrm{E}_0\,\bar{\psi}_2 = C_1 + \varepsilon^{2/5}\left(\bar{f}_1 - \frac{1}{2}x^{-1/2} + C_2\right)\;.$$

The function $\bar{f}_1$ is known to within an additive constant and, without changing the final result, we can take $\bar{f}_1(1) = 0$. Then, to satisfy the boundary condition $\bar{\psi}_2 = 0$ at $x = 1$ we have

$$C_1 = 0\;,\quad C_2 = \frac{1}{2}\;.$$

The inner expansion of $\bar{\psi}_2$ is

$$\mathrm{E}_1\,\bar{\psi}_2 = \bar{F}_1(X) + \varepsilon^{2/5}(\bar{f}_1(0) + \bar{F}_2(X))\;,$$

since the behaviour of $\bar{f}_1$ as $x \to 0$ is

$$\bar{f}_1 = \bar{f}_1(0) - \frac{8}{3}x^{1/4}\;,$$

where constant $\bar{f}_1(0) = 4.4221$ is obtained from the numerical solution of the equation for $\bar{f}_1$. The boundary condition $\bar{\psi}_2(0) = 0$ is satisfied by taking

$$\bar{F}_1(0) = 0\;,\quad \bar{f}_1(0) + \bar{F}_2(0) = 0\;.$$

Expansion of $\bar{\varphi}_3$. A regular expansion of $\bar{\varphi}_3$ to order 1 is

$$\bar{\varphi}_3 = \bar{g}_1(x) + \bar{G}_1(X) + \mathrm{o}(1)\;,\tag{S.2}$$

where $\bar{g}_1$ and $\bar{G}_1$ satisfy the equations

$$x^{1/4}\frac{\mathrm{d}\bar{g}_1}{\mathrm{d}x} - \bar{g}_1 = \bar{f}_1 - \frac{1}{2}x^{-1/2} + \frac{1}{2}\;,$$

$$\frac{\mathrm{d}\bar{G}_1}{\mathrm{d}X} = 0\;.$$

Function $\bar{G}_1$ is constant

$$\bar{G}_1 = C_3\;.$$

The outer and inner expansions of $\bar{\varphi}_3$ to order 1 are

$$\mathrm{E}_0\,\bar{\varphi}_3 = \bar{g}_1(x) + C_3\;,$$
$$\mathrm{E}_1\,\bar{\varphi}_3 = \bar{g}_1(0) + C_3\;.$$

In order to satisfy the condition $\bar{\varphi}_3(1) = 0$ and taking $\bar{g}_1(1) = 0$, we have

$$C_3 = 0 \ .$$

Expansion of $\bar{\psi}_3$. A regular expansion of $\bar{\Psi}_3$ to order 1 is

$$\bar{\Psi}_3 = \bar{H}_1(X) + o(1) \ , \tag{S.3}$$

where $\bar{H}_1(X)$ satisfies the equation

$$\frac{\mathrm{d}^2 \bar{H}_1}{\mathrm{d}X^2} + X^{1/4}\frac{\mathrm{d}\bar{H}_1}{\mathrm{d}X} = 0 \ .$$

The solution is

$$\bar{H}_1(X) = C_4 G_{5/4}(X) + C_5 \ .$$

The inner and outer expansions of $\bar{\Psi}_3$ are

$$\mathrm{E}_0\,\bar{\Psi}_3 = C_4 G_{5/4}(\infty) + C_5 \ ,$$
$$\mathrm{E}_1\,\bar{\Psi}_3 = C_4 G_{5/4}(X) + C_5 \ .$$

In order to satisfy the boundary conditions on $\bar{\Psi}_3$, we have

$$C_4 G_{5/4}(\infty) + C_5 = 0 \ ,$$
$$C_5 = -\bar{g}_1(0) \ .$$

The boundary conditions on $\bar{H}_1$ are

$$\bar{H}_1(0) = -\bar{g}_1(0) \ ,$$
$$\bar{H}_1 \to 0 \quad \text{as} \quad X \to \infty \ .$$

Identification with MMAE results. The results obtained above show that a regular expansion of $\bar{\Phi}_{a3}$, uniformly valid to order ε is

$$\bar{\Phi}_{a3} = \varphi_1 + \psi_1 - 1 + \varepsilon^{3/5}\bar{F}_1 + \varepsilon(\bar{f}_1 + \bar{g}_1 + \bar{F}_2 + \bar{H}_1) + o(\varepsilon) \ .$$

We define the following functions

$$F_2(X) = \bar{F}_1 + \frac{4}{3}X^{3/4} \ ,$$
$$f_2(x) = \bar{f}_1 + \bar{g}_1 - \frac{1}{2}x^{-1/2} + \frac{1}{2} \ ,$$
$$F_3(X) = \bar{F}_2 + \bar{H}_1 + \bar{f}_1(0) + \bar{g}_1(0) \ .$$

From the equations for $\bar{F}_1$, $\bar{f}_1$, $\bar{g}_1$ and $\bar{H}_1$ we deduce

$$\frac{\mathrm{d}^2 F_2}{\mathrm{d}X^2} + X^{1/4}\frac{\mathrm{d}F_2}{\mathrm{d}X} = \psi_1 \ ,$$
$$x^{1/4}\frac{\mathrm{d}f_2}{\mathrm{d}x} - f_2 = -\frac{\mathrm{d}^2\varphi_1}{\mathrm{d}x^2} \ ,$$
$$\frac{\mathrm{d}^2 F_3}{\mathrm{d}X^2} + X^{1/4}\frac{\mathrm{d}F_3}{\mathrm{d}X} = 0 \ ,$$

with the boundary conditions

$$F_2(0) = 0 \,,$$
$$F_2 \cong \frac{4}{3} X^{3/4} - \frac{1}{2} X^{-1/2} \quad \text{as} \quad X \to \infty \,,$$
$$f_2(1) = 0 \,,$$
$$F_3(0) = 0 \,,$$
$$F_3 \to \bar{f}_1(0) + \bar{g}_1(0) + \frac{1}{2} \quad \text{as} \quad X \to \infty \,.$$

Then, the following identifications hold

$$\psi_2 = F_2 \,,$$
$$\varphi_2 = f_2 \,,$$
$$\psi_3 = F_3 \,.$$

With the expression of φ_2, we obtain

$$\bar{f}_1(0) + \bar{g}_1(0) = 4 \,.$$

Finally, a regular expansion of Φ, uniformly valid to order ε is

$$\Phi = \varphi_1 + \psi_1 - 1 + \varepsilon^{3/5} \left(\psi_2 - \frac{4}{3} X^{3/4} \right) + \varepsilon \left(\varphi_2 + \frac{1}{2} x^{-1/2} - \frac{9}{2} + \psi_3 \right) + o(\varepsilon) \,.$$

This result is *exactly* the composite expansion (6.16) obtained with MMAE.
It is concluded that *SCEM results contain MMAE results.*
6-7. First, we have $\varphi_1 = \bar{\varphi}_1$ and, as $\varepsilon \to 0$, we obtain

$$\bar{\psi}_1 = \psi_1(X) - 1 + \text{EST} \,,$$

where $\varphi_1(x)$ and $\psi_1(X)$ are the functions appearing in MMAE.
Expansion of $\bar{\psi}_2$. Equation (6.34) for $\bar{\psi}_2$ is

$$\frac{\mathrm{d}^2 \bar{\psi}_2}{\mathrm{d}X^2} + X^{1/3} \frac{\mathrm{d}\bar{\psi}_2}{\mathrm{d}X} = -\varepsilon \frac{\mathrm{d}^2 \bar{\varphi}_1}{\mathrm{d}x^2} + \bar{\psi}_1 \,.$$

The boundary conditions are

$$\bar{\psi}_2(0, \varepsilon) = 0 \,, \quad \bar{\psi}_2(\varepsilon^{-3/4}, \varepsilon) = 0 \,.$$

It can be shown that

$$\bar{\psi}_2 = \bar{F}_1(X) - \varepsilon^{1/2} \ln \varepsilon \bar{F}_3^*(X) + \varepsilon^{1/2}(\bar{F}_2(X) + \bar{f}_1(x)) + \cdots \,. \tag{S.4}$$

From the equation for $\bar{\psi}_2$, we obtain the following equations

$$\frac{\mathrm{d}^2 \bar{F}_1}{\mathrm{d}X^2} + X^{1/3}\frac{\mathrm{d}\bar{F}_1}{\mathrm{d}X} = \frac{1}{3}X^{-4/3} + \psi_1 - 1 \,,$$

$$\frac{\mathrm{d}^2 \bar{F}_3^*}{\mathrm{d}X^2} + X^{1/3}\frac{\mathrm{d}\bar{F}_3^*}{\mathrm{d}X} = 0 \,,$$

$$\frac{\mathrm{d}^2 \bar{F}_2}{\mathrm{d}X^2} + X^{1/3}\frac{\mathrm{d}\bar{F}_2}{\mathrm{d}X} = -\frac{1}{2}X^{-2/3} \,,$$

$$x^{1/3}\frac{\mathrm{d}\bar{f}_1}{\mathrm{d}x} = -\frac{\mathrm{d}^2 \varphi_1}{\mathrm{d}x^2} - \frac{1}{3}x^{-4/3} + \frac{1}{2}x^{-2/3} \,.$$

The boundary conditions are deduced from the boundary conditions on $\bar{\psi}_2$ and from the calculation of $E_0\,\bar{\psi}_2$ and of $E_1\,\bar{\psi}_2$. We obtain

$$\bar{F}_1(0) = 0 \,,$$

$$\bar{F}_3^*(0) = 0 \,,$$

$$\bar{F}_2(0) + \bar{f}_1(0) = 0 \,,$$

$$\bar{f}_1(1) = 0 \,.$$

In fact, the condition $\bar{f}_1(1) = 0$ is chosen arbitrarily. Any other constant gives the same final result. Here, the condition $\bar{f}_1(1) = 0$ is chosen for convenience.

Calculating $E_0\,\bar{\psi}_2$ and applying the condition $\bar{\psi}_2 = 0$ at $x = 1$, we obtain as $X \to \infty$

$$\bar{F}_1 \cong -\frac{1}{2}X^{-2/3} \,,$$

$$\bar{F}_2 \cong \frac{1}{2} - \frac{1}{2}\ln X + \frac{3}{8}X^{-4/3} \,,$$

$$\bar{F}_3^* \to \frac{3}{8} \,.$$

The solution for $\bar{F}_3^*$ is

$$\bar{F}_3^* = \frac{3}{8}\frac{G_{4/3}(X)}{G_{4/3}(\infty)} \,.$$

In addition, we have the identification

$$\psi_2 = \bar{F}_1 + \frac{3}{2}X^{2/3} \,.$$

Indeed, it is easy to check that the functions ψ_2 and $\bar{F}_1 + \frac{3}{2}X^{2/3}$ satisfy the same equation and that the boundary conditions are identical. We note that the boundary conditions on ψ_2 result from the boundary conditions on $\bar{\psi}_2$ and not from the use of any matching principle.

Expansion of $\bar{\varphi}_3$. Equation (6.36) for $\bar{\varphi}_3$ is

$$x^{1/3}\frac{\mathrm{d}\bar{\varphi}_3}{\mathrm{d}x} - \bar{\varphi}_3 = \varepsilon^{-1/2}\bar{\psi}_2 \ ,$$

with

$$\bar{\varphi}_3(1,\varepsilon) = 0 \ .$$

It can be shown that

$$\bar{\varphi}_3 = -\frac{3}{8}\ln\varepsilon + \bar{G}_1(X) + \bar{g}_1(x) + \cdots \ . \tag{S.5}$$

The equations for $\bar{G}_1$ and $\bar{g}_1$ are

$$X^{1/3}\frac{\mathrm{d}\bar{G}_1}{\mathrm{d}X} = \bar{F}_1 \ ,$$

$$x^{1/3}\frac{\mathrm{d}\bar{g}_1}{\mathrm{d}x} - \bar{g}_1 = -\ln x + \frac{1}{2} + \bar{f}_1 \ ,$$

where the equation for $\bar{g}_1$ is obtained by considering $\mathrm{E}_0\,\bar{\psi}_2$.

The boundary conditions are deduced from the study of the outer expansion of $\bar{\varphi}_3$. From the condition $\bar{\varphi}_3 = 0$ at $x = 1$, we obtain

$$\bar{g}_1(1) = 0 \ ,$$

and the behaviour of $\bar{G}_1$ as $X \to \infty$ is

$$\bar{G}_1 \cong -\frac{1}{2}\ln X \ .$$

To a certain extent, the boundary conditions are arbitrary. The condition $\bar{g}_1(1) = 0$ has been chosen for convenience but any other constant can be chosen for $\bar{g}_1(1)$.

The equations for $\bar{f}_1$ and $\bar{g}_1$ are combined to give

$$x^{1/3}\frac{\mathrm{d}}{\mathrm{d}x}(\bar{f}_1 + \bar{g}_1) - (\bar{f}_1 + \bar{g}_1) = -\frac{\mathrm{d}^2\varphi_1}{\mathrm{d}x^2} - \frac{1}{3}x^{-4/3} + \frac{1}{2}x^{-2/3} - \ln x + \frac{1}{2} \ .$$

Consider the function

$$f_2 = \bar{f}_1 + \bar{g}_1 - \frac{1}{2}x^{-2/3} - \ln x + \frac{1}{2} \ .$$

The equation for f_2 is

$$x^{1/3}\frac{\mathrm{d}f_2}{\mathrm{d}x} - f_2 = -\frac{\mathrm{d}^2\varphi_1}{\mathrm{d}x^2} \ ,$$

and we obtain $f_2(1) = 0$.

Then, the functions φ_2 appearing with MMAE and f_2 are identical. We have

$$\bar{f}_1 + \bar{g}_1 = \varphi_2 + \frac{1}{2}x^{-2/3} + \ln x - \frac{1}{2} \ .$$

Expansion of $\bar{\psi}_3$. Equation (6.37) for $\bar{\psi}_3$ is

$$\frac{\mathrm{d}^2\bar{\psi}_3}{\mathrm{d}X^2} + X^{1/3}\frac{\mathrm{d}\bar{\psi}_3}{\mathrm{d}X} = -\varepsilon^{3/2}\frac{\mathrm{d}^2\bar{\varphi}_3}{\mathrm{d}x^2} ,$$

with the boundary conditions

$$\bar{\psi}_3(0,\varepsilon) = -\bar{\varphi}_3(0,\varepsilon) , \quad \bar{\psi}_3(\varepsilon^{-3/4},\varepsilon) = 0 .$$

It can be shown that

$$\bar{\psi}_3 = \bar{H}_1(X) + \alpha G_{4/3}(X) + \beta + \cdots , \tag{S.6}$$

where $G_{4/3}(X)$ is solution of the equation

$$\frac{\mathrm{d}^2 G_{4/3}}{\mathrm{d}X^2} + X^{1/3}\frac{\mathrm{d}G_{\frac{4}{3}}}{\mathrm{d}X} = 0 .$$

The equation for $\bar{H}_1(X)$ is

$$\frac{\mathrm{d}^2\bar{H}_1}{\mathrm{d}X^2} + X^{1/3}\frac{\mathrm{d}\bar{H}_1}{\mathrm{d}X} = -\frac{\mathrm{d}^2\bar{G}_1}{\mathrm{d}X^2} .$$

The boundary conditions on $\bar{H}_1$ are deduced from the boundary conditions on $\bar{\psi}_3$ and from the study of inner and outer expansions of $\bar{\psi}_3$. The function $\alpha G_{4/3}(X) + \beta$ has been introduced in order that $\bar{H}_1$ is a function of X only and does not depend on ε. A possible solution is

$$\bar{\psi}_3 = \bar{H}_1(X) - \frac{3}{8}\ln\varepsilon\,\frac{G_{4/3}(X)}{G_{4/3}(\infty)} + \frac{3}{8}\ln\varepsilon ,$$

and the boundary conditions are such that

$$\bar{H}_1(0) + \bar{G}_1(0) + \bar{g}_1(0) = 0 ,$$
$$\bar{H}_1 \to 0 \quad \text{as} \quad X \to \infty .$$

Identification with MMAE results. We define the function F_3 by

$$F_3 = \bar{F}_2 + \bar{G}_1 + \bar{H}_1 + \frac{9}{8}X^{4/3} - \frac{3}{4} .$$

From the equations for $\bar{F}_2$, $\bar{G}_1$ and $\bar{H}_1$ we obtain the equation for F_3

$$\frac{\mathrm{d}^2 F_3}{\mathrm{d}X^2} + X^{1/3}\frac{\mathrm{d}F_3}{\mathrm{d}X} = \psi_2 .$$

The boundary conditions on F_3 are obtained from the boundary conditions on $\bar{F}_2$, $\bar{G}_1$ and $\bar{H}_1$; from the expression of φ_2, we also have

$$\bar{f}_1(0) + \bar{g}_1(0) = -\frac{3}{4} .$$

Finally, we have

$$F_3(0) = 0 \, ,$$

$$F_3 \cong \frac{9}{8} X^{4/3} - \ln X - \frac{1}{4} \quad \text{as} \quad X \to \infty \, .$$

It follows that the function ψ_3 appearing with MMAE and the function F_3 are identical. We have

$$\bar{F}_2 + \bar{G}_1 + \bar{H}_1 = \psi_3 - \frac{9}{8} X^{4/3} + \frac{3}{4} \, .$$

The SCEM regular expansion to order ε is therefore

$$\Phi = \varphi_1 + \psi_1 - 1 + \varepsilon^{1/2} \left(\psi_2 - \frac{3}{2} X^{2/3} \right) - \varepsilon \ln \varepsilon \left(\frac{3}{4} \frac{G_{\frac{4}{3}}(X)}{G_{\frac{4}{3}}(\infty)} \right)$$

$$+ \varepsilon \left(\varphi_2 + \frac{1}{2} x^{-2/3} + \ln x + \psi_3 - \frac{9}{8} X^{4/3} + \frac{1}{4} \right) \, ,$$

or

$$\Phi = \varphi_1 + \psi_1 - 1 + \varepsilon^{1/2} \left(\psi_2 - \frac{3}{2} X^{2/3} \right) - \varepsilon \ln \varepsilon \left(\frac{3}{4} \frac{G_{4/3}(X)}{G_{4/3}(\infty)} - \frac{3}{4} \right)$$

$$+ \varepsilon \left(\varphi_2 + \frac{1}{2} x^{-2/3} + \ln X + \psi_3 - \frac{9}{8} X^{4/3} + \frac{1}{4} \right) \, .$$

This expansion is identical to the MMAE composite expansion (6.31).

Chapter 7

7-1. We have $\psi_0 = y$, $\Delta_0 = \varepsilon$, $\phi_0 = \sqrt{2x} f(\eta)$ with $\eta = \overline{Y}/\sqrt{2x}$ and $\overline{Y} = Y - F(x)$. The equation to solve is the same if written with $\overline{Y}$ or with Y. We have

$$\frac{\partial^4 \phi_0}{\partial Y^4} - \left(\frac{\partial \phi_0}{\partial Y} \frac{\partial}{\partial x} - \frac{\partial \phi_0}{\partial x} \frac{\partial}{\partial Y} \right) \frac{\partial^2 \phi_0}{\partial Y^2} = 0 \, .$$

This leads us to solve

$$f''' + f f'' = 0 \, ,$$

with $f(0) = f'(0) = 0$ and $f'(\infty) = 1$.

Moreover, we have $\delta_1 = \varepsilon$.

The outer and inner expansions of ψ to order ε are

$$\mathrm{E}\,\psi = y + \varepsilon \psi_1(x, y) \, ,$$

$$\mathrm{I}\,\psi = \varepsilon \phi_0(x, Y) \, .$$

Matching to order ε requires to evaluate the behaviour of $\phi_0(x, Y)$ as $Y \to \infty$

$$\phi_0(x, Y) \underset{Y \to \infty}{\cong} \sqrt{2x} \left[\frac{Y}{\sqrt{2x}} - \beta_0 \right] ,$$

$$\underset{Y \to \infty}{\cong} Y - F(x) - \beta_0 \sqrt{2x} ,$$

$$\underset{Y \to \infty}{\cong} \frac{y}{\varepsilon} - F(x) - \beta_0 \sqrt{2x} .$$

Applying the MVDP, we obtain

$$y + \varepsilon \psi_1(x, 0) = y - \varepsilon (F(x) + \beta_0 \sqrt{2x}) ,$$

whence

$$\psi_1(x, 0) = -(F(x) + \beta_0 \sqrt{2x}) .$$

To order 1, the equation of the streamline $\psi = 0$ is

$$y = 0 .$$

To order ε, the equation of the streamline $\psi = 0$ is obtained from a Taylor series expansion of ψ_1 in the neighbourhood of $y = 0$

$$y + \varepsilon \psi_1(x, 0) = 0 .$$

Taking into account the matching results, we obtain the equation of the streamline $\psi = 0$

$$y = \varepsilon \left(F(x) + \beta_0 \sqrt{2x} \right) .$$

This equation takes into account the wall deformation and also the influence of the boundary layer (displacement effect).

7-2. The reduced equations are the Euler equations

$$\frac{\partial u_\theta}{\partial \theta} + \frac{\partial}{\partial r}(r u_r) = 0 ,$$

$$\frac{u_\theta}{r} \frac{\partial u_\theta}{\partial \theta} + u_r \frac{\partial u_\theta}{\partial r} + \frac{u_\theta u_r}{r} = -\frac{1}{r} \frac{\partial p}{\partial \theta} ,$$

$$\frac{u_\theta}{r} \frac{\partial u_r}{\partial \theta} + u_r \frac{\partial u_r}{\partial r} - \frac{u_\theta^2}{r} = -\frac{\partial p}{\partial r} .$$

These equations are satisfied by

$$u_{\theta 1} = \sin \theta \left(1 + \frac{1}{r^2} \right) ,$$

$$u_{r1} = \cos \theta \left(-1 + \frac{1}{r^2} \right) ,$$

$$p_1 = p_\infty + \frac{1}{2} \left[1 - (u_{\theta 1}^2 + u_{r1}^2) \right] ,$$

374   Solutions of Problems

with the boundary conditions

$$u_{r1}(\theta, r = 1) = 0 \ , \quad u_{\theta 1} \xrightarrow[r \to \infty]{} \sin \theta \ , \quad u_{r1} \xrightarrow[r \to \infty]{} - \cos \theta \ .$$

The first condition expresses the slip condition at the wall; the other two conditions express the condition of uniform velocity at infinity.

As $\theta \to 0$ and $r \to 1$, we have

$$u_{\theta 1} = 2\theta + \cdots \ ,$$
$$u_{r1} = -2(r - 1) + \cdots \ ,$$
$$p_1 = p_\infty + \frac{1}{2} \left[1 - 4(\theta^2 + (r - 1)^2)\right] + \cdots \ ,$$

or, with the inner variables,

$$u_{\theta 1} = 2\varepsilon \Theta + \cdots \ ,$$
$$u_{r1} = -2\varepsilon R + \cdots \ ,$$
$$p_1 = p_\infty + \frac{1}{2} - 2\varepsilon^2(\Theta^2 + R^2) + \cdots \ .$$

The inner expansion is written in the form

$$u_\theta = \varepsilon U_{\theta 1}(\Theta, R) + \cdots \ ,$$
$$u_r = \varepsilon U_{r1}(\Theta, R) + \cdots \ ,$$
$$p = P_0 + \varepsilon^2 P_1(\Theta, R) + \cdots \ ,$$

and we have the matching conditions

$$\lim_{\substack{\Theta \to \infty \\ R \to \infty}} \frac{U_{\theta 1}}{\Theta} = 2 \ ,$$

$$\lim_{\substack{\Theta \to \infty \\ R \to \infty}} \frac{U_{r1}}{R} = -2 \ ,$$

$$P_0 = p_\infty + \frac{1}{2} \ ,$$

$$\lim_{\substack{\Theta \to \infty \\ R \to \infty}} \frac{P_1}{\Theta^2 + R^2} = -2 \ .$$

The equations for $U_{\theta 1}, U_{r1}, P_1$ are

$$\frac{\partial U_{\theta 1}}{\partial \Theta} + \frac{\partial U_{r1}}{\partial R} = 0 \ ,$$

$$U_{\theta 1}\frac{\partial U_{\theta 1}}{\partial \Theta} + U_{r1}\frac{\partial U_{\theta 1}}{\partial R} = -\frac{\partial P_1}{\partial \Theta} + \frac{\partial^2 U_{\theta 1}}{\partial \Theta^2} + \frac{\partial^2 U_{\theta 1}}{\partial R^2} \ ,$$

$$U_{\theta 1}\frac{\partial U_{r1}}{\partial \Theta} + U_{r1}\frac{\partial U_{r1}}{\partial R} = -\frac{\partial P_1}{\partial R} + \frac{\partial^2 U_{r1}}{\partial \Theta^2} + \frac{\partial^2 U_{r1}}{\partial R^2} \ .$$

At the wall, we have the no-slip conditions

$$R = 0 : \quad U_{\theta 1} = 0 , \quad U_{r1} = 0 .$$

We observe that the equations have the same form as the Navier-Stokes equations in cartesian coordinates. The resolution of the so-obtained equations is the Hiemenz problem. In the case of a flow impinging a flat wall, we obtain an exact solution of Navier-Stokes equations. Here, this is only an approximation since the equations result from seeking an approximate solution to Navier-Stokes equations.

We seek the solution in the form

$$U_{\theta 1} = \Theta \varphi'(R) , \quad U_{r1} = -\varphi(R) , \quad P_1 = -2(\Theta^2 + \Phi(R)) .$$

The equations are

$$\varphi'^2 - \varphi \varphi'' = 4 + \varphi''' ,$$
$$\varphi \varphi' = 2\Phi' - \varphi'' .$$

The first equation is also written as

$$\varphi''' + \varphi \varphi'' - \varphi'^2 + 4 = 0 ,$$

with the boundary conditions

$$R = 0 : \quad \varphi = 0 , \quad \varphi' = 0 ; \quad R \to \infty : \quad \varphi' \to 2 .$$

This is a standard boundary layer problem which belongs to the general class of Falkner-Skan's problem.

The function Φ is deduced from the integration of the equation in the R-direction

$$\Phi - \Phi(0) = \frac{\varphi^2}{4} + \frac{\varphi'}{2} .$$

The stagnation pressure in the boundary layer is

$$p_i = P_0 + \varepsilon^2 P_1 + \frac{\varepsilon^2}{2} \left[U_{\theta 1}^2 + U_{r1}^2 \right] ,$$

where

$$P_0 = p_\infty + \frac{1}{2} .$$

We obtain

$$p_i = P_0 + \varepsilon^2 \left[-2\Theta^2 - \varphi' - 2\Phi(0) + \frac{\Theta^2}{2} \varphi'^2 \right] ,$$

whence

$$p_i(0,0) = P_0 + \varepsilon^2 \left[-2\Phi(0) \right] ,$$
$$p_i(\Theta, R \to \infty) = P_0 + \varepsilon^2 \left[-2 - 2\Phi(0) \right] .$$

Thus, the difference of stagnation pressure between the stagnation point and the boundary layer edge is

$$p_i(\Theta, R \to \infty) - p_i(0,0) = -2\varepsilon^2 \ .$$

Therefore, a stagnation pressure probe does not measure exactly the stagnation pressure of the flow in which the probe is placed. This phenomenon is known as Barker's effect.

The determination of constant $\Phi(0)$ demands a detailed discussion. If we assume that the stagnation pressure in the boundary layer must match the stagnation pressure in the inviscid flow $(p_i(\Theta, R \to \infty) = P_0)$, we obtain $\Phi(0) = -1$.

7-3. We have $\delta_1 = H(x)\theta$, whence

$$\frac{d\delta_1}{dx} = \frac{dH}{dx}\theta + H\frac{d\theta}{dx} \ .$$

The boundary layer equations,

$$(H_{32} - HH'_{32})\frac{d\theta}{dx} + H'_{32}\frac{d\delta_1}{dx} + 3\frac{\delta_3}{u_e}\frac{du_e}{dx} = 2C_D \ , \quad H'_{32} = \frac{dH_{32}}{dH} \ ,$$

$$\frac{d\theta}{dx} + \theta\frac{H+2}{u_e}\frac{du_e}{dx} = \frac{C_f}{2} \ ,$$

become

$$(H_{32} - HH'_{32})\frac{d\theta}{dx} + H'_{32}\left(\frac{dH}{dx}\theta + H\frac{d\theta}{dx}\right) + 3\frac{\delta_3}{u_e}\frac{du_e}{dx} = 2C_D \ ,$$

$$\frac{d\theta}{dx} + \theta\frac{H+2}{u_e}\frac{du_e}{dx} = \frac{C_f}{2} \ ,$$

These equations can be written as

$$H_{32}\frac{d\theta}{dx} + 3\frac{\delta_3}{u_e}\frac{du_e}{dx} = 2C_D - H'_{32}\frac{dH}{dx}\theta \ ,$$

$$\frac{d\theta}{dx} + \theta\frac{H+2}{u_e}\frac{du_e}{dx} = \frac{C_f}{2} \ .$$

The determinant is

$$\Delta = H_{32}\theta\frac{H+2}{u_e} - 3\frac{\delta_3}{u_e} = \frac{\delta_3}{u_e}(H-1) \ .$$

This determinant does not vanish in domain $H > 1$. The calculation is always possible.

7-4. For a laminar, two-dimensional boundary layer, the dimensionless continuity and momentum equations are

$$\frac{\partial u^*}{\partial x^*} + \frac{\partial v^*}{\partial y^*} = 0 \, ,$$

$$u^* \frac{\partial u^*}{\partial x^*} + v^* \frac{\partial u^*}{\partial y^*} = -\frac{1}{\varrho^*}\frac{\partial p^*}{\partial x^*} + \nu^* \frac{\partial^2 u^*}{\partial y^{*2}} \, .$$

The dimensionless quantities are defined by

$$x = \frac{x^*}{l} \, , \quad y = \frac{y^*}{l} R^{1/2} \, , \quad u = \frac{u^*}{u_0} \, , \quad v = \frac{v^*}{u_0} R^{1/2} \, , \quad p = \frac{p^*}{\varrho u_0^2} \, .$$

1. In dimensionless form, the boundary layer equations are

$$\frac{\partial u}{\partial x} + \frac{\partial v}{\partial y} = 0 \, ,$$

$$u \frac{\partial u}{\partial x} + v \frac{\partial u}{\partial y} = -\frac{\mathrm{d}p}{\mathrm{d}x} + \frac{\partial^2 u}{\partial y^2} \, .$$

2. In the neighbourhood of x_0, we assume that the pressure gradient is given in the form

$$-\frac{\mathrm{d}p}{\mathrm{d}x} = p_0 + p_1(x - x_0) + p_2(x - x_0)^2 + \cdots \, ,$$

where p_0, p_1, ... are constants. The velocity profile at x_0 is

$$u = a_1 y + a_2 y^2 + a_3 y^3 + \cdots \, ,$$

where a_1, a_2, ... are functions of x. The velocity component v is obtained from the continuity equation by taking into account the wall condition $v(0) = 0$

$$v = -\frac{\mathrm{d}a_1}{\mathrm{d}x}\frac{y^2}{2} - \frac{\mathrm{d}a_2}{\mathrm{d}x}\frac{y^3}{3} - \frac{\mathrm{d}a_3}{\mathrm{d}x}\frac{y^4}{4} - \cdots \, .$$

The expressions of p, u et v are substituted in the momentum equation. Equating coefficients of like powers of y at x_0, we obtain

$$2a_2 + p_0 = 0 \, ,$$

$$a_3 = 0 \, ,$$

$$a_1 \frac{\mathrm{d}a_1}{\mathrm{d}x} - 24a_4 = 0 \, ,$$

$$\frac{2}{3}a_1 \frac{\mathrm{d}a_2}{\mathrm{d}x} - 20a_5 = 0 \, .$$

3. The momentum equation is differentiated with respect to x

$$\left(\frac{\partial u}{\partial x}\right)^2 + u\frac{\partial^2 u}{\partial x^2} + \frac{\partial v}{\partial x}\frac{\partial u}{\partial y} + v\frac{\partial^2 u}{\partial x \partial y} = -\frac{\mathrm{d}^2 p}{\mathrm{d}x^2} + \frac{\partial^3 u}{\partial x \partial y^2}\,.$$

We substitute the expressions of u, v, p and we equate the coefficients of like powers of y

$$2\frac{\mathrm{d}a_2}{\mathrm{d}x} + p_1 = 0\,,$$

$$\frac{\mathrm{d}a_3}{\mathrm{d}x} = 0\,.$$

4. We obtain the following relations

$$2a_2 + p_0 = 0\,,$$

$$a_3 = 0\,,$$

$$a_1\frac{\mathrm{d}a_1}{\mathrm{d}x} - 24a_4 = 0\,,$$

$$5!a_5 + 2a_1 p_1 = 0\,.$$

In these equations, p_0, p_1, ... are given coefficients. Thus, coefficients a_2, a_3, ... are not free since they are determined as function of p_0, p_1.

If these compatibility conditions are not satisfied, singularities appear when solving the boundary layer equations for $x > x_0$.

Let us examine the case $a_1 = 0$ corresponding to boundary layer separation. Substituting the expansions of u, v and p in the momentum equation, we obtain

$$2a_2 + p_0 = 0\,,$$

$$a_3 = 0\,,$$

$$a_4 = 0\,,$$

$$a_5 = 0\,,$$

$$6!a_6 = 2p_0 p_1\,,$$

$$a_7 = 0\,.$$

We assume that only the condition $2a_2 + p_0 = 0$ is satisfied. We have

$$a_1\frac{\mathrm{d}a_1}{\mathrm{d}x} - 24a_4 = 0\,.$$

Suppose that $a_4 \neq 0$ at $x = x_0$. In the neighbourhood of $x = x_0$ we obtain

$$a_1^2 = 48a_4(x - x_0)\,.$$

If a solution exists when $x < x_0$, we must have $a_4 < 0$ and the solution does not exist when $x > x_0$.

7-5. With the outer variables, we have

$$\left.\frac{\partial}{\partial x}\right|_{y=\text{cst}} = \frac{1}{n}\xi^{1-n}\left.\frac{\partial}{\partial \xi}\right|_{y=\text{cst}} .$$

Using the continuity equation, with $v(0) = 0$, we obtain

$$v = -\frac{1}{n}\xi^{1-n}(F_1 + 2\xi F_2 + \cdots) ,$$

with

$$F_1(y) = \int_0^y F_1'(z)\, \mathrm{d}z , \quad F_2(y) = \int_0^y F_2'(z)\, \mathrm{d}z .$$

We have $F_1(0) = 0$, $F_2(0) = 0$.

The expressions of u and v are substituted in the momentum equation

$$(F_0' + \xi F_1' + \xi^2 F_2' + \cdots)\frac{1}{n}\xi^{1-n}(F_1' + 2\xi F_2' + \cdots)$$

$$-\frac{1}{n}\xi^{1-n}(F_1 + 2\xi F_2 + \cdots)(F_0'' + \xi F_1'' + \xi^2 F_2'' + \cdots)$$

$$= p_0 + p_1\xi^n + p_2\xi^{2n} + \cdots + F_0''' + \xi F_1''' + \xi^2 F_2''' + \cdots .$$

With $n > 1$, we obtain the equation for F_1

$$F_0'F_1' - F_1 F_0'' = 0 .$$

The solution is

$$F_1 = k F_0' ,$$

where k is a constant yet undetermined. This solution enables us to satisfy the condition $F_1(0) = 0$ ($v = 0$) at $y = 0$ but not $\dfrac{\partial u}{\partial y} = 0$ at $y = 0$. Therefore, an inner layer is required.

With the inner variables, the differentiation rules give

$$\left.\frac{\partial}{\partial x}\right|_{y=\text{cst}} = \frac{1}{n}\xi^{1-n}\left.\frac{\partial}{\partial \xi}\right|_{\eta=\text{cst}} - \frac{1}{n}\frac{\eta}{\xi^n}\left.\frac{\partial}{\partial \eta}\right|_{\xi=\text{cst}} ,$$

$$\left.\frac{\partial}{\partial y}\right|_{x=\text{cst}} = \frac{1}{n\xi}\left.\frac{\partial}{\partial \eta}\right|_{\xi=\text{cst}} .$$

From the continuity equation, written in variables (ξ, η), we obtain the expression of v in the inner layer by taking into account the condition $v(0) = 0$

$$v = \xi^{1-n}(\eta f_0' - f_0) + \xi^{2-n}(\eta f_1' - 2f_1) + \xi^{3-n}(\eta f_2' - 3f_2) + \cdots .$$

The momentum equation becomes

$$(f_0' + \xi f_1' + \xi^2 f_2' + \cdots)$$
$$\times \frac{1}{n} \left[-\eta \xi^{-n} f_0'' + \xi^{1-n}(f_1' - \eta f_1'') + \xi^{2-n}(2f_2' - \eta f_2'') + \cdots \right]$$
$$+ \left[\xi^{1-n}(\eta f_0' - f_0) + \xi^{2-n}(\eta f_1' - 2f_1) + \xi^{3-n}(\eta f_2' - 3f_2) + \cdots \right]$$
$$\times \frac{1}{n\xi}(f_0'' + \xi f_1'' + \xi^2 f_2'' + \cdots)$$
$$= p_0 + p_1 \xi^n + p_2 \xi^{2n} + \cdots + \frac{1}{n^2 \xi^2}(f_0''' + \xi f_1''' + \xi^2 f_2''' + \cdots) \ .$$

We could choose $n = 2$ in order to keep the viscous term. The equation for f_0 would be

$$f_0''' + 2f_0 f_0'' = 0 \ .$$

However, the matching conditions with the outer layer give $f_0 = 0$. Then, we take $n = 3$. The equation for f_1 is

$$\frac{1}{3} f_1''' + 2f_1 f_1'' - f_1'^2 = 0 \ .$$

The solution must be such that $f_1(0) = 0$ and $f''(0) = 0$. In the neighbourhood of $\eta = 0$, the solution has the form

$$f_1 = \beta \eta + \frac{\beta^2}{2} \eta^3 - \frac{3}{20} \beta^3 \eta^5 + \cdots \ .$$

Coefficient β is calculated to satisfy the matching with the outer solution. Let us express this matching to order ξ. The outer solution is

$$\mathrm{E}_0 \, u = F_0' + \xi F_1' = a_1 y + a_2 y^2 + \cdots + \xi k(a_1 + 2a_2 y + \cdots) \ ,$$

that is, with the variable η

$$\mathrm{E}_0 \, u = 3a_1 \eta \xi + a_2 (3\eta \xi)^2 + \cdots + \xi k(a_1 + 6a_2 \eta \xi + \cdots) \ .$$

Therefore, we obtain

$$\mathrm{E}_1 \, \mathrm{E}_0 \, u = 3a_1 \eta \xi + a_1 k \xi \ .$$

Moreover, we have

$$\mathrm{E}_1 \, u = \xi f_1'(\eta) \ .$$

To obtain $\mathrm{E}_0 \, \mathrm{E}_1 \, u$, it is required to know the behaviour of f_1' as $\eta \to \infty$. From the equation for f_1, we have

$$f_1 \underset{\eta \to \infty}{\cong} \alpha \eta^2 + \mathrm{EST} \ ,$$

and

$$f_1' \underset{\eta\to\infty}{\cong} 2\alpha\eta + \text{EST} .$$

We obtain

$$\alpha = \frac{3}{2}a_1 , \quad k = 0 .$$

The first condition determines the function f_1 completely; the second condition gives $F_1 = 0$.

7-6. With the outer variables, we have

$$\left.\frac{\partial}{\partial x}\right|_{y=\text{cst}} = \frac{1}{n}\xi^{1-n}\left.\frac{\partial}{\partial\xi}\right|_{y=\text{cst}} .$$

Using the continuity equation, with $v(0) = 0$, we obtain

$$v = -\frac{1}{n}\xi^{1-n}(F_1 + 2\xi F_2 + \cdots) ,$$

with

$$F_1(y) = \int_0^y F_1'(z)\,\mathrm{d}z , \quad F_2(y) = \int_0^y F_2'(z)\,\mathrm{d}z .$$

We have $F_1(0) = 0$, $F_2(0) = 0$.

The expressions of u and v are substituted in the momentum equation

$$(F_0' + \xi F_1' + \xi^2 F_2' + \cdots)\frac{1}{n}\xi^{1-n}(F_1' + 2\xi F_2' + \cdots)$$

$$-\frac{1}{n}\xi^{1-n}(F_1 + 2\xi F_2 + \cdots)(F_0'' + \xi F_1'' + \xi^2 F_2'' + \cdots)$$

$$= p_0 + p_1\xi^n + p_2\xi^{2n} + \cdots + F_0''' + \xi F_1''' + \xi^2 F_2''' + \cdots .$$

With $n > 1$, we obtain the equation for F_1

$$F_0'F_1' - F_1 F_0'' = 0 .$$

The solution is

$$F_1 = kF_0' ,$$

where k is a constant yet undetermined. This solution does not enable us to satisfy, in particular, the wall condition $F_1(0) = 0$ ($v = 0$) if $k \neq 0$ since $a_0 \neq 0$. Therefore, an inner layer is required.

With the inner variables, the differentiation rules give

$$\left.\frac{\partial}{\partial x}\right|_{y=\text{cst}} = \frac{1}{n}\xi^{1-n}\left.\frac{\partial}{\partial\xi}\right|_{\eta=\text{cst}} - \frac{1}{n}\frac{\eta}{\xi^n}\left.\frac{\partial}{\partial\eta}\right|_{\xi=\text{cst}} ,$$

$$\left.\frac{\partial}{\partial y}\right|_{x=\text{cst}} = \frac{1}{n\xi}\left.\frac{\partial}{\partial\eta}\right|_{\xi=\text{cst}} .$$

From the continuity equation, written in variables ξ, η, we obtain the expression of v in the inner layer by taking into account the condition $v(0) = 0$

$$v = \xi^{1-n}(\eta f_0' - f_0) + \xi^{2-n}(\eta f_1' - 2f_1) + \xi^{3-n}(\eta f_2' - 3f_2) + \cdots .$$

The momentum equation becomes

$$(f_0' + \xi f_1' + \xi^2 f_2' + \cdots)$$
$$\times \frac{1}{n}\left[-\eta \xi^{-n} f_0'' + \xi^{1-n}(f_1' - \eta f_1'') + \xi^{2-n}(2f_2' - \eta f_2'') + \cdots\right]$$
$$+ \left[\xi^{1-n}(\eta f_0' - f_0) + \xi^{2-n}(\eta f_1' - 2f_1) + \xi^{3-n}(\eta f_2' - 3f_2) + \cdots\right]$$
$$\times \frac{1}{n\xi}(f_0'' + \xi f_1'' + \xi^2 f_2'' + \cdots)$$
$$= p_0 + p_1\xi^n + p_2\xi^{2n} + \cdots + \frac{1}{n^2\xi^2}(f_0''' + \xi f_1''' + \xi^2 f_2''' + \cdots) .$$

We choose $n = 2$ in order to keep the viscous term. The equation for f_0 is

$$f_0''' + 2f_0 f_0'' = 0 .$$

Let us examine the matching of u between the outer and inner layers. To order ξ^0, we have

$$\mathrm{E}_0\, u = F_0' = a_0 + a_1 y + a_2 y^2 + \cdots .$$

With the variable η, this expression becomes

$$\mathrm{E}_0\, u = a_0 + 2a_1\eta\xi + a_2(2\eta\xi)^2 + \cdots .$$

Therefore, we obtain
$$\mathrm{E}_1\, \mathrm{E}_0\, u = a_0 .$$

Moreover, we have
$$\mathrm{E}_1\, u = f_0'(\eta) ,$$

and therefore
$$\mathrm{E}_0\, \mathrm{E}_1\, u = \lim_{\eta \to \infty} f_0'(\eta) .$$

Then, the matching expresses that

$$\lim_{\eta \to \infty} f_0'(\eta) = a_0 .$$

The equation for f_1 is

$$f_1''' + 2f_0 f_1'' - 2f_0' f_1' + 4f_0'' f_1 = 0 .$$

We express the matching of u between the outer and inner layers to order ξ. We have

$$\mathrm{E}_0\, u = F_0'(y) + \xi F_1'(y) = a_0 + a_1 y + a_2 y^2 + \cdots + \xi k(a_1 + 2a_2 y + \cdots) ,$$

or, with the variable η

$$E_0 \, u = a_0 + 2a_1\eta\xi + a_2(2\eta\xi)^2 + \cdots + \xi k(a_1 + 4a_2\eta\xi + \cdots) \,,$$

whence

$$E_1 \, E_0 \, u = a_0 + 2a_1\eta\xi + ka_1\xi \,.$$

Moreover, we have

$$E_1 \, u = f_0'(\eta) + \xi f_1'(\eta) \,.$$

To obtain $E_0 \, E_1 \, u$, it is required to know the behaviour of f_1' as $\eta \to \infty$. We know that

$$f_0' \underset{\eta\to\infty}{\cong} a_0 + \text{EST} \,.$$

We deduce that we must have

$$f_1' \underset{\eta\to\infty}{\cong} 2a_1\eta + ka_1 + \cdots \,.$$

We have the following behaviours

$$f_0 \underset{\eta\to\infty}{\cong} A_0\eta + B_0 + \text{EST} \,,$$

$$f_1 \underset{\eta\to\infty}{\cong} A_1\eta^2 + B_1\eta + C_1 + \cdots \,.$$

It has already been seen that $A_0 = a_0$. Using the behaviours of f_0 and f_1 in the equation for f_1, we obtain

$$B_1 = 2\frac{a_1}{a_0}B_0 \,.$$

Taking into account that we must have

$$f_1' \underset{\eta\to\infty}{\cong} 2a_1\eta + ka_1 + \cdots \,,$$

we deduce

$$A_1 = a_1 \,, \quad k = 2\frac{B_0}{a_0} = -1.72a_0^{-1/2} \,.$$

7-7. We consider for example the operators E_0 and E_1 to order ξ^2. We have

$$E_1 \, u = 2 \left[f_0'(\eta) + \xi f_1'(\eta) + \xi^2 f_2'(\eta) \right] \,,$$
$$E_0 \, u = \chi_0'(y) + \xi\chi_1'(y) + \xi^2\chi_2'(y) \,,$$

with

$$\chi_0' = a_0 + a_1 y + a_2 y^2 + \cdots \,.$$

Assuming that χ_i' can be expanded in Taylor series in the neighbourhood of $y = 0$, we have

$$E_0 \, u = a_0 + a_1 y + a_2 y^2 + \cdots$$
$$+ \xi \left[\chi_1'(0) + y\chi_1''(0) + \cdots \right] + \xi^2 \left[\chi_2'(0) + y\chi_2''(0) + \cdots \right] \,.$$

With $y = 2^{1/2}\xi\eta$, we obtain

$$E_1\, E_0\, u = a_0 + \xi\left[a_1 2^{1/2}\eta + \chi_1'(0)\right] + \xi^2\left[a_2(2^{1/2}\eta)^2 + 2^{1/2}\eta\chi_1''(0) + \chi_2'(0)\right] .$$

With

$$E_1\, u = 2\left[f_0'(\eta) + \xi f_1'(\eta) + \xi^2 f_2'(\eta)\right] ,$$

the matching is expressed by

$$\lim_{\eta\to\infty} \frac{f_r'}{\eta^r} = \frac{a_r}{2} 2^{r/2} .$$

With the inner variables, the differentiation rules give

$$\left.\frac{\partial}{\partial x}\right|_{y=\text{cst}} = \frac{1}{n}\xi^{1-n}\left.\frac{\partial}{\partial \xi}\right|_{\eta=\text{cst}} - \frac{1}{n}\frac{\eta}{\xi^n}\left.\frac{\partial}{\partial \eta}\right|_{\xi=\text{cst}} ,$$

$$\left.\frac{\partial}{\partial y}\right|_{x=\text{cst}} = \frac{1}{2^{1/2}\xi}\left.\frac{\partial}{\partial \eta}\right|_{\xi=\text{cst}} .$$

The momentum equation writes

$$-2(f_0' + \xi f_1' + \xi^2 f_2' + \xi^3 f_3' + \xi^4 f_4' + \cdots)$$
$$\times 2\frac{\xi^{-n}}{n}\left[-\eta f_0'' + \xi(f_1' - \eta f_1'') + \xi^2(2f_2' - \eta f_2'')\right.$$
$$\left.+\xi^3(3f_3' - \eta f_3'') + \xi^4(4f_4' - \eta f_4'') + \cdots\right]$$
$$+\frac{2^{3/2}}{n}\xi^{1-n}\left[f_0 - \eta f_0' + \xi(2f_1 - \eta f_1') + \xi^2(3f_2 - \eta f_2')\right.$$
$$\left.+\xi^3(4f_3 - \eta f_3') + \xi^4(5f_4 - \eta f_4') + \cdots\right]$$
$$\times 2^{1/2}\xi^{-1}(f_0'' + \xi f_1'' + \xi^2 f_2'' + \xi^3 f_3'' + \xi^4 f_4'' + \cdots)$$
$$= -(1 + p_1\xi^n + p_2\xi^{2n}) + \xi^{-2}(f_0''' + \xi f_1''' + \xi^2 f_2''' + \xi^3 f_3''' + \xi^4 f_4''' + \cdots) .$$

As $a_0 = 0$ and $a_1 = 0$, we have

$$\lim_{\eta\to\infty} f_0' = 0 , \quad \lim_{\eta\to\infty} f_1' = 0 .$$

Then, we obtain $f_0 = 0$ and $f_1 = 0$; in addition, we are led to take $n = 4$. The equations for f_2, f_3 and f_4 are

$$f_2''' - 3f_2 f_2'' + 2f_2'^2 = 1 ,$$
$$f_3''' - 3f_2 f_3'' + 5f_2' f_3' - 4f_2'' f_3 = 0 ,$$
$$f_4''' - 3f_2 f_4'' + 6f_2' f_4' - 5f_2'' f_4 = 4f_3 f_3'' - 3f_3'^2 .$$

The no-slip condition at the wall ($u = 0$, $v = 0$) is expressed by

$$f_2(0) = f_3(0) = f_4(0) = 0 \ , \quad f_2'(0) = f_3'(0) = f_4'(0) = 0 \ .$$

The solutions are

$$f_2 = \frac{\eta^3}{6} \ ,$$
$$f_3 = \alpha_1 \eta^2 \ ,$$
$$f_4 = \alpha_2 \eta^2 - \frac{\alpha_1^2}{15} \eta^5 \ .$$

The conditions

$$\lim_{\eta \to \infty} \frac{f_r'}{\eta^r} = \frac{a_r}{2} 2^{r/2}$$

yield

$$a_2 = \frac{1}{2} \ , \quad a_3 = 0 \ , \quad a_4 = -\frac{\alpha_1^2}{6} \ .$$

We must have $a_4 \leq 0$ in order that the solution exists upstream of the separation point.

With the outer variables, we have

$$\left. \frac{\partial}{\partial x} \right|_{y=\mathrm{cst}} = \frac{1}{4} \xi^{-3} \left. \frac{\partial}{\partial \xi} \right|_{y=\mathrm{cst}} \ .$$

The momentum equation writes

$$-(\chi_0' + \xi \chi_1' + \xi^2 \chi_2' + \xi^3 \chi_3' + \cdots) \frac{1}{4} \xi^{-3} (\chi_1' + 2\xi \chi_2' + 3\xi^2 \chi_3' + \cdots)$$
$$+ \frac{1}{4} \xi^{-3} (\chi_1 + 2\xi \chi_2 + 3\xi^2 \chi_3 + \cdots)(\chi_0'' + \xi \chi_1'' + \xi^2 \chi_2'' + \xi^3 \chi_3'' + \cdots)$$
$$= -(1 + p_1 \xi^4 + p_2 \xi^8 + \cdots) + \chi_0''' + \xi \chi_1''' + \xi^2 \chi_2''' + \xi^3 \chi_3''' + \cdots \ .$$

The equations for χ_1, χ_2, χ_3 are

$$\chi_0'' \chi_1 - \chi_1' \chi_0' = 0 \ ,$$
$$\chi_0'' \chi_2 - \chi_2' \chi_0' = 0 \ ,$$
$$\chi_0'' \chi_3 - \chi_3' \chi_0' = 0 \ .$$

The solutions are

$$\chi_1 = k_1 \chi_0' \ , \quad \chi_2 = k_2 \chi_0' \ , \quad \chi_3 = k_3 \chi_0' \ ,$$

and we have

$$\chi_0' = \frac{1}{2} y^2 - \frac{\alpha_1^2}{6} y^4 + \cdots \ .$$

The matching conditions, examined above, yield

$$2f_2' \underset{\eta\to\infty}{\cong} a_2(2^{1/2}\eta)^2 + 2^{1/2}\eta\chi_1''(0) + \chi_2'(0) \, ,$$

$$2f_3' \underset{\eta\to\infty}{\cong} a_3(2^{1/2}\eta)^3 + \frac{1}{2}(2^{1/2}\eta)^2\chi_1'''(0) + 2^{1/2}\eta\chi_2''(0) + \chi_3'(0) \, ,$$

$$2f_4' \underset{\eta\to\infty}{\cong} a_4(2^{1/2}\eta)^4 + \frac{1}{3!}(2^{1/2}\eta)^3\chi_1''''(0) + \frac{1}{2}(2^{1/2}\eta)^2\chi_2'''(0)$$
$$+ 2^{1/2}\eta\chi_3''(0) + \chi_4'(0) \, .$$

With

$$a_2 = \frac{1}{2} \, , \quad a_3 = 0 \, , \quad a_4 = -\frac{\alpha_1^2}{6} \, ,$$

we obtain

$$k_1 = 0 \, , \quad k_2 = 2^{3/2}\alpha_1 \, , \quad k_3 = 2^{3/2}\alpha_2 \, .$$

From the inner expansion, we calculate

$$\left(\frac{\partial u}{\partial y}\right)_0 = 2^{1/2}\xi f_2''(0) + 2^{1/2}\xi^2 f_3''(0) + \cdots$$
$$= 2^{3/2}x^{1/2}\alpha_1 + \cdots$$
$$= \sqrt{-48a_4}\,x^{1/2} + \cdots \, .$$

The wall shear-stress vanishes with a behaviour in the square root of the distance to the separation point.

From the outer expansion, we have

$$\frac{\partial u}{\partial x} = \frac{1}{2}\xi^{-2}\chi_2' + \cdots$$
$$= \xi^{-2}2^{1/2}\alpha_1\chi_0'' + \cdots$$
$$= \frac{2^{1/2}}{x^{1/2}}\alpha_1\left(y - \frac{2}{3}\alpha_1^2 y^3 + \cdots\right) + \cdots \, ,$$

The expression of v is

$$v = \frac{1}{2}\xi^{-2}\chi_2 + \cdots$$
$$= \frac{2^{1/2}}{x^{1/2}}\alpha_1\left(\frac{y^2}{2} - \frac{\alpha_1^2}{6}y^4 + \cdots\right) + \cdots \, .$$

Thus, $\dfrac{\partial u}{\partial x}$ and v tend towards infinity as $x \to 0$. This behaviour is at variance with the boundary layer hypotheses. However, it cannot be concluded that the boundary layer equations are not valid to describe separation. Indeed, this is the way in which the boundary layer equations are solved which is concerned. For example, the inverse methods are able to describe separation with a perfectly regular behaviour. In strong coupling methods, even when using the standard boundary layer equations, separation is described without any sign of singularity.

Chapter 8

8-1.

1. The dimensionless quantities are

$$\mathcal{U} = \frac{u}{V_\infty} , \quad \mathcal{V} = \frac{v}{V_\infty} , \quad \mathcal{P} = \frac{p}{\varrho V_\infty^2} , \quad x = \frac{x^*}{L} , \quad y = \frac{y^*}{L} .$$

The Navier-Stokes equations write

$$\frac{\partial \mathcal{U}}{\partial x} + \frac{\partial \mathcal{V}}{\partial y} = 0 ,$$

$$\mathcal{U}\frac{\partial \mathcal{U}}{\partial x} + \mathcal{V}\frac{\partial \mathcal{U}}{\partial y} = -\frac{\partial \mathcal{P}}{\partial x} + \varepsilon^2 \left(\frac{\partial^2 \mathcal{U}}{\partial x^2} + \frac{\partial^2 \mathcal{U}}{\partial y^2} \right) ,$$

$$\mathcal{U}\frac{\partial \mathcal{V}}{\partial x} + \mathcal{V}\frac{\partial \mathcal{V}}{\partial y} = -\frac{\partial \mathcal{P}}{\partial y} + \varepsilon^2 \left(\frac{\partial^2 \mathcal{V}}{\partial x^2} + \frac{\partial^2 \mathcal{V}}{\partial y^2} \right) .$$

2. The reduced equations give the equations for u_1, v_1, p_1 which are the Euler equations

$$\frac{\partial u_1}{\partial x} + \frac{\partial v_1}{\partial y} = 0 ,$$

$$u_1\frac{\partial u_1}{\partial x} + v_1\frac{\partial u_1}{\partial y} = -\frac{\partial p_1}{\partial x} ,$$

$$u_1\frac{\partial v_1}{\partial x} + v_1\frac{\partial v_1}{\partial y} = -\frac{\partial p_1}{\partial y} .$$

3. To write the equations for U_1, V_1, P_1 we expand u_1, v_1, p_1 in the neighbourhood of $y = 0$ since $y = \varepsilon Y$, therefore $y \ll 1$, and we seek a regular expansion

$$u_1 = u_{10} + y u_{1y0} + \cdots$$
$$= u_{10} + \varepsilon Y u_{1y0} + \cdots ,$$
$$v_1 = v_{10} + y v_{1y0} + \cdots$$
$$= v_{10} - \varepsilon Y u_{1x0} + \cdots .$$

It is shown later that $v_{10} = 0$. Using this result, the Navier-Stokes equations become

$$\frac{\partial U_1}{\partial x} + \frac{\partial V_1}{\partial Y} = 0 ,$$

$$U_1 u_{1x0} + U_1\frac{\partial U_1}{\partial x} + u_{10}\frac{\partial U_1}{\partial x} + V_1\frac{\partial U_1}{\partial Y} - Y u_{1x0}\frac{\partial U_1}{\partial Y} = \frac{\partial^2 U_1}{\partial Y^2} + \cdots ,$$

$$-\varepsilon U_1 Y u_{1xx0} + \varepsilon U_1\frac{\partial V_1}{\partial x} + \varepsilon u_{10}\frac{\partial V_1}{\partial x} + \varepsilon V_1\frac{\partial V_1}{\partial Y} - \varepsilon Y u_{1x0}\frac{\partial V_1}{\partial Y}$$

$$= -\frac{\Delta}{\varepsilon}\frac{\partial P_1}{\partial Y} + \varepsilon\frac{\partial^2 V_1}{\partial Y^2} + \cdots .$$

Since the regular form of SCEM is applied, the boundary conditions must be prescribed order by order. We have

$$\mathcal{U} = u_1(x, y) + U_1(x, Y) + \cdots ,$$
$$\mathcal{V} = v_1(x, y) + \varepsilon V_1(x, Y) + \cdots ,$$

whence

$$y = 0 : \quad u_1 + U_1 = 0 , \quad v_1 = 0 , \quad V_1 = 0 .$$

As $y \to \infty$ we have

$$u_1 \to 1 , \quad v_1 \to 0 .$$

We also have

$$Y \to \infty , \quad U_1 \to 0 .$$

The solution for the outer expansion is simply $u_1 = 1$ and $v_1 = 0$. Then, the equations for U_1 and V_1 are

$$\frac{\partial U_1}{\partial x} + \frac{\partial V_1}{\partial Y} = 0 ,$$
$$U_1 \frac{\partial U_1}{\partial x} + \frac{\partial U_1}{\partial x} + V_1 \frac{\partial U_1}{\partial Y} = \frac{\partial^2 U_1}{\partial Y^2} ,$$

and the momentum equation in the direction normal to the wall is

$$(1 + U_1) \frac{\partial V_1}{\partial x} + V_1 \frac{\partial V_1}{\partial Y} = -\frac{\partial P_1}{\partial Y} + \frac{\partial^2 V_1}{\partial Y^2} .$$

4. We set

$$U = 1 + U_1 ,$$
$$V = \varepsilon V_1 ,$$

and we obtain

$$\frac{\partial U}{\partial x} + \frac{\partial V}{\partial y} = 0 ,$$
$$U \frac{\partial U}{\partial x} + V \frac{\partial U}{\partial y} = \varepsilon^2 \frac{\partial^2 U}{\partial y^2} ,$$

with the boundary conditions

$$y = 0 : \quad U = 0 , \quad V = 0 ,$$

and

$$y \to \infty : \quad U \to 1 .$$

We recover exactly Prandtl's model.

With Blasius' solution, we obtain

$$V_1 \underset{Y \to \infty}{\longrightarrow} = \frac{\beta_0}{\sqrt{2x}} .$$

With the transverse momentum equation

$$(1 + U_1)\frac{\partial V_1}{\partial x} + V_1\frac{\partial V_1}{\partial Y} = -\frac{\partial P_1}{\partial Y} + \frac{\partial^2 V_1}{\partial Y^2} \; ,$$

we obtain the behaviour of P_1 as $Y \to \infty$

$$-\frac{\partial P_1}{\partial Y} = \frac{\partial V_1}{\partial x} = -\frac{\beta_0}{2\sqrt{2}} x^{-3/2} \; ,$$

whence

$$P_1 \underset{Y \to \infty}{\cong} \frac{\beta_0}{2\sqrt{2}} x^{-3/2} Y \; ,$$

or

$$P_1 \underset{Y \to \infty}{\cong} \frac{\beta_0}{2\sqrt{2}} x^{-3/2} \frac{y}{\varepsilon} \; ,$$

or

$$\varepsilon P_1 \underset{Y \to \infty}{\cong} \frac{\beta_0}{2\sqrt{2}} x^{-3/2} y \; .$$

5. The equations for u_2, v_2, p_2 are

$$\frac{\partial u_2}{\partial x} + \frac{\partial v_2}{\partial y} = 0 \; ,$$

$$\frac{\partial u_2}{\partial x} = -\frac{\partial}{\partial x}(p_2 + \varepsilon P_1) \; ,$$

$$\frac{\partial}{\partial x}(v_2 + V_1) = -\frac{\partial}{\partial y}(p_2 + \varepsilon P_1) \; .$$

In the momentum equations, the quantity P_1 must be understood as the behaviour of P_1 as $Y \to \infty$; likewise, the quantity V_1 must be understood as the behaviour of V_1 as $Y \to \infty$.

The boundary conditions to take into account are

$$y \to \infty : \quad u_2 = 0 \; ,$$

$$y \to \infty : \quad v_2 \to -\frac{\beta_0}{\sqrt{2x}} \; ,$$

$$y = 0 : \quad v_2 = 0 \; .$$

The last condition enables us to satisfy the wall condition of zero normal velocity. The no-slip condition at the wall on u_2 must be discarded.

Taking into account the behaviour of V_1 and P_1 as $Y \to \infty$, and with the following change of functions

$$u_2^* = u_2 \; ,$$

$$v_2^* = v_2 + \frac{\beta_0}{\sqrt{2x}} \; ,$$

$$p_2^* = p_2 + \frac{\beta_0}{2\sqrt{2}} x^{-3/2} y \; ,$$

the equations become

$$\frac{\partial u_2^*}{\partial x} + \frac{\partial v_2^*}{\partial y} = 0 \, ,$$

$$\frac{\partial u_2^*}{\partial x} = -\frac{\partial p_2^*}{\partial x} \, ,$$

$$\frac{\partial v_2^*}{\partial x} = -\frac{\partial p_2^*}{\partial y} \, ,$$

with the boundary conditions

$$y \to \infty : \quad u_2^* = 0 \, ,$$

$$y \to \infty : \quad v_2^* = 0 \, ,$$

$$y = 0 : \quad v_2^* = \frac{\beta_0}{\sqrt{2x}} \, .$$

Then, the solution is

$$u_2^* = -\frac{\beta_0}{2} \frac{y}{\sqrt{x^2 + y^2}\sqrt{x + \sqrt{x^2 + y^2}}} \, ,$$

$$v_2^* = \frac{\beta_0}{2} \frac{\sqrt{x + \sqrt{x^2 + y^2}}}{\sqrt{x^2 + y^2}} \, .$$

Chapter 9

9-1.

1. Substituting the outer expansion in the Navier-Stokes equations, to the second order we obtain

$$\frac{\partial u_1}{\partial x} + \frac{\partial v_1}{\partial y} = 0 \, ,$$

$$(1 + ay)\frac{\partial u_1}{\partial x} + av_1 = -\frac{\partial p_1}{\partial x} \, ,$$

$$(1 + ay)\frac{\partial v_1}{\partial x} = -\frac{\partial p_1}{\partial y} \, .$$

2. The first order boundary layer equations are

$$\frac{\partial U_1}{\partial x} + \frac{\partial V_1}{\partial Y} = 0 \, ,$$

$$U_1\frac{\partial U_1}{\partial x} + V_1\frac{\partial U_1}{\partial Y} = \frac{\partial^2 U_1}{\partial Y^2} \, .$$

The second order boundary layer equations are

$$\frac{\partial U_2}{\partial x} + \frac{\partial V_2}{\partial Y} = 0 \, ,$$

$$\Delta_2 \left(U_1 \frac{\partial U_2}{\partial x} + V_1 \frac{\partial U_2}{\partial Y} + U_2 \frac{\partial U_1}{\partial x} + V_2 \frac{\partial U_1}{\partial Y} \right)$$

$$= -\Delta_2^* \frac{\partial P_2}{\partial x} + \varepsilon^2 \frac{\partial^2 U_1}{\partial x^2} + \Delta_2 \frac{\partial^2 U_2}{\partial Y^2} + \cdots ,$$

$$\varepsilon \left(U_1 \frac{\partial V_1}{\partial x} + V_1 \frac{\partial V_1}{\partial Y} \right) = -\frac{\Delta_2^*}{\varepsilon} \frac{\partial P_2}{\partial Y} + \varepsilon \frac{\partial^2 V_1}{\partial Y^2} .$$

The boundary layer equations are solved by assuming that $U_1 = f'(\eta)$ with $\eta = \dfrac{Y}{\sqrt{2x}}$. We have

$$V_1 = \frac{1}{\sqrt{2x}} \left[\eta f' - f \right] ,$$

$$f''' + f f'' = 0 ,$$

with $f(0) = 0$, $f'(0) = 0$, $f'(\infty) = 1$. On the other hand, the behaviour of f as $\eta \to \infty$ is

$$f(\eta) \underset{\eta \to \infty}{\cong} \eta - \beta_0 + \mathrm{EST} ,$$

with $\beta_0 = 1.21678$.

We observe that the first order boundary layer solution is independent of a.

3. The matching of $\mathcal{V}$ implies $\delta_1 = \varepsilon$. Indeed, to order ε, we have

$$\mathrm{E}_1 \, \mathcal{V} = \varepsilon V_1 ,$$

$$\mathrm{E}_0 \, \mathrm{E}_1 \, \mathcal{V} = \varepsilon \frac{1}{\sqrt{2x}} \left[\frac{Y}{\sqrt{2x}} - \frac{Y}{\sqrt{2x}} + \beta_0 \right]$$

$$= \varepsilon \frac{\beta_0}{\sqrt{2x}} ,$$

$$\mathrm{E}_1 \, \mathrm{E}_0 \, \mathcal{V} = \varepsilon v_1(x, 0) .$$

We obtain

$$v_1(x, 0) = \frac{\beta_0}{\sqrt{2x}} .$$

4. Substituting the behaviours of u_1, v_1, p_1 in the outer equations, we obtain
$b_0 = 2^{-1/2} \beta_0 x^{-1/2}$, $c_0 = -2^{1/2} a \beta_0 x^{1/2}$, $c_1 = 2^{-3/2} \beta_0 x^{-3/2}$,
$b_2 = -3 \cdot 2^{-7/2} \beta_0 x^{-5/2}$, $c_2 = 2^{-5/2} a \beta_0 x^{-3/2}$, $a_1 = -2^{-3/2} \beta_0 x^{-3/2}$.

5. The matching of pressure gives $\Delta_2^* = \varepsilon$. Indeed, to order ε, we have

$$\mathrm{E}_1 \, \mathrm{E}_0 \, \mathcal{P} = -2^{1/2} \varepsilon a \beta_0 x^{1/2} ,$$

$$\mathrm{E}_0 \, \mathrm{E}_1 \, \mathcal{P} = \mathrm{E}_0 \left[\varepsilon P_2(x, Y) \right] .$$

We obtain

$$\lim_{Y \to \infty} P_2(x, Y) = -2^{1/2} a \beta_0 x^{1/2} .$$

6. The matching of velocity $\mathcal{U}$ gives $\Delta_2 = \varepsilon$. Indeed, to order ε, we have

$$\mathrm{E}_1\,\mathrm{E}_0\,\mathcal{U} = 1 + \varepsilon a\,Y\ ,$$
$$\mathrm{E}_0\,\mathrm{E}_1\,\mathcal{U} = \mathrm{E}_0\left[U_1 + \varepsilon U_2\right]\ .$$

We have

$$\lim_{Y\to\infty}\,U_1 = 1\ ,$$

and

$$U_2 \underset{Y\to\infty}{\cong} aY\ .$$

7. We have

$$\frac{\partial P_2}{\partial Y} = 0\ ,$$

and

$$P_2 = -2^{1/2}a\beta_0 x^{1/2}\ ,$$

whence

$$\frac{\partial P_2}{\partial x} = -2^{-1/2}a\beta_0 x^{-1/2}\ .$$

The second order boundary layer equations are

$$\frac{\partial U_2}{\partial x} + \frac{\partial V_2}{\partial Y} = 0\ ,$$

$$U_1\frac{\partial U_2}{\partial x} + V_1\frac{\partial U_2}{\partial Y} + U_2\frac{\partial U_1}{\partial x} + V_2\frac{\partial U_1}{\partial Y} = 2^{-1/2}a\beta_0 x^{-1/2} + \frac{\partial^2 U_2}{\partial Y^2}\ .$$

Chapter 10

10-1.

1. The reduced equation is

$$\frac{\mathrm{d}y_0}{\mathrm{d}x} + y_0 = 0\ .$$

With the condition $y(1) = \beta$, the solution is

$$y_0 = \beta\,\mathrm{e}^{1-x}\ .$$

With the change of variable $\bar{x} = x/\delta(\varepsilon)$, the initial equation becomes

$$\frac{\varepsilon^3}{\delta^2}\frac{\mathrm{d}^2 y}{\mathrm{d}\bar{x}^2} + \delta^2\bar{x}^3\frac{\mathrm{d}y}{\mathrm{d}\bar{x}} + (\delta^3\bar{x}^3 - \varepsilon)y = 0\ .$$

To restore the boundary layer at $x = 0$, we must keep the term containing the second derivative. Comparing the order of magnitude of this term with the other terms shows that we must take $\delta = \varepsilon$. Then, the reduced equation is

$$\frac{\mathrm{d}^2\bar{y}_0}{\mathrm{d}\bar{x}^2} - \bar{y}_0 = 0\ ,$$

whence

$$\bar{y}_0 = A\,\mathrm{e}^{-\bar{x}} + B\,\mathrm{e}^{\bar{x}} \ .$$

The condition $y(0) = \alpha$ gives

$$A + B = \alpha \ .$$

To order 1, we have

$$\bar{\mathrm{E}}\,\mathrm{E}\,y = \beta\,\mathrm{e} \ ,$$

where $\bar{\mathrm{E}}$ and E are the expansion operators corresponding to the lower layer and to the upper layer respectively. We also have

$$\bar{\mathrm{E}}y = A\,\mathrm{e}^{-\bar{x}} + B\,\mathrm{e}^{\bar{x}} = A\,\mathrm{e}^{-x/\varepsilon} + B\,\mathrm{e}^{x/\varepsilon} \ ,$$

and

$$\mathrm{E}\,\bar{\mathrm{E}}y = \lim_{\varepsilon \to 0,\ x\,\text{fixed}} \left[A\,\mathrm{e}^{-x/\varepsilon} + B\,\mathrm{e}^{x/\varepsilon} \right] \ .$$

We conclude that $B = 0$ in order to have a possible matching, whence $A = \alpha$. Then $\mathrm{E}\,\bar{\mathrm{E}}y = 0$ and the matching is impossible since $\bar{\mathrm{E}}\,\mathrm{E}\,y \neq \mathrm{E}\,\bar{\mathrm{E}}y$.

We introduce an intermediate layer whose thickness is $\nu(\varepsilon)$. With the change of variable $\tilde{x} = x/\nu$, the initial equation becomes

$$\frac{\varepsilon^3}{\nu^2}\frac{\mathrm{d}^2 y}{\mathrm{d}\tilde{x}^2} + \nu^2\tilde{x}^3\frac{\mathrm{d}y}{\mathrm{d}\tilde{x}} + (\nu^3\tilde{x}^3 - \varepsilon)y = 0 \ .$$

Comparing the orders of magnitude of different terms shows that a distinguished limit is obtained by taking $\nu = \varepsilon^{1/2}$. The thickness of the corresponding layer is between the thicknesses of the upper and lower layers. The initial equation reduces to

$$\tilde{x}^3\frac{\mathrm{d}\tilde{y}_0}{\mathrm{d}\tilde{x}} - \tilde{y} = 0 \ .$$

The solution is

$$\tilde{y}_0 = C\,\mathrm{e}^{-1/(2\tilde{x}^2)} \ .$$

The matching between the upper layer and the intermediate layer to order 1 yields

$$\tilde{\mathrm{E}}\,\mathrm{E}\,y = \beta\,\mathrm{e} \ ,$$

and

$$\mathrm{E}\,\tilde{\mathrm{E}}y = C \ ,$$

whence $C = \beta\,\mathrm{e}$.

The matching between the intermediate layer and the lower layer to order 1 yields

$$\bar{\mathrm{E}}\tilde{\mathrm{E}}y = 0 \ .$$

Therefore, we have necessarily $B = 0$ and, with the condition $A + B = \alpha$, we have $A = \alpha$.

Then, the composite solution is

$$y_\mathrm{c} = \beta\,\mathrm{e}^{1-x} + \beta\,\mathrm{e}\,\mathrm{e}^{-1/(2\tilde{x}^2)} + \alpha\,\mathrm{e}^{-\bar{x}} - \beta\,\mathrm{e} \ .$$

2. To simplify, we assume that the triple layer structure is known but it would be possible to recover it with SCEM.

The outer approximation is the same as with MMAE

$$Y_0 = \beta\, e^{1-x} \ .$$

This approximation is complemented as

$$y = Y_0(x) + \widetilde{Y}_0(\tilde{x}, \varepsilon) \ .$$

The initial equation becomes

$$\varepsilon^2 \frac{d^2\widetilde{Y}_0}{d\tilde{x}^2} + \varepsilon^3 \frac{d^2 Y_0}{dx^2} + \varepsilon\tilde{x}^3 \frac{d\widetilde{Y}_0}{d\tilde{x}} - \varepsilon Y_0 + \varepsilon^{3/2}\tilde{x}^3\widetilde{Y}_0 - \varepsilon\widetilde{Y}_0 = 0 \ .$$

In this equation, $\varepsilon^3 \dfrac{d^2 Y_0}{dx^2}$ is of order ε^3 in domain $0 < A \le \tilde{x} \le 1$ where A is a constant independent of ε and Y_0 writes

$$Y_0 = \beta\, e^{1-x} = \beta\, e^{1-\varepsilon^{1/2}\tilde{x}} = \beta\, e(1 + \cdots) \ .$$

Then, the equation for $\widetilde{Y}_0$ is

$$\tilde{x}^3 \frac{d\widetilde{Y}_0}{d\tilde{x}} - \widetilde{Y}_0 = \beta\, e \ .$$

The solution is

$$\widetilde{Y}_0 = -\beta\, e + C\, e^{-1/(2\tilde{x}^2)} \ .$$

The condition $y(1) = \beta$ is satisfied by $Y_0(1) = \beta$. Then, we have

$$\tilde{x} = \frac{1}{\varepsilon^{1/2}} : \quad \widetilde{Y}_0 = 0 \ ,$$

whence

$$C = \beta\, e^{1+\varepsilon/2} \ .$$

We seek a UVA in the form

$$y_a = Y_0(x) + \widetilde{Y}_0(\tilde{x}, \varepsilon) + \overline{Y}_0(\bar{x}, \varepsilon) \ ,$$

The equation for $\overline{Y}_0$ is

$$\frac{d^2\overline{Y}_0}{d\bar{x}^2} - \overline{Y}_0 = 0 \ .$$

The solution is

$$\overline{Y}_0 = A\, e^{-\bar{x}} + B\, e^{\bar{x}} \ .$$

The boundary conditions are

$$\bar{x} = 0 : \quad Y_0 + \widetilde{Y}_0 + \overline{Y}_0 = \alpha \ ; \quad \bar{x} = \frac{1}{\varepsilon} : \quad Y_0 + \widetilde{Y}_0 + \overline{Y}_0 = \beta \ ,$$

whence

$$A = \frac{\alpha}{1 - \mathrm{e}^{-2/\varepsilon}} \ , \quad B = -\frac{\alpha\,\mathrm{e}^{-2/\varepsilon}}{1 - \mathrm{e}^{-2/\varepsilon}} \ ,$$

and the UVA is

$$y_\mathrm{a} = \beta\,\mathrm{e}^{1-x} - \beta\,\mathrm{e} + \beta\,\mathrm{e}^{1+\varepsilon/2}\,\mathrm{e}^{-1/(2\tilde{x}^2)} + \frac{\alpha}{1 - \mathrm{e}^{-2/\varepsilon}}\,\mathrm{e}^{-\tilde{x}} - \frac{\alpha\,\mathrm{e}^{-2/\varepsilon}}{1 - \mathrm{e}^{-2/\varepsilon}}\,\mathrm{e}^{\tilde{x}} \ .$$

The regular form of SCEM gives

$$y_\mathrm{a} = \beta\,\mathrm{e}^{1-x} - \beta\,\mathrm{e} + \beta\,\mathrm{e}\,\mathrm{e}^{-1/(2\tilde{x}^2)} + \alpha\,\mathrm{e}^{-\tilde{x}} \ .$$

3. The outer approximation is again the same

$$f_0 = \beta\,\mathrm{e}^{1-x} \ .$$

We seek a UVA in the form

$$y_\mathrm{a} = f_0 + \bar{f}_0(\bar{x},\varepsilon) \text{ with } \bar{x} = \frac{x}{\varepsilon} \ .$$

The initial equation becomes

$$\varepsilon^3 \frac{\mathrm{d}^2 f_0}{\mathrm{d}x^2} + \varepsilon \frac{\mathrm{d}^2 \bar{f}_0}{\mathrm{d}\bar{x}^2} + \varepsilon^2 \bar{x}^3 \frac{\mathrm{d}\bar{f}_0}{\mathrm{d}\bar{x}} - \varepsilon f_0 + \varepsilon^3 \bar{x}^3 \bar{f}_0 - \varepsilon \bar{f}_0 = 0 \ .$$

To cast the intermediate layer and the lower layer in a single approximation, we neglect the $\mathrm{O}(\varepsilon^3)$ terms

$$\frac{\mathrm{d}^2 \bar{f}_0}{\mathrm{d}\bar{x}^2} + \varepsilon \bar{x}^3 \frac{\mathrm{d}\bar{f}_0}{\mathrm{d}\bar{x}} - \bar{f}_0 = f_0 \ .$$

It is possible to write f_0 in the form

$$f_0 = \beta\,\mathrm{e}^{1-x} = \beta\,\mathrm{e}^{1-\varepsilon\bar{x}} = \beta\,\mathrm{e}(1 - \varepsilon\bar{x} + \cdots) \ .$$

The boundary conditions for $\bar{f}_0$ are

$$\bar{x} = 0: \quad f_0 + \bar{f}_0 = \alpha \ ; \quad \bar{x} = \frac{1}{\varepsilon}: \quad f_0 + \bar{f}_0 = \beta \ .$$

10-2. The main deck equations are

$$\frac{\partial \overline{U}_2}{\partial X} + \frac{\partial \overline{V}_2}{\partial \overline{Y}} = 0 \ ,$$

$$U_0 \frac{\partial \overline{U}_2}{\partial X} + \overline{V}_2 \frac{\mathrm{d}U_0}{\mathrm{d}\overline{Y}} + f \frac{\mathrm{d}f}{\mathrm{d}X}\left[\left(\frac{\mathrm{d}U_0}{\mathrm{d}\overline{Y}}\right)^2 - U_0 \frac{\mathrm{d}^2 U_0}{\mathrm{d}\overline{Y}^2}\right] = -\frac{\partial \overline{P}_2}{\partial X} \ ,$$

$$U_0 \frac{\partial \overline{V}_2}{\partial X} + \frac{\mathrm{d}}{\mathrm{d}X}\left(f \frac{\mathrm{d}f}{\mathrm{d}X}\right) U_0 \frac{\mathrm{d}U_0}{\mathrm{d}\overline{Y}} = -\frac{\partial \overline{P}_2}{\partial \overline{Y}} \ .$$

The lower deck equations are

$$\frac{\partial \widetilde{U}_1}{\partial X} + \frac{\partial \widetilde{V}_1}{\partial \widetilde{Y}} = 0 \, ,$$

$$(\lambda \widetilde{Y} + \widetilde{U}_1)\frac{\partial \widetilde{U}_1}{\partial X} + \widetilde{V}_1\left(\lambda + \frac{\partial \widetilde{U}_1}{\partial \widetilde{Y}}\right) = -\frac{\partial \widetilde{P}_1}{\partial X} + \frac{\partial^2 \widetilde{U}_1}{\partial \widetilde{Y}^2} \, ,$$

$$\frac{\partial \widetilde{P}_1}{\partial \widetilde{Y}} = 0 \, .$$

The boundary conditions at infinity are

$$\overline{Y} \to \infty: \quad \overline{U}_2 = 0 \, , \quad \overline{V}_2 = 0 \, .$$

The conditions at the wall are

$$\widetilde{Y} = 0: \quad \widetilde{U}_1 = 0 \, , \quad \widetilde{V}_1 = 0 \, .$$

The matching of velocities between the two decks yields

$$\lim_{\widetilde{Y}\to\infty} \widetilde{U}_1 = \lambda f(X) \, , \quad \overline{V}_2(X,0) = \lim_{\widetilde{Y}\to\infty}\left(\widetilde{V}_1 + \lambda \widetilde{Y}\frac{\mathrm{d}f}{\mathrm{d}X}\right) \, .$$

Moreover, in the lower deck, the pressure is constant along a normal to the wall

$$\widetilde{P}_1 = \overline{P}_1(X,0) \, .$$

If the shape of the hump is known, i.e. if the function $f(X)$ is known, it is not possible to determine directly the solution in the main deck or in the lower deck. The problems are coupled. There is no hierarchy between the decks. It is said that the interaction is strong.

10-3.

Zone 1. In the lower deck, the boundary conditions are

$$Y^* \to \infty: \quad U_1^* = 0 \, , \quad V_1^* = 0 \, .$$

In addition, the matching between the upper deck and the main deck yields

$$V_1^*(X,0) = 0 \, ,$$

because $\beta - \alpha < \beta - \dfrac{7\alpha}{3} + \dfrac{m}{2}$ as $\alpha < \dfrac{3m}{8}$ in zone 1. Then, in the upper deck, the solution is (see Appendix III)

$$U_1^* = 0 \, , \quad V_1^* = 0 \, , \quad P_1^* = 0 \, .$$

The matching conditions on pressure between the different decks yields

$$P_1^*(X,0) = \overline{P}_1(X,\overline{Y}) = \widetilde{P}_1(X,\widetilde{Y}) = 0 \, .$$

The main deck solution is

$$\overline{U}_1 = A(X)\frac{\mathrm{d}U_0}{\mathrm{d}\overline{Y}} \;, \quad \overline{V}_1 = -\frac{\mathrm{d}A}{\mathrm{d}X}U_0 \;.$$

The matching between the main deck and the lower deck yields also

$$\lim_{\widetilde{Y}\to\infty} \widetilde{U}_1 = \overline{U}_1(X,0) = \lambda A \;.$$

The lower deck equations are

$$\frac{\partial \widetilde{U}_1}{\partial X} + \frac{\partial \widetilde{V}_1}{\partial \widetilde{Y}} = 0 \;,$$

$$\lambda \widetilde{Y}\frac{\partial \widetilde{U}_1}{\partial X} + \lambda \widetilde{V}_1 = \frac{\partial^2 \widetilde{U}_1}{\partial \widetilde{Y}^2} \;.$$

The wall conditions are

$$\widetilde{Y} = 0 : \quad \widetilde{U}_1 = 0 \;, \quad \widetilde{V}_1 = V_w(X) \;.$$

At the boundary layer edge, we have

$$\widetilde{Y} \to \infty : \quad \frac{\partial \widetilde{U}_1}{\partial \widetilde{Y}} = 0 \;.$$

With these conditions, we can determine the lower deck solution and calculate A from

$$\lim_{\widetilde{Y}\to\infty} \widetilde{U}_1 = \lambda A \;.$$

Zone 2. The main deck solution is

$$\overline{U}_2 = A(X)\frac{\mathrm{d}U_0}{\mathrm{d}\overline{Y}} \;, \quad \overline{V}_2 = -\frac{\mathrm{d}A}{\mathrm{d}X}U_0 \;.$$

The matching between the upper deck and the main deck yields

$$V_2^*(X,0) = \lim_{\overline{Y}\to\infty} \overline{V}_2 = -\frac{\mathrm{d}A}{\mathrm{d}X} \;.$$

The problem is solved with a given distribution of $A(X)$. Then, the upper deck solution is calculated with

$$V_2^*(X,0) = -\frac{\mathrm{d}A}{\mathrm{d}X} \;.$$

Then, the value of $P_2^*(X,0)$ is obtained and $\widetilde{P}_1$ is known since the matching of pressure between the different decks gives

$$P_2^*(X,0) = \overline{P}_2(X,\overline{Y}) = \widetilde{P}_1(X,\widetilde{Y}) \;.$$

The lower deck equations are

$$\frac{\partial \widetilde{U}_1}{\partial X} + \frac{\partial \widetilde{V}_1}{\partial \widetilde{Y}} = 0 \,,$$

$$\lambda \widetilde{Y} \frac{\partial \widetilde{U}_1}{\partial X} + \lambda \widetilde{V}_1 = -\frac{\partial \widetilde{P}_1}{\partial X} + \frac{\partial^2 \widetilde{U}_1}{\partial \widetilde{Y}^2} \,.$$

The matching between the main deck and the lower deck gives in particular

$$\lim_{\widetilde{Y} \to \infty} \widetilde{U}_1 = 0 \,,$$

since $\beta - \dfrac{m}{2} < \beta + \dfrac{4\alpha}{3} - m$ as $\alpha > \dfrac{3m}{8}$.

Then, at the edge of the lower deck, the momentum equation gives

$$\lambda \widetilde{V}_{1e} = -\frac{\partial \widetilde{P}_1}{\partial X} \,,$$

with

$$\widetilde{V}_{1e} = \lim_{\widetilde{Y} \to \infty} \widetilde{V}_1 \,.$$

At the wall, we have

$$\widetilde{Y} = 0 : \quad \widetilde{U}_1 = 0 \,.$$

In the lower deck, a possible solution satisfying the boundary conditions is

$$\widetilde{U}_1 = 0 \,, \quad \widetilde{V}_1 = \widetilde{V}_{1e} \,,$$

and we obtain the value of V_w,

$$V_w = \widetilde{V}_{1e} = -\frac{1}{\lambda}\frac{\partial \widetilde{P}_1}{\partial X} \,.$$

Zone 3. The matching conditions between the upper deck and the main deck give in particular

$$V_1^*(X,0) = 0 \,,$$

since $\beta - \alpha < \dfrac{\beta - 3\alpha + m}{2}$. Now, the boundary conditions at infinity are

$$Y^* \to \infty : \quad V_1^* = 0 \,, \quad U_1^* = 0 \,.$$

We conclude that the upper deck solution is identically zero

$$U_1^* = 0 \,, \quad V_1^* = 0 \,, \quad P_1^* = 0 \,.$$

The matching conditions between the upper deck and the main deck give

$$\lim_{\overline{Y} \to \infty} \overline{U}_1 = 0 \,,$$

since $\dfrac{\beta - \alpha}{2} < \beta - \alpha.$

The matching between the main deck and the lower deck gives

$$\overline{V}_1(X,0) = 0 \; ,$$

since $\dfrac{\beta - 3\alpha + m}{2} < \dfrac{\beta - 3\alpha + 2m}{4} \; .$

The main deck solution is

$$\overline{U}_1 = A(X)\frac{\mathrm{d}U_0}{\mathrm{d}\overline{Y}} \; , \quad \overline{V}_1 = -\frac{\mathrm{d}A}{\mathrm{d}X}U_0 \; .$$

The matching between the main deck and the lower deck gives

$$\lim_{\widetilde{Y}\to\infty} \widetilde{U}_1 = \lambda A \; , \quad \widetilde{V}_1 \underset{\widetilde{Y}\to\infty}{\cong} -\lambda\frac{\mathrm{d}A}{\mathrm{d}X}\widetilde{Y} \; .$$

In addition, the matching of pressure between the different decks yields

$$\widetilde{P}_1 = \overline{P}_1 = P_1^*(X,0) = 0 \; .$$

Therefore, the lower deck equations are

$$\frac{\partial \widetilde{U}_1}{\partial X} + \frac{\partial \widetilde{V}_1}{\partial \widetilde{Y}} = 0 \; ,$$

$$\widetilde{U}_1\frac{\partial \widetilde{U}_1}{\partial X} + \widetilde{V}_1\frac{\partial \widetilde{U}_1}{\partial \widetilde{Y}} = \frac{\partial^2 \widetilde{U}_1}{\partial \widetilde{Y}^2} \; ,$$

with the boundary conditions at the wall

$$\widetilde{Y} = 0 : \quad \widetilde{U}_1 = 0 \; , \quad \widetilde{V}_1 = V_w(X) \; .$$

With the condition

$$\lim_{\widetilde{Y}\to\infty} \widetilde{U}_1 = \lambda A \; ,$$

the momentum equation, as $\widetilde{Y} \to \infty$, gives

$$A\frac{\mathrm{d}A}{\mathrm{d}X} = 0 \; .$$

With $A \to 0$ as $X \to -\infty$, we obtain $A(X) = 0$, whence

$$\lim_{\widetilde{Y}\to\infty} \widetilde{U}_1 = 0 \; .$$

Then, the lower deck solution is

$$\widetilde{U}_1 = 0 \; , \quad \widetilde{V}_1 = V_w(X) \; .$$

To the considered order, the effect of blowing is restricted to the lower deck in which the component of velocity normal to the wall remains unchanged.

Zone 4. The lower deck equations are

$$\frac{\partial \widetilde{U}_1}{\partial X} + \frac{\partial \widetilde{V}_1}{\partial \widetilde{Y}} = 0 \,,$$

$$\widetilde{U}_1 \frac{\partial \widetilde{U}_1}{\partial X} + \widetilde{V}_1 \frac{\partial \widetilde{U}_1}{\partial \widetilde{Y}} = -\frac{\partial \widetilde{P}_1}{\partial X} + \frac{\partial^2 \widetilde{U}_1}{\partial \widetilde{Y}^2} \,,$$

$$\frac{\partial \widetilde{P}_1}{\partial \widetilde{Y}} = 0 \,.$$

Now, the matching between the lower deck and the main deck yields

$$\lim_{\widetilde{Y} \to \infty} \widetilde{U}_1 = 0 \,,$$

since $0 < \beta - \dfrac{m}{2} < 2\beta + \alpha - \dfrac{3m}{2}$. Then, as $\widetilde{Y} \to \infty$, the X-momentum equation gives

$$\frac{\partial \widetilde{P}_1}{\partial X} = 0 \,.$$

As $\dfrac{\partial \widetilde{P}_1}{\partial \widetilde{Y}} = 0$, using the condition $\widetilde{P}_1 \to 0$ as $X \to -\infty$, we obtain $\widetilde{P}_1 = 0$.
In addition, at the wall, we must have

$$\widetilde{Y} = 0 : \quad \widetilde{U}_1 = 0 \,, \quad \widetilde{V}_1 = V_w(X) \,.$$

A possible solution in the lower deck is

$$\widetilde{U}_1 = 0 \,, \quad \widetilde{V}_1 = V_w(X) \,.$$

The matching of the pressure between the different decks yields

$$P_2^*(X,0) = \overline{P}_2 = \widetilde{P}_1 = 0 \,.$$

Taking into account the conditions

$$Y^* \to \infty : U_2^* = 0 \,, \quad V_2^* = 0 \,,$$

the upper deck solution is identically zero

$$U_2^* = 0 \,, \quad V_2^* = 0 \,, \quad P_2^* = 0 \,.$$

The main deck solution has the form

$$\overline{U}_2 = A(X)\frac{\mathrm{d}U_0}{\mathrm{d}\overline{Y}} \,, \quad \overline{V}_2 = -\frac{\mathrm{d}A}{\mathrm{d}X}U_0 \,.$$

The matching between the main deck and the upper deck gives

$$\lim_{\overline{Y} \to \infty} \overline{V}_2 = V_2^*(X,0) = 0 \,.$$

With $A \to 0$ as $X \to -\infty$, we obtain $A(X) = 0$.

Chapter 11

11-1.

1. The equation for f_0 is

$$\frac{d^2 f_0}{dy^2} + \frac{1}{y}\frac{df_0}{dy} + f_0\frac{df_0}{dy} = 0 \ ,$$

with

$$\lim_{y\to\infty} f_0 = 1 \ .$$

The solution of this non linear equation is $f_0 = 1$. Then, the equation for f_1 is

$$\frac{d^2 f_1}{dy^2} + \frac{1}{y}\frac{df_1}{dy} + \frac{df_1}{dy} = 0 \ ,$$

with

$$\lim_{y\to\infty} f_1 = 0 \ .$$

The linear equation for f_1 has the solution

$$f_1 = A\int_y^\infty \frac{e^{-t}}{t}\, dt \ ,$$

which satisfies the condition at infinity.

2. The equation for $\bar{f}_0$ is

$$\frac{d^2 \bar{f}_0}{d\bar{y}^2} + \frac{1}{\bar{y}+1}\frac{d\bar{f}_0}{d\bar{y}} = 0 \ ,$$

with $\bar{f}_0(0) = 0$. The solution is

$$\bar{f}_0 = K\ln(\bar{y} + 1) \ .$$

3. We have

$$E_1\, E_0\, u = 1 + \delta_1 A(-\ln\varepsilon\bar{y} - \gamma) \ ,$$
$$E_0\, E_1\, u = \bar{\delta}_1 K\ln\bar{y} \ ,$$

whence

$$\delta_1 A = \frac{1}{\gamma + \ln\varepsilon} \ , \qquad \bar{\delta}_1 K = -\delta_1 A \ .$$

The composite solution is

$$u = 1 + \frac{1}{\gamma + \ln\varepsilon}\int_y^\infty \frac{e^{-t}}{t}\, dt - \frac{1}{\gamma + \ln\varepsilon}\ln(\bar{y} + 1) + \frac{1}{\gamma + \ln\varepsilon}\ln\bar{y} \ ,$$

or

$$u = 1 + \frac{1}{\gamma + \ln\varepsilon}\left[\int_y^\infty \frac{e^{-t}}{t}\, dt + \ln\frac{\bar{y}}{\bar{y} + 1}\right] \ .$$

11-2.

1. With $y' = y + \varepsilon$, the initial equation becomes

$$\frac{\mathrm{d}^2 u}{\mathrm{d}y'^2} + \frac{1}{y'}\frac{\mathrm{d}u}{\mathrm{d}y'} + u\frac{\mathrm{d}u}{\mathrm{d}y'} = 0 \; ,$$

with the boundary conditions

$$u(\varepsilon) = 0 \; , \qquad \lim_{y \to \infty} u = 1 \; .$$

We obtain exactly the Stokes-Oseen's flow model proposed by Lagerstrom.

2. The equation for F_1 is

$$\frac{\mathrm{d}^2 F_1}{\mathrm{d}y^2} + \frac{1}{y}\frac{\mathrm{d}F_1}{\mathrm{d}y} + \frac{\mathrm{d}F_1}{\mathrm{d}y} = 0 \; .$$

With the condition $F_1(\infty) = 0$ which satisfies the condition at infinity for u, the solution is

$$F_1 = A \int_y^\infty \frac{\mathrm{e}^{-t}}{t}\,\mathrm{d}t \; ,$$

but the condition at $y = 0$ cannot be satisfied since F_1 becomes infinite as $y \to 0$. The equation for $\overline{F}_1$ is

$$\frac{\mathrm{d}^2 \overline{F}_1}{\mathrm{d}\bar{y}^2} + \frac{1}{\bar{y}+1}\frac{\mathrm{d}\overline{F}_1}{\mathrm{d}\bar{y}} = \frac{1}{\bar{y}(\bar{y}+1)}\varepsilon\frac{\mathrm{d}F_1}{\mathrm{d}y} \; .$$

The term $\varepsilon\dfrac{\mathrm{d}F_1}{\mathrm{d}y}$ must be simplified since $\overline{F}_1$ must be a function of $\bar{y}$ only. We have

$$\frac{\mathrm{d}F_1}{\mathrm{d}y} = -A\frac{\mathrm{e}^{-y}}{y} = -\frac{A}{\varepsilon}\frac{\mathrm{e}^{-\varepsilon\bar{y}}}{\bar{y}} = -\frac{A}{\varepsilon}\frac{1}{\bar{y}} + \cdots \; .$$

The equation for $\overline{F}_1$ becomes

$$\frac{\mathrm{d}^2 \overline{F}_1}{\mathrm{d}\bar{y}^2} + \frac{1}{\bar{y}+1}\frac{\mathrm{d}\overline{F}_1}{\mathrm{d}\bar{y}} = -\frac{A}{\bar{y}^2(\bar{y}+1)} \; .$$

After mutliplying by $\bar{y}+1$, the equation can be integrated easily. We have

$$\overline{F}_1 = A\ln\bar{y} - A\ln(\bar{y}+1) + B\ln(\bar{y}+1) + C \; ,$$

whence

$$u = 1 + \delta_1 A \int_y^\infty \frac{\mathrm{e}^{-t}}{t}\,\mathrm{d}t + \delta_1 A\ln\bar{y} - \delta_1 A\ln(\bar{y}+1) + \delta_1 B\ln(\bar{y}+1) + \delta_1 C \; .$$

The condition $u = 1$ as $y \to \infty$ gives $B = 0$ and $C = 0$. After expanding $\int_y^\infty \frac{\mathrm{e}^{-t}}{t}\,\mathrm{d}t$ as $y \to 0$, the condition at $y = 0$ yields

$$1 - \delta_1 A\gamma - \delta_1 A\ln\varepsilon = 0 \; ,$$

that is

$$\delta_1 A = \frac{1}{\gamma + \ln \varepsilon} \, .$$

Finally, the solution is

$$u = 1 + \frac{1}{\gamma + \ln \varepsilon} \left[\int_y^\infty \frac{e^{-t}}{t} \, dt + \ln \frac{\bar{y}}{\bar{y}+1} \right] \, .$$

The same result as with MMAE (Problem 11-1) is recovered.

3. The equation for $\bar{f}_1$ is

$$\frac{d^2 \bar{f}_1}{d\bar{y}^2} + \frac{1}{\bar{y}+1} \frac{d\bar{f}_1}{d\bar{y}} = \frac{1}{\bar{y}(\bar{y}+1)} \varepsilon \frac{d f_1}{dy} \, .$$

With the equation for f_1,

$$\frac{d^2 f_1}{dy^2} + \frac{1}{y} \frac{d f_1}{dy} + \frac{d f_1}{dy} = 0 \, ,$$

we form the equation for $g = f_1 + \bar{f}_1$

$$\frac{d^2 g}{dy^2} + \frac{1}{y+\varepsilon} \frac{dg}{dy} = -\frac{d f_1}{dy} \, .$$

With $f_1 = A \displaystyle\int_y^\infty \frac{e^{-t}}{t} \, dt$, we obtain

$$\frac{d^2 g}{dy^2} + \frac{1}{y+\varepsilon} \frac{dg}{dy} = A \frac{e^{-y}}{y} \, .$$

This equation is integrated after multiplying by $y + \varepsilon$

$$g = A \int_y^\infty \frac{e^{-t}}{t+\varepsilon} \, dt + A\varepsilon \int_y^\infty \left[\frac{1}{\xi+\varepsilon} \int_\xi^\infty \frac{e^{-t}}{t} \, dt \right] d\xi + \alpha \ln(y+\varepsilon) + \beta \, .$$

The condition $u = 1$ as $y \to \infty$ gives $\alpha = 0$ and $\beta = 0$. The condition at $y = 0$ yields

$$\delta_1 A \left\{ \int_0^\infty \frac{e^{-t}}{t+\varepsilon} \, dt + \varepsilon \int_0^\infty \left[\frac{1}{\xi+\varepsilon} \int_\xi^\infty \frac{e^{-t}}{t} \, dt \right] d\xi \right\} = -1 \, .$$

Note 5.2. Observing that in the vicinity of $y = 0$, it is not appropriate to approximate $y + \varepsilon$ by y, we can take for f_1 the following equation

$$\frac{d^2 f_1}{dy^2} + \frac{1}{y+\varepsilon} \frac{d f_1}{dy} + \frac{d f_1}{dy} = 0 \, .$$

Then, the solution satisfying $f_1(\infty) = 0$ is

$$f_1 = A \int_{y+\varepsilon}^{\infty} \frac{e^{-t}}{t}\, dt\,.$$

The condition $u(0) = 0$ can be satisfied by taking

$$\delta_1 A = -\frac{1}{\displaystyle\int_{\varepsilon}^{\infty} \frac{e^{-t}}{t}\, dt}\,.$$

We note that $f_1 = f_1(x, \varepsilon)$, which is pertinent with the generalized form of SCEM. In addition, it is easily shown that $\bar{f}_1 = 0$. The solution

$$u = 1 - \frac{\displaystyle\int_{y+\varepsilon}^{\infty} \frac{e^{-t}}{t}\, dt}{\displaystyle\int_{\varepsilon}^{\infty} \frac{e^{-t}}{t}\, dt}$$

is an excellent approximation of the exact solution.

11-3.

In wall variables, the inner region equation writes

$$\frac{du^+}{dy^+} + F_c^2 \ell^{+2} \left(\frac{du^+}{dy^+}\right)^2 = 1\,.$$

At $y^+ = 0$, we have $u^+ = 0$. For $y^+ \gg 1$, this equation becomes

$$\chi y^+ \frac{du^+}{dy^+} = 1\,,$$

or

$$u^+ = \frac{1}{\chi} \ln y^+ + C\,.$$

The plot of the law of the wall (Fig. S.4) exhibits a logarithmic zone for values of y^+ greater than 75. Constant C of the logarithmic law is about $C = 5.28$.

In the close vicinity of the wall, the velocity profile is linear $u^+ = y^+$ but the region where this law is satisfied is very narrow ($y^+ < 3$). Between this region and the logarithmic law, there exists a buffer layer.

11-4. We express the matching to order $\dfrac{u_\tau}{u_e}$. We have

$$\mathrm{E}_0\, \frac{u}{u_e} = 1 + \frac{u_\tau}{u_e} \left[\frac{1}{\chi} \ln \eta - \frac{B}{\chi} \{2 - w(\eta)\}\right]\,,$$

and

$$\mathrm{E}_1\, \mathrm{E}_0\, \frac{u}{u_e} = 1 + \frac{u_\tau}{u_e} \left[\frac{1}{\chi} \ln y^+ + \frac{\nu}{\delta u_\tau} - \frac{2B}{\chi}\right]\,,$$

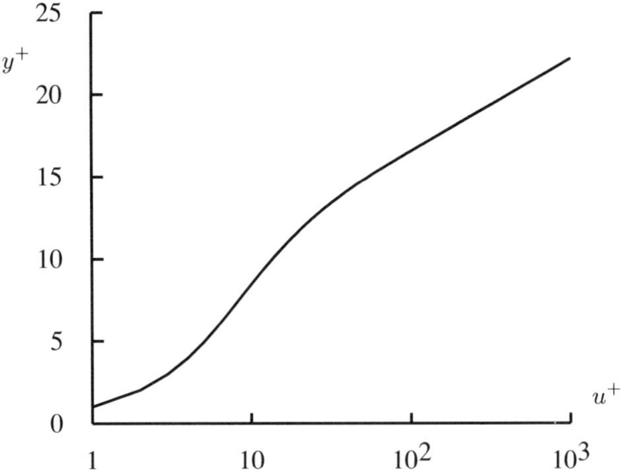

Fig. S.4. Velocity profiles in semi-log coordinates in the inner region

where, in this last expression, the function between square brackets must be understood as being its behaviour, y^+ being kept fixed, as $\dfrac{u_\tau}{u_e} \to 0$, that is as $\dfrac{u_e\delta}{\nu} \to \infty$; therefore, we have $\eta = y^+ \dfrac{\nu}{u_\tau\delta} \to 0$.

In addition, we have

$$\mathrm{E}_1 \frac{u}{u_e} = \frac{u_\tau}{u_e} f(y^+) = \frac{u_\tau}{u_e} f\left(\frac{y}{\delta}\frac{\delta u_\tau}{\nu}\right) .$$

The behaviour of f as $\dfrac{u_\tau\delta}{\nu} \to \infty$, $\dfrac{y}{\delta}$ being kept fixed, is given by the behaviour of f as $y^+ \to \infty$, whence

$$\mathrm{E}_0\,\mathrm{E}_1 \frac{u}{u_e} = \frac{u_\tau}{u_e}\frac{1}{\chi} \ln \frac{y}{\delta}\frac{\delta u_\tau}{\nu} + C\frac{u_\tau}{u_e} .$$

Then, the matching condition $\mathrm{E}_0\,\mathrm{E}_1 \dfrac{u}{u_e} = \mathrm{E}_1\,\mathrm{E}_0 \dfrac{u}{u_e}$ yields

$$1 + \frac{u_\tau}{u_e}\left[\frac{1}{\chi}\ln y^+ \frac{\nu}{\delta u_\tau} - \frac{2B}{\chi}\right] = \frac{u_\tau}{u_e}\frac{1}{\chi}\ln\frac{y}{\delta}\frac{\delta u_\tau}{\nu} + C\frac{u_\tau}{u_e} ,$$

This condition can be written as

$$1 - \frac{u_\tau}{u_e}\frac{1}{\chi}\ln\frac{u_\tau\delta}{\nu} - \frac{2B}{\chi}\frac{u_\tau}{u_e} = C\frac{u_\tau}{u_e} .$$

We deduce

$$\frac{u_e}{u_\tau} = \frac{1}{\chi}\ln\left(\frac{u_\tau}{u_e}\frac{u_e\delta}{\nu}\right) + \frac{2B}{\chi} + C ,$$

and we have

$$\frac{u_\tau}{u_e} = \sqrt{\frac{C_f}{2}} \; .$$

Then, the skin-friction law tells us that $\dfrac{u_\tau}{u_e} \to 0$ as $\dfrac{u_e\delta}{\nu} \to \infty$; we also have $\dfrac{u_\tau\delta}{\nu} \to \infty$.

A composite approximation gives a UVA in the whole boundary layer

$$\frac{u}{u_e} = \mathrm{E}_0 \, \frac{u}{u_e} + \mathrm{E}_1 \, \frac{u}{u_e} - \mathrm{E}_1 \, \mathrm{E}_0 \, \frac{u}{u_e} \; ,$$

whence

$$\frac{u}{u_e} = \frac{u_\tau}{u_e} \left[f(y^+) + \frac{B}{\chi}\omega(\eta) \right] \; ,$$

or

$$\frac{u}{u_e} = 1 + \frac{u_\tau}{u_e} \left[f(y^+) - \frac{1}{\chi}\ln\frac{\delta u_\tau}{\nu} - C - \frac{B}{\chi}\{2 - \omega(\eta)\} \right] \; .$$

11-5. In the overlap region, we have simultaneously

$$\frac{u_e - u}{u_\tau} = -\frac{1}{\chi}\ln\eta + D \quad \text{as} \quad \eta \to 0 \, ,$$

$$\frac{u}{u_\tau} = \frac{1}{\chi}\ln y^+ + C \quad \text{as} \quad y^+ \to \infty \, .$$

Adding member to member, we eliminate u and y

$$\frac{u_e}{u_\tau} = \frac{1}{\chi}\ln\frac{u_\tau\delta}{\nu} + C + D$$

$$= \frac{1}{\chi}\ln\frac{u_\tau}{u_e} + \frac{1}{\chi}\ln\frac{u_e\delta}{\nu} + C + D \; .$$

The properties of the logarithmic function give

$$\frac{u_\tau}{u_e} \to 0 \quad \text{as} \quad R_\delta \to \infty \, .$$

We make the change of variables

$$(x, y) \longmapsto \left(X = x, \eta = \frac{y}{\delta} \right) \; .$$

The differentiation rules give

$$\frac{\partial}{\partial x} = \frac{\partial}{\partial X} - \eta\frac{\delta'}{\delta}\frac{\partial}{\partial \eta} \; ,$$

$$\frac{\partial}{\partial y} = \frac{1}{\delta}\frac{\partial}{\partial \eta} \; .$$

Integrating the continuity equation yields

$$v = -\gamma u_e \delta'(\eta F' - F) - \delta u_e'(\eta - \gamma F) + \delta u_e \gamma' F \ .$$

The momentum equation becomes

$$\frac{\partial}{\partial \eta}\left(\frac{\tau}{\tau_w}\right) = 2\beta F' - \beta\gamma F'^2 + \beta\frac{u_e}{u_e'}\frac{\gamma'}{\gamma}(F' - \gamma F'^2 + \gamma F F'')$$
$$-\beta\left(1 + \frac{u_e}{u_e'}\frac{\delta'}{\delta}\right)(\eta F'' - \gamma F F'') \ .$$

The skin-friction law is

$$\frac{1}{\gamma} = \frac{1}{\chi}\ln\gamma\frac{u_e\delta}{\nu} + C + D \ .$$

This shows that $\gamma \to 0$ as $R_\delta \to \infty$. By differentiating with respect to x, we obtain

$$-\beta\frac{\gamma'}{\gamma}\frac{u_e}{u_e'} = \frac{\gamma/\chi}{1 + \gamma/\chi}\left(\beta - \frac{\delta'}{\gamma}\right) \ .$$

Then, we can say that $\beta\dfrac{\gamma'}{\gamma}\dfrac{u_e}{u_e'}$ tends to zero like γ.

The momentum equation reduces to

$$\frac{\partial}{\partial \eta}\left(\frac{\tau}{\tau_w}\right) = 2\beta F' - \beta\left(1 + \frac{u_e}{u_e'}\frac{\delta'}{\delta}\right)\eta F'' \ .$$

We integrate with respect to η from $\eta = 0$ where we have $\tau/\tau_w = 1$ and $F = 0$

$$\frac{\tau}{\tau_w} - 1 = 2\beta F - \beta\left(1 + \frac{u_e}{u_e'}\frac{\delta'}{\delta}\right)(\eta F' - F) \ .$$

At $\eta = 1$ we have $\tau = 0$, $F' = 0$ and $F = F_1$ whence

$$-1 = 2\beta F_1 + \beta\left(1 + \frac{u_e}{u_e'}\frac{\delta'}{\delta}\right)F_1 \ ,$$

and

$$-\beta\left(1 + \frac{u_e}{u_e'}\frac{\delta'}{\delta}\right) = \frac{1}{F_1} + 2\beta \ .$$

The momentum equation writes

$$\frac{\tau}{\tau_w} = 1 - \frac{F}{F_1} + \left(\frac{1}{F_1} + 2\beta\right)\eta F' \ .$$

11-6. The scales of velocity, length and time of small structures are $\boldsymbol{v}$, $\boldsymbol{\eta}$, $\boldsymbol{\tau}$ respectively. Between these scales, we have the relations

$$\varepsilon = \frac{\boldsymbol{v}^2}{\boldsymbol{\tau}} \ , \quad \frac{\boldsymbol{v}\boldsymbol{\eta}}{\nu} = 1 \ , \quad \boldsymbol{\tau} = \frac{\boldsymbol{\eta}}{\boldsymbol{v}} \ .$$

The first relation comes from the definition of dissipation (amount of energy transformed into heat by unit time); the second relation comes from the hypothesis that the characteristic Reynolds number of dissipative structures is of order unity; the third relation is simply the link between the scales of velocity, length and time. In particular, we obtain

$$\eta = \frac{\nu^{3/4}}{\varepsilon^{1/4}} \ .$$

Dimensional arguments gives the shape of the spectrum in the range of large structures and in the range of small structures

$$E = u^2 \ell F(\xi \ell) \ ,$$

$$E = \nu^{5/4} \varepsilon^{1/4} f\left(\xi \frac{\nu^{3/4}}{\varepsilon^{1/4}}\right) \ .$$

We assume that the spectrum follows a power law ξ^α in the overlap range. We have

$$\nu^{5/4} \varepsilon^{1/4} \xi^\alpha \frac{\nu^{3\alpha/4}}{\varepsilon^{\alpha/4}} = u^2 \ell \xi^\alpha \ell^\alpha \ .$$

The viscosity must disappear, so that we have

$$\alpha = -\frac{5}{3} \ ,$$

and we obtain

$$\varepsilon = \frac{u^3}{\ell} \ .$$

Therefore, we observe that the value of dissipation is independent of viscosity, but the physical mechanism remains a transformation of kinetic energy into heat due to the deformation power of viscous forces within the flow. The physical mechanism of dissipation is closely related to viscosity but the amount of dissipated energy does not depend on it.

Chapter 12

12-1.

Equations (12.1a–12.1c) become

$$\frac{\partial u}{\partial X} + \frac{\partial v}{\partial y} = 0 \ ,$$

$$u_0 \frac{\partial u}{\partial X} + v\frac{du_0}{dy} + \varepsilon^r \left(u\frac{\partial u}{\partial X} + v\frac{\partial u}{\partial y}\right) = -\varepsilon^{s-r}\frac{\partial p}{\partial X} + \varepsilon^{m+\alpha}\frac{\partial^2 u}{\partial X^2} + \varepsilon^{m-\alpha}\frac{\partial^2 u}{\partial y^2} \ ,$$

$$u_0 \frac{\partial v}{\partial X} + \varepsilon^r \left(u\frac{\partial v}{\partial X} + v\frac{\partial v}{\partial y}\right) = -\varepsilon^{s-r-2\alpha}\frac{\partial p}{\partial y} + \varepsilon^{m+\alpha}\frac{\partial^2 v}{\partial X^2} + \varepsilon^{m-\alpha}\frac{\partial^2 v}{\partial y^2} \ .$$

The case of interest occurs when, in a first approximation, Navier-Stokes equations reduce to a form leading to a singular perturbation problem. Here, first order differential equations are obtained, for high Reynolds numbers, when $m > |\alpha|$. This is a necessary condition for a small perturbation to be significant. This condition shows why we assume high Reynolds numbers.

In the boundary layer, the order of magnitude of u_0 is

$$u_0(y) = \mathrm{O}(\varepsilon^\beta) \,.$$

Thus, to obtain a significant degeneracy with possible separation, the streamwise velocity perturbation in the boundary layer is clearly $\mathrm{O_S}(\varepsilon^\beta)$.

Using the variables X and Y, (12.1a–12.1c) become

$$\frac{\partial U}{\partial X} + \frac{\partial V}{\partial Y} = 0 \,,$$

$$U\frac{\partial U}{\partial X} + V\frac{\partial U}{\partial Y} = \varepsilon^{m-\alpha-3\beta}\left[-\frac{\partial P}{\partial X} + \frac{\partial^2 U}{\partial Y^2}\right] + \varepsilon^{(m+\alpha-\beta)}\frac{\partial^2 U}{\partial X^2} \,,$$

$$U\frac{\partial V}{\partial X} + V\frac{\partial V}{\partial Y} = -\varepsilon^{(m-3\alpha-5\beta)}\frac{\partial P}{\partial Y} + \varepsilon^{(m-\alpha-3\beta)}\frac{\partial^2 V}{\partial Y^2} + \varepsilon^{(m+\alpha-\beta)}\frac{\partial^2 V}{\partial X^2} \,.$$

In the longitudinal momentum equation, the second term on the right hand side is negligible compared to the first term if $m - \alpha - 3\beta < m + \alpha - \beta$. This condition corresponds to a slope of the wall indentation such that $\beta + \alpha > 0$. In order to have viscous and inertia terms of the same order, we take $3\beta = m - \alpha$.

With $\beta = (m - \alpha)/3$, the regular expansions are

$$\mathcal{U} = \varepsilon^{(m-\alpha)/3}U_1(X,Y) + \cdots \,,$$
$$\mathcal{V} = \varepsilon^{(2m+\alpha)/3}V_1(X,Y) + \cdots \,,$$
$$\mathcal{P} - p_c = \varepsilon^{(2m-2\alpha)/3}P_1(X,Y) + \cdots \,.$$

Then, we obtain the boundary layer equations

$$\frac{\partial U_1}{\partial X} + \frac{\partial V_1}{\partial Y} = 0 \,,$$

$$U_1\frac{\partial U_1}{\partial X} + V_1\frac{\partial U_1}{\partial Y} = -\frac{\partial P_1}{\partial X} + \frac{\partial^2 U_1}{\partial Y^2} \,,$$

$$\frac{\partial P_1}{\partial Y} = 0 \,.$$

In the boundary layer, we have $\dfrac{\partial P_1}{\partial Y} = 0$. Then, the pressure expansion is such that

$$\mathcal{P} - p_c = \varepsilon^{2(m-\alpha)/3}P_1(X) + \cdots \,.$$

In the middle deck, we have

$$\mathcal{P} - p_c = -2\varepsilon^{m-\alpha}X + \varepsilon^s p_1(X,y) + \cdots \,.$$

The condition $\beta > 0$, i.e. $m - \alpha > 0$, leads to $m - \alpha > 2(m - \alpha)/3$. Then, to order $\varepsilon^{2(m-\alpha)/3}$, the MVDP leads to

$$s = \frac{2}{3}(m - \alpha) \ .$$

The matching condition on pressure gives

$$\lim_{y \to -1/2} p_1(X, y) = P_1(X) \ .$$

The core flow equations (or middle deck equations) are

$$\frac{\partial u_1}{\partial X} + \frac{\partial v_1}{\partial y} = 0 \ ,$$

$$u_0 \frac{\partial u_1}{\partial X} + v_1 \frac{du_0}{dy} = -\varepsilon^{2(m-\alpha)/3-r} \frac{\partial p_1}{\partial X} \ ,$$

$$u_0 \frac{\partial v_1}{\partial X} = -\varepsilon^{2(m-4\alpha)/3-r} \frac{\partial p_1}{\partial y} \ .$$

From the longitudinal momentum equation, a non trivial solution is obtained if and only if

$$r \leq \frac{2}{3}(m - \alpha) \ .$$

If $r < 2(m - \alpha)/3$, the longitudinal momentum equation is

$$u_0 \frac{\partial u_1}{\partial X} + v_1 \frac{du_0}{dy} = 0 \ .$$

The solution of the continuity equation and longitudinal momentum equation is

$$u_1 = A(X) \frac{du_0}{dy} \ ,$$

$$v_1 = -\frac{dA}{dX} u_0 \ .$$

Using the expansion operators M and I, we obtain
- in the middle deck, to order ε^r: $M\mathcal{U} = u_0 + \varepsilon^r u_1$,
- in the boundary layer, to order $\varepsilon^{(m-\alpha)/3}$: $I\mathcal{U} = \varepsilon^{(m-\alpha)/3}U_1$.

In order to apply the matching condition $MI\mathcal{U} = IM\mathcal{U}$ with the MVDP, the operators M and I must be considered to the same order.

If $r < (m - \alpha)/3$, the matching of the longitudinal component of velocity obtained with M and I to order r gives the trivial solution $A(X) = 0$. Then, we are led to take $r \geq (m-\alpha)/3$. With $u_0 = 1/4 - y^2$, the matching condition $MI\mathcal{U} = IM\mathcal{U}$ gives
- if $r = (m - \alpha)/3$, with M and I taken to order r,

$$\lim_{Y \to \infty} (U_1 - Y) = A(X) \ ,$$

- if $r > (m - \alpha)/3$, with M and I taken to order $(m - \alpha)/3$,

$$\lim_{Y \to \infty} (U_1 - Y) = 0 \ .$$

In addition, with the transverse momentum equation in the middle deck, we find that the case $r < 2(m - 4\alpha)/3$ is impossible. Then, we have $r \geq 2(m - 4\alpha)/3$. The transverse momentum equation gives

- if $r > 2(m - 4\alpha)/3$,

$$\frac{\partial p_1}{\partial y} = 0 \ ,$$

- if $r = 2(m - 4\alpha)/3$,

$$u_0 \frac{\partial v_1}{\partial X} = -\frac{\partial p_1}{\partial y} \ .$$

The above results show that the couple (r, α) is in the hatched triangle in Fig. S.5 and $\alpha > 0$. Smith's theory is obtained for the least degenerated case, $r = 2m/7$ and $\alpha = m/7$. To a certain extent, this theory is equivalent to the triple deck theory for external flows.

In the general non symmetric case, for a given value of α, the most significant degeneracies are given by the smallest value of r. The corresponding indentation induces the largest perturbation in the streamwise velocity. Then, for $0 < \alpha < m$, we have

- $r = (m - \alpha)/3$ if $m/7 \leq \alpha < m$,
- $r = 2(m - 4\alpha)/3$ if $0 \leq \alpha \leq m/7$.

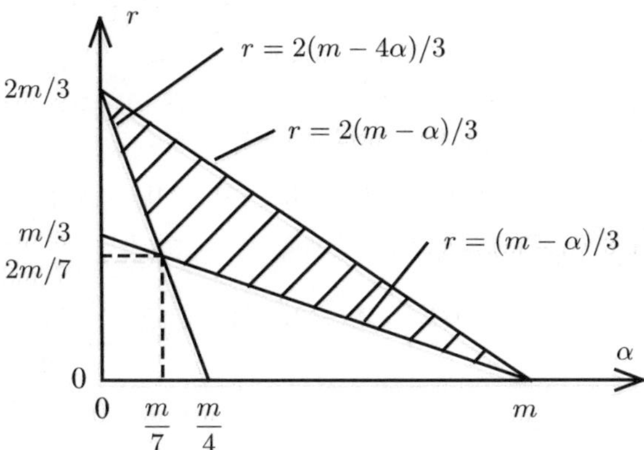

Fig. S.5. Domain covered by the study in the (r, α)-plane

12-2.

If $0 < \alpha \leq m/7$, we have $r = 2(m - 4\alpha)/3$. The transverse momentum equation gives

$$\frac{\partial p_1}{\partial y} = -u_0 \frac{\partial v_1}{\partial X} = u_0^2 A''(X) \ .$$

To calculate the pressure p_1, we know that, for the lower wall

$$\lim_{y \to -1/2} p_1(X, y) = P_1(X) .$$

In the same way, for the upper wall, we have

$$\lim_{y \to 1/2} p_1(X, y) = \overline{P}_1(X) .$$

We also know that

$$u_0 = \frac{1}{4} - y^2 .$$

Finally, we have

$$p_1(X, y) = P_1(X) + \frac{A''(X)}{60} \left(12y^5 - 10y^3 + \frac{15}{4}y + 1 \right) .$$

We obtain

$$\overline{P}_1(X) = P_1(X) + \frac{A''(X)}{30} .$$

With Prandtl's transformation, the lower wall boundary layer equations become

$$\frac{\partial U_1}{\partial X} + \frac{\partial W}{\partial Z} = 0 ,$$

$$U_1 \frac{\partial U_1}{\partial X} + W \frac{\partial U_1}{\partial Z} = -\frac{\partial P_1}{\partial X} + \frac{\partial^2 U_1}{\partial Z^2} ,$$

$$\frac{\partial P_1}{\partial Z} = 0 .$$

The boundary conditions at the wall are

$$Z = 0 : \quad U_1 = 0 , \quad W = 0 ,$$

and the matching condition gives

$$\lim_{Z \to \infty} (U_1 - Z) = B(X) + F(X) .$$

We have
- $B(X) = A(X)$ if $r = (m - \alpha)/3$,
- $B(X) = 0$ if $r > (m - \alpha)/3$.

In the upper wall boundary layer, the equations for $\overline{U}_1, \overline{V}_1, \overline{P}_1$ are

$$\frac{\partial \overline{U}_1}{\partial X} + \frac{\partial \overline{V}_1}{\partial Y} = 0 ,$$

$$\overline{U}_1 \frac{\partial \overline{U}_1}{\partial X} + \overline{V}_1 \frac{\partial \overline{U}_1}{\partial Y} = -\frac{\partial \overline{P}_1}{\partial X} + \frac{\partial^2 \overline{U}_1}{\partial Y^2} ,$$

$$\frac{\partial \overline{P}_1}{\partial Y} = 0 .$$

With Prandtl's transformation, these equations become

$$\frac{\partial \overline{U}_1}{\partial X} + \frac{\partial \overline{W}}{\partial \overline{Z}} = 0 \, ,$$

$$\overline{U}_1 \frac{\partial \overline{U}_1}{\partial X} + \overline{W} \frac{\partial \overline{U}_1}{\partial \overline{Z}} = -\frac{\partial \overline{P}_1}{\partial X} + \frac{\partial^2 \overline{U}_1}{\partial \overline{Z}^2} \, ,$$

$$\frac{\partial \overline{P}_1}{\partial \overline{Z}} = 0 \, .$$

The boundary conditions at the wall are

$$\overline{Z} = 0 : \quad \overline{U}_1 = 0 \, , \quad \overline{W} = 0 \, .$$

The matching condition gives

$$\lim_{\overline{Z} \to \infty} (\overline{U}_1 - \overline{Z}) = -B(X) + G(X) \, ,$$

and we have
- $B(X) = A(X)$ if $r = (m - \alpha)/3$,
- $B(X) = 0$ if $r > (m - \alpha)/3$.

If $m/7 < \alpha < m$, we have $r = (m - \alpha)/3$. The transverse momentum equation is

$$\frac{\partial p_1}{\partial y} = 0 \, ,$$

which gives

$$P_1(X) = \overline{P}_1(X) \, .$$

The two canonical problems for the boundary layers are the same and the pressure distributions in the lower and upper wall boundary layers are the same. This leads to

$$A = \frac{1}{2}[G(X) - F(X)] \, .$$

The slope δ of the wall indentation is given by

$$\delta = \varepsilon^{(m+2\alpha)/3} \, .$$

We obtain the classification of the different studied cases
- if $0 < \alpha < m/7$, we have

$$1 \prec L \prec \mathcal{R}^{1/7} \, ,$$
$$\mathcal{R}^{-1/3} \succ \delta \succ \mathcal{R}^{-3/7} \, ,$$

- if $m/7 < \alpha < m$, we have

$$\mathcal{R}^{1/7} \prec L \prec \mathcal{R} \, ,$$
$$\mathcal{R}^{-3/7} \succ \delta \succ \mathcal{R}^{-1} \, .$$

12-3. From Problem 12-1, we have $\beta = 1$ and $m - \alpha = 3$. Then, the lines bounding the domain of study shown in Fig. S.5 are determined below.

- The line $r = (m - \alpha)/3$ becomes $r = 1$.
- The line $r = 2(m - 4\alpha)/3$ becomes $r = 2(1 - \alpha)$.
- The line $r = 2(m - \alpha)/3$ becomes $r = 2$.

We obtain the domain shown in Fig S.6.

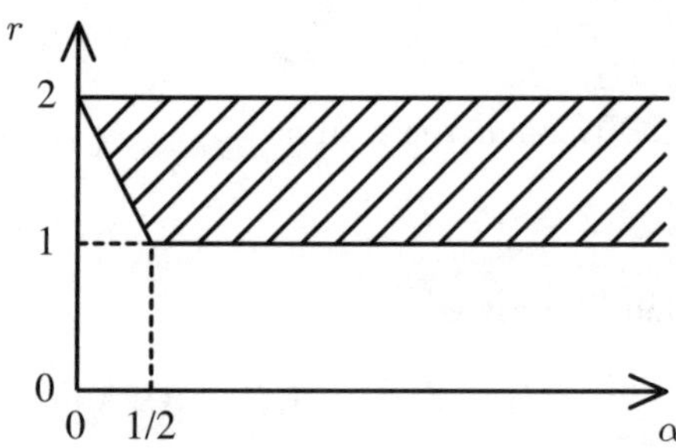

Fig. S.6. Domain covered by the study in the (r, α)-plane $(\beta = 1)$

From the relation

$$\mathcal{R} = \varepsilon^{-m} \, ,$$

we obtain

$$\mathcal{R} = \varepsilon^{-(3+\alpha)} \, .$$

From the relation

$$\delta = \varepsilon^{(m+2\alpha)/3} \, ,$$

we obtain

$$\delta = \varepsilon^{1+\alpha} \, .$$

The core flow momentum equations become

$$u_0 \frac{\partial u}{\partial X} + v \frac{du_0}{dy} + \varepsilon^r \left(u \frac{\partial u}{\partial X} + v \frac{\partial u}{\partial y} \right) = -\varepsilon^{2-r} \frac{\partial p}{\partial X} + \mathrm{O}(\varepsilon^3) \, ,$$

$$u_0 \frac{\partial v}{\partial X} + \varepsilon^r \left(u \frac{\partial v}{\partial X} + v \frac{\partial v}{\partial y} \right) = -\varepsilon^{2(1-\alpha)-r} \frac{\partial p}{\partial y} + \mathrm{O}(\varepsilon^3) \, .$$

The first approximation is given by the first significant perturbation, i.e. the smallest value fo r for a given value of α. Then, for $\alpha > 0$, we have $r < 2$ and the first order core flow equations give

$$\frac{\partial u_1}{\partial X} + \frac{\partial v_1}{\partial y} = 0 \, ,$$

$$u_0 \frac{\partial u_1}{\partial X} + v_1 \frac{du_0}{dy} = 0 \, .$$

The solution is

$$u_1 = A_1(X)\frac{\mathrm{d}u_0}{\mathrm{d}y} \ ,$$

$$v_1 = -A_1'(X)u_0 \ .$$

For the second order, two cases must be considered.

1. $\alpha \geq 1/2$, $r = 1$. The core flow momentum equations are

$$u_0\frac{\partial u}{\partial X} + v\frac{\mathrm{d}u_0}{\mathrm{d}y} + \varepsilon\left(u\frac{\partial u}{\partial X} + v\frac{\partial u}{\partial y}\right) = -\varepsilon\frac{\partial p}{\partial X} + \mathrm{O}(\varepsilon^3) \ ,$$

$$u_0\frac{\partial v}{\partial X} + \varepsilon\left(u\frac{\partial v}{\partial X} + v\frac{\partial v}{\partial y}\right) = -\varepsilon^{1-2\alpha}\frac{\partial p}{\partial y} + \mathrm{O}(\varepsilon^3) \ .$$

Then, we are led to take the AEs of u and v as

$$u = u_1 + \varepsilon u_2 + \cdots \ ,$$

$$v = v_1 + \varepsilon v_2 + \cdots \ .$$

We obtain the equations

$$\frac{\partial u_2}{\partial X} + \frac{\partial v_2}{\partial y} = 0 \ ,$$

$$u_0\frac{\partial u_2}{\partial X} + v_2\frac{\mathrm{d}u_0}{\mathrm{d}y} = -\frac{\partial p_1}{\partial X} - u_1\frac{\partial u_1}{\partial X} - v_1\frac{\partial u_1}{\partial y} \ .$$

For the case $\alpha > 1/2$, we have

$$0 = \frac{\partial p_1}{\partial y} \ .$$

For the case $\alpha = 1/2$, we have

$$u_0\frac{\partial v_1}{\partial X} = -\frac{\partial p_1}{\partial y} \ .$$

2. $0 < \alpha < 1/2$, $r = 2(1 - \alpha)$. The core flow momentum equations are

$$u_0\frac{\partial u}{\partial X} + v\frac{\mathrm{d}u_0}{\mathrm{d}y} = -\varepsilon^{2\alpha}\frac{\partial p}{\partial X} + \mathrm{O}(\varepsilon^{2(1-\alpha)}) \ ,$$

$$u_0\frac{\partial v}{\partial X} = -\frac{\partial p}{\partial y} + \mathrm{O}(\varepsilon^{2(1-\alpha)}) \ .$$

Then, we are led to take the AEs of u and v as

$$u = u_1 + \varepsilon^{2\alpha}u_2 + \cdots \ ,$$

$$v = v_1 + \varepsilon^{2\alpha}v_2 + \cdots \ .$$

which gives the equations

$$\frac{\partial u_2}{\partial X} + \frac{\partial v_2}{\partial y} = 0 \,,$$

$$u_0 \frac{\partial u_2}{\partial X} + v_2 \frac{\mathrm{d}u_0}{\mathrm{d}y} = -\frac{\partial p_1}{\partial X} \,,$$

$$u_0 \frac{\partial v_1}{\partial X} = -\frac{\partial p_1}{\partial y} \,.$$

In fact, the AE of $\mathcal{U}$ valid for both cases $(\alpha > 0)$ is

$$\mathcal{U} = u_0 + \varepsilon^r u_1 + \varepsilon^2 u_2 + \cdots$$

which shows that the gauge of the second order is ε^2. This corresponds to the line $r = 2$ in Fig. S.6. The AE of $\mathcal{V}$ is

$$\mathcal{V} = \varepsilon^{r+\alpha} v_1 + \varepsilon^{2+\alpha} v_2 + \cdots$$

For $\alpha > 1/2$, the second order equations give

$$\frac{\partial u_2}{\partial X} + \frac{\partial v_2}{\partial y} = 0 \,,$$

$$u_0 \frac{\partial u_2}{\partial X} + v_2 \frac{\mathrm{d}u_0}{\mathrm{d}y} = -\frac{\partial p_1}{\partial X} - u_1 \frac{\partial u_1}{\partial X} - v_1 \frac{\partial u_1}{\partial y} \,,$$

and we have also

$$0 = \frac{\partial p_1}{\partial y} \,.$$

We obtain

$$\frac{\partial}{\partial y} \left(\frac{v_2}{u_0} \right) = \frac{1}{u_0^2} \frac{\partial p_1}{\partial X} - A_1 \frac{\mathrm{d}A_1}{\mathrm{d}X} \frac{\mathrm{d}}{\mathrm{d}y} \left(\frac{1}{u_0} \frac{\mathrm{d}u_0}{\mathrm{d}y} \right) \,,$$

and the solution is

$$u_2(X, y) = 2 \left[y \ln \left(\frac{1/2 + y}{1/2 - y} \right)^2 - 2 \right] p_1 + \frac{A_1^2}{2} \frac{\mathrm{d}^2 u_0}{\mathrm{d}y^2} + A_2(X) \frac{\mathrm{d}u_0}{\mathrm{d}y} \,,$$

$$v_2(X, y) = \left[\left(\frac{1}{2} + y \right) \left(\frac{1}{2} - y \right) \ln \left(\frac{1/2 + y}{1/2 - y} \right)^2 + 2y \right] \frac{\partial p_1}{\partial X}$$

$$- A_1 \frac{\mathrm{d}A_1}{\mathrm{d}X} \frac{\mathrm{d}u_0}{\mathrm{d}y} - \frac{\mathrm{d}A_2}{\mathrm{d}X}(X) u_0 \,.$$

For the lower wall boundary layer, we have

$$\mathcal{U} = \varepsilon U_1 + \cdots \,,$$

$$\mathcal{V} = \varepsilon^{2+\alpha} V_1 + \cdots \,,$$

where the boundary layer variable is

$$Y = \frac{1/2 + y}{\varepsilon} \ .$$

Two cases are considered.
1. For $r = 1$, $\alpha \geq 1/2$, we obtain

$$\lim_{Y \to \infty} (U_1 - Y) = A_1(X) \ .$$

On the other hand, we have

$$\mathrm{I}\mathcal{V} = \varepsilon^{2+\alpha} V_1 \ ,$$
$$\mathrm{M}\mathcal{V} = \varepsilon^{1+\alpha} v_1 + \varepsilon^{2+\alpha} v_2 \ .$$

To the same order $\varepsilon^{2+\alpha}$, we obtain

$$\lim_{Y \to \infty} (V_1 + Y A_1') = -\frac{\partial p_1}{\partial X} - A_1 \frac{\mathrm{d}A_1}{\mathrm{d}X} \ .$$

2. For $r = 2(1 - \alpha)$, $0 < \alpha < 1/2$, we obtain

$$\lim_{Y \to \infty} (U_1 - Y) = 0 \ .$$

On the other hand, we have

$$\mathrm{M}\mathcal{V} = \varepsilon^{(2-\alpha)} v_1 + \varepsilon^{(2+\alpha)} v_2 \ ,$$

and

$$\lim_{Y \to \infty} V_1 = -\frac{\partial p_1}{\partial X} \ .$$

For the symmetric channel, two cases are considered.
1. For $0 < \alpha < 1/2$, $r = 2(1 - \alpha)$, we have

$$p_1 = P_1 + \frac{A_1''}{60} \left(12 y^5 - 10 y^3 + \frac{15}{4} y + 1 \right) \ ,$$

so that the upper and lower walls pressures are related by

$$\overline{P}_1 = P_1 + \frac{A_1''}{30} \ .$$

2. For $\alpha > 1/2$, $r = 1$, we have

$$A_1(X) = \frac{1}{2}(G - F) \ ,$$
$$\overline{P}_1 = P_1 \ .$$

If the channel is symmetric, $F = G$.

In both cases, for a symmetric channel, we obtain $A_1 = 0$ for $\alpha > 0$. We have

$$\mathcal{U} = u_0 + \varepsilon^2 u_2 + \cdots ,$$
$$\mathcal{V} = \varepsilon^{2+\alpha} v_2 + \cdots ,$$

and

$$u_2(X, y) = 2 \left[y \ln \left(\frac{1/2 + y}{1/2 - y} \right)^2 - 2 \right] p_1 + A_2 \frac{\mathrm{d} u_0}{\mathrm{d} y} ,$$

$$v_2(X, y) = \left[\left(\frac{1}{2} + y \right) \left(\frac{1}{2} - y \right) \ln \left(\frac{1/2 + y}{1/2 - y} \right)^2 + 2y \right] \frac{\partial p_1}{\partial X} - \frac{\mathrm{d} A_2}{\mathrm{d} y} u_0 .$$

Note 5.3. If the function F (and the function G) depends on ε and can be expanded at least to first order as

$$F(x, \varepsilon) = F_1(X) + \varepsilon F_2(X) + \cdots$$

then, it is possible that $A_2 \neq 0$. The condition $F_1 = G_1$ leads to $A_1 = 0$. The condition $F_2 = G_2$ leads to $A_2 = 0$.

References

1. Computation of viscous-inviscid interactions. AGARD CP 291, 1981. [185]
2. Numerical and physical aspects of aerodynamic flows IV, edited by Tuncer Cebeci. Springer Verlag, Berlin, Heidelberg, 1990. [185]
3. F. Álvarez, N. Moraga, and J. Cousteix. An updated interactive boundary layer method for high reynolds number flows. *Int. J. Numer. Meth. Engng.*, 2006. [187]
4. W. Appel. *Mathématiques pour la physique et les physiciens.* H&K Éditions, Paris, 2005. 3e édition. [53, 312]
5. B. Aupoix, J.Ph. Brazier, and J. Cousteix. Asymptotic Defect Boundary-Layer Theory Applied to Hypersonic Flows. *AIAA Journal*, 30(5):1252–1259, May 1992. [79, 200]
6. C.M. Bender and S.A. Orszag. *Advanced Mathematical Methods for Scientists and Engineers – Asymptotic Methods and Perturbation Theory.* Springer-Verlag, Inc., New York, NY, 1999. [15]
7. J.Ph. Brazier. Étude asymptotique des équations de couche limite en formulation déficitaire, 1990. [79, 200, 201]
8. J.Ph. Brazier, B. Aupoix, and J. Cousteix. Étude asymptotique de la couche limite en formulation déficitaire. *Comptes Rendus de l'Académie des Sciences*, t. 310 Série II:1583–1588, 1990. [79, 200]
9. J.E. Carter. A new boundary layer inviscid iteration technique for separated flow. In *AIAA Paper 79-1450.* 4 th Computational fluid dynamics conf., Williamsburg, 1979. [134, 156, 169, 175, 185]
10. J.E. Carter and S.F. Wornom. Solutions for incompressible separated boundary-layers including viscous-inviscid interaction. Aerodynamic analysis requiring advanced computers, 1975. NASA SP-347 125–150. [134, 156, 185]
11. G. Casalis. Méthodes mathématiques en mécanique des fluides. Cours SUPAERO, 2000. [63]
12. D. Catherall and W. Mangler. The integration of a two-dimensional laminar boundary-layer past the point of vanishing skin friction. *J. Fluid. Mech.*, 26(1):163–182, 1966. [133, 152, 167]
13. T. Cebeci. *An Engineering Approach to the Calculation of Aerodynamic Flows.* Horizons Publishing Inc, Long Beach, California – Springer-Verlag, Berlin, Heidelberg, New York, 1999. [134, 157, 169, 175, 185, 187, 191, 192, 193]
14. T. Cebeci and J. Cousteix. *Modeling and Computation of Boundary-Layer Flows, Second Edition.* Horizons Publishing Inc, Long Beach, California – Springer-Verlag, Berlin, Heidelberg, New York, 2005. [186, 191, 192, 238]

15. L.Y. Chen, N. Goldenfeld, and Y. Oono. Renormalization group and singular perturbations: Multiple scales, boundary layers, and reductive perturbation theory. *Phys. Rev. E*, 54(1):376–394, july 1996. [24, 119]

16. F.H. Clauser. The turbulent boundary layer. *Advances Appl. Mech.*, 4:1–51, 1956. [256]

17. J.D. Cole. *Perturbation methods in applied mathematics.* Pure and applied mathematics. Blaisdell Publishing Company, Waltham, MA, 1968. [16, 63, 95, 99, 102]

18. D. Coles. The law of the wake in the turbulent boundary layer. *J. Fluid Mech.*, 1:191, 1956. [256]

19. D. Coles. Remarks on the equilibrium turbulent boundary layer. *J. Aero. Sci.*, 24, 1957. [256]

20. J. Cousteix. *Couche limite laminaire.* Cepadues, Toulouse, 1988. [148, 149, 188]

21. J. Cousteix. *Turbulence et Couche Limite.* Cepadues, Toulouse, 1989. [238, 239]

22. J. Cousteix, J.Ph. Brazier, and J. Mauss. Perturbation tridimensionnelle d'une couche limite de Blasius. *C. R. Acad. Sci. Paris*, t. 329, Série II b:213–219, 2001. [142, 155, 230]

23. J. Cousteix and R. Houdeville. Singularities in three-dimensional turbulent boundary layer calculations and separation phenomena. *AIAA Journal*, 19(8), August 1981. [134]

24. J. Cousteix, J.C. Le Balleur, and R. Houdeville. Calcul des couches limites turbulentes instationnaires en mode direct ou inverse, écoulements de retour inclus. *La Recherche Aérospatiale*, 3:147–157, Mai-Juin 1980. [134]

25. J. Cousteix and J. Mauss. Approximations of the Navier-Stokes equations for high Reynolds number flows past a solid wall. *Jour. Comp. and Appl. Math.*, 166(1):101–122, 2004. [172]

26. E.M. de Jager and Jiang Furu. *The theory of singular perturbations*, volume 42 of *North-Holland series in applied mathematics and mechanics.* Elsevier, Amsterdam, 1996. [19, 54, 82, 102]

27. A. Dechaume. Analyse asymptotique et numérique des équations de Navier-Stokes : cas du canal, Doctoral thesis, Université Paul Sabatier, Toulouse, 2006. [290]

28. A. Dechaume, J. Cousteix, and J. Mauss. An interactive boundary layer model compared to the triple deck theory. *Eur. J. of Mechanics B/Fluids*, 24:439–447, 2005. [267]

29. F.R. DeJarnette and R.A. Radcliffe. Matching Inviscid/Boundary-Layer Flowfields. *AIAA Journal*, 34(1):35–42, january 1996. [176, 197]

30. M. Drela and M.B. Giles. Ises: a two-dimensional viscous aerodynamic design and analysis code. AIAA Paper 87–1118, 1987. [185]

31. H. Dumitrescu and V. Cardoş. Improved formulation of boundary-layer-type flows. *AIAA Journal*, 40(4):794–796, 2002. [176]

32. L.F. East. A representation of second-order boundary layer effects in the momentum integral equation and in viscous-inviscid interactions. Technical Report 81002, Royal Aircraft Establishment, 1981. [79]

33. W. Eckhaus. *Matched asymptotic expansions and singular perturbations.* North Holland, Amsterdam and London; American Elsevier, New York, 1973. [16, 43, 49, 59, 65, 79, 89]

34. W. Eckhaus. *Asymptotic analysis of singular perturbations*. Studies in mathematics and its applications, 9. North-Holland, 1979. [16, 77]

35. L.E. Fraenkel. On the matched asymptotic expansions. *Proc. Camb. Phil. Soc.*, 65(Part I, II and III):209–284, 1969. [80]

36. K.O. Friedrichs. *Fluid Dynamics*. Brown University, 1942. [14]

37. F. Gaible, J. Cousteix, and J. Mauss. Analyse asymptotique de l'écoulement au voisinage d'un point d'arrêt. *C.R. Acad. Sci., Paris*, t. 313, Série II:145–150, 1991. [158]

38. P. Germain. Méthodes asymptotiques en mécanique des fluides. In R. Balian et J.L. Peube, editor, *Fluid Dynamics*. Gordon and Breach Publishers, 1973. École d'été de Physique Théorique, Les Houches. [15, 90]

39. S. Goldstein. Concerning some Solutions of the Boundary Layer Equations in Hydrodynamics. *Proc. Camb. Phil. Soc.*, XXVI(Part I):1–30, 1930. [133, 160, 161, 163, 190]

40. S. Goldstein. On laminar boundary-layer flow near a position of separation. *Quarterly J. Mech. and Appl. Math.*, 1:43–69, 1948. [133, 140, 152, 165]

41. J.P. Guiraud. Going on with asymptotics. In R. Gatignol P.A. Bois, E. Dériat and A. Rigolot, editors, *Asymptotic Modelling in Fluid Mechanics*, number 442 in Lectures Notes in Physics, pages 257–307. Springer-Verlag, Heidelberg, 1995. Proceedings of a symposium in honour of Professor Jean-Pierre Guiraud held at the Université Pierre et Marie Curie, Paris, France, 20-22 april 1994. [90, 134]

42. E.J. Hinch. *Perturbation methods*. Applied mathematics. Cambridge University Press, 1991. [15, 53, 63, 81, 92, 99, 121]

43. M.H. Holmes. *Introduction to perturbation methods*, volume 20 of *Texts in Applied Mathematics*. Springer, 1998. [15]

44. Z. Janour. Resistance of a plate in parallel flow at low Reynolds numbers. NACA TM 1316, 1924. [191]

45. S. Kaplun. *Fluid mechanics and singular perturbations*. Academic Press, New York, 1967. [16, 63, 67]

46. S. Kaplun and P.A. Lagerstrom. Asymptotic expansions of Navier-Stokes solutions for small Reynolds numbers. *J. Math. Mech.*, 6:585–93, 1957. [67]

47. P.A. Lagerstrom. Méthodes asymptotiques pour l'étude des équations de Navier-Stokes. Lecture Notes, translated by T.J. Tyson, California Institute of Technology, Pasdena, CA, 1965. [16, 133]

48. P.A. Lagerstrom. *Matched asymptotic expansions, ideas and techniques*. Applied mathematical sciences, 76. Springer-Verlag, 1988. [16, 63, 78, 85, 89, 99, 118, 121]

49. P.Y. Lagrée, E. Berger, M. Deverge, C. Vilain, and A. Hirschberg. Characterization of the pressure drop in a 2d symmetrical pipe: some asymptotical, numerical and experimental comparisons. *ZAMM Z. Ange. Math. Mech.*, 85(1):1–6, 2005. [185, 267, 275, 279]

50. P.Y. Lagrée and S. Lorthois. The rns/prandtl equations and their link with other asymptotic descriptions: Application to the wall shear stress in a constricted pipe. *International Journal of Engineering Science*, 43:352–378, 2005. [185, 267]

51. L. Landau and E. Lifschitz. *Mécanique des fluides*. Ellipses, 1994. 3ème édition en français, titre original : Téorétitcheskaïa fizika v 10 tomakh Tom VI Guidrodinamika, traduit du russe par S. Medvédev,. [133, 140, 303]

52. M. Lazareff and J.C. Le Balleur. Méthode de couplage fluide parfait fluide visqueux en tridimensionnel avec calcul de la couche limite par méthode multi-zones. In *T.P. ONERA n 1986-134*. 28ème réunion ss grpe AAAS, groupe sectoriel Franco-Soviétique Aéronautique, 24-29 Août Moscou 1986. [185]

53. J.C. Le Balleur. Couplage visqueux-non visqueux : analyse du problème incluant décollements et ondes de choc. *La Recherche Aérospatiale*, 6:349–358, Novembre-Décembre 1977. [134, 156, 169, 175, 185]

54. J.C. Le Balleur. Couplage visqueux-non visqueux : méthode numérique et applications aux écoulements bidimensionnels transsoniques et supersoniques. *La Recherche Aérospatiale*, 1978(2):65–76, Mars-Avril 1978. [134, 156, 185]

55. J.C. Le Balleur. Calcul par couplage fort des écoulements visqueux transsoniques incluant sillages et décollements. Profils d'ailes portants. *La Recherche Aérospatiale*, 3:161–185, Mai-Juin 1981. [79, 185]

56. J.C. Le Balleur and P. Girodroux-Lavigne. Calculation of fully three-dimensional separated flows with an unsteady viscous-inviscid interaction method. In *T.P. ONERA no 1992-1*, California State University, Long Beach CA (USA), January 13-15 1992. 5th Int. Symp. on numerical and physical aspects of aerodynamical flows. [134, 156, 185]

57. J.C. Le Balleur and M. Lazareff. A multi-zonal marching integral method for 3D boundary-layer with viscous-inviscid interaction. In *T.P. ONERA n 1984-67*. 9ème Congrès International des Méthodes Numériques en Mécanique des Fluides, Saclay 25-29 Juin 1984. [185]

58. M.J. Lighthill. On boundary-layer and upstream influence: II. Supersonic flows without separation. *Proc. R. Soc.*, Ser. A 217:478–507, 1953. [133, 140, 147]

59. M.J. Lighthill. On displacement thickness. *J. Fluid Mech.*, 4:383–392, 1958. [139]

60. R.C. Lock. A review of methods for predicting viscous effects on aerofoils and wings at transonic speed. In AGARD Conference Proceedings No. 291, editor, *Computation of viscous-inviscid interactions,* ref. 2. North Atlantic Treaty Organization, 1981. [134]

61. J. Luneau and A. Bonnet. *Théories de la dynamique des fluides.* Cepadues, Toulouse, 1989. [305, 308]

62. J.J. Mahony. An expansion method for singular perturbation problems. *J. Austral. Math. Soc.*, 2:440–463, 1962. [20, 81]

63. J. Mauss. Un théorème d'estimation pour un problème de perturbation singulière lié à une équation différentielle du second ordre. *C.R. Acad. Sci., Paris*, t. 278, Série A:1177–1180, 1974. [102, 103]

64. J. Mauss. Asymptotic modelling for separating boundary layers. In R. Gatignol P.A. Bois, E. Dériat and A. Rigolot, editors, *Asymptotic Modelling in Fluid Mechanics*, number 442 in Lectures Notes in Physics, pages 239–254. Springer-Verlag, Heidelberg, 1995. Proceedings of a symposium in honour of Professor Jean-Pierre Guiraud held at the Université Pierre et Marie Curie, Paris, France, 20-22 april 1994. [71, 72, 143, 229]

65. J. Mauss, A. Achiq, and S. Saintlos. Sur l'analyse conduisant à la théorie de la triple couche. *C.R. Acad. Sci., Paris*, t. 315, Série II:1611–1614, 1992. [143, 229]

66. J. Mauss and J. Cousteix. Uniformly valid approximation for singular perturbation problems and matching principle. *C. R. Mécanique*, 330, issue 10:697–702, 2002. [81]

67. G.L. Mellor. The large Reynolds number asymptotic theory of turbulent boundary layers. *Int. J. Eng. Sci.*, 10:851–873, 1972. [238]

68. G.L. Mellor and D.M. Gibson. Equilibrium turbulent boundary layer. FLD No 13, Princeton University, 1963. [256]

69. A.F. Messiter. Boundary-layer flow near the trailing edge of a flat plate. *SIAM J. Appl. Math.*, 18:241–257, 1970. [134, 169, 191]

70. R. Michel, C. Quémard, and R. Durant. Application d'un schéma de longueur de mélange à l'étude des couches limites turbulentes d'équilibre. N.T. 154, ONERA, 1969. [254, 256, 257]

71. L.M. Milne-Thomson. *Theoretical Aerodynamics*. Dover Publications, Inc, New York, fourth edition, 1966. [193, 305]

72. A.H. Nayfeh. *Perturbation methods*. Pure and applied mathematics. John Wiley & sons, 1973. [15, 82, 99]

73. A.H. Nayfeh. Triple-deck structure. *Computers and Fluids*, 20(3):269–292, 1991. [143, 229]

74. V.Ya. Neyland. Theory of laminar boundary-layer separation in supersonic flow. *Izv. Akad. Nauk SSSR Mekh. Zhidk. Gaza.*, 4:53–57, 1969. Engli. transl. Fluid Dyn. 4 (4), pp 33–35. [134, 169]

75. R.E. O'Malley, Jr. *Introduction to singular perturbations*, volume 14 of *Applied Mathematics and Mechanics*. Academic Press, New York, London, 1974. [19, 82, 102]

76. R.L. Panton. Review of Wall Turbulence as Described by Composite Expansions. *Applied Mechanics Reviews*, 58(1):1–36, january 2005. [260]

77. S.B. Pope. *Turbulent Flows*. Cambridge University Press, Cambridge, UK, 2000. [237]

78. L. Prandtl. Über Flüßigkeitsbewegung bei sehr kleiner Reibung. Proceedings 3rd Intern. Math. Congr., Heidelberg, pages 484–491, 1904. [14, 133]

79. C. Roget. Structures asymptotiques et calculs d'écoulements sur des obstacles bi et tridimensionnels, Doctoral thesis, Université Paul Sabatier, Toulouse, 1996. [142, 186, 230]

80. C. Roget, J.Ph. Brazier, J. Mauss, and J. Cousteix. A contribution to the physical analysis of separated flows past three-dimensional humps. *Eur. J. Mech. B/Fluids*, 3:307–329, 1998. [142, 146, 155, 230]

81. A.P. Rothmayer and F.T. Smith. *The handbook of Fluid Dynamics, R.W. Johnson ed.*, chapter 25, Numerical Solution of Two-Dimensional, Steady Triple-Deck Problems. CRC Press, Boca Raton, FL, USA and Springer-Verlag, Heidelberg, Germany, 1998. [134, 185, 279]

82. A.P. Rothmayer and F.T. Smith. *The handbook of Fluid Dynamics, R.W. Johnson ed.*, chapter 23, Incompressible Triple-Deck Theory. CRC Press, Boca Raton, FL, USA and Springer-Verlag, Heidelberg, Germany, 1998. [135, 142]

83. A.P. Rothmayer and F.T. Smith. *The handbook of Fluid Dynamics, R.W. Johnson ed.*, chapter 24, Free Interactions and Breakaway Separation. CRC Press, Boca Raton, FL, USA and Springer-Verlag, Heidelberg, Germany, 1998. [135]

84. J.C. Rotta. *Turbulent boundary layers in incompressible flows*, volume 2 of *Progress in Aeronautical Sciences*. Pergamon Press, 1962. [256]

85. S. Saintlos and J. Mauss. Asymptotic modelling for separating boundary layers in a channel. *Int. J. Engng. Sci.*, 34(2):201–211, 1996. [267]

424   References

86. H. Schlichting and K. Gersten. *Boundary Layer Theory*. Springer-Verlag, Berlin, Heidelberg, 8^e edition, 2000. [134, 135, 142, 190, 207, 303]

87. Dennis S.C.R. and Chang G.A. Numerical integration of the Navier-Stokes equations for steady two-dimensional flow. *Phys. Fluid Suppl.*, II:88–93, 1969. [191]

88. Dennis S.C.R. and Dunwoody J. The steady flow of a viscous fluid past a flat plate. *J. Fluid Mech.*, 24:577–595, 1966. [191]

89. F.T. Smith. Flow through constricted or dilated pipes and channels: part 1. *Quarterly Journal of Mechanics and Applied Mathematics*, XXIX(Pt 3):343–364, 1976. [267, 283]

90. F.T. Smith. Flow through constricted or dilated pipes and channels: part 2. *Quarterly Journal of Mechanics and Applied Mathematics*, XXIX(Pt 3):365–376, 1976. [267, 283]

91. F.T. Smith. Upstream interactions in channel flows. *J. Fluid Mech.*, 79(part 4):631–655, 1977. [267, 283]

92. F.T. Smith. On the high Reynolds number theory of laminar flows. *IMA J. Appl. Math.*, 28(3):207–281, 1982. [135, 267]

93. F.T. Smith, P.W.M. Brighton, P.S. Jackson, and J.C.R. Hunt. On boundary-layer flow past two-dimensional obstacles. *J. Fluid Mech.*, 113:123–152, 1981. [229]

94. I.J. Sobey. *Introduction to interactive boundary layer theory*. Oxford Applied and Engineering Mathematics. Oxford University Press, Oxford, 2000. [192, 267, 275]

95. K. Stewartson. On the flow near the trailing edge of a flat plate II. *Mathematika*, 16:106–121, 1969. [134, 191]

96. K. Stewartson. Is the singularity at separation removable? *J. Fluid Mech.*, 44(2):347–364, 1970. [154]

97. K. Stewartson. Multistructured boundary-layers of flat plates and related bodies. *Adv. Appl. Mech.*, 14:145–239, 1974. [134]

98. K. Stewartson. D'Alembert's Paradox. *SIAM Review*, 23(308–343):145–239, 1981. [135, 147]

99. K. Stewartson, F.T. Smith, and K. Kaups. Marginal separation. *Studies in Applied Mathematics*, 67:45–61, 1982. [190]

100. K. Stewartson and P.G. Williams. Self induced separation. *Proc. Roy. Soc.*, A 312:181–206, 1969. [134, 169]

101. V.V. Sychev. Concerning laminar separation. *Izv. Akad. Nauk. SSSR Mekh. Zhidk. Gaza*, 3:47–59, 1972. [134, 147]

102. V.V. Sychev, A.I. Ruban, Vic.V. Sychev, and G.L. Korolev. *Asymptotic theory of separated flows*. Cambridge University Press, Cambridge, U.K., 1998. [147, 156, 169]

103. J.P.K. Tillett. On the laminar flow in a free jet of liquid at high Reynolds numbers. *J. Fluid Mech.*, 32(part 2):273–292, 1968. [207, 208]

104. E.R. Van Driest. On turbulent flow near a wall. *Jour Aeron Sci*, 23:1007, 1956. [256]

105. M. Van Dyke. Higher approximations in boundary-layer theory. Part 1. General analysis. *J. of Fluid Mech.*, 14:161–177, 1962. [133, 169, 185, 196, 217]

106. M. Van Dyke. Higher approximations in boundary-layer theory. Part 2. Application to leading edges. *J. of Fluid Mech.*, 14:481–495, 1962. [201, 211]

107. M. Van Dyke. *Perturbation methods in fluid mechanics*. Academic Press, New York, 1964. [16, 62, 71, 81, 133, 139, 169, 191]

108. M. Van Dyke. *Perturbation methods in fluid mechanics.* Parabolic Press, Stanford, CA, 1975. [15, 27, 28, 54, 71, 80, 90, 130]

109. A.E.P. Veldman. New, quasi-simultaneous method to calculate interacting boundary layers. *AIAA Journal*, 19(1):79–85, January 1981. [134, 156, 157, 169, 175, 185, 186]

110. A.E.P. Veldman. Matched asymptotic expansions and the numerical treatment of viscous-inviscid interaction. *Journal of Engineering Mathematics*, 39:189–206, 2001. [134, 185, 191]

111. A.E.P. Veldman. Viscous-Inviscid Interaction: Prandtl's Boundary Layer challenged by Goldstein's Singularity. In J. Cousteix and J. Mauss, editors, *Proc. BAIL2004 Conf. on Boundary and Interior Layers*, 2004. [169, 175]

112. F. Verhulst. *Methods and Applications of Singular Perturbations – Boundary Layers and Multiple Timescale Dynamics*, volume 50 of *Texts in Applied Mathematics*. Springer, New York, NY, 2005. [15, 82]

113. M.I. Višik and L.A. Lyusternik. Regular degeneration and boundary layer for linear differential equations with a small parameter. *Uspekki Mat. Nauk 12, 1957. Amer. Math. Soc. Transl. Serv. 2, 20*, pages 239–364, 1962. [41]

114. D.C. Wilcox. *Perturbation methods in the computer age.* DCW Industries, Inc., La Cañada, CA, 1995. [54]

115. K.S. Yajnik. Asymptotic theory of turbulent shear flows. *J. Fluid Mech.*, 42, Part 2:411–427, 1970. [238]

116. R.Kh. Zeytounian. *Asymptotic modelling of fluid flow phenomena.* Kluwer Academic Publishers, Dordrecht, The Netherlands, 2002. [90, 134]

Subject index

Scientific Computation

Scientific Computation

Printing: Krips bv, Meppel
Binding: Stürtz, Würzburg